Fascism, Aviation and Mythical Modernity

Fascism, Aviation and Mythical Modernity

Fernando Esposito
Universität Tübingen, Germany

Translated by Patrick Camiller

Originally published as Mythische Moderne. Aviatik, Faschismus und die Sehnsucht nach Ordnung in Deutschland und Italien, © 2011 by Oldenbourg Wissenschaftsverlag GmbH, Munich

English edition first published 2015 by
PALGRAVE MACMILLAN

Palgrave Macmillan in the UK is an imprint of Macmillan Publishers Limited, registered in England, company number 785998, of Houndmills, Basingstoke, Hampshire RG21 6XS.

Palgrave Macmillan in the US is a division of St Martin's Press LLC,
175 Fifth Avenue, New York, NY 10010.

Palgrave Macmillan is the global academic imprint of the above companies and has companies and representatives throughout the world.

Palgrave® and Macmillan® are registered trademarks in the United States, the United Kingdom, Europe and other countries.

ISBN 978–1–137–36298–8

This book is printed on paper suitable for recycling and made from fully managed and sustained forest sources. Logging, pulping and manufacturing processes are expected to conform to the environmental regulations of the country of origin.

A catalogue record for this book is available from the British Library.

A catalog record for this book is available from the Library of Congress.

Typeset by MPS Limited, Chennai, India.

The translation of this work was funded by Geisteswissenschaften International – Translation Funding for Humanities and Social Sciences from Germany, a joint initiative of the Fritz Thyssen Foundation, the German Federal Foreign Office, the collecting society VG WORT and the Börsenverein des Deutschen Buchhandels (German Publishers & Booksellers Association).

Contents

List of Illustrations

Preface

In his foreword to *Totemism and Exogamy* the Scottish anthropologist James George Frazer voices a principle of scientific integrity that seems quite suitable for the following: "That my conclusions on these difficult questions are final, I am not so foolish as to pretend. I have changed my views repeatedly, and I am resolved to change them again with every change of the evidence, for like a chameleon the candid enquirer should shift his colours with the shifting colours of the ground he treads."[1] Has the ground that I had trodden in my dissertational thesis *Mythische Moderne. Aviatik, Faschismus und die Sehnsucht nach Ordnung in Deutschland und Italien* shifted colors in the four years since it was published in Germany? By and large, I believe it hasn't. Nonetheless, the field of fascism studies has seen quite a number of important publications and the centennial of the outset of the First World War has brought forth a plethora of books on the "seminal catastrophe." So while it seems as if there were no need for this chameleon to change its hue or for the book to be "revised," a concise and selective account of the most influential and interesting research on the diverse subjects covered in this book since 2011 seems appropriate.

As the reader will learn in Chapter I.1, *Idea non Vincit*, I've chosen the art historian and cultural scientist Aby Warburg and his postage stamp design *Idea vincit* as paradigmatic embodiments of a liberal project of modernity and accordingly liberal contributions to the aviation discourse. Whereas the last year saw the London Warburg Institute highly publicized for quite regrettable reasons, the afterlife of Warburg himself and the Hamburg School that arose around him has continued to be vibrant.[2] From this wide array of works it is Emily Levine's *Dreamland of Humanists* that has especially caught my attention, as she not only emphasizes the enlightened, cosmopolitan, republican and transatlantic vein of this Hanseatic circle, but also stresses the importance that a liberal version of palingenesis had for them:[3] It was an aspiration towards a revival of the humanist Renaissance that underlay Warburg's interest in the *Nachleben* of classical antiquity and in the Renaissance itself. This only hints at the centrality of the palingenetic topos for *all* modern ideologies and not only for fascism proper, which I believe should be looked into more closely in future research.

Warburg's antipode, the poet and aviator Gabriele D'Annunzio, who was felicitously named Mussolini's "John the Baptist" by the British journalist Sisley

---

[1] James George Frazer, *Totemism and Exogamy. A Treatise on Certain Early Forms of Superstition and Society*. London 1910, p. xiii.

[2] For an ongoing bibliography see Björn Biester's blog: http://aby-warburg.blogspot.de/.

[3] Emily J. Levine, *Dreamland of humanists. Warburg, Cassirer, Panofsky, and the Hamburg school*. Chicago, IL 2013. See also: Id., "The Other Weimar: The Warburg Circle as Hamburg School," *Journal of the History of Ideas* 74/2013, pp. 307–30.

Huddleston and who is the main protagonist of my second chapter has received a popular new biography by Lucy Hughes-Hallet.[4] Albeit John Woodhouse's *Defiant Archangel* had extensively presented this central figure of Italian culture and politics to the English speaking public, Hughes-Hallet has supplied a highly readable account of the adventures of this cocaine-driven seducer, war hero and bard and called new attention to the *tremendum et faszinosum* that he emanates.

The commemoration of the outbreak of the First World War has been accompanied by so many new books on such diverse aspects of the war that it seems foolhardy to attempt an overview at this point. One tendency though is quite obvious: as with so many other historical subjects, the First World War has become global. Two of the most outstanding works give proof of this: in the Anglo-American landscape it is the three volumes of *The Cambridge History of the First World War* edited by Jay Winter,[5] while in Germany this applies first and foremost to Jörn Leonhard's *Die Büchse der Pandora*, which articulately shows the demise of the old liberal and bourgeois world and the birth of a new one that was not least characterized by the expansion and multiplication of the "spaces of violence."[6]

Even though the globalization of our perspectives on the First World War has partially shifted the focus of our attention away from the "storms of steel" on the Western Front, Ernst Jünger – not unlike D'Annunzio – continues to fascinate and repel his readers as well as the scientific community. Recent years have seen the publication of the historico-critical edition of his classic depiction of the carnage along with Helmut Kiesel's very helpful concordance covering the many metamorphoses the book went through over the ages.[7] Furthermore, we have seen a very familial double biography of Ernst and his brother Friedrich Georg Jünger which at times shows the banality of the life of this erstwhile protofascist.[8] Further aesthetes of the war and fascists of the first hour, the Italian Futurists, received a major retrospective at New York's Guggenheim Museum that was accompanied by a very informative catalogue.[9]

But, what about the field of fascism itself? In Germany fascism studies have been – aside from a few exceptions – largely dormant. Yet, in 2012 the *Institut für Zeitgeschichte* in Munich organized a major conference which was aimed both at synthesizing the existing research and at discussing new fields for further research, especially fascist internationalism or rather the joint objective of a European, anti-Semitic, racist, anti-communist and anti-liberal new order.[10] From the much

---

[4] Lucy Hughes-Hallett, *The Pike. Gabriele d'Annunzio, Poet, Seducer and Preacher of War*, London, 2013.

[5] Jay Winter, *The Cambridge History of the First World War*, 3 vols., Cambridge, 2014.

[6] Jörn Leonhard, *Die Büchse der Pandora. Geschichte des Ersten Weltkriegs*, Munich, 2014.

[7] Ernst Jünger, *In Stahlgewittern*, 2 vols. (ed. Helmut Kiesel), Stuttgart, 2013. See also: Matthias Schöning (ed.), *Ernst Jünger-Handbuch. Leben – Werk – Wirkung*, Stuttgart, 2014.

[8] Jörg Magenau: *Brüder unterm Sternenzelt. Friedrich Georg und Ernst Jünger. Eine Biographie.* Klett-Cotta Verlag, Stuttgart, 2012.

[9] Vivien Greene (ed.), *Italian Futurism 1909–1944. Reconstructing the Universe*, New York, NY, 2014.

[10] See: Thomas Schlemmer and Hans Woller (ed.), *Der Faschismus in Europa. Wege der Forschung*, Munich, 2014.

wider range of English publications on fascism five relatively recent works seem particularly to stand out: Michael R. Ebner's *Ordinary Violence in Mussolini's Italy*, Paul Corner's *The Fascist Party and Popular Opinion* and Kate Ferris's *Everyday Life in Fascist Venice* may be summed up under the keyword "consensus," as they offer new insights into the relationship between violence, coercive measures and the "manufacture of consent" during Mussolini's dictatorship.[11] Joshua Arthurs' *Excavating Modernity* and Aristotle Kallis's *The Third Rome*, on the other hand, look into fascism's relationship to the Roman past and examine the nexus between fascism, temporality and modernity.[12] As will hopefully become clear in the course of the following book, the concepts of modernity and historical temporality are pivotal to the understanding of fascism. At a workshop organized by Sven Reichardt and myself at the Villa Vigoni, we had the opportunity of discussing this central role of temporality amongst others with Joshua Arthurs, Martin Baumeister, Ruth Ben-Ghiat, Raul Cârstocea, Claudio Fogu, Peter Fritzsche and Roger Griffin, and our (I believe) quite interesting findings were recently published.[13]

This brings me to the acknowledgments: I would especially like to thank Roger Griffin for making the publication of this book at Palgrave Macmillan possible. He regarded it as "one long Germanism," yet in a positive way. I hope that further readers judge it similarly. My special thanks goes to Patrick Camiller. German dissertational theses are certainly rarely a treat to translate and mine was probably no exception. Nonetheless, I believe he's done a wonderful job. This translation would also never have been possible without the help of Cordula Hubert and the *Oldenbourg Verlag*, Jenny McCall and Jade Moulds from Palgrave Macmillan and the generous prize *Geisteswissenschaften International* awarded by the *Börsenverein des deutschen Buchhandels*. As all further acknowledgments made in the original book are still valid, I'll refer anyone interested to those pages and just thank you, Anne Ulrich, again for everything.

---

[11] Michael R. Ebner, *Ordinary Violence in Mussolini's Italy*, Cambridge, 2011; Paul Corner, *The Fascist Party and Popular Opinion in Mussolini's Italy*, Oxford, 2012; Kate Ferris, *Everyday life in Fascist Venice, 1929–1940*, Basingstoke, 2012.

[12] Joshua Arthurs, *Excavating Modernity. The Roman Past in Fascist Italy*, Ithaca, NY, 2012 and Aristotle Kallis, *The Third Rome. 1922–1943. The Making of the Fascist Capital*, Basingstoke, 2014.

[13] See the special issue on Fascist Temporalities of the *Journal of Modern European History* 13 (February 2015) ed. by Fernando Esposito.

Introduction

In a recently published text from his literary estate, the Austrian writer Thomas Bernhard argued that "the cold increases with the clarity."[1] This was the leitmotif of his address on receiving the Bremen Literature Prize in 1975, and it defines the view of "modernity" adopted in the present work.

The trope of coldness or a "cold heart" had already established itself in the Romantic movement, as in Wilhelm Hauff's fairytale "Das kalte Herz,"[2] and served to describe the modern age and the prevailing attitude to life.[3] But today's perspective is more disillusioned than that of Romanticism: it is guided neither by the desire for a brighter, warmer world nor by the longing for a "homeland" or "community" that supposedly existed in the past. Clarity, it would seem, has increased. For it has clearly become impossible to escape coldness or "transcendental homelessness," to flee from chaos into order.[4] Such flights give rise to those crimes and madness that the Marxist philosopher and literary theorist Georg Lukács described in his *The Theory of the Novel* (first published in 1916) as "objectivations of transcendental homelessness" and a new coldness. For Bernhard too, "the clarity increases with the cold." In his address of 1975, the misanthropic writer explained:

> I had to live through a long war and I saw hundreds of thousands die, and others who went on right over them; everyone went on, in reality, everything changed, in truth; in the five decades during which everything turned to revolt and everything changed, during which a thousand-year-old fairy tale gave way to *the* reality and *the* truth, I felt myself getting colder and colder while a new world and a new nature arose from the old. It is harder to live without fairy tales, that is why it is so hard to live in the twentieth century.[5]

---

[1] Thomas Bernhard, *My Prizes*, New York, NY 2010, p. 120.

[2] First published in the *Märchen Almanach auf das Jahr 1826* (Fairytale Almanac of 1826).

[3] See Manfred Frank, "Das Motiv des 'kalten Herzens' in der romantisch-symbolischen Dichtung," in Frank (ed.), *Kaltes Herz, unendliche Fahrt, neue Mythologie. Motiv-Untersuchungen zur Pathogenese der Moderne*, Frankfurt/Main 1989, pp. 11–49.

[4] On "transcendental homelessness," see Lukács, *The Theory of the Novel. A Historico-Philosophical Essay on the Forms of Great Epic Literature*, London 1971 [orig. 1920], p. 41.

[5] Bernhard, *My Prizes*, p. 118.

The underlying assumption of this study is that "knowledge" of the fairytale quality of the world is both a blessing and a curse. It defines the "postmodern condition" that spread out in the last decades of the twentieth century, emanating not least from a re-reading of the philosophy of Friedrich Nietzsche.[6] Modernity owes to this knowledge its capacity for self-criticism, for the deconstruction of reigning orders, and for the renewed emancipation of human beings from their "self-incurred immaturity."[7]

This knowledge, however, also spawns fear of anomie and chaos and a longing for order; it is the foundation of the dialectic of freedom and order, of modernity itself. And it is through this optic that we shall focus on fascism, understanding it as an attempt to repress such knowledge, to escape "transcendental homelessness" through myth, to replace anomie with a *nomos*,[8] chaos with order, contingency with necessity, and ambivalence with unequivocal clarity.[9] Fascism arose out of modernity and was itself modern. At the heart of this study will be the longing for order radicalized by the First World War, and the fascist blueprint for order as it found expression in aviation discourse.

Aviation, fascism and the longing for order

After the First World War, fascists sought to escape the "state of emergency" through the establishment of a "perfect" order.[10] The airplane and the aviator served as a symbol of its aspiration to an eternal order and highlighted the dynamism of its movements. Airplanes aroused admiration and surrounded those who flew them with a bold, vital and youthful aura of impetuosity. "Flying," Wulf Bley, the author of numerous books on aviation, wrote in 1936

> is more than forward movement in the air. The airplane is more than a means to that end. Flying is the supreme enhancement of life. But life is moving and being moved, forming and being formed. So the aviator becomes the form of a new age, on which he impresses his stamp. [...] The aviator is a new type of man. Not only is the German aviator a new type of German; the German type of aviator as such has a formative impact in the world.[11]

---

[6] See Jean-François Lyotard, *The Postmodern Condition. A Report on Knowledge*, Minneapolis, MN 1984.

[7] Immanuel Kant, "An Answer to the Question: 'What Is Enlightenment?'" in Hans Reiss (ed.), *Political Writings*, 2nd edn, Cambridge 1991, p. 58.

[8] On the concept of *nomos*, see Peter L. Berger, *The Sacred Canopy. Elements of a Sociological Theory of Religion*, Garden City, NY 1967.

[9] See Zygmunt Bauman, *Modernity and Ambivalence*, Cambridge 1991.

[10] On "perfect order," see Zygmunt Bauman, *Liquid Modernity*, Cambridge 2000, p. 29; and on the "state of exception" or "emergency, " Carl Schmitt, *Political Theology. Four Chapters on the Concept of Sovereignty*, Chicago, IL 1985 [orig. 1922], pp. 13–15. Cf. Michael *Makropoulos*, "Haltlose Souveränität. Benjamin, Schmitt und die Klassische Moderne in Deutschland," in Manfred Gangl/Gérard Raulet (eds.), *Intellektuellendiskurse in der Weimarer Republik*, Darmstadt 1994, pp. 197–211, 200.

[11] Wulf Bley, "Vorwort," in Bley (ed.), *Deutschland zur Luft*, Stuttgart 1936, p. 9.

Bley's Italian colleague Guido Mattioli emphasized in his book *Mussolini aviatore*:

> No machine requires as much concentration of the human mind, as much human will power, as the flying machine does. The pilot really knows what it means to govern. Hence there appears to be a necessary, inner spiritual affinity between aviation and fascism. Every aviator is a born fascist.[12]

This work will consider the idea of a "necessary, inner spiritual affinity between aviation and fascism."[13] Its guiding assumption is that the link created between

---

[12] Guido Mattioli, *Mussolini aviatore e la sua opera per l'aviazione*, Rome 1935–36, p. 2. From 1926 on, the Fascist calendar starting with the October 1922 "March on Rome" applied in Italy; thus, Year I began on October 29, 1922 and ended on October 28, 1923, and each successive year ran from October 29 until the next October 28. This system was also used by publishers, and so both the first and the second year are given here when it has not been possible to say for sure in which of the two a book appeared. See Luca Scuccimara, "Era fascista," in Victoria De Grazia/Sergio Luzzatto (eds.), *Dizionario del fascismo*, vol. 1, Turin 2002, pp. 480–81.

[13] The nexus between fascism and aviation has previously been investigated only by Mario Isnenghi, and even then not in detail or in relation to the question of modernity. In a chapter of his *L'Italia del Fascio*, he points out how the development of aviation may be read as paralleling that of fascism, and he even calls the aviation of the time a "metaphor of fascism." (Mario Isnenghi, *L'Italia del Fascio*, Florence 1996, pp. 233–51, esp. 233.) The desirability of a study of this nexus is suggested by the works of Peter Fritzsche and Robert Wohl, which stress the great symbolic power of aviation: see Peter Fritzsche, *A Nation of Fliers. German Aviation and the Popular Imagination*, Cambridge, MA 1992; Robert Wohl, *A Passion for Wings. Aviation and the Western Imagination 1908–1918*, New Haven, CT 1994; and Robert Wohl, *The Spectacle of Flight. Aviation and the Western Imagination 1920–1950*, New Haven, CT 2005. When the manuscript for this book was being completed, the works of Christian Kehrt (*Moderne Krieger. Die Technikerfahrungen deutscher Militärpiloten 1910–1945*, Paderborn 2010) and Stefanie Schüler-Springorum (*Krieg und Fliegen. Die Legion Condor im Spanischen Bürgerkrieg*, Paderborn 2010) – which do much to fill this gap – had not yet been published. On aviation in the Soviet Union, see Scott W. Palmer, *Dictatorship of the Air. Aviation Culture and the Fate of Modern Russia*, Cambridge 2006. On the connection between aviation and modernity in Britain and Germany, see Bernhard Rieger, *Technology and the Culture of Modernity in Britain and Germany, 1890–1945*, Cambridge 2005. Also especially worthy of mention are the studies by Aribert Reimann and René Schilling, which focus on the fascination of aviation, the significance of the flier as a new technological hero, and his roots in the context of war: see René Schilling, *"Kriegshelden." Deutungsmuster heroischer Männlichkeit in Deutschland 1813–1945*, Paderborn 2002, and Aribert Reimann, *Der große Krieg der Sprachen. Untersuchungen zur historischen Semantik in Deutschland und England zur Zeit des Ersten Weltkrieges*, Essen 2000. Further insights are provided in Detlef Siegfried's study of the Junkers airplane works: *Der Fliegerblick. Intellektuelle, Radikalismus und Flugzeugproduktion bei Junkers 1914–1934*, Bonn 2001. Modris Eksteins' discussion of Charles Lindbergh and his perception as a "new Christ" is of special value: see *Rites of Spring. The Great War and the Birth of the Modern Age*, New York, NY 2000 [repr.]. A number of interesting contributions on air travel under Fascism are contained in Massimo Ferrari (ed.), *Le ali del ventennio. L'aviazione italiana dal 1923 al 1945. Bilanci storiografici e prospettive di giudizio*, Milan 2005. There is a noteworthy bibliography on aviation in Felix Philipp Ingold, *Literatur und Aviatik. Europäische Flugdichtung 1909–1927*, Frankfurt/Main 1980, followed a few years later by Laurence Goldstein, *The Flying Machine and Modern Literature*, Bloomington, IN 1986. The significance

the two may be understood as an attempt to "form" an alternative modernity. A look at some rival worldviews of aviation will help to bring out this dawning of "another modernity," with its promises and its claim to superiority.[14] In particular, the fascist encoding of aviation symbols will be set against the background of an ideal-typical liberal alternative.

"It is an anomaly worth considering," wrote Paul Fussell, "that such a myth-ridden world could take shape in the midst of a war [WWI] representing a triumph of modern industrialism, materialism, and mechanism."[15] Taking this as our point of departure, we shall argue that fascism grew out of a longing for order radicalized by the First World War, and that it may be understood as an attempt to assuage this longing through the construction of a mythical modernity. Our aim is to illuminate this complex and to draw out the conceptions of order that underlie it. The focus is aviation discourse in Germany and Italy between roughly 1909 and 1939, and thus the slightly extended "crisis years of classical modernity."[16] But flying *per se* is not the center of attention. Rather, it stands for the mythical modernity and order that are our chief cognitive concern.

The question of the constitution of modernity and the social order to be constructed is debated implicitly and explicitly in aviation discourse; implicitly in that each text reflects the order that brings it forth and that it continually

---

of flying in art history is explored in Christoph Asendorf, *Super Constellation. Flugzeug und Raumrevolution. Die Wirkung der Luftfahrt auf Kunst und Kultur der Moderne*, Vienna 1997. Finally, some of the most important articles on aspects of the subject are: Yaron Jean, "'Mental Aviation' – Conquering the Skies in the Weimar Republic," *Tel Aviver Jahrbuch für deutsche Geschichte* 28/1999, pp. 429–58; Christian Kehrt, "'Schneid, Takt und gute Nerven.' Der Habitus deutscher Militärpiloten und Beobachter im Kontext technisch strukturierter Handlungszusammenhänge, 1914–1918," *Technikgeschichte* 72/2005, pp. 177–201; John H. Morrow Jr., "Knights of the Sky. The Rise of Military Aviation," in Frans Coetzee/Marylin Shevin-Coetzee (eds.), *Authority, Identity, and the Social History of the Great War*, Providence, RI 1995, pp. 305–24; George L. Mosse, "The Knights of the Sky and the Myth of the War Experience," in Robert A. Hinde/Helen E. Watson (eds.), *War. A Cruel Necessity? The Bases of Institutionalized Violence*, London 1995, pp. 132–42; Michael Paris, "The Rise of Airmen. The Origins of Air Force Elitism, 1890–1918," *Journal of Contemporary History* 28/1993, pp. 123–41; Stefanie Schüler-Springorum, "Vom Fliegen und Töten. Militärische Männlichkeit in der deutschen Fliegerliteratur, 1914–1939," in Karen Hagemann/Stefanie Schüler-Springorum (eds.), *Heimat-Front. Militär und Geschlechterverhältnisse im Zeitalter der Weltkriege*, Frankfurt/ Main 2002, pp. 208–33.

[14] On the "other modernity," see Thomas Rohkrämer, *Eine andere Moderne? Zivilisationskritik, Natur und Technik in Deutschland 1880–1933*, Paderborn 1999.

[15] Paul Fussell, *The Great War and Modern Memory*, New York, NY 2000 [repr.], p. 115.

[16] The concept of discourse used in this study will be explained in Section 1.2 below, "Theoretical and Methodological Approach." The choice of chronology roughly situates the discourse in question between 1909 the Brescia air show and the outbreak of the Second World War in Europe. On the "crisis years of classical modernity," see Detlev J. K. Peukert, *The Weimar Republic. The Crisis of Classical Modernity*, London 1991; and on the concept of crisis and its usefulness in interpreting the Weimar Republic, Moritz Föllmer/Rüdiger Graf (eds.), *Die "Krise" der Weimarer Republik. Zur Kritik eines Deutungsmusters*, Frankfurt/Main 2005.

updates;[17] and explicitly, since people at the time saw and instrumentalized the airplane and the aviator as symbols of "modernity." Modern societies represented themselves in this collective symbolism and announced their project for the modern age. Therefore the airplane and the aviator may be read as totems of the fascist, Marxist and liberal blueprints for modernity.[18]

This book enquires into the "substructures of thought," the interpretive patterns and tropes, that defined the fascist blueprint for modernity, and traces them at the level of aviation discourse.[19] The fascist conception of order, originating in the war and developing in the interwar years, define our investigative horizon: the temporal order and the "order of the community" that underlay the fascist vision of modernity stand at the center of our attention. These chronopolitical and social conceptions of order will become clearer from narratives of aviator heroes, and from the vision of the new man that fascism sought to make a reality through an "anthropological revolution."[20] We shall look at the origins of these blueprints during the First World War and follow their evolution in the postwar period, showing the context in which they acquired a meaning and tracing the media channels through which they spread and increased their effect. Hero images will be understood as models or prototypes that people were urged to use for the "writing" of their own lives. The media-conveyed norms and values that such heroes embodied were inscribed in the reality of society, and the world was interpreted in accordance with "prescribed" patterns.

We shall also examine how the desirability of a fascist blueprint for modernity grew over the years, whereas the liberal vision of man, society and the future became less and less convincing. One supposition is that, already during the First World War, the imagery of war and heroes that the popular and highbrow media helped to shape displayed a tendency toward a model of society and the future that had no place for a narrative of emancipation and progress; order, it seems, was preferred to freedom in Germany and Italy. In this respect, for all the differences between National Socialism and Italian Fascism, they may be seen as one and the same alternative to liberal and Stalinist-Marxist modernity.[21]

---

[17] "Text" refers here to any readable objectification of meaning – hence to semiotic or symbolic systems in the broadest sense of the term.

[18] Emile Durkheim, *The Elementary Forms of the Religious Life*, New York, NY 1965 [orig. 1915], pp. 244–45. Following Durkheim's definition of totem, the aviator and the airplane would be society itself, "objectified and represented in the mind." On the liberal codification and instrumentalization of aviation, see Chapter I.1 below: *Idea non vincit*; and, on the Soviet equivalent, Palmer, *Dictatorship of the Air*.

[19] On the "substructures of thought," see Hans Blumenberg, *Paradigms for a Metaphorology*, Ithaca, NY 2010, p. 5.

[20] On the politics of time, see Peter Osborne, *The Politics of Time. Modernity and the Avant-garde*, London 1995. On fascism and anthropological revolution, see *inter alia*: Emilio Gentile, *Politics as Religion*, Princeton, NJ 2006, pp. 46–7; and "Il uomo nuovo del fascismo. Riflessioni su un esperimento totalitario di rivoluzione antropologica," in Gentile, *Fascismo. Storia e interpretazione*, Rome 2005, pp. 235–64.

[21] See my definition of fascism in Section 1.3 below: "Definition of the Central Analytic Categories." It has not been possible to examine here the Stalinist-Marxist blueprint for

It will be shown that the aviator and his airplane were paradigmatic symbols of the mythical modernity that is the focus of this study. That which appears paradoxical to modernization theory – the conceptual linkage of myth and modernity – is harmonized in the collective symbol of aviation. The seemingly contradictory solidifies there into unity; the supposedly non-synchronous proves to be synchronous, and the merely juxtaposed comes together in a "form."[22]

Mythical modernity, grounded on a palingenetic myth of the nation or *Volk*, was perceived as an answer both to the crisis of reason and to the crisis of historicism. It would fulfill the longing for perfect order and satisfy the need for a sacral and aesthetic *nomos*. It was supposed to overcome the contingency and ambivalence of pluralist standpoints, the value relativism arising from historicism, the critical self-examination associated with rationalism and the uncertainty of a provisional, context-dependent world. In place of chaos and disorientation, the new order would have its center in the nation, race or *Volk*, or anyway in an unshakable Absolute beyond history and self-questioning.

Theoretical and methodological approach

The present work is a contribution to a *new* history of ideas.[23] Theoretically it is based upon the philosophical hermeneutics of Hans-Georg Gadamer,[24] and methodologically on the history of experience,[25] discourse history,[26] and the history of

---

modernity. On the comparison between Nazi Germany and Stalinist Russia, and for a survey of the literature, see Robert Gellately, *Lenin, Stalin, and Hitler. The Age of Social Catastrophe*, London 2007; Michael Geyer/Sheila Fitzpatrick (eds.), *Beyond Totalitarianism. Stalinism and Nazism Compared*, Cambridge 2009; Richard J. Overy, *The Dictators. Hitler's Germany and Stalin's Russia*, London 2004. Wolfgang Schivelbusch contrasts Germany and Italy with the United States in his *Three New Deals. Reflections on Roosevelt's America, Mussolini's Italy, and Hitler's Germany, 1933–1939*, New York, NY 2006. On Germany and Italy, see Richard Bessel (ed.), *Fascist Italy and Nazi Germany. Comparisons and Contrasts*, Cambridge 2000 [repr.] and Armin Nolzen/Sven Reichardt (eds.), *Faschismus in Italien und Deutschland. Studien zu Transfer und Vergleich*, Göttingen 2005.

[22] Hans-Georg Soeffner, "Flying Moles (Pigeon-Breeding Miners in the Ruhr District). The Totemistic Enchantment of Reality and the Technological Disenchantment of Longing," in Soeffner (ed.), *The Order of Rituals. The Interpretation of Everyday* Life, New Brunswick, NJ 1997, pp. 95–116; here p. 96.

[23] See Lutz Raphael, "Ideen als gesellschaftliche Gestaltungskraft im Europa der Neuzeit. Bemerkungen zur Bilanz eines DFG-Schwerpunktprogramms," in Lutz Raphael/Heinz-Elmar Tenorth (eds.), *Ideen als gesellschaftliche Gestaltungskraft im Europa der Neuzeit. Beiträge für eine erneuerte Geistesgeschichte*, Munich 2006, pp. 11–27.

[24] Hans-Georg Gadamer, *Truth and Method*, 2nd rev. edn, London 1989.

[25] Nikolaus Buschmann/Horst Carl, "Zugänge zur Erfahrungsgeschichte des Krieges: Forschung, Theorie, Fragestellung," in Buschmann/Carl (eds.), *Die Erfahrung des Krieges. Erfahrungsgeschichtliche Perspektiven von der Französischen Revolution bis zum Zweiten Weltkrieg*, Paderborn 2001, pp. 11–26.

[26] For a survey of discourse history, see Achim Landwehr, *Geschichte des Sagbaren. Einführung in die historische Diskursanalyse*, Tübingen 2001; Jürgen Martschukat, "Geschichte schreiben mit Foucault – eine Einleitung," in Martschukat (ed.), *Geschichte schreiben mit Foucault*, Frankfurt/Main 2002, pp. 7–26; and Philipp Sarasin, "Geschichtswissenschaft und

concepts and metaphors.[27] At the heart of this new history of ideas are the "sub-structures of thought," symbolic-semantic nexuses, terminological or conceptual compounds, and ideas of order through which reality is structured and meaning produced. The basic assumption is that, although "men make their own history," they do so

> not of their own free will; not under circumstances they themselves have chosen but under the given and inherited circumstances with which they are directly confronted. The tradition of the dead generations weighs like a night-mare on the minds of the living. And, just when they appear to be engaged in the revolutionary transformation of themselves and their material surround-ings, in the creation of something which does not yet exist, precisely in such epochs of revolutionary crisis they timidly conjure up the spirits of the past to help them; they borrow their names, slogans, and costumes so as to stage the new world-historical scene in this venerable disguise and borrowed language.[28]

To take account of the fact that human beings make their history under the weight of "dead generations," language will be at the center of this study. With Gadamer, we shall consider language as "Being that can be understood."[29] It stores "the given and inherited circumstances with which men are directly confronted," the "tradition" or knowledge of a society, its current ways of perceiving the world and the concepts, ideas and rules on which contemporaries base the order of real-ity. It is in language that reality first becomes understandable, both to the histori-cal actor and to the historian.[30] For the symbolic-semantic order is the system of coordinates within which an intelligible world is constructed and interests and actions first acquire a meaning.[31]

Diskursanalyse," in Sarasin (ed.), *Geschichtswissenschaft und Diskursanalyse*, Frankfurt/Main 2003, pp. 10–60.

[27] For a survey of conceptual history and the history of metaphor, see Hans-Erich Bödeker (ed.), *Begriffsgeschichte, Diskursgeschichte, Metapherngeschichte*, Göttingen 2002; Hans Ulrich Gumbrecht, *Dimensionen und Grenzen der Begriffsgeschichte*, Munich 2006, esp. pp. 7–36. See also the "founding texts" of German conceptual history: Reinhart Koselleck, "Einleitung," in Otto Brunner/Werner Conze/Reinhart Koselleck (eds.), *Geschichtliche Grundbegriffe. Historisches Lexikon zur politisch-sozialen Sprache in Deutschland*, vol. 1, Stuttgart 1972, pp. XIII–XXVII, and Rolf Reichardt, "Einleitung," in Rolf Reichardt/Eberhard Schmitt (eds.), *Handbuch politisch-sozialer Grundbegriffe in Frankreich 1680–1820*, vol. 1/2, Munich 1985, pp. 39–148.

[28] Karl Marx, "The Eighteenth Brumaire of Louis Bonaparte," in Marx, *Surveys from Exile*, London 1973, p. 146.

[29] Gadamer, *Truth and Method*, p. 474.

[30] Ibid., p. 443. "Not only is the world world insofar as it comes into language, but language, too, has its real being only in the fact that the world is presented in it."

[31] In this sense, language really does present itself as the "house of being." See Martin Heidegger, "Letter on 'Humanism' " (1946), in William McNeill (ed.), *Pathmarks*, Cambridge 1998, pp. 239–76; here 239.

Language, then, passes on the order of reality into which human beings are born and "thrown." It is the cognitive storehouse of society, determining and making possible meaningful actions on the part of historical players.[32] But neither knowledge nor its repository is static; it is subject to constant change that is at the center of the historian's interest. For in the encounter with the world, in the use of language and the application of its rules, our cognitive repository not only reproduces itself but adapts to the changing world. By integrating ever new objects and situations, it is updated and modified as the bearer of knowledge. As Wittgenstein put it, the language game and its rules adapt to the situation and to those who play it.[33] The transformation of the linguistic and mental ordering of reality, or of the "socialized interpretive categories," in the period around the First World War will be traced here through a reconstruction and deconstruction of aviation discourse.[34] But what is a discourse? And what can be expected from its investigation?

---

[32] See Peter L. Berger and Thomas Luckmann, *The Social Construction of Reality*, Baltimore, MD 1966, pp. 52ff. We read there: "Language is capable of becoming the objective repository of vast accumulations of meaning and experience, which it can then preserve in time and transmit to following generations. [...] As a sign system, language has the quality of objectivity. I encounter language as a facticity external to myself and it is coercive in its effect on me. Language forces me into its patterns. [...] Language provides me with a ready-made possibility for the ongoing objectification of my unfolding experience. Put differently, language is pliantly expansive so as to allow me to objectify a great variety of experiences coming my way in the course of my life. Language also typifies experiences, allowing me to subsume them under broad categories in terms of which they have meaning not only to myself but also to my fellow men." The relationship between language and history that informs the present study is also made clear in the following quotation from Algirdas Greimas: "Language is a global system of signs that permeates a culture and gives it expression. It is not just a repository of words, from which a mere few can serve as witnesses to a history that otherwise takes place in a sphere beyond language; rather, as a symbolic system, it is itself the locus where history takes place. It forms an autonomous social space, which surpasses individuals and imposes on them certain patterns of feeling and behavior. Words organized into structural "ensembles" – vocabularies – mutually define one another and form an objective, inescapable plane of language in which the historian can discover mentalities and models of collective sensibility. [...] That is the level where social roles are allocated and social frameworks develop for models of feeling and norms of morality." Algirdas J. Greimas, "Histoire et linguistique," *Annales. Économies, Sociétés, Civilisations* 13/1958, pp. 110–14; here p. 111; quoted in Peter Schöttler, "Sozialgeschichtliches Paradigma und historische Diskursanalyse," in Jürgen Fohrmann/Harro Müller (eds.), *Diskurstheorien und Literaturwissenschaft*, Frankfurt/Main 1988, pp. 159–99; here p. 163.

[33] See Ludwig Wittgenstein, *Philosophical Investigations*, 2nd edn, Oxford 1958.

[34] See Buschmann/Carl, "Zugänge zur Erfahrungsgeschichte des Krieges," pp. 18–19: "Whereas, from the player's viewpoint, [experiences] always appear as unique elements in his or her biography, their interpretation involves socialized categories whose common basis is language, qua 'positive condition and conduit of experience itself.' Language should thus be regarded as the basic condition of experience, in so far as it gives the semantic apparatus that makes experiences possible in consciousness and communication."

A discourse will here be understood as a linguistic context in which actions are embedded. They acquire their meaning or signification through this insertion in a symbolic-semantic order; this meaning, which historical players attach to their actions and recognize in their world, is at the center of studies in the history of ideas.[35] "Discourses," according to Philipp Sarasin, "are historically specific thematic speech contexts, which define possibilities and limits of meaningful speech and coherent social action."[36] A discourse therefore represents the reference system, or "involvement context [*Bewandtniszusammenhang*]," within which the object – in our case: aviation, airplane, aviator and flying – is classified and hence constituted in the first place.[37] However, the discourse does not represent a thing in, above or behind the world; it is a heuristic concept, with whose help the place of an object within its semantic network or context can be reconstructed. It serves to answer the question why, "at a given historical moment, only a limited number

---

[35] For Foucault's concept of discourse, see in particular his "text on method:" Michel Foucault, *The Archaeology of Knowledge*, London 1972. The following definition by the sociologist Rainer Diaz-Bone should serve as an initial guide for this study: "Discourse may be more precisely defined as a declarative system in which the facts that are being 'talked about' are put forward as cognitive elements. [...] In so far as facts are not only treated as things but also combined with evaluations, classified in relation to one another, and associated with concepts, they acquire a meaning and hence the cognitive status of 'things' that discourse communities experience as prediscursive but are anything but simply given or capable of being naively experienced. In their interconnection, the statements of a discourse exhibit a system of rules that achieves the formation of 'concepts,' 'objects,' speaker positions, and thematic choices. The discursive order is thus an inherently systematic practice that generates the cognitive order. The declarative system is an enabling context for the individual statement, and conversely the flow of statements reproduces the regularity of the discourse. One might say, in the manner of Bourdieu: discursive practice is both structured (by the system of rules) and structuring (of the cognitive order and the system of rules)." See Rainer Diaz-Bone, "Zur Methodologisierung der Foucaultschen Diskursanalyse," *Historical Social Research* 31/2006, pp. 243–74; here pp. 251–52.

[36] Philipp Sarasin, "Subjekte, Diskurse, Körper. Überlegungen zu einer diskursanalytischen Kulturgeschichte," in Wolfgang Hardtwig/Hans-Ulrich Wehler (eds.), *Kulturgeschichte Heute*, Göttingen 1996, pp. 131–64; here p. 142.

[37] On the "involvement context," see Martin Heidegger, *Being and Time*, New York, NY 1962, §18 ("Involvement and Significance"), p. 114. See also his introduction to philosophy given in Freiburg in the 1928–29 winter semester: "Einleitung in die Philosophie," in *Gesamtausgabe*, vol. 27, ed. by Otto Saame and Ina Saame Speidel, Frankfurt/Main 1996, pp. 81–2: "But the area of the translucent involvement contexts, the perspective of what is just now apparent to us, is changeable and does constantly change. If we say chalk, sponge, table, and auditorium, we constrict ourselves, as it were, to this particular space. But the auditorium itself is directly inside the university building – this building on this city square, the city of Freiburg, in certain environs beneath the sky, by day, by night, in a certain kind of weather. This whole context is actually hidden from us when we say that this piece of chalk lies here on the desk. All these ambient contextual elements of beings have no fixed boundaries. They are not placed next to one another; rather, those further away shine as a single whole in and through those that are closer."

of statements can be made about a particular issue, although in purely linguistic terms an infinite number of statements are possible."[38]

For example, what the flier "is" at a particular time becomes clear when his place within the semantic structure has been worked out. Which terms and concepts do certain speakers associate him with at time t_1 and which images do these generate? How do these associations, and the rules of their perception, change? Which ideas are present at time t_2 by virtue of such change? What does it mean for people's understanding of "flier" if the significance of nobility, for example, changes? (Figure I.1)

The aviator therefore acquires his "identity" and significance both from everything he is *not* (i.e. through difference) and from the terms in whose context and structural proximity he is classified. This changing semantic context will be examined in the course of this study. It has two consequences. First, detailed accounts are required of the linguistic cosmos in which aviation was located; only these will make it clear how aviation symbolism functioned. Second, it is often not aviation *per se* but its connotations, not the aviator but conceptions prevalent at the time (of a new nobility, for example), that will be the main focus of attention. Symbols or metaphors continually establish a connection; they bring something to the fore *as something*. As George Lakoff and Mark Johnson put it, metaphorical notions represent "ways of partially structuring one experience in terms of another."[39] The

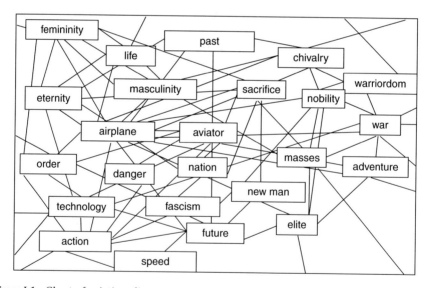

Figure I.1 Chart of aviation discourse

[38] Landwehr, *Geschichte des Sagbaren*, p. 7. In Foucault the question is: "how is it that one particular statement appeared rather than another?" See Foucault, *The Archaeology of Knowledge*, p. 30.

[39] George Lakoff/Mark Johnson, *Metaphors We Live By*, Chicago, IL 1980, p. 77.

meaning that arises through this connection, as well as the "experiential space" out of which it is created and the "horizon of expectation" onto which it is projected, are a central concern of the history of ideas.[40] So, in order to draw out the perception and interpretation of aviation at that time, it is necessary "to range freely, with no fear of border guards," supplementing aviation sources with the most diverse texts.[41]

The body of sources is thus quite heterogeneous. The media material for the period in question is nearly endless, but account has been taken of the widest and most representative possible spectrum relating to aviation, together with a large number of published and archive sources that complete or generally assist our understanding of the subject. The study examines not only aviation magazines, but also Aby Warburg's design for a postage stamp, Futurist *aeropittura* paintings, Leni Riefenstahl's film *Triumph of the Will* and the Italian Aeronautics Exhibition held in Milan in 1934. Events at the level of high culture are considered along with products of popular culture. The books on war in the air that appeared in Germany as part of the literary mobilization during the First World War form a central part of the corpus,[42] providing us with information about different layers of mass society. In Italy, because of the still high rate of illiteracy, such books did not play such an important role; they began to gain influence only in the wake of the propaganda efforts of the Fascist regime. Great significance is accorded to the writings of Gabriele D'Annunzio and the Futurists, which, alongside those of Ernst Jünger and Aby Warburg, should be seen as paradigmatic statements. Prominent speakers in the discourse were capable of operating more agilely within the symbolic-semantic network, and their forcefulness and powers of interpretation meant that what they said received a greater hearing. The following section will explain the central categories of this book – fascism, modernity and myth – and define the research field in which it should be located.

Definition of the central analytical categories

In his *Reactionary Modernism*, published in 1984, Jeffrey Herf pointed to a paradox that recalls Paul Fussell's previously noted "anomaly:"[43] on the one hand, German

---

[40] Reinhart Koselleck, "'Space of Experience' and 'Horizon of Expectation' – Two Historical Categories," in Koselleck, *Futures Past. On the Semantics of Historical Time*, Cambridge, MA 1985, pp. 267–88.

[41] Aby Warburg, "Italian Art and International Astrology in the Palazzo Schifanoia, Ferrara (1912)," in Warburg, *The Renewal of Pagan Antiquity. Contributions to the Cultural History of the European Renaissance*, Los Angeles, CA 1999, pp. 563–91; here p. 585.

[42] On the literary mobilization, see Chapter II.1 below, "Don Quixote of the Air." Cf. Marieluise Christadler, *Kriegserziehung im Jugendbuch. Literarische Mobilmachung in Deutschland und Frankreich vor 1914*, Frankfurt/Main 1978.

[43] Since the state of research on particular themes will be mentioned in the relevant chapters of this study, the following section mainly covers titles that concern it in its entirety. Having been completed in late summer 2009, the text refers only in exceptional cases to literature that has appeared since then.

nationalism and Nazi Germany spurned "modernity" (i.e. the political values deriving from the French Revolution and social values born out of the Industrial Revolution), but on the other hand, despite their antimodern, Romantic and irrationalist roots, they welcomed modern technology.[44] With the help of his concept of "reactionary modernism," Herf showed that this paradox of technology and antimodernism was intelligible if not actually resolvable, and that Germany and its intellectuals suffered from a clear lack of reason, liberalism and "modernity."[45]

The mythical modernity at issue in this study refers back to, while also rejecting, Herf's "reactionary modernism." Modernism may be understood, in line with Herf's analysis, as a "cultural response to technological advance which incorporated technology into a romantic, fundamentalist revolt against processes of rationalization."[46] But we shall dispute Herf's normative statement that modernism can be equated with reason, liberalism and enlightenment.[47] And we shall challenge the assumption of modernization theory that there is a single modernity, a single straight line leading to liberal-democratic Western-style modernity based upon the Enlightenment and the French Revolution. A deviation from this path, which thinkers from Max Weber to Hans-Ulrich Wehler and Jürgen Habermas all hold up as uniquely valid, will be regarded neither as a reaction or *Sonderweg* nor as premodern or antimodern.

Whereas previous investigations (mostly in the field of social history) have revolved around the question of the modernity of National Socialism or Italian Fascism, taking liberal modernity as their (normative) yardstick, the focus here will be fascist modernity as such.[48] The main supporting texts will be those of

---

[44] Jeffrey Herf, *Reactionary Modernism. Technology, Culture, and Politics in Weimar and the Third Reich*, Cambridge, MA 1984, p. 1. See also Jeffrey Herf, "The Engineer as Ideologue. Reactionary Modernists in Weimar and Nazi Germany," *Journal of Contemporary History* 19/1984, pp. 631–48, and "Der nationalsozialistische Technikdiskurs. Die deutschen Eigenheiten des reaktionären Modernismus," in Wolfgang Emmerich/Carl Wege (eds.), *Der Technikdiskurs in der Hitler-Stalin-Ära*, Stuttgart 1995, pp. 72–93; and Thomas Rohkrämer, "Antimodernism, Reactionary Modernism and National Socialism. Technocratic Tendencies in Germany, 1890–1945," *Contemporary European History* 8/1999, pp. 29–50.

[45] Herf, *Reactionary Modernism*, p. 234.

[46] Herf, "The Engineer as Ideologue," p. 646. Here and on the following, see also Riccardo Bavaj, *Die Ambivalenz der Moderne im Nationalsozialismus. Eine Bilanz der Forschung*, Munich 2003, pp. 48–52.

[47] Cf. Rohkrämer, *Eine andere Moderne?*, pp. 15–16.

[48] For surveys of the question of the modernity of National Socialism and Fascism, see: Bavaj, *Die Ambivalenz der Moderne*; Peter Fritzsche, "Nazi Modern," *Modernism/Modernity* 3/1996, pp. 1–21; and Stanley G. Payne, *A History of Fascism 1914–1945*, 2nd edn, London 1997, pp. 471–86. We should already mention here some of the key texts in the debate: Ralf Dahrendorf, *Society and Democracy in Germany*, New York, NY 1979; David Schoenbaum, *Hitler's Social Revolution. Class and Status in Nazi Germany, 1933–1939*, New York, NY 1966; Timothy W. Mason, "The Origins of the Law on the Organization of National Labour of 20 January 1934. An Investigation into the Relationship between 'Archaic' and 'Modern' Elements in Recent German History," in Timothy W. Mason (ed.), *Nazism, Fascism and the Working Class*, Cambridge 1995, pp. 77–103; Detlev J. K. Peukert, *Inside Nazi Germany. Conformity, Opposition, and Racism in Everyday Life*, London 1987; Detlev J. K. Peukert, *Max Webers Diagnose der Moderne*,

authors who, inspired by the 1970s critique of progress and skepticism about modernity, have seen fascism, and especially Nazi extermination policies, as a genuine expression of Janus-faced or ambivalent modernity.[49] In his *Modernism and Fascism* (2007) Roger Griffin pointed out – in an allusion to Horkheimer – that anyone not prepared to talk about modernity is in danger of misunderstanding the nature of fascism.[50] The present study too assumes that (1) the nexus of fascism and modernity is central and that fascism can therefore be understood only in the context of modernity; and (2) the concept of myth is of the greatest significance for an understanding of modernity. In the rest of this section, these three terms will be delimited separately from one another, but since they belong closely together they will be reunited at the end of the investigation.

Fascism

The term fascism has changed its content in accordance with the particular interests and approaches of its researchers.[51] Three phases of comparative study may be distinguished from one another, although the boundaries between the second and

---

Göttingen 1989; Rainer Zitelmann, *Hitler. The Policies of Seduction*, London 1999; Zygmunt Bauman, *Modernity and the Holocaust*, Ithaca, NY 1989; Michael Prinz/Rainer Zitelmann, *Nationalsozialismus und Modernisierung*, Darmstadt 1991. On Italian Fascism and its modernity, see: Andrew Hewitt, *Fascist Modernism. Aesthetics, Politics, and the Avant-Garde*, Stanford, CA 1993; Emilio Gentile, "The Conquest of Modernity. From Modernist Nationalism to Fascism," *Modernism/Modernity* 1/1994, pp. 55–87; Walter L. Adamson, "The Culture of Italian Fascism and the Fascist Crisis of Modernity. The Case of il Selvaggio," *Journal of Contemporary History* 30/1995, pp. 555–75; Timothy W. Mason, "Italy and Modernization. A Montage," *History Workshop Journal* 25/1998, pp. 127–47; Ruth Ben-Ghiat, *Fascist Modernities. Italy, 1922–1945*, Berkeley, CA 2004 [repr.]; Nicola Tranfaglia, *Fascismi e modernizzazione in Europa*, Turin 2001; Wolfgang Schieder, "Die Geburt des Faschismus aus der Krise der Moderne," in Christof Dipper (ed.), *Deutschland und Italien 1860–1960*, Munich 2005, pp. 159–79.

[49] Bauman, *Modernity and the Holocaust* and *Modernity and Ambivalence*; Peukert, *Inside Nazi Germany* and "Die Genesis der 'Endlösung' aus dem Geist der Wissenschaft," in *Max Webers Diagnose der Moderne*, Göttingen 1989, pp. 102–21. The works of both Bauman and Peukert are in the tradition of Horkheimer's and Adorno's *Dialectic of Enlightenment*, against which Herf positioned himself so decisively: Theodor Adorno/Max Horkheimer, *Dialectic of Enlightenment* (orig. 1944), New York, NY 1972.

[50] Horkheimer argued: "Whoever is not willing to talk about capitalism should keep quiet about fascism". See Max Horkheimer, "The Jews in Europe," in, S. E. Bonner/D. M. Kellner (eds.), *Critical Theory and Society. A Reader*, New York, NY 1989, pp. 77–94; here p. 78. Griffin, for his part, writes: "Less dogmatically, this book has argued that anyone not prepared to talk about modernity or modernism may by all means speak of fascism, but risks remaining oblivious of important aspects of its dynamics, and incapable of resolving the many paradoxes posed by its relationship to liberal capitalist modernity. In particular, it means underestimating the causal significance in its genesis and its appeal attributable to the crucial role played by the 'disembedding,' anomy-generating impact of modernization in fuelling countless revolts against 'decadence' of which fascism was but one example – albeit one which changed the course of modern history in a way that cost millions of lives." Roger Griffin, *Modernism and Fascism. The Sense of a New Beginning under Mussolini and Hitler*, Basingstoke 2007, p. 344.

[51] The following remarks are not an exhaustive survey of fascism research but will serve to introduce the approach adopted in this book.

the third have turned out to be rather fluid.[52] Only in the climate of the 1990s, which was more favorable to a "cultural history of fascism," did aspects already developed in the 1970s by George L. Mosse or Emilio Gentile begin to have a major impact in the English-speaking world.[53]

Analysis of the fascist phenomenon began with the emergence of Italian Fascism itself.[54] The Marxist theorists who first addressed the subject saw it as essentially an "instrument of monopoly capital." In December 1933, the 13th Plenum of the Executive Committee of the Communist International (ECCI) defined it as "the open, terrorist dictatorship of the most reactionary, most chauvinist, and most imperialist elements of finance capital."[55] The bourgeoisie, it was claimed, used the fascist movement to protect its property from the revolutionary proletariat. Fateful in this regard was the thesis of social fascism, that is, the equals sign that the Comintern placed between Social Democracy and fascism, on the grounds that both were "instruments of capitalist dictatorship."[56] The idea of fascism as an agency of "state monopoly capitalism" survived the Second World War and became official doctrine in the GDR, the self-styled bastion of antifascism. The nexus of capitalism and fascism, highlighted by Max Horkheimer too in 1939, was taken up again in West Germany by the student protest movement of 1968.[57] Meanwhile, as conservative-liberal opinion distanced itself from totalitarianism theory, a second phase of fascism studies developed there in the 1960s.

---

[52] On the phases of fascism studies, with a review of the relevant literature, see Sven Reichardt, "Neue Wege der vergleichenden Faschismusforschung," *Mittelweg 36* 16/2007, pp. 9–25; and "Was mit dem Faschismus passiert ist. Ein Literaturbericht zur internationalen Faschismusforschung seit 1990, Teil 1," *Neue Politische Literatur* 49/2004, 3, pp. 385–406.

[53] Conversely, we may say that the third period of research in the English-speaking countries – which was geared to cultural history – largely overlooked the socially oriented studies of fascism published by Wolfgang Schieder and his circle since the 1970s, which had emphasized such aspects as the transfer between Italian Fascism and National Socialism. See Wolfgang Schieder, *Faschistische Diktaturen. Studien zu Italien und Deutschland*, Göttingen 2008; Christof Dipper/Rainer Hudemann/Jens Petersen (eds.), *Faschismus und Faschismen im Vergleich. Wolfgang Schieder zum 60. Geburtstag*, Cologne 1998; Christof Dipper/Wolfgang Schieder (eds.), *Faschismus und Gesellschaft in Italien. Staat – Wirtschaft – Kultur*, Cologne 1998; Wolfgang Schieder, "Die NSDAP vor 1933. Profil einer faschistischen Partei," *Geschichte und Gesellschaft* 19/1993, pp. 141–54; Wolfgang Schieder (ed.), *Faschismus als soziale Bewegung. Deutschland und Italien im Vergleich*, Hamburg 1976.

[54] On the following, see: Arnd Bauerkämper, *Der Faschismus in Europa 1918–1945*, Stuttgart 2006, pp. 18–24; Renzo De Felice, *Le interpretazioni del fascismo*, Rome 2001 [repr.]; Payne, *A History of Fascism*, pp. 441–50; Roger Griffin (ed.), *Fascism. Critical Concepts in Political Science*, vol. 1, *The Nature of Fascism*, London 2004; and Wolfgang Wippermann, *Faschismustheorien. Die Entwicklung der Diskussion von den Anfängen bis heute*, Darmstadt 1997[7], pp. 11–57.

[55] *The Communist International 1919–1943. Documents*, vol. 3, *1929–1943*, ed. by Jane Degras, London 1965, p. 296.

[56] Ibid., vol. 2, p. 139.

[57] See Norbert Frei, *1968. Jugendrevolte und globaler Protest*, Bonn 2008, pp. 84ff.

By virtue of its exculpatory and relativistic connotations, totalitarianism theory acquired more or less official status in the Federal Republic.[58] In the conflicts of the 1920s and 1930s, it had already been the conservative-liberal response to the challenges of Communism and fascism, both of which were seen to threaten liberty and to be seeking the establishment of a totalitarian structure of rule. The Cold War then gave fresh impetus to such theories, especially those of Carl J. Friedrich, Zbigniew Brzezinski and Hannah Arendt,[59] but in the 1960s the concept lost much of its explanatory power. After the collapse of the Soviet Union, however, it experienced another revival. Thus, in association with Eric Voegelin's concept of "political religion," it became a central element in Emilio Gentile's analysis of fascism and made a considerable contribution to our understanding of the phenomenon.[60] The key point was no longer whether a regime was totalitarian or not – that is, whether its rule encompassed the whole of society – but that totalitarianism was an "experiment in political rule."[61]

In his *Three Faces of Fascism*, first published in German in 1963, Ernst Nolte disputed the explanatory power of the concept of totalitarianism, insisting that it was impossible to grasp the "modification" if "it is subsumed in the general concept."[62] But even then he understood fascism in terms of its relationship to Marxism or Bolshevism. Without Marxism, he argued, there is no fascism. Nolte's phenomenologically derived definition was as follows: "Fascism is anti-Marxism which seeks to destroy the enemy by the evolvement of a radically opposed yet related ideology and by the use of almost identical and yet typically

---

[58] Wolfgang Wippermann, "The Post-War German Left and Fascism," in *Journal of Contemporary History* 11/1976, pp. 185–219; here p. 193. See also Abbott Gleason, *Totalitarianism. The Inner History of the Cold War*, New York, NY 1995.

[59] See, among others: Eckhard Jesse (ed.), *Totalitarismus im 20. Jahrhundert. Eine Bilanz der internationalen Forschung*, 2nd edn, Baden-Baden 1999; Hans Maier (ed.), *Totalitarianism and Political Religions. Concepts for the Comparison of Dictatorships. Theory and History of Interpretation*, 3 vols., New York, NY 2004–2007; Wolfgang Wippermann, *Totalitarismustheorien. Die Entwicklung der Diskussion von den Anfängen bis heute*, Darmstadt 1998.

[60] On the concept of political religion and the connection with totalitarianism theory, see Chapter I.2 below.

[61] "By the term 'totalitarianism' we mean an experiment in political domination implemented by a revolutionary movement that has been organized by a party with military discipline and an all-absorbing concept of politics aimed at the monopoly of power, which on taking power by legal or illegal means destroys or transforms the previous regime and builds a new state founded on a single-party regime with the principal objective of conquering society, that is, the subjugation, integration, and homogenization of the ruled on the basis of the totally political nature of existence, whether individual or collective, as interpreted by the categories, myths, and values of an institutionalized ideology in the form of a political religion, with the intention of molding individuals and masses through an anthropological revolution, in order to regenerate the essence of humanity and to create a new man devoted body and soul to the realization of the revolutionary and imperialist projects of the totalitarian party, and thus a new civilization of a supranational nature." Gentile, *Politics as Religion*, p. 46.

[62] Ernst Nolte, *Three Faces of Fascism. Action Française, Italian Fascism, National Socialism*, New York, NY 1969, Appendix A, p. 571.

modified methods, always, however, within the unyielding framework of national self-assertion and autonomy."[63] In addition, he set out a "fascist minimum," which, though often criticized for being purely negative,[64] has been taken up by a number of writers, not least Stanley Payne, Roger Griffin and Roger Eatwell.[65] For Nolte, then, apart from anti-Marxism, central features of fascism were anti-liberalism and anticonservatism, the *Führerprinzip*, the party-army, and the claim to totality (which recurred in Gentile's definition).[66] Illuminating as this "fascist minimum" appeared to be, the further "philosophical" definition of fascism by this one-time disciple of Heidegger proved to be obscure and mysterious. At the most fundamental level, fascism was supposed to be a "metapolitical phenom-enon" involving "resistance to transcendence."[67] As Thomas Nipperdey pointed out in his review in the *Historische Zeitschrift*, the word transcendence "was not a happy choice, because it gave rise to misunderstandings" – a defect that Nolte himself had admitted.[68] But the very definition of fascism as a "metapolitical phe-nomenon," on a higher level than the realm of politics, seems to have been more pioneering than the necessarily essentialist "fascist minimum."

"Transcendence," Nolte writes, "looking back on what has been and forward to what is coming, reaches out toward the whole."[69] And resistance to such tran-scendence means resistance to the human striving beyond self to the eternal, resistance to "theoretical transcendence" qua "abstraction of thought."[70] On the other hand, resistance to "practical transcendence" qua "abstraction of life" is resistance to the process of estrangement or disengagement between man and the world. Fascists and their supporters no longer understand this estrangement process as "release and differentiation," but rather as a "torture and curse,"[71] as a symptom of decay.

This idea of "resistance to transcendence" might be seen as an early allusion to the awareness of order as a task – a dimension of fascism that will be considered further in this section, as it underlies Zygmunt Bauman's conception of moder-nity.[72] Nolte's process of estrangement and disenchantment not only releases people from a supposedly "natural" and "given" order, but also sows doubt about

---

[63] Ibid., p. 46.

[64] See Payne, *A History of Fascism*, p. 5.

[65] See Roger Eatwell, "On Defining the 'Fascist Minimum.' The Centrality of Ideology," *Journal of Political Ideologies* 1/1996, pp. 303–19; and Roger Griffin, *The Nature of Fascism*, London 1993 [repr.], p. 38.

[66] Ernst Nolte, *Die Krise des liberalen Systems und die faschistischen Bewegungen*, Munich 1968, p. 385.

[67] Nolte, *Three Faces of Fascism*, p. 537.

[68] Thomas Nipperdey, "Der Faschismus in seiner Epoche. Zu den Werken von Ernst Nolte zum Faschismus," *Historische Zeitschrift* 210/1970, pp. 620–38. See Nolte, *Three Faces of Fascism*, p. 539.

[69] Nolte, *Three Faces of Fascism*, p. 540.

[70] Ibid., p. 542.

[71] Ibid., p. 540.

[72] Bauman, *Modernity and Ambivalence*, esp. pp. 1–17.

whether such an order exists at all. Fascism rose up against the dissolution of "naturalness," against "progress," against the currents of practical and theoretical transcendence that postulated human beings as the center point of order – and thus against the process of cognitive emancipation. Human emancipation from the divine order had destroyed "fairy-tales" and bequeathed a longing for order as well as a dark foreboding of its impossibility. The emancipation process had generated "forces which are born of [that] process and then turn against their own origin"; yet these forces remained a prisoner of their origin in so far as they simply chose a human construct, the nation or *Volk*, as the center point of the order.[73] To put it more sharply, we might say that Nolte seems to assume in advance that fascism can be understood as resistance to the "transcendental homelessness" that first prompted the endeavors of bourgeois society toward theoretical and practical transcendence.

In the United States, George L. Mosse's *The Crisis of German Ideology* (1964) appeared more or less in parallel to studies of fascism inspired by Nolte.[74] Pointing to the *völkisch* roots of Nazi ideology, Mosse underlined the religious dimension of fascist ideology and practice and the purposes of mass mobilization served by its style of politics, its aesthetics, symbolism, rituals, myths, festivals and buildings. In his later *The Fascist Revolution* (1990), he argued that fascism should be understood as a cultural – hence "metapolitical" – movement and revolution:

> Fascism considered as a cultural movement means seeing fascism as it saw itself and as its followers saw it, to attempt to understand the movement in its own terms. Only then, when we have grasped fascism from the inside out, can we truly judge its appeal and its power. For fascism created a political environment which attempted to encompass the entire man or woman, to address, above all, the senses and emotions, and at the same time to make the abstract concrete as something uplifting and familiar which can be seen and touched.[75]

For Mosse, the following were the core of fascism studies: (1) a certain perception of human beings and an understanding of fascist self-presentations as a reflection of that perception; (2) nationalism and racism as a belief system; (3) the birth of fascism out of the First World War, leading to an emphasis on experiences of war, comradeship, and virility; and (4) the dialectic of leader and people.[76] Mosse not only anticipated the turn in cultural history; his work on the "nationalization of the masses," the cult of fallen soldiers and the construction of masculinity

---

[73] Nolte, *Three Faces of Fascism*, p. 567.

[74] George L. Mosse, *The Crisis of German Ideology. Intellectual Origins of the Third Reich*, London 1966 [repr.].

[75] George L. Mosse, *The Fascist Revolution. Towards a General Theory of Fascism*, New York, NY 1999, p. x.

[76] Ibid., pp. xi–xvii.

established the approach to fascism (the one also adopted here) that centered on the way in which it was perceived and interpreted by the contemporaries.[77]

One writer who made especially productive use of Mosse's new impetus was Emilio Gentile, the disciple of Mussolini's biographer Renzo De Felice. Already in 1975 Gentile's book on the origins of fascist ideology had pointed to the centrality of the myths of renewal and the new man.[78] And in his *The Sacralization of Politics in Fascist Italy*, first published in Italian in 1993, he showed how the Fascists aestheticized and sacralized politics, and argued that fascism may be understood as a political religion. Recent works by Gentile have constantly emphasized the modernity of Fascism:[79]

> Born from the experience of the Great War, fascism was a manifestation of political modernism, an ideology that accepted modernization and thought it possessed the formula by which to give human beings, swept by the whirl of modernity, the power to face the challenges of history and create a new civilization. Fascism was not anti-modern, but rather had its own vision of modernity which opposed the visions of liberalism, socialism, and communism, and which claimed the right to impose its own form of modernity on the twentieth century.[80]

Along with Mosse and Gentile, Stanley Payne forms a "bridge" between the second and third periods of fascism research. Having specialized in Spanish fascism since the early 1960s, Payne's two later books, *Fascism. Comparison and Definition* (1980) and a considerably more extensive *History of Fascism* (1995), took up and developed Nolte's attempt to establish a "fascist minimum." His own list of attributes was meant to make it easier to compare the many different manifestations of fascism; Nolte's minimum was expanded to include quite a number of categories, giving a more precise typology. However, Payne did not want fascism to be seen as a "monolithic, reified, taxonomic category," a kind of Platonic Idea.[81] He also grouped his list of criteria under three headings: (1) ideology and goals of the fascists; (2) negations; and (3) style and organization. His definition of fascism was "a form of revolutionary ultranationalism for national rebirth that is based on a

---

[77] See esp. George L. Mosse, *The Nationalization of the Masses. Political Symbolism and Mass Movements in Germany from the Napoleonic Wars through the Third Reich*, New York, NY 2001; *Fallen Soldiers. Reshaping the Memory of the World Wars*, New York, NY 1990; and *The Image of Man. The Creation of Modern Masculinity*, New York, NY 1996.

[78] Emilio Gentile, *The Origins of Fascist Ideology, 1918–1925*, New York, NY 2005 [repr.].

[79] See Emilio Gentile, *The Sacralization of Politics in Fascist Italy*, Cambridge, MA 1996; *Storia del partito fascista, 1919–1922. Movimento e milizia*, Rome 1989; *The Conquest of Modernity*, op. cit.; and *The Struggle for Modernity. Nationalism, Futurism, and Fascism*, Westport, CT 2003.

[80] Emilio Gentile, "The Myth of National Regeneration in Italy. From Modernist Avant-Garde to Fascism," in Matthew Affron/Mark Antliff (eds.), *Fascist Visions. Art and Ideology in France and Italy*, Princeton, NJ 1997, pp. 25–45; here p. 41.

[81] Stanley G. Payne, *Fascism. Comparison and Definition*, Madison, WI 1980; and *A History of Fascism*, p. 14.

primarily vitalist philosophy, is structured on extreme elitism, mass mobilization, and the *Führerprinzip*, positively values violence as an end as well as means, and tends to normatize war and/or the military values."[82]

Furthermore, in his "epignostic theory of fascism," Payne adds to this working definition by clarifying the circumstances in which major fascist movements arose. These necessary conditions comprise:

> [a] strong influence from the cultural crisis of the fin de siècle in a situation of perceived mounting cultural disorientation; the background of some form of organized nationalism before World War I; an international situation of perceived defeat, status humiliation, or lack of dignity; a state system comparatively new that was entering or had just entered a framework of liberal democracy; a situation of increasing political fragmentation; large sectors of workers, farmers, or petit bourgeois that were either not represented or had lost confidence in the existing parties; and an economic crisis perceived to stem in large measure from foreign defeat or exploitation.[83]

Four years earlier, following on from George L. Mosse and Emilio Gentile, Roger Griffin had defined fascism as "a genus of political ideology whose mythic core in its various permutations is a palingenetic form of populist ultra-nationalism."[84]

The shift of ideological fronts after the end of the Soviet empire formed the background for the third phase of comparative research on fascism, which Sven Reichardt dates from the publication of Griffin's book.[85] Once freed from any suspicion that they were offering a form of political legitimacy, historians were able to use the concept of fascism for an analysis of the "age of extremes," which had now seemingly come to an end.[86] Griffin, for example, in agreement with the results of Emilio Gentile's research, was already arguing in the 1990s that fascism had been embedded in modernism.[87] In his *Modernism and Fascism* (2007) he actually defined fascism as

> a form of programmatic modernism that seeks to conquer political power in order to realize a totalizing vision of national or ethnic rebirth. Its ultimate end is to overcome the decadence that has destroyed a sense of communal belonging and drained modernity of meaning and transcendence and usher in a new era of cultural homogeneity and health.[88]

---

[82] Payne, *A History of Fascism*, p. 14.

[83] Ibid., p. 494.

[84] Griffin, *The Nature of Fascism*, p. 26.

[85] Reichardt, "Neue Wege der vergleichenden Faschismusforschung," p. 11.

[86] On the use of this term, see Eric Hobsbawm, *The Age of Extremes. The Short Twentieth Century, 1914–1991*, London 1994.

[87] See Roger Griffin, "Introduction," in Roger Griffin (ed.), *International Fascism. Theories, Causes, and the New Consensus*, London 1998, pp. 1–20; here p. 14.

[88] Griffin, *Modernism and Fascism*, p. 182.

This approach had significant heuristic advantages. Drawing on Peter Osborne's *The Politics of Time*, Griffin showed that fascism had attempted to establish a new temporal order.[89] It had had a future-oriented dynamic, which Griffin, referring back to the conservative revolutionary Arthur Moeller van den Bruck, called one of "reconnection forwards" [*Wiederanknüpfung nach Vorwärts*].[90]

Building on these insights, the present study will demonstrate that neither the glorification of a mythical past nor the construction of a mythical modernity was "reactionary." Rather, they served to "root" a disoriented and transcendentally homeless society. In Nietzschean terms, fascists may be seen as "energetic and powerful" souls who placed the past in the service of life. Fascist "monumentalist history" served the palingenesis of the nation or *Volk*.[91] Fascism offered a new interpretation of the present as a breakthrough to the future of a revived eternal nation, which had decayed under liberalism and found its cohesion threatened by Marxism. Fascism was thus hailed by numerous intellectuals, as well as by sections of the masses, as an alternative modernity and the beginning of a new age. The core of this alternative modernity was the nation or people; it was the bedrock on which the new community rested. Before we turn to the concept of modernity, we should once again explain the context in which the idea of fascist modernism or the modernity of fascism is being considered.

Fascism and modernity

Although the common denominator of the many works in the third period of comparative fascism studies is not yet fully distinct, four characteristic features may be identified.[92] *First*, the turn from social to cultural history marks even works that are rather dismissive of "cultural studies."[93] *Second*, such works appear

---

[89] See Osborne, *The Politics of Time*; and Roger Griffin, "'I Am No Longer Human. I Am a Titan. A God!' The Fascist Quest to Regenerate Time," in Roger Griffin (ed.), *A Fascist Century. Essays*, New York, NY 2008, pp. 3–23.

[90] Griffin, *Modernism and Fascism*, pp. 132 and 177ff.

[91] On "monumentalist history," see Chapter III.1 below; and Friedrich Nietzsche, "On the Uses and Disadvantages of History for Life (1874)," in *Untimely Meditations*, Cambridge 1983, pp. 57–123, esp. p. 72, which also foregrounds the "great and powerful" (*Thätige und Mächtige*).

[92] In this connection, Roger Griffin spoke of a "new consensus" – although not everyone he included was in agreement. See Griffin's introduction to "The Primacy of Culture. The Current Growth (or Manufacture) of Consensus within Fascist Studies," *Journal of Contemporary History* 37/2002, pp. 21–43. Here (p. 24) the consensus is formulated as follows: "Fascism is a genus of modern politics which aspires to bring about a revolution in the political and social culture of a particular national or ethnic community. While extremely heterogenous in the specific ideology of its many permutations, in its social support, in the form of organization it adopts as an anti-systemic movement, and in the type of political system, regime, or homeland it aims to create, generic fascism draws its internal cohesion and affective driving force from a core myth that a period of perceived decadence and degeneracy is imminently or eventually to give way to one of rebirth and rejuvenation in a postliberal order."

[93] Even Robert Paxton – who believes that "the ideas that underlie fascist actions [...] are best deduced from those actions" and seeks to grasp fascism as a dynamic process – cannot

concerned to avoid essentialist definitions of fascism, although they are not always successful. Whether fascism is conceived dynamically (as in Robert Paxton[94]), sociologically, in terms of its membership (as in Michael Mann[95]), or praxeologically, through an analysis of its use of violence (as in Sven Reichardt[96]), there is factual agreement that the various fascisms were different and partly contradicted one another, but also that the same term can nevertheless be applied to them. *Third,* some 20 years after the beginning of this new period of fascism research inaugurated by the "victory" of the Western liberal system and the collapse of the Eastern bloc, it is possible to see in a new light the pan-European dimension of the fascist phenomenon and to contextualize the respective national narratives of antifascist resistance. *Fourth* – the most important point for our purposes – changes in the understanding of modernity have marked new ways of looking at fascism.

Initially, in the English-speaking countries, a newly critical approach to modernity in general asserted itself as the optimism of planning and progress disappeared, modernization theory faded away, and the rivalry between the two systems came to an end. A new reception of Horkheimer and Adorno, as well as impulses stemming from the French poststructuralists, also played a role in this development. Modernity itself was increasingly historicized. Today, in Germany too, it is possible to question the "project of modernity" or to expose its normative content without being labeled as conservative or even right-wing and revisionist.[97] To situate fascism in general or Nazism in particular within the context of modernity does not entail an exculpatory "normalization" of German or

---

avoid an ideal-typical definition that displays many of the "cultural" aspects mentioned by Griffin in the list he claims to be consensual. See Robert O. Paxton, *The Anatomy of Fascism,* New York, NY 2004, pp. 219–20.

[94] Paxton argues that the main focus should be on fascism as a process rather than on its inherent character: see *The Anatomy of Fascism,* as well as "The Five Stages of Fascism," *Journal of Contemporary History* 70/1998, pp. 1–23.

[95] Michael Mann, *Fascists,* Cambridge 2004.

[96] On the "praxeological" approach, see Sven Reichardt, "Praxeologie und Faschismus. Gewalt und Gemeinschaft als Elemente eines praxeologischen Faschismusbegriffs," in Karl H. Hörning/Julia Reuter (eds.), *Doing Culture. Neue Positionen zum Verhältnis von Kultur und sozialer Praxis,* Bielefeld 2004, pp. 129–53; and Sven Reichardt, *Faschistische Kampfbünde. Gewalt und Gemeinschaft im italienischen Squadrismus und in der deutschen SA,* Cologne 2002. In the latter work (p. 717) we read: "It is not ideology or a coherent political program that made fascism an autonomous, clearly distinguishable phenomenon, but rather its combination of forms of political practice with certain political attitudes. [...] Violence, organizational ceremony, and fervor of belief were general hallmarks of the fascist movement." The following five features, "in an ideal-typical sense," are for Reichardt characteristic of the ascending phase of Italian and German fascism: "1. Both movements were marked by a high degree of violence and militancy" (p. 719). "2. Neither developed stable internal institutions for the formation of the party's will, or a bureaucratic party structure" (p. 720). "3. Both movements claimed a monopoly over male youth and saw themselves as the 'organized will of the youth,' directed against the 'gerontocratic' and 'feminine' democracies" (p. 722). "4. Both Italian and German fascism displayed a characteristic form of politics" (p. 722). And "5. Both movements avoided the setting of clearcut programmatic objectives" (p. 724).

[97] Jürgen Habermas, "Modernity. An Unfinished Project," in Maurizio P. d'Entrèves/Seyla Benhabib (eds.), *Habermas and the Unfinished Project of Modernity,* Cambridge, MA 1997, pp. 38–55.

Italian history, nor does it involve downplaying the Holocaust. On the contrary, as Zygmunt Bauman showed back in 1989, anyone who speaks of the Holocaust may not remain silent about modernity.[98] As Detlev Peukert stated in 1982, fascist crimes represent "one of the pathological forms of development of modernity."[99] In 1996 Peter Fritzsche emphasized that National Socialism built a totalitarian version of modernity; its most conspicuous artifact, however, was not the gas mask (as Fritzsche suggested) but, as we will show, the aircraft.[100] To place fascism in the context of modernity is to see it as the product of a long-term development and to anchor it in a *longue durée* of contemporary history.

The relationship of National Socialism to modernity was raised in the second half of the 1960s, by Ralf Dahrendorf and David Schoenbaum, before it was taken up more widely in the course of the 1980s.[101] Since modernization theory and the *Sonderwegsthese* (the thesis of Germany's "special path") shaped how the question was posed, fascism appeared as a case of antimodern regression. But why?

The modernization theory associated with the US economist Walt W. Rostow was a normative theory of historical development. Arising during the Cold War, against the background of decolonization and the growing attractiveness of the Soviet model,[102] it offered a blueprint for the development of Western-style democracies based on the history of Britain and the United States. Certain basic assumptions – for example, a free market and a bourgeois middle class – had to be met for the transition from traditional to modern society to be successful. However, modern society was not only (highly) industrialized, grounded on science and technology, and uniquely oriented to economic growth; it was also conceived as a liberal-democratic capitalist society functioning in accordance with the British and American model. The liberal system of the Western democracies was thus equated with "modernity": it was *the* goal of history, in a teleological scheme of things. The course of Anglo-American history, in particular, was the measure of progress for the "underdeveloped" countries, but also for Germany, Italy and Japan. The Eurocentrism, or Atlantic-centrism, of the model needs no further explanation.[103]

---

[98] Bauman, *Modernity and the Holocaust*.

[99] Peukert, *Inside Nazi Germany*, p. 183.

[100] Fritzsche, "Nazi Modern."

[101] Dahrendorf, *Society and Democracy in Germany*; Schoenbaum, *Hitler's Social Revolution*. See also Henry A. Turner Jr., "Fascism and Modernization," *World Politics* 24/1972, pp. 547–64; Bavaj, *Die Ambivalenz der Moderne*; and Payne, *A History of Fascism*, pp. 471–86.

[102] On modernization theory and the Cold War, see Michael Latham, *Modernization as Ideology. American Social Science and 'Nation Building' in the Kennedy Era*, Chapel Hill, NC 2000. On modernization theory as the "organizing model of the postwar consensus," see Anselm Doering-Manteuffel/Lutz Raphael, *Nach dem Boom. Perspektiven auf die Zeitgeschichte seit 1970*, Göttingen 2008, pp. 21–7, 60–1.

[103] Dipesh Chakrabarty, *Provincializing Europe. Postcolonial Thought and Historical Difference*, Princeton, NJ 2000. Claude Lévi-Strauss and others pointed as early as the 1950s to the normative character of the Western model of progress: see his *Race and History*, Paris (Unesco) 1952.

The *Sonderwegsthese* too arose against the background of modernization theory.[104] An underdeveloped bourgeoisie, the failed revolution of 1848, a weak liberalism, the dominance of traditional elites, an arrogant sense of superiority over Western "civilization": all these played a role in Germany's road to the "Third Reich." According to advocates of the thesis, it was those deviations from the Western norm that led to the break with civilization represented by the Holocaust; despite the bureaucratic planning and organization, the underlying eugenic ideas, the praxis of social engineering and the industrial execution of genocide involving millions of people, the Shoah was excluded from modernity. In the early 1980s, however, Geoff Eley was already arguing convincingly against this interpretation. In his view, the origins of German fascism lay less in supposed "preindustrial continuities" with the Empire than in the specific contradictions of a society that had undergone rapid capitalist transformation. Germany's susceptibility to fascism stemmed not from a "preindustrial blockage of modernization," but from an accelerated irruption of modernity.[105]

But the admission of a nexus between National Socialism and modernity would have meant questioning the hard-won self-understanding of the Bundesrepublik, which, notwithstanding the break with civilization, appeared to have completed the "long road to the West."[106] This self-image began to be questioned anyway in the early 1980s. But the discussion of the modernity of National Socialism took place against the background of: (1) a new trend in political-historical discourse and in the politics of memory that came with the Kohl era; (2) the *Historikerstreit* over Ernst Nolte's "causal nexus" thesis; and (3) the reorganization of Europe following the end of the Cold War. It was in this context, that Rainer Zitelmann, who was associated with Nolte, wrote a dissertation that tried to present Hitler as a "revolutionary" committed to the modernization of Germany.[107] And a few years later, Zitelmann and Michael Prinz edited a work on "National Socialism and

---

[104] Thomas Mergel, "Die Modernisierungstheorie auf dem Weg zu einer Theorie der Moderne," in Thomas Mergel/Thomas Welskopp (eds.), *Geschichte zwischen Kultur und Gesellschaft. Beiträge zur Theoriedebatte*, Munich 1997, p. 232. See also Hans-Ulrich Wehler, *Modernisierungstheorie und Geschichte*, Göttingen 1975; and *Deutsche Gesellschaftsgeschichte*, vol. 5, *Von der Gründung der beiden deutschen Staaten bis zur Vereinigung 1949–1990*, Munich 2008. On p. 782 of the latter, Wehler argues that the decisive influence was not Rostow's modernization theory but "the analysis and interpretation developed by Adam Smith, Karl Marx, Max Weber, and many others of the evolutionary process that led to Western modernity."

[105] Geoff Eley, "The British Model and the German Road. Rethinking the Course of German History Before 1914," in David Blackbourn/Geoff Eley (eds.), *The Peculiarities of German History. Bourgeois Society and Politics in Nineteenth-Century Germany*, Oxford 1984, pp. 37–155, pp. 154–55.

[106] On the *Sonderweg* as the "master narrative" of the Federal Republic, see Thomas Welskopp, "Identität *ex negativo*. Der 'deutsche Sonderweg' als Metaerzählung in der bundesdeutschen Geschichtswissenschaft der siebziger und achtziger Jahre," in Konrad H. Jarausch (ed.), *Die historische Meistererzählung. Deutungslinien der deutschen Nationalgeschichte nach 1945*, Göttingen 2002, pp. 109–39.

[107] See Zitelmann, *Hitler*.

modernization" that once more ran into the danger of abetting a rehabilitation of National Socialism.[108] Still, if we abstract from these initiatives (which may have had a motivation in the politics of the day), the anchoring of fascism in the context of modernity brought a clear heuristic gain. We should therefore now look briefly at the powerful works by Detlev Peukert and Zygmunt Bauman, and once again at those of Jeffrey Herf.

In his *Inside Nazi Germany*, originally published in German in 1982, Peukert – one of the first German historians to take the work of Foucault seriously[109] – pointed out that "the practice of terror by the National Socialists" harked back to the "repressive aspect of discipline and standardization inherent in social-political conceptions around the turn of the century." He also noted:

> Consistent in its rejection of the legacy of 1789, National Socialism envisaged a society with modern institutions and technologies but owing nothing to the ideals of equal rights, emancipation, self-determination, and fraternity. It pushed the utopian belief in all-embracing final solutions of social problems to the ultimate logical extreme, pursuing a bureaucratic racial-biological design and ultimately eradicating all sources of maladjustment and friction.[110]

Peukert not only traced the "genesis of the Final Solution out of the spirit of science,"[111] but in distinguishing between "modern institutions and technologies" and the models of 1789 he took the first decisive step toward a more accurate understanding of modernity, foreshadowed by Max Weber in his *Science as a Vocation*.[112]

While Peukert recognized that fascism should be seen as "one of the possibilities of modern civilization in crisis,"[113] and that "modern institutions and technologies" by no means necessarily entail the desirable goals or values of

---

[108] Prinz/Zitelmann (eds.), *Nationalsozialismus und Modernisierung*. For critiques of Prinz and Zitelmann, see Bavaj, *Die Ambivalenz der Moderne*, pp. 42ff. and Christof Dipper, "Zwischen 'Historikerstreit' und der Debatte über 'Nationalsozialismus und die Moderne,'" in Gertraud Diendorfer/Gerhard Jagschitz/Oliver Rathkolb (eds.), *Zeitgeschichte im Wandel*, Innsbruck 1998, pp. 110–21; Norbert Frei, "Wie modern war der Nationalsozialismus?," *Geschichte und Gesellschaft* 19/1993, pp. 367–87; Hans Mommsen, "Nationalsozialismus und Modernisierung," *Geschichte und Gesellschaft* 21/1995, pp. 391–402; Axel Schildt, "NS-Regime, Modernisierung und Moderne. Anmerkungen zur Hochkonjunktur einer andauernden Diskussion," in Dan Diner/Frank Stern (eds.), *Nationalsozialismus aus heutiger Perspektive*, Göttingen 1994, pp. 3–22.

[109] See Detlev J. K. Peukert, "Die Unordnung der Dinge. Michel Foucault und die deutsche Geschichtswissenschaft," in François Ewald/Bernhard Waldenfels (eds.), *Spiele der Wahrheit. Michel Foucaults Denken*, Frankfurt/Main 1991, pp. 320–39.

[110] Peukert, *Inside Nazi Germany*, p. 248 (translation modified). On Peukert's understanding of modernity, see also his *The Weimar Republic*, esp. pp. 275–81.

[111] Peukert, *Max Webers Diagnose der Moderne*, pp. 102–21. In this book (pp. 81–2) he also clearly distanced himself from the *Sonderwegsthese*.

[112] Max Weber, "Science as a Vocation," in H. H. Gerth/C. Wright Mills (eds.), *From Max Weber. Essays in Sociology*, Boston, MA 1948, pp. 129–56.

[113] Peukert, *Max Webers Diagnose der Moderne*, pp. 81 and 104.

liberty, equality and fraternity, the American historian Jeffrey Herf took as his starting point a necessary unity of means and ends. His widely accepted interpretation of "reactionary modernism" based itself upon a normative conception of modernity:[114] antimodernist or even reactionary were those who rejected the ideals of the French Revolution and the social-economic results of the Industrial Revolution, and so too were the Romanticism, antimaterialism and antirationalism of the "conservative revolutionaries" and National Socialists; modernist, on the other hand, was their incorporation of technology. "Reactionary modernists," Herf argued, succeeded in cutting the nexus between technology and "civilization." They aestheticized technology, interpreting it as an instrument of the will to power, and concluded that it was indispensable in the Darwinian struggle for survival. In this way, "modernist" technology was integrated into the "reactionary" sphere of German culture.[115] "Reactionary modernists" therefore did precisely what Herf and other historians working in the context of modernization theory had abstained from; they differentiated between means and ends.

This distinction was also central to the work of a sociologist standing in the tradition of Horkheimer and Adorno, namely Zygmunt Bauman.[116] Like Peukert, he argued in his *Modernity and the Holocaust* (1989) that the Holocaust was "a rare, yet significant and reliable, test of the hidden possibilities of modern society."[117] The aetiological – that is, foundational – myth of Western societies spoke of their ascent from barbarism.[118] According to the "myth of the Enlightenment," reproduced by scores of sociological theories and historical narratives alike, the Holocaust could be understood only in terms of a failed civilizing process, an "incomplete project of modernity." But the exact opposite was the case:

> The "Final Solution" did not clash at any stage with the rational pursuit of efficient, optimal goal-implementation. On the contrary, it arose out of a genuinely rational concern, and it was generated by bureaucracy true to its form and purpose. [...] The Holocaust was not an irrational outflow of the not-yet-fully eradicated residues of pre-modern barbarity. It was a legitimate

---

[114] See Herf, *Reactionary Modernism*; and his "Der nationalsozialistische *Technikdiskurs*." In the latter article (p. 72) he writes: "By modernity we should not exclusively or principally understand the scientific and technological progress of the last two centuries. It certainly includes this, but it is much more. It also encompasses the formation of nations and states, a certain notion of human dignity irrespective of social or ethnic status, economic markets regulated by the stable rule of law, and conceptual ideas of individual freedom, social equality, and liberal democracy."

[115] Herf, *Reactionary Modernism*, pp. 224–25.

[116] Anson Rabinbach, "Nationalsozialismus und Moderne. Zur Technik-Interpretation im Dritten Reich," in Wolfgang Emmerich/Carl Wege (eds.), *Der Technikdiskurs in der Hitler-Stalin-Ära*, Stuttgart 1995, pp. 94–113, here p. 98. With *The Dialectic of Enlightenment* in mind, Rabinbach points out that for Bauman "a modernity defined by the rationalization of means" ultimately presupposes "the irrationality of ends."

[117] Bauman, *Modernity and the Holocaust*, p. 12.

[118] Ibid.

resident in the house of modernity; indeed, one who would not be at home in any other house.[119]

In any event, the rules of instrumental reason could not prevent such a phenomenon as the Holocaust:

> there is nothing in those rules which disqualifies the Holocaust-style methods of "social engineering" as improper or, indeed, the actions they served as irrational. [...] it was the spirit of instrumental rationality, and its modern, bureaucratic form of institutionalization, which made the Holocaust-style solutions not only possible, but eminently "reasonable" – and increased the probability of their choice.[120]

For Bauman, the spirit of instrumental reason and modernity gave birth to a "gardening state," which treated nature and society as objects to be ruled, improved and organized.[121] The genocidal side of the fascist "gardening state" will not be considered in the present study, but it should be noted here that the construction of the new man was only the other face of the modern Janus head. To remain with Bauman's metaphor, we may say that the same gardener who eradicated "weeds" also cultivated the "useful plants" he wished to have. Cultivation and eradication were two sides of the same coin.

Before we look more closely at the concept of modernity in the light of Bauman's theoretical work, mention should be made of some special studies of the fascism–modernity nexus that have appeared in recent years. Their main focus has been the question of the cultural modernity of Italian Fascism in particular,[122] that is, "modernist" aspects in the narrower sense of the term: the links between Fascism and *avant garde* or other artists and the self-presentation of the regime and its dealings with the mass media,[123] as well as the character of Fascist warfare.[124]

---

[119] Ibid., p. 17.

[120] Ibid., p. 18.

[121] Cf. Bauman, *Modernity and Ambivalence*; and James C. Scott, *Seeing Like a State. How Certain Schemes to Improve the Human Condition Have Failed*, New Haven, NJ 1998.

[122] There is a survey of the literature in Paul Betts, "The New Fascination with Fascism. The Case of Nazi Modernism," *Journal of Contemporary History* 37/2002, pp. 541–58. See also JCH 3/1996, *Fascism and Culture*, on modernism and modernity; JCH 31/1996, *The Aesthetics of Fascism*; and *Stanford Italian Review* (1990) on *Fascism and Culture*. Here we shall mention only the most important monographs on the subject; further references may be found in individual chapters.

[123] See among others: Mark Antliff, *Avant-Garde Fascism. The Mobilization of Myth, Art, and Culture in France, 1909–1939*, Durham, NC 2007; Ben-Ghiat, *Fascist Modernities*; Emily Braun, *Mario Sironi and Italian Modernism*, New York, NY 2000; Patricia Chiantera-Stutte, *Von der Avantgarde zum Traditionalismus. Die radikalen Futuristen im italienischen Faschismus von 1919 bis 1931*, Frankfurt/Main 2002; Simonetta Falasca-Zamponi, *The Fascist Spectacle. The Aesthetics of Power in Mussolini's Italy*, Berkeley, CA 2000 [repr.]; Claudia Lazzaro/Roger J. Crum (eds.), *Donatello among the Blackshirts. History and Modernity in the Visual Culture of Fascist Italy*, Ithaca, NY 2005; Marla Susan Stone, *The Patron State. Culture and Politics in Fascist Italy*, Princeton, NJ 1998; Frank Vollmer, *Die politische Kultur des Faschismus. Stätten totalitärer Diktatur in Italien*, Cologne 2007.

[124] Giulia Brogini-Künzi, *Italien und der Abessinienkrieg 1935/36. Kolonialkrieg oder Totaler Krieg*, Paderborn 2006; Aram Mattioli, *Experimentierfeld der Gewalt. Der Abessinienkrieg und seine internationale Bedeutung 1935–1941*, Zurich 2005.

Similar studies of National Socialism remain much to be desired,[125] as the exception of Peter Fritzsche's pioneering essay "Nazi Modern" confirms.[126]

Modernity

What modernity is – an attitude or *ethos*,[127] or even an epoch – and when it began and perhaps ended, are questions that have been more intensely debated since the end of the postwar boom and the rise of theories of "post," "late," "second" or "reflexive" modernity.[128] Among historians, however, as we have just noted, the widespread use of the term does not at all mean that there is clarity about its content.[129] A clear-cut and above all definitive answer is no more possible than in the case of "fascism"; nor will any attempt be made to provide one. But it needs to be explained how the word will be understood and used. The present study supports itself on Zygmunt Bauman's theory of modernity and its underlying concept of order, complementing these with a temporal dimension.

Modernity is first of all a historiographical category, developed out of the investigative horizon that marked classical sociology in the second half of the nineteenth century.[130] Theorists such as Marx and Weber, as well as Durkheim, Simmel and Tönnies, were all concerned with the transformation of "traditional" societies and the rise of new formations. But it was in the 1950s and 1960s that the varying collection of features attributed to "modern" societies found entry into modernization theory as such and its teleological narrative of development. It set out certain "basic processes" that distinguished modern from traditional society: for example, rationalization and secularization, the dominance of science and technology, industrialization and urbanization, social differentiation and individualization, increased social and spatial mobility, and the rise of mass culture and a public space accessible to the masses.[131] This list, comprising a kind of

---

[125] Not the least indication of this is that, since the monographs by Reichel and Michaud (originally published in German and French respectively, in 1991 and 1996), scarcely any comparable works have appeared. See Eric Michaud, *The Cult of Art in Nazi Germany*, Stanford, CA 2004; Peter Reichel, *Der schöne Schein des Dritten Reichs. Gewalt und Faszination des deutschen Faschismus*, Hamburg 2006 [repr.]. A notable exception is: Uwe Hebekus, *Ästhetische Ermächtigung. Zum politischen Ort der Literatur im Zeitraum der Klassischen Moderne*, Munich 2009.

[126] Fritzsche, "Nazi Modern."

[127] On the ethos of modernity, see Eva Erdmann/Rainer Forst/Axel Honneth (eds.), *Ethos der Moderne. Foucaults Kritik der Aufklärung*, Frankfurt/Main 1990.

[128] See Doering-Manteuffel/Raphael, *Nach dem Boom*.

[129] See Ute Schneider, "Spurensuche. Reinhart Koselleck und die 'Moderne,'" in Lutz Raphael/ Ute Schneider (eds.), *Dimensionen der Moderne. Festschrift für Christof Dipper*, Frankfurt/Main 2008, pp. 61–71; and Christoph Cornelißen, "Ein ständiges Ärgernis? Die Moderne in der (west-)deutschen Geschichtsschreibung," in ibid., pp. 235–48.

[130] On the following, see Thorsten Bonacker/Andreas Reckwitz (eds.), *Kulturen der Moderne. Soziologische Perspektiven der Gegenwart*, Frankfurt/Main 2007.

[131] For a "dichotomy alphabet" of traditional and modern societies, see Wehler, *Modernisierungstheorie und Geschichte*, pp. 14–15; also Ulrich Herbert, "Europe in High Modernity. Reflections on a Theory of the Twentieth Century," *Journal of Modern European History* 5/2007, pp. 5–21. For a critique of the concept of high modernity, see Lutz Raphael, "Ordnungsmuster der 'Hochmoderne?' Die Theorie der Moderne und die Geschichte der

"modernist minimum," will be borne in mind in what follows, but the main aim here is to underpin it with a more elementary definition that does justice to the possibility of "multiple modernities," including variants of modernity and their narratives that deviate from the Western liberal norm.[132]

The most serious and comprehensive account of modernity and the transformation of "traditional" societies goes back to Max Weber's theory of Western rationalization and "disenchantment" of the world,[133] which, he argued, led to the overcoming of magical and mythical world-views, extensive secularization, and a capacity "in principle, [to] master all things by calculation."[134] Capitalist economy too was grounded on it.[135] Although Weber himself was skeptical about the scale of this Western rationalization process and the possibility of rational legitimation of ethical-normative judgments, the Weberian account of the transformation of traditional societies as a process of rationalization became the foundation for the normative definition of modernity characteristic of modernization theory. Theorists in this tradition believed that the Western evolutionary narrative of progress, based on an understanding of supposedly universal (instrumental) reason, could be consistently maintained and transposed to societies outside Europe. It was an attitude that persisted beyond its peak in the 1960s, involving a belief that the whole world could be calculated, planned and systematized.[136]

This book does not share the interpretation of Weber advanced by modernization theory. It will, to be sure, accept that the questioning of "traditional" metaphysical-religious world-views was the gateway to modernity, that modern attitudes were a result of the "disenchantment of the world." But it will regard rationalization, and especially the secularization of world-views, as an incomplete process – even one that can never be successfully completed.[137] Despite their "rational apologetics,"

---

europäischen Gesellschaften im 20. Jahrhundert," in Raphael/Schneider (eds.), *Dimensionen der Moderne*, pp. 73–91.

[132] See Shmuel N. Eisenstadt, "Multiple Modernities," *Daedalus* 129/2000, pp. 1–29.

[133] See Max Weber, "The 'Rationalism' of Western Civilization," in Stephen Kalberg (ed.), *Readings and Commentaries on Modernity*, Malden, MA 2005, pp. 53–64; and *The Protestant Ethic and the Spirit of Capitalism* [1930], New York, NY 1992.

[134] Weber, "Science as a Vocation," p. 139.

[135] On the role of this process in Weber's theory, see Jürgen Habermas, *The Theory of Communicative Action*, vol. 1, *Reason and the Rationalization of Society*, Boston, MA 1984, pp. 143–242; and Wolfgang Schluchter, *The Rise of Western Rationalism. Max Weber's Developmental Theory*, Berkeley, CA 1981.

[136] Anselm Doering-Manteuffel, "Konturen von Ordnung in den Zeitschichten des 20. Jahrhunderts," in Thomas Etzemüller (ed.), *Die Ordnung der Moderne. Social Engineering im 20. Jahrhundert*, Bielefeld 2009, pp. 41–64; Gabriele Metzler, "'Geborgenheit im gesicherten Fortschritt.' Das Jahrzehnt von Planbarkeit und Machbarkeit," in Matthias Frese/Julia Paulus/Karl Teppe (eds.), *Demokratisierung und gesellschaftlicher Aufbruch. Die sechziger Jahre als Wendezeit der Bundesrepublik*, Paderborn 2003, pp. 777–97; and Dirk van Laak, "Planung. Geschichte und Gegenwart des Vorgriffs auf die Zukunft," *Geschichte und Gesellschaft* 34/2008, pp. 305–26.

[137] See *inter alia* Michael Burleigh, *Earthly Powers. The Clash of Religion and Politics in Europe from the French Revolution to the Great War*, New York, NY 2005, and also (for the United States)

modern orders of reality too remain rooted in metaphysics and the transcendental. In this sense, as Bruno Latour rightly said, "we have never been modern."[138]

Zygmunt Bauman links up, via Horkheimer and Adorno, with aspects of Weber's legacy and weaves them into his theory of modernity. But before we look more closely at his definition, we should stress again that our own vantage point on modern life is floating and fluctuating; the present itself is marked by "volatility" and "liquidity."[139] Our relationship to modernity is double-edged, and the modern age itself was ambivalent in so far as it involved both solidification and its opposite: "all that is solid melts into air" (to quote Marx and Engels). Modernity was (and is) breakup and consolidation. It was and is marked by the dialectic of freedom and order.

Freedom. The threshold of the modern age was crossed – according to the Kantian–Weberian perspective – when "man emerged from his self-incurred immaturity" and freed himself from the traditional religious-metaphysical order.[140] The "modern spirit" turned against the prevailing order, profaning "the sacred," disavowing the past, and dethroning tradition. But

> all this was done not in order to do away with the solids once and for all and make the brave new world free of them for ever, but to clear the site for *new and improved solids*; to replace the inherited set of deficient and defective solids with another set, which was much improved and preferably perfect, and for that reason no longer alterable.[141]

The wish to do away with the old order appeared reasonable because everything was

> already rusty, mushy, coming apart at the seams, and altogether unreliable. Modern times found the pre-modern solids in a fairly advanced state of disintegration; and one of the most powerful motives behind the urge to melt them was the wish to discover or invent solids of – for a change – *lasting* solidity.[142]

Bauman has in mind the dissolution of the Ancien Régime, the "dual revolution" which, based on the Enlightenment and the rise of capitalism, reached a climax in the French and the Industrial Revolution.[143] Here we see the initial

---

Michael Hochgeschwender, *Amerikanische Religion. Evangelikalismus, Pfingstlertum und Fundamentalismus*, Frankfurt/Main 2007.

[138] Bruno Latour, *We Have Never Been Modern*, Cambridge, MA 1993. See also Hartmut Böhme, *Fetischismus und Kultur. Eine andere Theorie der Moderne*, Reinbek 2006, pp. 22–3.

[139] Here and on the following see Bauman, *Liquid Modernity*, esp. pp. 1–15.

[140] Kant, "What Is Enlightenment?," p. 54.

[141] Bauman, *Liquid Modernity*, p. 3. Emphases in the original.

[142] Ibid. Emphases in the original.

[143] See Eric J. Hobsbawm, *The Age of Revolution. Europe 1789–1848*, London 1962. Elsewhere, Bauman refers to the impossibility of periodizing modernity. But he calls modernity the "historical period that began in Western Europe with a series of profound social-structural and intellectual transformations of the seventeenth century and achieved its maturity: (1) as

positive face of Janus-headed modernity; the critical relationship to the past, the reflexive relationship to the present, the hope in a different and better future.

Order. But once the encrusted old order has been laid waste, the past obliterated, and the entry to a different future begun, Janus rears his other face. As the longed-for freedom threatens to become experienced as ambivalence, contingency and anomie, a struggle against the latter is waged in the name of order. Thus, if the order-dissolving spirit brings forth chaos, its positive connotation passes instead to the anomie-removing, order-creating spirit of modernity.[144] Order becomes a task:

> Among the multitude of impossible tasks that modernity set itself and that made modernity into what it is, the task of order (more precisely and most importantly, of *order as a task*) stands out. [...] We can think of modernity as of a time when order – of the world, of the human habitat, of the human self, and of the connection between all three – is *reflected upon*, a matter of thought, of concern, of a practice that is aware of itself, conscious of being a conscious practice, and wary of the void it would leave were it to halt or merely relent.[145]

Liberation from order has an unsettling effect, since there is no longer any *nomos* to protect against the "terror" or "absolutism of reality."[146] The answers to the questions of what I can know, what I should do and what I may hope lose their supposed clarity. Ambivalence and contingency become more widespread, calling forth disorientation and resistance. The need for meaning grows, and that means the need for order. If the breakdown of order leaves behind chaos, enlightenment reverts to mythology.[147] The order-dissolving spirit of modernity becomes order-generating. The open horizon of expectation gets closed and the shape of the future defined. The present becomes the site of the conscious realization of the vision of order, of the overcoming of ambivalence and of the generation of a new clarity. The past becomes a reserve from which any traditions can be selected to root and legitimize the planned order.

We cannot here consider in detail whether, or how, the liberal, enlightening, "disenchanting" spirit became itself a source of mythical enchantment. But we will show how fascism – which rolled together many supposedly antimodern currents of the high industrial period that began around 1890 – rebelled against the failed "gerontocratic" liberal order and promised a new departure with

---

a cultural project – with the growth of Enlightenment; (2) as a socially accomplished form of life – with the growth of industrial (capitalist, and later also communist) society." Bauman, *Modernity and Ambivalence*, p. 4.

[144] This latter spirit itself becomes negatively connoted if the new order is experienced as inhuman and inflexible – not least because chaos is rediscovered where order was supposed to have been established. See Bauman, *Modernity and Ambivalence*, esp. pp. 13ff.

[145] Ibid., pp. 4–5.

[146] See Peter L. Berger, *The Sacred Canopy*, esp. pp. 22–6. On the absolutism of reality, see Hans Blumenberg, *Work on Myth*, Cambridge, MA 1985.

[147] Adorno/Horkheimer, *Dialectic of Enlightenment*, p. 8.

positive connotations for many people living at the time.[148] Aviation was made a metaphor for this awakening and for the stated objective.

The striving for order and solidity was so prominent in fascism that its promise of a new dawn, with its revolutionary, order-dissolving dynamic, has mostly been left out of account or interpreted as a form of reaction.[149] Yet, in negating the past and shaping the present in the name of a "better" future, the fascists fulfilled the essential criterion of modernity. Liberal modernity appeared chaotic and meaningless to many of their contemporaries, and a wish to overcome the existing order began to grow at the latest in the years after the First World War. For again everything was "rusty, musty, coming apart at the seams and altogether unreliable," and "one of the most powerful motives behind the urge to melt [the premodern solids] was the wish to discover or invent solids of lasting solidity."[150] Both Communism/Bolshevism and fascism promised a stable order which, being "perfect" and "definitive," was of "lasting solidity." Only a mythical order – whose mythical character will be explained in the following section – was capable of satisfying this longing. Only a mythical order was capable of satisfying this longing for something eternally valid.

In any case, we should take Bauman at his word: "order and chaos are *modern twins*."[151] The order-generating spirit is unthinkable without the order-dissolving spirit, which creates and sharpens the awareness of chaos. Where chaos is revealed and order is already conceived as a task, the spirit or ethos of modernity is already present. Then there is no turning back. Man has already emerged out of order and knows in so far as he knows of the contingency of the metaphysical-religious order. This awareness of the historicity and foundations of all order can be suppressed and denied, but not forgotten. Once in the world, historical knowledge shows itself to be universal; in principle, nothing is protected against historicization. Nietzsche expressed this ambiguity as early as 1889: "Once you discovered me, it was no great feat to find me: the difficulty now is to lose me."[152] The difficulty was to find something unhistorical or suprahistorical. The attempt to build a stable, definitive order is thus driven by a longing to forget the historicity and contingency of existence and to lose the spirit of modernity. But this longing is itself intrinsically modern; it is a logical consequence of the awareness of chaos and the discovery of order as a task.

Modernity, then, turns out to be a dialectical process, involving the delegitimation and destruction of order through historical thinking, as well as the quest for and construction of order on the basis of a putative Eternal and Absolute.

---

[148] See Gunther Mai, "Agrarische Transition und industrielle Krise. Anti-Modernismus in Europa in der ersten Hälfte des 20. Jahrhunderts," *Journal of Modern European History* 4/2006, pp. 5–37.

[149] The most notable exception is Roger Griffin: see his *Modernism and Fascism*.

[150] Bauman, *Liquid Modernity*, p. 3.

[151] Bauman, *Modernity and Ambivalence*, p. 4. Emphasis in the original.

[152] Letter to Georg Brandes, January 4, 1889, in Christopher Middleton (ed.), *Selected Letters of Friedrich Nietzsche*, Chicago, IL 1969, p. 345. Cf. Bauman, *Modernity and Ambivalence*, pp. 4, 234.

Modernity is characterized by a dialectic of freedom and order that bears within itself the seeds of radicalization, since life in freedom is perceived as chaotic and meaningless, and the longed-for order is impossible. Awareness of chaos and discovery of order as a task, knowledge of the non-natural character of the "god-given" order, are what differentiate modernity from premodernity.[153] And it is belief in the possibility of order as such, together with hope that ambivalence can be overcome with the help of a non-historicizable foundation of order, which distinguishes solid modernity from liquid modernity. The spirit of liquid modernity is marked by a truly all-encompassing historical knowledge, which does not spare even the historicizing subject; everything is made, everything is conditional.

The basic hypothesis of this book is that where the foundation of order is not historicized, where belief in the Unhistorical or Suprahistorical prevails, man still lives within a mythical horizon. Despite the unwelcome teleological implication, it may be argued that modernity is an intermediate phase, wedged between a world "ignorant" of the problem of order and chaos and one that has lost faith in the possibility of order (even if the longing for it has not disappeared, despite the knowledge of its catastrophic side-effects).

Before a temporal dimension is added to this definition, we should again emphasize that modernity is a spirit, attitude or ethos rather than an epoch. It could be the latter only if the spirit of modernity had been all-encompassing. We should start, however, from the idea of the "simultaneity of the non-simultaneous," which treats neither the Before nor the After as being of greater value. Knowledge or belief that the project to delineate order and remove ambivalence is an impossibility coexists with that very project and with knowledge of, or belief in, the "natural," "god-given" order. The birth of the problem of order as the beginning of modernity, or of doubt about the possibility of order as the end of modernity,[154] cannot be dated in a clear-cut manner, since there are various intersections and overlapping attitudes to order. Nevertheless, in the Euro-Atlantic area the spirit of modernity has known definite highpoints; the clearest is the age following the Enlightenment.

Reinhart Koselleck's conception of a "bridge period" (*Sattelzeit*) "between 1770 and 1830" may be integrated into this definition of "modernity";[155] the fact that he does not explicitly speak of a "modern" age, preferring the term *Neuzeit*, is

---

[153] See Bauman, *Modernity and Ambivalence*, p. 6: "The discovery that order was *not natural* was discovery of *order as such*. The *concept* of order appeared in consciousness only simultaneously with the *problem* of order, of order as a matter of *design* and *action*, order as an obsession. To put it yet more bluntly, order as a problem emerged in the wake of the ordering flurry, as a reflection on ordering practices. Declaration of the 'non-naturalness of order' stood for an order already coming out of hiding, out of non-existence, out of silence."

[154] See Gianni Vattimo, *The End of Modernity. Nihilism and Hermeneutics in Postmodern Culture*, Cambridge 1988.

[155] Reinhart Koselleck, "*Neuzeit* – Remarks on the Semantics of Modern Concepts of Movement," in Koselleck, *Futures Past. On the Semantics of Historical Time*, pp. 222–54; here p. 247. See also Lynn Hunt, *Measuring Time, Making History*, Budapest 2008.

no obstacle to this.[156] In short, Koselleck's theses imply that modern concepts of movement, and especially of progressive advance, took shape at a time when a gulf was opening between "space of experience" and "horizon of expectation": that is, when future expectations were superseding what all past experiences offered:

> The peasant world, which two hundred years ago comprised up to 80 per cent of all persons in many parts of Europe, lived within the cycle of nature. [...] the expectations cultivated in this peasant-artisan world (and no other expectations could be cultivated) subsisted entirely on the experiences of their predecessors, experiences which in turn became those of their successors. [...] As long as the Christian doctrine of the Last Things set an immovable limit to the horizon of expectation (roughly speaking, until the mid-seventeenth century), the future remained bound to the past.[157]

As the pace of change quickened, however, this old temporal order broke up. A new experience of time then gave rise to a new temporal order, which harnessed and stabilized that very experience. Koselleck's thinking may here be harmonized with Bauman's: the new experience of time – or the new temporality of accelerated change and its accompanying loss of past experience – generated the need to reorder the three levels of time: past, present and future.

The modern experience of time, including the development of historicism and the reorganization of these three temporal dimensions, cannot be elucidated here. Despite Koselleck's theory of historical times, it remains an acute task for research to clarify this whole complex web of questions. But all we can do here is refer to the work of Koselleck, and to Peter Osborne's little noticed book *The Politics of Time*.[158] Osborne, like Koselleck, points out that modernity is characterized by a distinctive temporality: "Modernity is a form of historical time which valorizes the new as the product of a constantly self-negating temporal dynamic." It is a special kind of temporalization of history, "through which the three dimensions of phenomenological or lived time (past, present and

---

[156] Cf. Schneider, *Spurensuche*; and Christof Dipper, "Die 'Geschichtlichen Grundbegriffe.' Von der Begriffsgeschichte zur Theorie der historischen Zeiten," *Historische Zeitschrift* 270/2000, pp. 281–308.

[157] Koselleck, "'Space of Experience' and 'Horizon of Expectation,'" pp. 263–64.

[158] Now also see Aleida Assmann, *Ist die Zeit aus den Fugen? Aufstieg und Fall des Zeitregimes der Moderne*, Munich 2013; Hans Ulrich Gumbrecht, *Unsere breite Gegenwart*, Berlin 2010; François Hartog, *Régimes d'historicité. Présentisme et expériences du temps*, Paris 2003; Hunt, *Measuring Time*; Stephen Kern, *The Culture of Time and Space, 1880–1918*, London 1983; Achim Landwehr, *Die Geburt der Gegenwart. Eine Geschichte der Zeit im 17. Jahrhundert*, Frankfurt/Main 2014; Chris Lorenz/Berber Bevernage (eds.), *Breaking up time. Negotiating the Borders between Present, Past and Future*, Göttingen 2013 and especially Koselleck's *Futures Past* and Osborne, *The Politics of Time*.

future) are linked together within the dynamic and eccentric unity of a single historical view."[159]

The acceleration of change leads to a dislocation of the three temporal dimensions; the nexus of past, present and future appears broken.[160] Permanent denial of, and breaks with, the past lead to a hiatus between past and present and a devaluation of accumulated experience. The less experience functions as a guide to the future, argues Koselleck, the more this future has to be envisaged on the basis of ungrounded expectations.[161] At the same time, by virtue of the acceleration of change, all that can be expected of the future is that it will be different from the present. The relationship between present and future is therefore disturbed. Yet while the widening gulf between past, present and future was unsettling, it also created the opportunity to design the future and to interpret the present in the light of the most diverse visions of the past.

In any event, modern temporality is marked by hiatuses. This discontinuity produced by the order-dissolving spirit of modernity results in a sense of disorientation. The modern experience of time, especially in a present felt to be beset with crisis, is also an experience of chaos that requires order. The vanishing force of the "envisioned past" and the openness of the "envisioned future" have an effect that varies between emancipatory and threatening.[162] The openness of tomorrow changes into the contingency of today. Past, present and future require order. In turn, the ordering of time means the generation of a continuity beyond the hiatuses that legitimizes action in the here and now. This legitimation may be achieved through rejection of yesterday or else derived from it, but always with regard to a malleable future.

Thus, according to Hartmut Rosa, "the interconnection of the three levels of time [...] always follows narrative patterns":

> Everyday time, biographical time, and historical time are related to each other, and mutually criticized and justified, in cultural and individual narratives. The meaning and relative weight of the past, present, and future and thus also the relevance and relative weight of tradition and change are determined simultaneously in narrative schemata [*Entwürfe*]. In them every present appears grounded upon a past and related to a future. Cultural and institutional forms of change and persistence are legitimated, and sometimes criticized, through the narrative interrelation of everyday time, life history, and world history,

---

[159] Osborne, *The Politics of Time*, pp. xi and ix. See also Hans Ulrich Gumbrecht, "Modern, Modernität, Moderne," in Otto Brunner/Werner Conze/Reinhart Koselleck (eds.), *Geschichtliche Grundbegriffe. Historisches Lexikon zur politisch-sozialen Sprache in Deutschland*, vol. 4, Stuttgart 1978, pp. 93–131.

[160] On acceleration, see Hartmut Rosa, *Social Acceleration. A New Theory of Modernity*, New York, NY 2013.

[161] Koselleck, "*Neuzeit*," p. 249.

[162] Koselleck, " 'Space of Experience' and 'Horizon of Expectation,' " pp. 257ff.

though of course the balance between dynamic and stabilizing forces, between movement and inertia, varies historically.[163]

Rosa notes that the task of interconnecting or reconciling the three time levels is solved by the introduction of a fourth level, "sacred time":

> This "holy time" overarches the linear time of life and history, establishes its beginning and end, and sublates life history and world history in a common, higher, and, so to speak, timeless time. [...] This temporal concordance is by no means always already secured. Instead it has to be produced in political and social processes of contestation.[164]

Here we can do no more than postulate three distinct periods since the eighteenth century in which the "politics of time" has played a preponderant role. That means three phases when temporal concordance was a key issue, or when the nexus of past, present and future was fought over: (1) the "bridge period"; (2) the "crisis of classical modernity," beginning in the 1890s and becoming especially virulent after the First World War; and (3) the decades since the 1970s. In each case, as Koselleck remarked of the "bridge period," "time itself becomes a title of legitimation open to occupation from all sides."[165]

In the liberal order that established itself in the wake of the Enlightenment, historicism produced continuity, temporal concordance and order with the help of the narrative of progress.[166] The concept of progress, and people's faith in it, organized the chaos resulting from the ever wider gap between experience and expectation.

> Terminologically, the spiritual *profectus* was either displaced or dissolved by a worldly *progressus*. The objective of possible completeness, previously attainable only in the Hereafter, henceforth served the idea of improvement on earth and made it possible for the doctrine of the Last Things to be superseded by the hazards of an open future. [...] Henceforth history could be regarded as a long-term process of growing fulfilment, which, despite setbacks and deviations, was ultimately planned and carried out by men themselves. The objectives were then transferred from one generation to the next, and the effects anticipated by plan or prognosis became the titles of legitimation of political action.[167]

Progress became the central category of a "secular religion of self-assurance" – secular in so far as, "despite the many historical-intellectual links between progress and Christian hopes in the future," it "aimed at active transformation of this

---

[163] Rosa, *Social Acceleration*, pp. 10–11.
[164] Ibid., p. 11.
[165] Koselleck, "*Neuzeit*," p. 248.
[166] Cf. Bedrich Loewenstein, *Der Fortschrittsglaube. Geschichte einer europäischen Idee*, Göttingen 2009.
[167] Koselleck, "'Space of Experience' and 'Horizon of Expectation,'" pp. 265–66.

world, not at a world beyond."[168] History, with its "narrative schemas," became the main medium in and through which the temporal order was produced.[169]

In the last third of the nineteenth century, however, the temporal concordance bound up with the historicist concept of progress began to fall apart, not least as a result of full-scale industrialization. Once more the acceleration of change widened the gulf between experience and expectation, until the First World War finally turned it into a virtually unbridgeable chasm. Furthermore, the war shattered hopes in perfectibility and "progress," so thoroughly that the whole idea fell into disrepute – as did the historicism that Nietzsche had excoriated in the second of his *Untimely Meditations*, "On the Uses and Disadvantages of History for Life" (1874).[170]

Often hitching their mast to Nietzsche, *avant garde* intellectuals in the years after the First World War tended to regard historicism – that is, the insight that all being and thought, and especially all values, have come about historically – as the fundamental problem. For historicity made it "impossible to discover substantive ultimate justifications of reality."[171] Many of these intellectuals, from the right as well as the left of the political spectrum (as the example of Walter Benjamin shows), longingly searched for a transhistorical bedrock to underpin reality and values.[172] But it looks as if some of them already vaguely sensed that any such foundation for a New Age and a New Order would rest on shaky ground. The spirit of modernity had destroyed the transhistorical and transcendental in principle, killing God (as Nietzsche put it) and profaning the highest

---

[168] Koselleck, "Fortschritt," in Brunner/Conze/Koselleck (eds.), *Geschichtliche Grundbegriffe. Historisches Lexikon zur politisch-sozialen Sprache in Deutschland*, vol. 2, Stuttgart 1975, pp. 351–423, here p. 411; cf. Koselleck, " 'Space of Experience' and 'Horizon of Expectation,'" p. 266.

[169] See Koselleck, "Fortschritt."

[170] On the crisis of historicism and the whole complex of antihistoricism, see *inter alia* Anselm Doering-Manteuffel, "Mensch, Maschine, Zeit. Fortschrittsbewusstsein und Kulturkritik im ersten Drittel des 20. Jahrhunderts," in *Jahrbuch des Historischen Kollegs 2003*, Munich 2004, pp. 91–119, and "Die antihistoristische Revolution im ersten Drittel des 20. Jahrhunderts. Eine Fallstudie zu den Spielräumen und Erkenntnismöglichkeiten von 'Ideengeschichte,'" MS. See also Otto Gerhard Oexle, "Krise des Historismus – Krise der Wirklichkeit. Eine Problemgeschichte der Moderne," in Oexle (ed.), *Krise des Historismus – Krise der Wirklichkeit. Wissenschaft, Kunst und Literatur 1880–1932*, Göttingen 2007, pp. 11–116.

[171] Michael Makropoulos, *Modernität als ontologischer Ausnahmezustand? Walter Benjamins Theorie der Moderne*, Munich 1989, p. 148, and "Crisis and Contingency. Two Categories of the Discourse of Classical Modernity," *Thesis Eleven* 111/2012, pp. 9–18.

[172] On Walter Benjamin and his "antihistoricist" theses on the philosophy of history, see Osborne, *The Politics of Time*, pp. 138–50. Cf. Griffin, *Modernism and Fascism*, pp. 177–78: "even though they remained poles apart in their reaction to Nazism, both these intellectual giants [Heidegger und Benjamin] applied their philosophical powers to the diagnosis of modernity's all-consuming decadence after the cataclysm of the First World War with a view to *transcending* it, each looking to a mythicized past as the source of the inspiration needed to inaugurate a new, revitalized, nomic society."

values.[173] Although most contemporaries knew nothing of the epistemological subtleties of historicity and relativism, their disorientation was no less great for all that. They too had become transcendentally homeless, and at the level of everyday action they too were confronted with the consequences of the hiatus in the temporal order.

Further development and improvement of human existence through the spread of universalist reason had in any case proved an illusory expectation. At the latest, by the end of the war, the course of history since the Enlightenment and the French Revolution appeared to numerous people as evidence, not of progress and ascent, but of decline and degradation. The actual or perceived defeat compelled a rethinking of history in both Germany and Italy.[174] The past that had led to national ruin was to be erased, while the aim was to link up with the eternal realm, already revealed in the past, that promised victory, salvation, community and future greatness. The present was seen as a workshop, in which this nexus of past and future would be constructed.[175] What counted now was to produce a new temporal concordance, a New Age.

This aspiration to a New Age, together with delegitimation of the prevailing (temporal) order and an attempt to establish a new (temporal) order, is intrinsically modern. For it negates the past and revalues the new. The fact that what the fascists sought to establish was nevertheless an imagined eternal realm, one offering a stable orientation and supposedly already present at some time in the past, shows not only the depth of the crisis and of the problem of historicity, but also the kind of paradoxes to which the temporal dynamic of modernity could lead. As if the past could be simply swept away, they thought they could reproduce a legitimizing continuity only through a break with the recent past. Peter Osborne, differing with the view of Jeffrey Herf, explains as follows the temporal logic underlying the antihistoricist revolution:

> [A]s a counter-revolutionary ideology, conservative revolution is modernist in the full temporal sense [...] of affirming the temporality of the new. Its image of the future may derive from the mythology of some lost origin or suppressed national essence, but its temporal dynamic is rigorously futural. [...] Conservative revolution is a form of revolutionary reaction. It understands that what it would "conserve" is already lost (if indeed it ever existed, which is doubtful), and hence must be created anew. It recognizes that under such circumstances the chance presents itself fully to realize this "past" for the

---

[173] This impossibility in principle of any transhistorical foundation informed those theorists of a liquid modernity who considered it "finally" settled that all being and thought was historical.

[174] Reinhart Koselleck, "On the Anthropological and Semantic Structure of *Bildung*," in Koselleck, *The Practice of Conceptual History. Timing History, Spacing Concepts*, Stanford, CA 2002, pp. 170–207.

[175] See Peter Fritzsche, "Historical Time and Future Experience in Postwar Germany," in Wolfgang Hardtwig (ed.), *Ordnungen in der Krise. Zur politischen Kulturgeschichte Deutschlands 1900–1933*, Munich 2007, pp. 141–64, here p. 141.

first time. The fact that the past in question is primarily imaginary is thus no impediment to its political force, but rather its very condition (myth).[176]

Fascist antihistoricism is a long way from the intellectual flights of fancy of certain "conservative revolutionaries" or a Walter Benjamin. For the great majority of fascists in Italy and Germany, the mythical past and the realm of eternity with which they sought to link up were neither imagined nor constructed. For some of them, indeed, these were a "scientifically" demonstrable reality. And all considered that reality to be powerful and effective.[177]

What conservative-revolutionary intellectuals such as Martin Heidegger, Ernst Jünger and Carl Schmitt shared with the fascists was a rejection of liberal progress and a longing for a stable, definitive order with a suprahistorical basis.[178] Purposive advance from the old into a rationally directed future was no longer capable of legitimizing action; that required a timeless quantity, rooted in the past and valid for the future. For most fascists, the newness and historical contingency of their vision of order remained hidden. They believed that their order had set free and reaffirmed an eternal truth buried in the past; that seemed the only way to avoid the problem of contingency and to establish an order whose legitimacy was assured. This entity, promising a way out of the relativism of the historical process, was to be found in the Nation or *Volk*, the absolute, suprahistorical core of a stable, definitive and completed order. The task was none other than to restore the Nation or *Volk*.

The longed-for rebirth of the Nation or *Volk* had already begun in the "experience of August 1914," as well as in the trenches of the Western front and the battles on the Isonzo and in the Alps. These had revealed the suprahistorical core of the legitimate order that now had to be set free. In a way, the August experience, the *maggio radioso* of 1915 (which saw Italy enter the war), and the wartime sense of community represented the Holy Night that would now be celebrated again. Selective glorification of the past was not regressive or reactionary, but served to "root" a disoriented and transcendentally homeless society. Fascists conjured up such pasts, in which the suprahistorical bedrock of their order shone through. They used history "monumentally," as an "active and powerful" force,[179] intending it to bring about the rebirth of the Nation or *Volk*. They placed the past "in

---

[176] Osborne, *The Politics of Time*, p. 164. Cf. Fritzsche, "Nazi Modern," p. 16, and Griffin, *Modernism and Fascism*, pp. 180–81.

[177] See Frank Lothar Kroll, *Utopie als Ideologie. Geschichtsdenken und politisches Handeln im Dritten Reich*, Paderborn 1998.

[178] On Heidegger's longing for order and his early active involvement in the construction of a National Socialist order, see the informative, if apologetic, book: Phillipe Lacoue-Labarthe, *Heidegger and the Politics of Poetry*, Chicago, IL 2007, and Daniel Morat, *Von der Tat zur Gelassenheit. Konservatives Denken bei Martin Heidegger, Ernst Jünger und Friedrich Georg Jünger 1920–1960*, Göttingen 2007. On Lacoue-Labarthe, see Rabinbach, *Nationalsozialismus und Moderne*, pp. 105–10.

[179] See Nietzsche, "On the Uses and Disadvantages of History for Life," p. 67.

the service of life," bringing forth a temporal harmony between past, present and future that incorporated them into the context of a higher, sacred time.

The palingenesis that fascism aimed to achieve can easily be understood as corresponding to the temporal dynamic of modernity, but it may also be interpreted as an attempt to overcome it. Rebirth will complete the self-negating temporal dynamic, the break with a past from which the fascists themselves emerged. Henceforth, as in the Christmas festival, everything will be new. At the same time, that which is new has already existed before and is legitimized by that very fact. The means to produce this legitimacy of order is *myth*.

Myth

The concept of myth and its derivatives are often used in connection with fascism, but they are seldom properly defined.[180] Even when "myth" is not employed imprecisely, the sense of "pseudonarrative" – hence the opposition between myth and truth, and the connotations of primitive and archaic, illusory and instrumental – plays an important role. Such uses are deeply normative, however, since myth becomes "untrue, primitive narrative" only in comparison with logos. This has the effect of positioning fascism, and the author in question, either outside or inside the Enlightenment narrative of progress and modernization theory; myth is seen as that which has been overcome by Reason and Science.[181] Here, we shall offer a different definition, one based on philosophical theories of myth.[182]

This work rests upon a functional understanding of myth. Following Hans Blumenberg, it can be established that myth operates as a means to ward off "the

---

[180] Griffin, for example, devotes only a few pages to the concept of myth (*The Nature of Fascism*, pp. 27–36). Emilio Gentile, who also postulates a mythic core of Italian Fascism, has a short chapter on the subject in his *The Struggle for Modernity* (2003), but it takes the meaning and function of myth as given. Moreover, myth appears in Gentile as a demagogic instrument, not as a mode of order: see *The Struggle for Modernity*, pp. 77–88. Cf. Herfried Münkler, *Die Deutschen und ihre Mythen*, Berlin 2009.

[181] In line with Weberian modernization theory, mythical fascism becomes the primitive, archaic Other of modernity, a reactionary relapse into something already superseded. See Weber, "Science as a Vocation," p. 139: "The increasing intellectualization and rationalization [...] indicate [...] knowledge or belief that [...] one can, in principle, master all things by calculation. This means that the world is disenchanted. One need no longer have recourse to magical means in order to master or implore the spirits, as did the savage, for whom such mysterious powers existed. Technical means and calculations perform the service."

[182] For surveys of philosophical myth-theory, see Emil Angehrn, *Die Überwindung des Chaos. Zur Philosophie des Mythos*, Frankfurt/Main 1996; Wilfried Barner/Anke Detken/Jörg Wesche (eds.), *Texte zur modernen Mythentheorie*, Stuttgart 2003; Christoph Jamme, *Einführung in die Philosophie des Mythos. Neuzeit und Gegenwart*, Darmstadt 1991. Solidly grounded analyses of the phenomenon of myth and modernity may be found especially in Karl Heinz Bohrer (ed.), *Mythos und Moderne. Begriff und Bild einer Rekonstruktion*, Frankfurt/Main 1983; Manfred Fuhrmann (ed.), *Terror und Spiel. Probleme der Mythenrezeption [Poetik und Hermeneutik, Arbeitsergebnisse einer Forschungsgruppe IV]*, Munich 1971; and Christoph Jamme, *"Gott an hat ein Gewand." Grenzen und Perspektiven philosophischer Mythos-Theorien*, Frankfurt/Main 1999.

absolutism of reality."[183] As Emil Angehrn puts it, summarizing Blumenberg's position: "Myth serves to weaken archaic fears, by means of naming, organizing, and structuring that make reality clear and manageable."[184] Myth operates at this fundamental anthropological level:

> as a kind of historical memory, which paints a picture of the world's Becoming and the origins of a community; as a description of the existing world that allows some understanding of its decisive forces and laws; as a description of the world that allows for an understanding of determining powers and laws; myth is both constitution and means of appropriation. Qua description, myth is not simply a likeness of the world but organizes and systematizes it; it gives the world a certain profile; it facilitates orientation through a penetrative understanding. Qua interpretation, myth produces the world and at the same time is a means enabling human beings to describe and understand themselves.[185]

In what follows, we shall understand by myth certain narratives that serve a socially important function of producing order and ground reality on a suprahistorical realm.[186] A mythical order is the totality of these narratives. Myth is a social practice of a linguistic nature:

> Myths serve to accredit with supreme value the existence and constitution of a society. [...] The, or one, achievement of myth [...] lies in the normative realm and has to do with justifying the conditions of life in social contexts. [...] In mythical narratives, something existing in nature or humanity is related to and established by a holy sphere. "Established by" means here: derived from – not in the sense of a simple causal relation, such as one finds in the natural sciences, but in the sense of a justification.[187] But to "justify" or "accredit" something means to relate it to a value that is indisputably intersubjective. And the only radically indisputable thing for human subjects is what they regard as holy – as unassailable, ubiquitous, and omnipotent.[188]

---

[183] See Blumenberg, *Work on Myth*.

[184] Angehrn, *Die Überwindung des Chaos*, p. 38.

[185] Ibid., p. 37.

[186] Here and below, see Manfred Frank, *Gott im Exil. Vorlesungen über die Neue Mythologie*, Frankfurt/Main 1988, pp. 15ff. On myth as statement and (secondary) semiological system, see Roland Barthes, *Mythologies*, New York, NY 1972, pp. 109 and passim. See also Münkler, *Die Deutschen und ihre Mythen*.

[187] Cf. Angehrn, *Die Überwindung des Chaos*, pp. 264–65, where we read: "Mythology, cosmology, religion, natural science, and historiography put in perspective various spheres of reality and various ways of interconnecting particularity. Strict laws are only one variant, although they are often proposed as an ideal form of systematization."

[188] Frank, *Gott im Exil*, p. 16.

So, myth is narrative that constitutes community by overcoming chaos and generating order within a realm of the holy or suprahistorical.[189] Myth is a kind of authentication of knowledge.[190] A community authenticates its knowledge – what it regards as true, good, just and beautiful – by associating it with a supreme value or being. The knowledge thereby acquires a context within which it proves meaningful – indeed, is able to develop its meaning. It receives an unshakable, unquestionable foundation: its direction and purpose.

Myth as a linguistic or discursive form is synthetic: it creates a unitary, "organic" order, instead of dissolving order into its individual components. Myth would seem to be inductive, whereas *logos* is deductive and requires general premises on which to base itself. Myth creates an interpretive context in which everyone and everything is legitimized or delegitimized by reference to a sacred Absolute outside history.[191] This sacred sphere is the Archimedean point of an order that, through the presence of the Absolute, achieves stability, escapes change or becoming, and is therefore "eternal."

The quest for a different temporal harmony analyzed in the previous section, in which the self-negating temporal dynamic of modernity is both completed and superseded, can be achieved by way of a mythical order. Myth establishes an order in which (linear) time is sublated in a sacral time that is both new and eternal. The "origins" are repeated in the myth. But, writes Emil Angehern, the mythical recollection

> does not seek only to conjure up an event that took place at a certain point in the past, to represent it in contemporary imagery, narrative, or gestures as something that once existed. Rather, the original happening is experienced as something happening today; the original occurrence is celebrated as an action

---

[189] Angehrn, *Die Überwindung des Chaos*, pp. 259–320. On the parallels between this philosophical concept of myth and the one that (especially in Mussolini) can be traced back to Sorel, see Zeev Sternhell/Mario Sznajder/Maia Asheri, *The Birth of Fascist Ideology. From Cultural Rebellion to Political Revolution*, Princeton, NJ 1994, pp. 78–91; and Michael Tager, "Myth and Politics in the Works of Sorel and Barthes," *Journal of the History of Ideas* 47/1986, pp. 625–39. Tager writes (pp. 626–27): "A pragmatic rather than an analytical attitude characterized Sorel's study of myth. What concerned him was not whether an event like the resurrection actually occurred but only its capacity to evoke sacrifice and heroism among its believers. [...] Rather than examining the psychological or sociological aspects of myth, Sorel insistently asked a more immediate question: can it provoke a reformation of man and society?" The key difference is therefore that Sorel regards myths as instrumental narratives for the manipulation of the masses. See Georges Sorel, *Reflections on Violence* [orig. 1906], New York, NY 1975, p. 126: "Myth must be judged as a means of acting *on the present*" (emphasis in the original).

[190] Hans-Georg Gadamer, "Mythos und Vernunft," in *Gesammelte Werke*, vol. 8, *Ästhetik und Poetik*, Tübingen 1993, pp. 163–69; here p. 165. On the legitimation of knowledge through narrative, see Lyotard, *The Postmodern Condition*, esp. pp. 27–36.

[191] In terms of the model of discourse presented here, this means that the aviator is constituted by – and thus acquires meaning from – his association with the nation.

effective here and now; the original conflict is registered as one that has never been resolved. What is summoned up is not times of old but a deeper dimension of the Now – a stratum where the Now transcends its chronological place and is no longer just one in a succession of moments. The mark of the mythical conception of time is simultaneity of the non-simultaneous, co-presence in a time of origins. People alive today become contemporaneous with the original happening. Sacred history permeates the secular world and constitutes its foundation.[192]

The fascist yearning for rebirth of the nation corresponds to this temporal schema. The reborn Nation or *Volk* is not something from the past that is now visualized anew, but an eternal being that returns from the ruins left behind by progress.

Fascism – a working definition

We should now state more precisely the concept of fascism used in this book and spell out how it differs from some other recent studies. First of all, it focuses on fascism as an ideology. This does not imply, however – as in Zeev Sternhell's work, for example – that fascism is a closed political structure or theory, comparable to Marxism in its coherence and consistency.[193] The exact opposite is the case. We shall take seriously its own conception of itself as a *movement*: which means, first, a fluid and volatile "worldview," and, second (the position of George L. Mosse) a "cultural revolution" expressing a particular ordering of the world.[194] Worldview here signifies – in an allusion to Heidegger's concept of *Weltbild* – a conception of beings as a whole.[195] Order, on the other hand, is understood with Foucault as

> that which is given in things as their inner law, the hidden network that determines the way they confront one another, and also that which has no existence except in the grid created by a glance, an examination, a language; and

---

[192] Angehrn, *Die Überwindung des Chaos*, pp. 66–7.

[193] Sternhell/Sznajder/Asheri, *The Birth of Fascist Ideology*; cf. David D. Robert, "How Not to Think about Fascism and Ideology. Intellectual Antecedents and Historical Meaning," *Journal of Contemporary History* 35/2000, pp. 185–211. A closer look at the actual diversity of historical "Marxism" shows that it was by no means as coherent as it has long been assumed.

[194] Mosse, *The Fascist Revolution*, p. xi: "Culture in our case must not be narrowly defined as a history of ideas, or as confined to popular culture, but instead understood as dealing with life seen as a whole – a totality, as indeed the fascist movement sought to define itself. Cultural history centers above all upon the perceptions of men and women, and how these are shaped and enlisted in politics at a particular place and time." On Mosse's interpretation of fascism as "cultural revolution," see Payne, *A History of Fascism*, pp. 450–51.

[195] Martin Heidegger, "The Age of the World Picture" (1938), in Heidegger (ed.), *Off the Beaten Track* [*Holzwege*], Cambridge 2001, pp. 67–85. "Understood in an essential way, 'world picture' does not mean 'picture of the world' but, rather, the world grasped as picture. Beings as a whole are not taken in such a way that a being is first only in being in so far as it is set in place by representing-producing (*vorstellend-herstellend*) humanity. Whenever we have a world picture, an essential decision occurs concerning beings as a whole. The being of beings is sought and found in the representedness of beings."

it is only in the blank spaces of this grid that order manifests itself in depth as though already there, waiting in silence for the moment of its expression.[196]

Order is a level between practice and theory, a "middle region" between "the already 'encoded' eye and reflexive knowledge." This region, however,

> can be posited as the most fundamental of all: anterior to words, perceptions, and gestures, which are then taken to be more less exact, more or less happy, expressions of it [...]; more solid, more archaic, less dubious, always more "true" than the theories that attempt to give those expressions explicit form, exhaustive application, or philosophical foundation. Thus, in every culture, between what one might call the ordering codes and reflections upon order itself, there is the pure experience of order and of its modes of being.[197]

Fascism, then, will be understood as a way of thinking and seeing, or a specific order of the world prior to statements and actions that determines those very statements and actions. It will be treated neither as an "ism" or political ideology in the strict sense of the word (as in Sternhell), nor in terms of its establishment of rule or its exercise of violence (as in Robert O. Paxton and Sven Reichardt). No attempt will be made at a sociology of fascist movements, such as we find in Michael Mann. The choice of a different focus on ideology or the history of ideas, linking up with Mosse, Gentile and Griffin, does not at all mean that we reject Paxton's developmental model of fascism, Reichardt's praxeological analysis of fascist combat organizations, or Mann's wider sociological approach. Rather, we shall propose keeping a distance from absolutist claims or essentialist definitions, preferring instead to allow different approaches to exist alongside one another. Only such an acceptance of multiple perspectives can do justice to the inherent fluidity of fascism. For the phenomenon includes a complex, and partly contradictory, array of thought patterns, interpretive models and concepts, and of behavioral motives, ambitions, yearnings and fears (as well as corresponding modes of appropriation, interpretation, and objectification) among individual fascists. There were as many fascisms as there were fascists.[198] Fascism satisfied a

---

[196] Michel Foucault, *The Order of Things. An Archeology of the Human Sciences*, New York, NY 1994, pp. xx–xxi. Cf. Manfred Frank, *Was ist Neostrukturalismus?*, Frankfurt/Main 1984, pp. 138ff.

[197] Foucault, *The Order of Things*, p. xxii.

[198] Hans Frank, the former governor and "butcher" of Poland, noted this "plurality": "The formula: National Socialism is exclusively what so-and-so says or does – by which each representative meant himself – appeared in place of the assumption in the Party program that National Socialism is the realization of that program. In the beginning, many names were pressed into that formula: Hitler, Göring, Strasser, Röhm, Goebbels, Hess, Rosenberg and others. *Basically there were as many 'National Socialisms' as there were leading men.* [Emphases added] From January 30, 1933 on, Hitler's decisive position meant that his name alone was valid in the formula – at least officially." See Hans Frank, *Im Angesicht des Galgens. Deutung Hitlers und seiner Zeit auf grund eigener Erlebnisse und Erkenntnisse. Geschrieben im Nürnberger Justizgefängnis*, 2nd edn, Neuhaus bei Schliersee 1955, pp. 176–77. I am grateful to Frank Reichherzer for referring me to Hans Frank's statement;

longing; it did not offer a theory, but promised a clear, non-contingent world that people would be able to understand. This basis for its success under the political conditions of mass society is not the least of the reasons why it cannot be grasped in an essentialist manner.

Contemporary conceptions of fascism were fluid both synchronically and dia-chronically. Not only did understandings of fascism diverge at the same point in time; the conceptions of the self-same fascists varied over the course of time. Their understanding of fascism resulted from the web of concepts in which it was enmeshed, from the ever-changing links to the ideas and images surrounding it. The surrounding concepts that constituted fascism were not the same for Filippo Tommaso Marinetti and for a big landowner in the Po valley. And they changed even for the "leading men" themselves, depending on whether fascists were try-ing to conquer power or to consolidate the state power they had already won, on whether they were feigning a desire for peace or conducting a war of annihi-lation.[199] But, given its variety of manifestations, how can fascism be subsumed under a generic concept?

Rather than think of a generic concept as a Platonic universal, it seems more meaningful to understand it heuristically as the expression of a "family resemblance."[200] The "kinship relations" among fascisms in a single country or across national boundaries may then be thought of as "a complicated network of similarities overlapping and criss-crossing," "sometimes overall similarities, sometimes similarities of detail."[201] These various fascisms – to remain with Wittgenstein's illuminating image – may be seen as individual fibers spun into a thread: "And the strength of the thread does not reside in the fact that some one fiber runs through its whole length, but in the overlapping of many fibers. [...] Something runs through the whole thread – namely, the continuous overlapping of those fibers."[202]

---

see also Frank Reichherzer, "'Das Wehr-Denken ist deutsch, nationalsozialistisch.' Zum Verhältnis von wehrwissenschaftlichem Denken und nationalsozialistischer Ideologie in der Zwischenkriegszeit," in Käte Meyer-Drawe/Kristin Platt (eds.), *Wissenschaft im Einsatz*, Munich 2007, pp. 243–67.

[199] On this conception of fascism as a process, see Paxton, *Anatomy of Fascism*.

[200] On "family resemblances," see Ludwig Wittgenstein's *Philosophical Investigations*, I: 59ff., Oxford 1968, pp. 31ff.

[201] Ibid., I: 66. "For if you look at them you will not see something that is common to *all*, but similarities, relationships, and a whole series of them at that" (I: 67). Wittgenstein con-tinues: "I can think of no better expression to characterize these similarities than 'family resemblances;' for the various resemblances between members of a family: build, features, color of eyes, gait, temperament, etc. etc. overlap and criss-cross in the same way." And he chooses numbers as an example of what he has in mind: "Why do we call something a 'number?' Well, perhaps because it has a – direct – relationship with several things that have hitherto been called number; and this can be said to give it an indirect relationship to other things we call the same name. And we extend our concept of number as in spinning a thread we twist fiber on fiber."

[202] Ibid., I: 67.

Some generic feature common to species *x*, National Socialism, and species *y*, the Romanian Legion of Archangel Michael, may stand in a contradictory relationship to some other common feature that makes it seem justifiable to subsume species *w*, Italian Fascism, and species *z*, the Spanish Falange, under the generic concept of fascism. Nevertheless, there is a closeness between species *z* and species *x* that runs through the overlapping of both with species *a*, for example, the Hungarian Arrow Cross. The same goes for definitions of the generic concept themselves. The strength of the thread lies in "the overlapping of many fibers."

This understanding of generic concepts is not at all based on some exaggerated "postmodern" relativism. It is an attempt to implement the "knowledge" that worldviews and social realities are even less static than texts. In order to grasp these context-dependent and constantly mutating visions of reality, we need "fluid concepts" as well as an awareness that, although texts seek to fix change and movement, they are bound to fail. If a working definition of fascism is given nonetheless, it is with a realization of its limited scope and its dependence on a particular investigative horizon.

Fascism was a metapolitical phenomenon and a cultural revolution. Its objective was to escape transcendental homelessness by bringing about a mythical order based upon a sacral, absolutized Nation or *Volk*. The fascist blueprint for a mythical order differed from conservative and liberal or Marxist visions, in respect both of the temporal order at which it aimed and of the underlying "sacred reality."[203] It diverged from the conservative blueprint for order also in the role it ascribed to "the masses," but what divided the two was mainly the future-oriented temporal dynamic of fascism. The fascist mythical order was further opposed to the idea of progress by means of which liberalism and Marxism organized time. Fascism posited instead a narrative of eternity that satisfied the yearning for both dynamization and "deceleration," in so far as it promised renewal of the Nation or *Volk* as a transhistorical entity. Whereas the fascist mythical order involved an absolutization of the Nation or *Volk*, the liberal and Marxist conceptions of order started from a different Archimedean point.

Structure of the work

It should be made clear that this work is not centrally concerned with particular "aviation heroes" or the formation of national air forces, nor with the building of the Italian Fascist or German Nazi Party. Rather, using aviation discourse as its source material, it focuses on the conceptions of order and related ideas through which perceived chaos was structured into a meaningful world during and after the First World War. This ostensibly stable and eternal order is presented as mythical modernity.

The investigation is divided into three parts. Part I deals with the widespread *longing for order*, which took a more radical form in the First World War. Taking Aby Warburg as the intellectual paradigm, Chapter I.1 identifies an ideal-typical liberal conception of order as a counterpoint to the fascist blueprint for modernity.

---

[203] On the "sacred *realissimum*," see Berger, *The Sacred Canopy*, p. 32.

Warburg's postage stamp featuring an aircraft borne aloft by the inscription *Idea vincit* is a compressed objectification of the liberal vision of modernity based on reason and progress. Chapter I.2 then counterposes to this Gabriele D'Annunzio's political-religious encoding of the aviator and his martial interpretations of the Icarus myth. It shows how D'Annunzio, who, not least because of his flight to Vienna in 1918, became one of the most popular "heroes of the air," integrated the aviator into his nationalist "theologeion of salvation" and constructed a mythical order with the sacralized nation at its center.[204]

Part II is devoted to the *fractured order* in the First World War. The focus shifts from intellectuals to mass society and the channels along which mythical order and the "curious literariness of real life" (noted by Paul Fussell) spread into it.[205] Chapter II.1 first demonstrates, from the example of the Brescia airshow, how an image of the aviator as conqueror and superman took shape in both high and popular culture, before turning to the production and reception of mass reading material arising out of the wartime literary mobilization. Chapter II.2 then discusses the hero images and interpretive models present in this material, as well as their functions in the context of the First World War. The aviator-hero was a countervailing figure to the industrialized ground war and its tendencies toward stasis, offering a heroic matrix for readers to imitate. Chapter II.3 looks at the transitional capacity of this aviation hero narrative; it shows that fusion with the nation was the main virtue of such heroes, but also that this contributed to the unity of society understood as a structured community. Aviation heroes were transitional figures who made it easier to overcome the fragile, declining order and who came to embody a new social nobility versed in technology. They provided a narrative that transcended profane time and embedded war, suffering, sacrifice and death in a horizon of meaning.

Finally, Part III describes the *eternal order* that fascists sought to achieve. Drawing especially on Ernst Jünger's conception of "monumentalist history," Chapter III.1 shows how war became the beginning of a new future. The Futurists too waged war on the past: they saw the aviator as the symbol of a new dawn, as well as of a new man born out of the heroism of the world war. The Futurists, but also the Esposizione dell' Aeronautica Italiana (the Italian aeronautical exhibition of 1934), illustrate how aviation became a techno-totem of the fascist order. In conclusion, Chapter III.2 explains the extent to which the new postwar order envisaged by fascists should be understood as mythical, and how it functioned as an answer to both the crisis of reason and the crisis of historicism. Having considered the concepts of fascism, modernity and myth separately from one another, the text now fits them together again and demonstrates the complementarity of myth and modernity.

---

[204] On D'Annunzio's "theologeion of salvation," see Hans Ulrich Gumbrecht, "I redentori della vittoria. Über Fiumes Ort in der Genealogie des Faschismus," in Hans Ulrich Gumbrecht/Friedrich Kittler/Bernhard Siegert (eds.), *Der Dichter als Kommandant. D'Annunzio erobert Fiume*, Munich 1996, pp. 83–115.

[205] Fussell, *The Great War and Modern Memory*, p. ix.

Part I
Longing for Order

1

Idea non vincit: Warburg and the Crisis of Liberal Modernity

Four reasons suggest themselves why this study of aviation discourse and its links with war, fascism and modernity should begin with the art historian and cultural theorist Aby Warburg. First, Warburg deserves attention from the point of view of theory and methodology; he sets an example for a "cultural history" that escapes interdisciplinary "border guards."[1]

Second, in our present examination of the character of modernity, Warburg plays a dual role *vis-à-vis* his contemporaries. On the one hand, he is a source who needs to be interrogated concerning an earlier historical period; on the other, he clearly protrudes forward into the present day. As a theorist who himself sought an answer to the question of modernity, he influences the understanding of modernity that underlies our own study. This makes it essential to historicize his own person as much as the impulses we have received from him.

Third, as the scion of a leading Jewish banking family, Warburg described himself as *"ebreo di sangue, Amburghese di cuore, d'anima Fiorentino"* ("a Jew by blood, a Hamburger at heart, a Florentine in spirit").[2] We portray him as a paradigmatic representative of liberal modernity, central aspects of which may be identified

---

[1] Aby Warburg, "Italian Art and International Astrology in the Palazzo Schifanoia, Ferrara" (1912), in Warburg, *The Renewal of Pagan Antiquity. Contributions to the Cultural History of the European Renaissance*, Los Angeles, CA 1999, pp. 563–91; here p. 585. See Ulrich Raulff, "Von der Privatbibliothek des Gelehrten zum Forschungsinstitut. Aby Warburg, Ernst Cassirer und die neue Kulturwissenschaft," *Geschichte und Gesellschaft* 23/1997, pp. 28–43; and Bernd Roeck, "Psychohistorie im Zeichen Saturns. Aby Warburgs Denksystem und die moderne Kulturgeschichte," in Wolfgang Hardtwig/Hans-Ulrich Wehler (eds.), *Kulturgeschichte heute*, Göttingen 1996, pp. 231–54. It may be thought illuminating in this context that Warburg studied in Bonn under Karl Lamprecht and others. But it is his nomadic, sometimes thoroughly eccentric, habits and especially his methodological rejection of anything absolute or cut-and-dried that account for the current interest in his work and ensure him a place in this book. On his academic career path, see Ernst H. Gombrich, *Aby Warburg. An Intellectual Biography*, Oxford 1986, pp. 25–66; Karen Michels, *Aby Warburg. Im Bannkreis der Ideen*, Munich 2007, pp. 27–34; and Bernd Roeck, *Der junge Aby Warburg*, Munich 1997, pp. 41–53.

[2] Gertrud Bing, "Aby M. Warburg. Vortrag," in Aby M. Warburg, *Ausgewählte Schriften und Würdigungen*, ed. Dieter Wuttke, Baden-Baden 1980, pp. 455–64; here p. 464.

with reference to him as an ideal type. This background serves as a contrasting foil to emphasize the mythical modernity that is the subject of this book.[3]

Fourth, Warburg is a speaker in the discourse of aviation. In 1926 he designed a postage stamp and handed it to the foreign minister of the day, Gustav Stresemann. Its main graphic element, an airplane whose wings are adorned with the motto *Idea vincit*, will be interpreted here as a condensed symbol of liberal modernity, though not strictly in the sense of Warburg's or Panofsky's iconography. The stamp, on which Warburg commissioned Otto Heinrich Strohmeyer to do further work, met with no greater success than the republic it was supposed to represent. But it is significant as one of the voices in a discourse that would no longer be given a hearing. This story will be told here in stages, though not always chronologically.

Bourgeois, citizen of the world, *Bildungsbürger*

"But the influence of the Warburg Institute, if profound, was narrow."[4] Although Warburg is better known outside the circle of art historians than he was when Gay's book first appeared in 1968, his life and theoretical contribution still require contextualization. He was born in 1866 into a German-Jewish family, whose bank M. M. Warburg & Co. had been founded in 1798 and had become one of the leading houses in Hamburg, or indeed Germany, active on the international stage.[5] Despite the distinctive *haut bourgeois* milieu of Wilhelmine financiers from which he stemmed, Warburg's habitus was more in keeping with that of the *Bildungsbürger*, the cultured middle classes.[6] In parallel with the description of himself as "a Jew by blood, a Hamburger at heart, a Florentine in spirit," we might therefore say that Warburg was a bourgeois by blood, a cosmopolitan at heart and a *Bildungsbürger* in spirit. These three categories help to place him socially, but above all in a "space of experience." The nexus of bourgeoisie, culture and

---

[3] Peter Gay has already taken Warburg and his circle as a contrast to fascist mythical modernity. "The austere empiricism and scholarly imagination of the Warburg style were the very antithesis of the brutal anti-intellectualism and vulgar mysticism threatening to barbarize German culture in the 1920s; this was Weimar at its best." Peter Gay, *Weimar Culture. The Outsider as Insider,* Harmondsworth 1974, p. 35.

[4] Gay, *Weimar Culture,* p. 35.

[5] On the M. M. Warburg & Co. banking house, see Eduard Rosenbaum/Ari J. Sherman, *M.M. Warburg and Co., 1938–1978, Merchant Bankers of Hamburg,* Boston, MA 1979; and on the Warburg family, Ron Chernow, *The Warburgs. The Twentieth-Century Odyssey of a Remarkable Jewish Family,* New York, NY 1994.

[6] For an insight into the milieu of German-Jewish high finance before the First World War, see Boris Barth, "Weder Bürgertum noch Adel – Zwischen Nationalstaat und kosmopolitischem Geschäft. Zur Gesellschaftsgeschichte der deutsch-jüdischen Hochfinanz vor dem Ersten Weltkrieg," *Geschichte und Gesellschaft* 25/1999, pp. 94–122. There we read (on p. 115) that, "with their conspicuously lavish lifestyle," top financiers "sought to emphasize their distance from upwardly mobile layers of the bourgeoisie." "Especially in Prussia, financiers who recognized their Jewishness stood in between the nobility and the big bourgeoisie, without actually being socially integrated."

liberalism, also including Warburg's "Jewishness" and rational, progressive world-view, can here be only intimated rather than exhaustively explored.[7]

A central concern here is the role of scientific rationality in strivings for a liberal order. The stress on an approach to the world guided by reason had its origins in the Enlightenment and its promise of an end to man's "self-incurred immaturity."[8] At an individual level, this impetus joined up with the ideal of a cultural-educative formation (*Bildung*). This in turn, understood as a project for the cultivation of mature human beings, was a prerequisite and reflection of progress in history. It was a liberal Enlightenment conception, found in the person of Warburg and given objective form in his design for a postage stamp.[9]

The story was told in the family that Warburg renounced his rights of primogeniture at the age of 13. His younger brother Max promised in return "to buy him all the books he wanted."[10] From 1905 on, the cultural studies library he founded in Hamburg, the Kulturwissenschaftliche Bibliothek Warburg (K.B.W.), was run on a semipublic basis; it comprised more than 16,000 volumes when it and its staff emigrated to London in December 1933, four years after Warburg's death.[11] Warburg may not have entered the family bank, but the K.B.W. – which both embodied and facilitated his thinking – could never have existed without it. One would not be doing justice to the relationship between bankers and scholars, however, if one described Warburg as a supplicant. It was customary among Wilhelmine industrialists, financiers and so-called *Kaiserjuden* to make a social commitment and to act as sponsors of learning. According to Ulrich Raulff, the K.B.W. bestowed legitimacy on the Jewish banking house, and with that behind him Warburg managed to build his "model institute" as a home for the history of modern art and culture.[12]

But Warburg distanced himself from his brothers' bourgeois lifestyle and "ostentatious consumption."[13] He thought of himself as bourgeois only "by blood," that is, by origin. Since scholars or intellectuals "are hard to accommodate in the house of

---

[7] Despite an abundance of literature on the bourgeoisie, there is a need for research to demonstrate fully on an empirical basis, and in relation to the history of ideas, the nexus of scientific rationality, education, progress and liberalism. See the recently published: Franco Moretti, *The Bourgeois. Between History and Literature*, London 2013.

[8] Immanuel Kant, "An Answer to the Question: 'What Is Enlightenment?'" in Hans Reiss (ed.), Political Writings, 2nd edn, Cambridge 1991, p. 58. p. 54.

[9] See George L. Mosse, "Das deutsch-jüdische Bildungsbürgertum," in Reinhart Koselleck (ed.), *Bildungsbürgertum im 19. Jahrhundert. Teil II. Bildungsgüter und Bildungswissen*, Stuttgart 1990, pp. 168–80.

[10] Max Warburg, Rede zur Gedenkfeier, May 5, 1929, quoted in Gombrich, *Aby Warburg*, p. 22.

[11] Martin Warnke, "'Ich bin ein wissenschaftlicher Privatbankier, dessen Credit so gut ist wie der der Reichsbank.' Aby Warburg und die Warburg Bank. Vorwort," in Michels, *Aby Warburg*, pp. 11–19. On the "emigration" of the library, see Dieter Wuttke, "Die Emigration der Kulturwissenschaftlichen Bibliothek Warburg und die Anfänge des Universitätsfaches Kunstgeschichte in Großbritannien," in Horst Bredekamp/Michael Diers/Charlotte Schoell-Glass (eds.), *Aby Warburg. Akten des internationalen Symposions Hamburg 1990*, Weinheim 1991, pp. 141–63.

[12] Raulff, "Von der Privatbibliothek des Gelehrten," pp. 36f.

[13] Ibid., p. 37.

social stratification,"[14] it anyway seems more important to situate Warburg in the context of the Wilhelmine and Weimar intelligentsia. Although he was rooted socio-economically in the world of high finance, his cosmopolitan and *Bildungsbürger* aspects open up the main perspective in which his testimony should be read.

The concept of cosmopolitanism allows us to situate Warburg in the society of the Wilhelmine Empire and the Weimar Republic, for it rises above the mostly negative connotations of things foreign during both periods, as well as the self-perception resulting from them. As we have seen, Warburg came from a Jewish family. He himself did not practice the Jewish faith – although or precisely because he had been brought up in strict orthodoxy – and he married a Protestant woman, Mary Hertz. His Jewishness consisted "only" of a cultural heritage, in which he participated by virtue of his upbringing. To the majority German society, however, he remained "a Jew" – or at best a "German Jew." The ambivalence of such multiple identities, and of the (possibly plural) solidarities bound up with them, was perceived as a danger; it seemed to place a question mark over claims to an exclusive, homogenous national identity.[15] A similar threat seemed to come from Ultramontane Catholics and "rootless" Social Democrats "without a fatherland." But, at the latest with the *Judenzählung* of 1916 – the "head count" of 1916 in the army, designed to confirm accusations that Jews were underrepresented at the front – the anti-Semitism always latent in the country came to the fore in the trope of Jewish "internationalism."[16] The unwillingness of the majority to allow room for plurality or ambivalence forced those facing exclusion to revalue the niche existence and identity imposed on them. Many affirmed the complementarity of cosmopolitanism and (German) national sentiment; Friedrich Meinecke, for example, indulged "bearers of German culture" with the idea that "the true, the best German national feeling also includes the cosmopolitan ideal of a humanity beyond nationality and that it is 'un-German to be merely German'." Only "so-called public opinion" saw a contradiction between the two.[17]

As the example of Warburg plainly shows, there was indeed no contradiction. For assimilated Jews, who saw themselves as part of German culture, the accusation that they were insufficiently patriotic was a baseless and unpardonable insult; there was no reason to doubt their patriotism.[18] Thus, on May 6, 1915, shortly

---

[14] Ralf Dahrendorf, *Society and Democracy in Germany*, New York, NY 1969, p. 103. Quoted in Klaus Vondung, "Probleme einer Sozialgeschichte der Ideen," in Vondung (ed.), *Das wilhelminische Bildungsbürgertum. Zur Sozialgeschichte seiner Ideen*, Göttingen 1976, p. 9.

[15] See here Zygmunt Bauman, *Modernity and Ambivalence*, Cambridge 1991 and *Modernity and the Holocaust*, Ithaca, NY 1989.

[16] On the *Judenzählung*, see Ulrich Sieg, *Jüdische Intellektuelle im Ersten Weltkrieg. Kriegserfahrungen, weltanschauliche Debatten und kulturelle Neuentwürfe*, Berlin 2001, pp. 87–96. On Warburg's reaction to the "census" in particular and anti-Semitism in general, see Charlotte Schoell-Glass, *Aby Warburg und der Antisemitismus. Kulturwissenschaft als Geistespolitik*, Frankfurt/Main 1998.

[17] Friedrich Meinecke, *Cosmopolitanism and the National State* (orig. 1908), Princeton, NJ 1970, pp. 21f.

[18] On the patriotism and nationalism of German Jews, and their "August 1914 experience," see Sieg, *Jüdische Intellektuelle im Ersten Weltkrieg*, pp. 53–87, and Ulrich Sieg, *Jüdische Intellektuelle und die Krise der bürgerlichen Welt im Ersten Weltkrieg*, Stuttgart 2000, p. 15.

before his 49th birthday, Warburg wrote to his disciple Wilhelm Waetzoldt, the future director-general of the state museums in Berlin:

> Should Italy really opt for betrayal, the question will arise for me as to whether I can somehow make myself militarily useful. Physical incapacity (you know I have been seriously afflicted for years) means no direct service at the front; can I not sit the interpreters' exam, so that I can be used in our country or in Italy? Or might my real-life experiences in Italy spare me this exam, even though I did my year's service in 1894 only as an NCO (unqualified, of course, because unbaptized)? Best of all, I would like to go to preach in an Italian prison camp, though not as an NCO.[19]

Warburg remained a staunch nationalist at least until 1918, and as late as 1917 he still fervently believed in a victorious peace.[20] Warburg's nationalism softened after the war, however. Like many other Jewish intellectuals, who increasingly saw nationalism as a dead end in the face of war and growing anti-Semitism, he felt proud of belonging to the German *Kulturnation*. The more tangible the drive to exclude them became, the more clearly German Jews (referring to Lessing, Goethe, Schiller and Humboldt) invoked the liberal, (neo)humanist and universalist ideals of the German tradition. Warburg packed this bourgeois-universalist utopianism into the bridge-builder image that is often found in his correspondence. In a letter from 1918 to the art historian Gustav Pauli, he writes:

> Until now Germany has had no solid bourgeoisie capable of standing on its own two feet and advancing criticisms of its own. Natural aristocrats, temperamentally if reluctantly believing in freedom, were lacking as leaders, and there were too few bridge-builders. [...] Prince Max is such a bridge-builder in his moral character. If only he had come sooner! But this Pontifex Max will be a

---

[19] WIA, GC, Aby Warburg to Wilhelm Waetzoldt, May 6, 1915.

[20] WIA, GC, Aby Warburg to Selma Fliess, May 8, 1917. "The provisional triumph of the world's enslavement by England and the phraseology of the French Revolution is for me by far the most Satanic development of the war. To see Balfour and Viviani together at Washington's grave, speaking of a freedom that America wrested for itself through resistance to English tyranny, is one of the craziest things reality has ever come up with. Despite everything, the most enslaved nation must now show that this war of the machine is in reality a war of the idea. The next few weeks will, I think, bring a decision in our favor. If our incredibly brave men in the West continue to hold out as a wall on which the enemy's bodies bleed, and if the submersibles continue to clear the sea, then even the American panders of war fever will no longer be able to do a thing. I admit that for a long time the destruction of works of art has made no impression on me. Every artillery observation that the enemy can make from a cathedral tower is an outrage if it could have been stopped with a couple of grenades. At stake now is something quite different from memories of the past; if the Entente wins, Germany will at best become an agency in the Anglo-American human slaughterhouse. To be sure, we in Germany are exposed to the fits of unscrupulous politicians, who have been enslaving the masses through the press, but the struggle against them is beginning and must be carried through. To be able to help us in proportion to my sadly declining powers is my greatest wish."

leader of the *Pontifice minimi* of poor bridge-builders, who are supposed to save us. For my poor self, long before the war I set out in this image the rest of my mission in life as a scholar and human being.[21]

Warburg's bridges, scholarly in nature, stretched across national as well as disciplinary boundaries.[22] Thus, when his brother Paul married the New York banker's daughter Nina Loeb in October 1895, Warburg crossed the Atlantic for the occasion and undertook the cultural-ethnological studies among the Hopi Indians that will be discussed later in this chapter. Before traveling to New Mexico, however, he got in touch with some "pioneers of native research," including Cyrus Adler and Franz Boas.[23] These Warburg bridges indicate the networks that grew up among Jews in the Diaspora and exposed them to the charge of internationalism.[24] But, above all, they illustrate Warburg's commitment to the universalism of science, which involves verifying and – unless they are refuted – accepting claims to truth, irrespective of the nationality, religious affiliation, class or "race" of the person who makes them.[25] Warburg did not feel that he belonged to the Jewish community, but rather to a scholarly community that paid homage to science and reason.

---

[21] WIA, GC, Aby Warburg to Gustav Pauli, October 10, 1918. [A few days earlier, on October 3, 1918, Prince Max of Baden had been appointed chancellor of the Reich and prime minister of Prussia; on October 5 he proposed an armistice on the basis of Woodrow Wilson's fourteen-point program. *Trans. note.*]

[22] On Warburg's far-flung scholarly contacts, see Heinz Paetzold, *Ernst Cassirer. Von Marburg nach New York. Eine philosophische Biographie*, Darmstadt 1995. Paetzold lists some of the scholars who shared Warburg's interests and collaborated with the K.B.W. The circles around Hamburg University included, apart from Ernst Cassirer (pp. 71f.): "Gustav Pauli and Erwin Panofsky (art history), Karl Reinhardt (classical philology), Richard Salomon (Byzantine history), Hellmut Ritter (Oriental languages), and later Bruno Snell (classical philology) and Heinrich Junker (Indo-European studies)." But there were also various German and international guests: "the classical philologists and historians of religion Franz Dornseiff, Franz Joseph Dölger, Robert Eisler, Eduard Fraenkel, Hugo Gressmann, Hans Leitzmann, Joseph Kroll, Eduard Norden, Richard Reitzenstein, Hans Heinrich Schaeder, Ulrich von Wilamowitz-Moellendorf and Karl-Ludwig Schmidt; the art historians Adolf Goldschmidt, Jaques Mesnil (Belgium), Wolfgang Stechow, Hubert Schrade and Julius von Schlosser; the Romance scholars Ernst Robert Curtius and Fritz Schalk; the literary theorists André Jolles (Netherlands), Clemens Lugowski and Arturo Farinelli (Italy); the historians Alfred Doren, Hubert Pruckner and Percy Ernst Schramm; the historians of astronomy and astrology Franz Boll and Wilhelm Gundel; the Orientalist Richard Hartmann; and the Egyptologist Hermann Kees."

[23] Aby Warburg, draft of the Kreuzling lecture, quoted from Gombrich, *Aby Warburg*, p. 88; cf. Ulrich Raulff, "Nachwort," in Aby Warburg, *Schlangenritual. Ein Reisebericht*, Berlin 1988 (orig. 1923), pp. 63–94; here esp. pp. 66–71.

[24] According to Dan Diner, it is precisely this transnationality, transterritoriality, or non-territorial disposition of the Jews that could make Jewish history a new paradigm for historiography in the postnational age. See Dan Diner, "Geschichte der Juden. Paradigma einer europäischen Geschichtsschreibung," in Dan Diner, *Gedächtniszeiten. Über jüdische und andere Geschichten*, Munich 2003, pp. 246–62.

[25] On universalism as an "institutional imperative" of science, see Robert K. Merton, "The Ethos of Science," in Piotr Sztompka (ed.), *On Social Structure and Science*, Chicago, IL 1996, pp. 267–76, 268f.

In scholarship, many German Jews hoped, the individual would be judged by his talents and achievements, not by his origins.[26] Warburg, then, stood for an unprejudiced cosmopolitanism guided by reason, with such vehemence that George L. Mosse saw him and many other Jewish scholars as "mirroring the Enlightenment ideal of *Bildung*;" they thought of themselves as "guardians of that ideal," since *Bildung* enabled them to assimilate and break with the Jewish faith.[27] *Bildung*, science and reason therefore presented themselves as connecting links in a chain – indeed, as the bridges that made intellectuals members of a universal community and offered Jews a way of integrating into German society.

As nationalism grew more radical and acquired *völkisch* hues in the Wilhelmine Empire, the original nexus of liberalism, nation, enlightenment, emancipation, *Bildung* and progress began to come apart.[28] Visions of a cosmopolitan community of scholars and a European humanism were now overshadowed by dreams of a homogenized Nation or *Volk*; nationalism turned to the vital and elemental as mainstays of support, not to reason and education. "Blood ties" and "roots" in the people supplanted the ideal of educating individuals as rational beings and citizens of the world. And Jews, accused of being ethereal *Luftmenschen* with no roots in the soil,[29] were excluded from membership in this "imagined community" – not least because they clung all the more firmly to the original ideals of the Enlightenment.

These ideals found expression in Warburg's life as a scholar. But two further typical loci and objects of *Bildungsbürger* longing also met up there: Italy

---

[26] Shulamit Volkov, "Jewish Success in Science," in Volkov, *Germans, Jews, and Antisemites. Trials in Emancipation*, New York, NY 2006, pp. 224–7, esp. 228ff.

[27] Mosse, "Das deutsch-jüdische Bildungsbürgertum," p. 168. On what follows, see also Mosse, *Jüdische Intellektuelle in Deutschland. Zwischen Religion und Nationalismus*, Frankfurt/ Main 1992. Of the extensive literature on the concept of *Bildung* in general, see esp. Georg Bollenbeck, *Bildung und Kultur. Glanz und Elend eines deutschen Deutungsmusters*, Frankfurt/ Main 1994; Reinhart Koselleck, "Einleitung – Zur anthropologischen und semantischen Struktur der Bildung," in Koselleck (ed.), *Bildungsbürgertum im 19. Jahrhundert*, pp. 11–46; Rudolf Vierhaus, "Bildung," in Otto Brunner/Werner Conze/Reinhart Koselleck (eds.), *Geschichtliche Grundbegriffe. Historisches Lexikon zur politisch-sozialen Sprache in Deutschland*, Stuttgart 1972, vol. 1, pp. 508–51. On the nexus between *Bildungsbürgertum* and liberalism, see Dieter Langewiesche, "Bildungsbürgertum und Liberalismus im 19. Jahrhundert," in Jürgen Kocka (ed.), *Bildungsbürgertum im 19. Jahrhundert. Teil IV. Politischer Einfluß und gesellschaftliche Formation*, Stuttgart 1989, pp. 95–121.

[28] Both the bourgeoisie and liberalism underwent a number of mutations in the course of the nineteenth century. They therefore require more detailed consideration than they can be given here. See *inter alia* Geoff Eley, "Liberalism, Europe and the Bourgeoisie 1860–1914," in David Blackbourn/Richard J. Evans (eds.), *The German Bourgeoisie. Essays on the Social History of the German Middle Class from the Late Eighteenth to the Early Twentieth Century*, London 1991, pp. 293–317; Dieter Langewiesche, *Liberalism in Germany*, Princeton, NJ 1999; Jörn Leonhard, *Liberalismus. Zur historischen Semantik eines europäischen Deutungsmusters*, Munich 2001; James J. Sheehan, *German Liberalism in the Nineteenth Century*, Chicago, IL 1978.

[29] On the *Luftmensch* metaphor, see Nicolas Berg, *Luftmenschen. Zur Geschichte einer Metapher*, Göttingen 2008.

(especially Renaissance Italy) and classical antiquity. As we shall see, both would play an essential role in Warburg's encoding of the airplane. But the force behind him was not the kind of utopian antiquity associated with the art historian Johann Winckelmann, but Nietzsche's very different antiquity and Burckhardt's Renaissance. It was the survival of "Dionysian" passions and the ongoing quest for their "Apollonian" restraint in various forms of expression that aroused Warburg's attention.

In his obituary article, Warburg's disciple Erwin Panofsky concisely summarized this interest. Warburg's life's work, he wrote, had been driven by a will

> to see the history of human culture as a history of human passions, which in their terrible simplicity – wish to possess, to give, to kill, and to die – have remained constant in an existential stratum only seemingly covered over by civilization, and which the form-giving mind, for that very reason, must reveal and tame in ever new cultural patterns.[30]

If Warburg could write the history of human passions, not of human reason, this was only because he was convinced

> that Greek antiquity for the first time fought in full the dual spiritual battle for revelation and taming of man's primal agitation. [...] Both as bearer of holy symmetry and as demonic Medusa, antiquity had to be repeatedly forgotten, repeatedly gained, and repeatedly overcome.[31]

The conflict between "thought" and "arousal factors," "holy symmetry" and "demonic Medusa," reason and passion, shaped Warburg's scholarly work and his very existence; it is the antagonism that makes him a paradigmatic representative of his epoch. The conclusion that Warburg drew from it – that antiquity must be forgotten, regained and surmounted over and over again – was the result of a development that will be examined in a moment.

In any event, the starting point of Warburg's thinking was a liberal order of knowledge associated with the *Bildungsbürgertum* and the Enlightenment tradi-tion.[32] This system of thought, speech and action was already breaking up by the turn of the century. But with the First World War and its aftermath, the cracks widened into seemingly insurmountable chasms. This destruction of the liberal order left behind the widespread longing for order that is the theme of this study.

---

[30] Erwin Panofsky, "A. Warburg," in *Repertorium für Kunstwissenschaft*, vol. LI (1930), pp. 1–4, 1f; the obituary first appeared on October 28, 1929 in the *Hamburger Fremdenblatt*.

[31] Ibid., p. 2.

[32] Warburg's close colleague Gertrud Bing also saw him as an Enlightenment figure. In her lecture of 1958 at the unveiling of a Warburg bust in the Hamburg Kunsthalle, she said: "Warburg believed in the power of reason; he was a man of the Enlightenment, precisely because he knew so well the legacy of demonic antiquity. Lessing's *Laocoön* was the great influence of his youth, and he felt beholden to the eighteenth-century German Enlightenment." Bing, "Aby M. Warburg," p. 463.

Two key aspects of the bourgeois-liberal Enlightenment order will be especially important in what follows: (1) a belief in both the liberating-dissolving and the structuring-consolidating power of reason *qua* science, and its hegemony over other approaches and worldviews; and (2) a resulting conception of time as oriented to progress. Since the late eighteenth century, these had been central, almost *a priori*, articles of faith for the German *Bildungsbürgertum*, but also for the liberal bourgeoisie in general as well as the workers' movement. After all, they legitimized the elimination of hereditary privileges and the longed-for dissolution of the Estates as the basis of society, while also promoting the evaluation of individuals by their achievements and furthering the development of new knowledge and technologies. In place of the traditional order rooted in religion and metaphysics, a more just and perfect one based on reason alone seemed set to appear. But "reason" was thereby absolutized, and the order-dissolving spirit of modernity became an order-generating spirit. The idea that it was possible to eliminate chaos and to "master all things by calculation" proved to be as much of an illusion as the progressive course of history.[33]

As a *Bildungsbürger*, Warburg thought of his own life as a project to develop the human within himself under the guidance of reason, and hence as a mirror image of historical progress at the level of the individual. The course of time, as well as his own mental affliction and especially the First World War, led Warburg to a correction of this worldview that found expression in his scholarly work. However, the realization that progress and the unfolding of reason were insufficient, that "human passions [...] had remained constant in an existential stratum only seemingly covered over by civilization," did not lead him to give up the normative claims underlying his liberal attitude to the world. In this he differed from the great majority of his contemporaries.

Why Athens has to be constantly won back from Alexandria

As the character of nationalism changed, large sections of the *Bildungsbürgertum* and the bourgeoisie in general moved away from their Enlightenment roots.[34] To be a bourgeois in Germany, even a member of the educated middle classes, did not necessarily imply a liberal orientation. But Warburg became a loyal Republican and, as a member of the Deutsche Demokratische Partei (DDP), threw in his lot with the Weimar Republic.[35] The First World War, in shattering the optimism of reason and belief in progress, had shaken the foundations of his thinking. Yet as he revised his view of history and world, he maintained his adherence to the principles of the Enlightenment. In this section, we shall look in particular at how he corrected his idea of a progressive direction of history.

---

[33] Max Weber, "Science as a Vocation," in H. H. Gerth/C. Wright Mills (eds.), *From Max Weber. Essays in Sociology*, Boston, MA 1948, p. 139.

[34] Hans-Ulrich Wehler, *Deutsche Gesellschaftsgeschichte*, vol. 3, *Von der "Deutschen Doppelrevolution" bis zum Beginn des Ersten Weltkrieges 1849–1914*, Munich 1995, pp. 938–61.

[35] Ulrich Raulff, *Wilde Energien. Vier Versuche zu Aby Warburg*, Göttingen 2003, pp. 72f.

Like so many of his contemporaries, Warburg lost his "faith" because of the world war. After all, the idea of a "worldly *progressus*" that had "displaced or dissolved" the "spiritual *profectus*"[36] had been the religion of the age. As Stefan Zweig put it in his memoirs:

> This faith in an uninterrupted and irresistible "progress" truly had the force of a religion for that generation. One began to believe more in this "progress" than in the Bible, and its gospel appeared ultimate because of the daily new wonders of science and technology. In fact, at the end of this peaceful century, a general advance became more marked, more rapid, more varied.[37]

When "the lights went out" over Europe in 1914, as the British foreign secretary Sir Edward Grey put it,[38] central tenets of the religion of progress collapsed. The course of history now pointed to anything but a better future, and by no means did man seem as rational as had been previously supposed. Sigmund Freud, for instance, wrote in 1915:

> We expected that the great ruling nations of the white race, the leaders of mankind, [...] would find some other way of settling their differences and conflicting interests. [...] But the war in which we did not want to believe broke out and brought – disappointment. It is not only bloodier and more destructive than any foregoing war, as a result of the tremendous development of weapons of attack and defense, but it is at least as cruel, bitter, and merciless as any earlier war.[39]

Freud was here referring to the great disillusionment that gripped Europe during the war. But the sober simplicity of the term "disappointment" should not blind us to the fact that he was thoroughly aware of its depth and destructiveness.

> Caught in the whirlwind of these war times, without any real information or any perspective upon the great changes that have already occurred or are about to be enacted, lacking all premonition of the future, it is small wonder that we ourselves become confused as to the meaning of impressions which crowd in upon us or of the value of the judgments we are forming. It would seem as though no event had ever destroyed so much of the precious heritage of mankind, confused so many of the clearest intellects or so thoroughly debased what is highest.[40]

Warburg was one of these "clearest intellects" so confused by the war, and it was not least the "meaning of impressions" that made him fall ill. The war so

---

[36] Reinhart Koselleck, "'Space of Experience' and 'Horizon of Expectation' – Two Historical Categories," in Koselleck (ed.), *Futures Past: On the Semantics of Historical Time*, Cambridge, MA 1985 [orig. 1979], p. 265.

[37] Stefan Zweig, *The World of Yesterday. An Autobiography*, New York, NY 1943, p. 3.

[38] Edward Grey, *Twenty-Five Years. 1892–1916*, vol. 2, London 1925, p. 20.

[39] Sigmund Freud, *Reflections on War and Death*, New York, NY 1918, pp. 4, 11.

[40] Ibid., p. 1.

thoroughly undermined his already shaky sense of order that the whole "world of yesterday" seemed to him to be collapsing. But before this breakdown led him to various psychiatric clinics and finally to Ludwig Binswanger in Kreuzlingen, he was still endeavoring to make sense of the phenomenon of the war[41] – witness his extensive reading, his collection activities, the war archive and his efforts in the *Rivista*, on the strength of ideas and supposedly better arguments, to stop Italy joining the war on the side of the Entente.[42] Also his *Pagan and Antique Prophecy in Words and Images in the Age of Luther* (1920) and his serpent ritual lecture (1923) represent attempts to bring the light of scholarship to bear upon the chaos produced by the war and to calm the author's own fears. At the same time, they may be understood as interpretations of the "relapse" into irrational forms of mastering chaos. Thus, it is only logical to regard Warburg as a pioneer in the study of Europe during the "crisis of the liberal system" – a crisis that had its origins *inter alia* in the collapse of the axioms of reason and progress, which horrified Warburg like so many others.[43] In his piece on Luther, the examination of a distant yet kindred phenomenon – the co-presence of mathematics with fear of demons – created space in which he could think about, and thereby master, the situation brought about by the First World War.[44]

During his illness, Warburg reflected on his attempts to bring order into chaos intellectually. In a letter of 1922 to his assistant Fritz Saxl, he writes:

From those times [Warburg's childhood] comes the dread brought about by disjointed visual recollections or stimulations of the organs of smell and hearing; the fear brought forth by chaos, the attempt to bring order into this chaos intellectually – an attempt that may be described as the tragic childhood experiment of the thinking man – therefore began very early on, much too early for my nervous constitution.[45]

Warburg's "nervous constitution" went beyond the typical "nervousness" of the age.[46] His severe mental illness, which developed during the war, is of interest

---

[41] On Warburg's illness, see Ludwig Binswanger/Aby Warburg, *La guarigione infinita. Storia clinica di Aby Warburg*, ed. Davide Stimili, Vicenza 2005.

[42] See Dorothea McEwan, "Ein Kampf gegen Windmühlen. Warburgs pro-italienische publizistische Initiative," in Gottfried Korff (ed.), *Kasten 117. Aby Warburg und der Aberglaube im Ersten Weltkrieg*, Tübingen 2007, pp. 135–63; and Peter J. Schwartz, "Aby Warburgs Kriegskartothek. Vorbericht einer Rekonstruktion," in ibid., pp. 39–69.

[43] Nolte, *Die Krise des liberalen Systems und die faschistischen Bewegungen*.

[44] Warburg's most influential biographer, Ernst Gombrich, argued that in this text he "consciously or unconsciously" linked his research on the fear of devils in the Reformation with his daily political experiences on the wartime home front, and that he had "a deep spiritual sympathy" with the problem. The idea that the Luther piece was connected with Warburg's impending mental breakdown, and above all that it had a motivation in the period, appears to be a hermeneutic self-evidence. See Gombrich, *Aby Warburg*, pp. 207, 214.

[45] WIA, GC, Aby Warburg to Fritz Saxl, May 10, 1922.

[46] On the "nervous age," see Joachim Radkau, *Das Zeitalter der Nervosität. Deutschland zwischen Bismarck und Hitler*, Munich 1998.

here only in so far as Warburg seems to have drawn a parallel between it and the all-European malady; just like Europe at war, he was no longer capable of using reason to order chaos and to calm his fears. To put it in a nutshell, he started from the supposition that the world is first perceived as chaotic.[47] Chaos produces fear, and this fear drives people on to cultural and intellectual feats of order that serve to overcome the causes of fear.[48] There are various responses to the phobic cause: that is, specific forms of the ordering of chaos, each of which corresponds to a stage of human development.

As Blumenberg writes with reference to Franz Rosenzweig, a worldly object originally funds itself in the "chaos of the unnamed."[49] It triggers fear, since "as something nameless, it cannot be conjured up or appealed to or magically attacked."[50] The act of naming is the first step of a still teleologically conceived development, which peaks in rational mastery of the world and the supremacy of logical thinking.[51] The fetish or totem, in which fear is directly objectified, corresponds to the human childhood of "primitive culture." The "savage" banishes and exorcises his original fear of chaos and the "absolutism of reality" in an object.[52] To him the primal force of fear appears to reside in the magical object. Thus fear and object are still fused together. Since the "savage" does not realize his own effect in shaping and generating the world, he is still under the spell of the object, which he equates with his primal fear. He has still put only a small distance between himself and the cause of his fear, but this space nevertheless permits a ritual practice of invocation and magic.[53]

Right at the other end of the cultural scale, in phylogenetically "adult" human beings, Warburg finds logical reactions to the cause of fear, and their objectification in words and numbers. Whereas Warburg associated the "savage" magical-mythical reaction with "man as grasper," and the logical reaction with "man as thinker or conceptualizer."[54] In the latter, the distance (or "deliberative space") between subject and object is at its greatest. Warburg's "space for thought" may thus be regarded as a distance from the object gained through the *logos* of abstraction. The *logos*, especially in its most abstract mathematical form, first

---

[47] On the following, see Gombrich, *Aby Warburg*, pp. 216ff.; and Böhme, *Fetischismus und Kultur*, pp. 243ff.

[48] Warburg's starting point has many parallels with Hans Blumenberg's concept of the "absolutism of reality" that can be mastered through myth, as well as with Ernst Cassirer's view that cultural achievements have their origin in attempts to order chaos. See Hans Blumenberg, *Work on Myth*, Cambridge, MA 1985; and Ernst Cassirer, *The Myth of the State* (orig. 1946), New Haven, CT 1963, p. 15.

[49] Blumenberg, *Work on Myth*, pp. 15ff.

[50] Ibid., pp. 34–5.

[51] Cf. Gombrich, *Aby Warburg*, p. 217.

[52] On the "absolutism of reality," see Blumenberg, *Work on Myth*, esp. ch. 1.

[53] For an accessible "translation" of Warburg's notes, see Gombrich, *Aby Warburg*, pp. 217f.

[54] On this distinction, see Warburg, *Schlangenritual*, p. 25. [Cf. the variant English translation of Warburg's text: "A Lecture on Serpent Ritual," *Journal of the Warburg Institute*, vol. 2, no. 4 (April 1939); here p. 282.]

makes the world truly an object – an object both of reflection and of planning and manipulation.[55]

Between fetish and concept lies the intermediate stage of "myth;" the symbol and the image are the forms corresponding to this stage. "Between the primitive man […] who grasps the object directly before him" and abstract-logical man, "who plans and awaits the results of his action," is the man "who interposes symbols between himself and the world."[56] For Warburg, images and symbols are "configured affects," stores of surviving phobic energy. But they also create a distance from the original cause. The fear is objectified and given a form; its cause is named and thereby warded off. Admittedly the fear is present in the image or symbol, but unlike the totem or fetish it is not equated with the object. This is what Warburg means when he says: "You live and do me no harm."[57]

The evolutionary-teleological theories of nineteenth-century anthropology and ethnology, with their characteristic tinges of colonialism, had a strong influence on Warburg's stages model of the cultural process.[58] The First World War made a revision unavoidable, since it had become highly questionable to plot a linear evolution from archaic-magical "man the grasper" through mythical man "interposing symbols between himself and the world" to rational civilized man deploying science and technology. Warburg felt compelled to conclude that, as Panofsky put it, "human passions had remained the same" and were only covered over with a veneer of civilization. Instead of believing that superstitious, instinctually driven man was a thing of past, he now realized that he was a timeless form still very much alive in today's world. In his Luther text, Warburg therefore asserted that "that age when logic and magic blossomed, like trope and metaphor, in Jean Paul's words, 'grafted to a single stem,' is inherently timeless."[59] This "timeless" juxtaposition of magic and logic, of "grasping" and "reasoning" man, of myth and *logos*, is a basic analytic tool for an understanding of mythical modernity. For the latter becomes plausible only if the co-presence of myth and *logos* supplants the normative, because teleological, "simultaneity of the non-simultaneous."[60]

---

[55] Warburg still views positively the split between man and world, subject and object. Unlike Heidegger, for instance, he by no means thinks of it as estrangement. Myth, we may say at once, was seen by many of Warburg's contemporaries as a possible way of ending the split or estrangement between man and the world, restoring the unity of subject and object, and thereby overcoming the distortions of "modernity."

[56] Warburg, "A Lecture on Serpent Ritual," p. 282.

[57] This was the epigraph to Warburg's fragments on the psychology of art; quoted from Gombrich, *Aby Warburg*, p. 71.

[58] See Raulff, "Nachwort," pp. 74f. According to Gombrich (*Aby Warburg*, p. 59), Warburg's understanding of man's development was "still rooted in the optimistic creed of evolutionism. Progress led from savagery to higher and higher forms of control, and art had a share in this ascent." See also the recently published Nicola Gess (ed.), *Literarischer Primitivismus*, Berlin 2013.

[59] Aby Warburg, "Pagan and Antique Prophecy in Words and Images in the Age of Luther," in Warburg, *The Renewal of Pagan Antiquity*, pp. 597–667; here p. 599.

[60] On the simultaneity of the non-simultaneous, see Wilhelm Pinder, *Das Problem der Generation in der Kunstgeschichte Europas*, Berlin 1926. It was Ernst Bloch, though, not the

For Warburg, progress changed from a fact into a normative imperative. Thus, unlike most of his contemporaries – among whom concepts such as "the primitive," "the elemental" and "the primordial" were experiencing an unprecedented boom – he did not derive a vitalist philosophy of life from the "eternal return" of magical and mythical modes of thought. With the discovery of the Unconscious, *fin-de-siècle* Europe had embarked on a long journey into its "Inner Africa" in search of the "noble savage."[61] Until the war it had seemed that this had been finally left behind, but now – as that "savage" had cast off the thin veneer of civilization and revealed himself without disguise – it seemed as if all along he had been painfully missed.

That was not what Warburg felt. The grounds for his rejection of "the primitive" and "the archaic" may lie in his mental illness. Certainly he experienced this as the triumph of irrational Dionysian inner forces over his reason; he had to accept that magical or mythical reactions to fears persisted alongside logical attempts to master them and could not in the end be conquered and dispelled. They continually demanded their tribute, and the logical ordering of a chaotic reality had to be wrested from them again with difficulty. Yet it seems idle to engage in biographical speculation, when Warburg himself addressed the problem scientifically during the war years. As in the Reformation period – the analogy seems appropriate – Europe found itself in a new "age of Faust, in which the modern scientist – caught between magic practice and cosmic mathematics – was trying to insert the conceptual space of rationality between himself and the object. Athens has constantly to be won back again from Alexandria."[62]

How should we understand this? Faced with the reality of the war, Warburg felt compelled to revise his diachronic stages model of the cultural process. The war revealed that science and magic, *logos* and myth, technological advance and huge bloodbaths were not mutually exclusive. The enchanted man of Egyptian Alexandria, at home in a mythical *nomos*, had not returned; he had never really gone away. Hence the war and the magical-mythical practices it generated were not a relapse. The difference with the past was that the new-old Alexandrian man now had a vast potential for destruction, originating in the scientific-technological development that began its triumphal march in the Athens of Socrates.

As an "Athenian" Warburg argued that phobic causes again and again elicited magical and mythical, not only logical, reactions.[63] Disenchantment was not a once and for all process. "Conceptual man" had to free himself again and again from the clutches of "grasping man." Although the war forced Warburg to take leave of his phylogenetic-teleological ideas, he still clung to the hierarchical

---

art historian Pinder, who made the phrase more widely known. See Ernst Bloch, *Erbschaft dieser Zeit*, Zurich 1935.

[61] Jean Paul, *Selina, oder über die Unsterblichkeit der Seele*, in *Sämtliche Werke*, I, vol. 6, ed. Norbert Miller, Munich 1996, pp. 1105–236, 1182.

[62] Warburg, "Pagan and Antique Prophecy," p. 650.

[63] Cf. Peter Gay, *Freud, Jews, and Other Germans. Masters and Victims in Modernist Culture*, New York, NY 1978, pp. 129f.

model in which *logos* was superior to magic and myth, and the telos of humanity was rational development. Such progress, however, was an ongoing dialectical process – an imperative, not a one-off achievement.

Warburg's correction also affected the value of "primitive" practices – witness the quotation from Goethe with which he ends his Luther text. "Such abuses are forgivable in dark ages, when they are entirely in character. Superstition is simply the use of false means to a true end, and is therefore neither so reprehensible as it is believed to be, nor so rare in so-called enlightened centuries and among enlightened people."[64] As we shall see in the next section, Warburg built on this insight in his serpent ritual lecture, where he offered a further means of understanding the crisis that had gripped Europe. The war itself, marking a rift in civilization, was surely the main cause of the chaos that now called for the re-establishment of order. But another phobic cause was also destroying the space for thought and had to be overcome: technology.

> The forces of nature are no longer seen in anthropomorphic shapes; they are conceived as an endless succession of waves, obedient to the touch of the human hand. With these waves the civilization of the mechanical age is destroying what natural science, itself emerging out of myth, had won with such vast effort – the sanctuary of devotion, the remoteness needed for contemplation. The modern Prometheus and the modern Icarus, Franklin and the Wright Brothers who invented the aeroplane, are those fateful destroyers of our sense of distance who threaten to lead the world back into chaos.[65]

Disenchantment of the serpent

In Warburg's eyes, the "modern Prometheus," "modern Icarus" or even technology in general was destructive of the realm of thought. As is well known, such a skeptical – or positively hostile – attitude to technology was not uncommon among German intellectuals.[66] But Warburg's position was by no means as one-sided as the preceding quotation would seem to suggest; though certainly critical of technology and its consequences, he was at the same time captivated by it. This is apparent in, among other places, the K.B.W. library in Hamburg.[67] Warburg's fascination over many years with aviation and the "modern Icarus" strengthens this impression. He nursed great enthusiasm for the German airship pioneer

---

[64] Johann Wolfgang Goethe, *Materialien zur Geschichte der Farbenlehre*; quoted from Warburg, "Pagan and Antique Prophecy," p. 651.

[65] Warburg, "A Lecture on Serpent Ritual," p. 292.

[66] See Joachim Radkau, *Technik in Deutschland. Vom 18. Jahrhundert bis zur Gegenwart*, Frankfurt/Main 1989; and Thomas Rohkrämer, *Eine andere Moderne? Zivilisationskritik, Natur und Technik in Deutschland 1880–1933*, Paderborn 1999

[67] The new library building in Hamburg was fitted with a pneumatic dispatch system, a book and a passenger lift, conveyor belts, and lifting tables, as well as numerous telephones with outside lines and for communication inside the building. See Tilmann von Stockhausen, *Die Kulturwissenschaftliche Bibliothek Warurg. Architektur, Einrichtung und Organisation*, Hamburg 1992.

Hugo Eckener, who in October 1924 crossed the Atlantic in a Zeppelin.[68] And as early as 1913, the art historian Warburg had published an essay on "Airship and Submarine in the Medieval Imagination."[69]

Two "Northern European tapestries" or *arrazzi* that adorned the Palazzo Doria in Rome had attracted his attention. They show the heroic exploits of Alexander the Great, who, according to legend, traveled to the bottom of the sea in a kind of glass barrel and rode up into the sky in a metal structure drawn by four griffins. In discussing these fabulous scenes on the wall tapestries, Warburg had noted the juxtaposition of mythological and scientific worldviews, or the "dichotomy" in the "spiritual construction," of Renaissance man. Not unlike Paul Fussell, who was filled with wonder at the role of myth in mechanized industrial warfare,[70] Warburg addressed the paradox that an "uncritical faith in griffins" existed alongside a spirit "that put fire to practical use, in the siege artillery of Duke Philip the Good of Burgundy [1396–1467]."[71]

> This "Burgundian Antique," like its Italian counterpart, had a role of its own to play in the creation of modern man, with his determination to conquer and rule the world. While continuing to visualize the elemental sphere of fire as inaccessible even to the preternatural strength of fabulous oriental beasts, man himself, through firearms, had already tamed the fiery element and pressed it into his own service. It seems to me by no means far-fetched to tell the modern aviator, as he considers the "up-to-the-minute" problem of motor cooling systems, that his intellectual pedigree stretches back in line direct – by way of Charles the Bold, trying to cool the burning feet of his heaven-storming griffins with wet sponges – to *le grand Alixandre*.[72]

In 1913 Warburg still felt the existence of *logos* and *mythos* alongside each other to be an anomaly. Whereas the modern aviator supposedly faced the problem of motor cooling systems with nothing other than his technical knowhow and scientific worldview, his spiritual ancestors had solved it in a symbolic-pictorial, indeed mythical, manner. As this study will make clear, however, that ostensible "dichotomy" did not mark modern man only *in statu nascendi*.

On April 21, 1923 – ten years after his "Airship" essay, and nearly five after the end of the war – Warburg gave his famous lecture on serpent ritual to demonstrate

---

[68] This is confirmed in a letter from Warburg to Felix von Eckhardt. "As you perhaps know," he wrote, "I have greatly warmed to the idea that Eckener should become President of the Reich, and I should therefore like to know whether the person in question, about whom the note also says that he has recently been enjoying high regard in America, was Eckener." WIA, GC, A. Warburg to Felix von Eckhardt, 2.9.1925.

[69] Aby Warburg, "Airship and Submarine in the Medieval Imagination" (1913), in Warburg, *The Renewal of Pagan Antiquity*, pp. 333–42.

[70] Fussell, *The Great War and Modern History*, p. 115.

[71] "Airship and Submarine," p. 336.

[72] Ibid., p. 337.

his recovery and his renewed capacity for work.[73] This "travel report," in which Warburg the ethnologist presented his impressions from a trip among the Pueblo Indians 27 years earlier, dealt with a modern, seemingly paradoxical, phenomenon:

> What interested me as a cultural historian was that, in the middle of a country that had made technology an admirable precision weapon in the hands of intellectual man, an enclave of primitive pagan humanity could still persist – one which unshakably pursues [...] in agriculture and hunting certain magical practices that we are accustomed to judge as symptomatic of a thoroughly backward form of human existence. [...] This juxtaposition of fantastic magic and sober purposive action seems to us a symptom of splitting.[74]

In "Airship and Submarine" Warburg had considered the coexistence of a belief in griffins with advances in pyro-weaponry, while in his text on Luther he had examined the presence of science alongside fear of demons in Reformation Europe. Now, in the lecture on serpent ritual, the juxtaposition of a magical-mythical and a scientific-technological relationship to the world shifts to the American continent, where Warburg attributes it to the native inhabitants. It is almost as if he was seeking to externalize the "splitting." As we shall see in the following chapters, however, it was not necessary to cross the ocean or to reach back into ancient times; Alexandria and Athens, myth and enlightenment, the order-generating and order-dissolving spirit of modernity were indigenous to the Europe of his time. It would seem that, although Warburg could feel the "split" in Europe seismographically, he still ascribed it to "the Other." In this way, he gained the "level-headedness" that he needed to understand the contemporary handling of war, technology and chaos.

The juxtaposition of mythical and scientific practice was especially clear in the case of technology. This "admirable precision weapon in the hands of intellectual man," this solid token of "sober purposive action," had become a "second nature" that threatened to engulf human beings in the same way as the first, real nature.[75] For Warburg, just as space for thought had been wrested from first nature, the task now was to conquer it in the space between man and second nature.

The serpent of Warburg's lecture is a symbol of both first and second nature, even if in the case of technology its manifestations are more abstract than "Edison's copper snake" or "the lightning held captive in electric wire."[76] Warburg's travel report, whose snake theme may at first seem remote or even aberrant in relation to our central concern with aviation, proves to be a lead worth following up, not least because the snake operates as a metaphor for technology

---

[73] On the context of the lecture, see Raulff, "Nachwort," pp. 63ff.

[74] Warburg, *Schlangenritual*, p. 10; cf. "A Lecture on Serpent Ritual," p. 282.

[75] The conception of technology, or all human cultural achievements, as a "second nature" runs from Hegel, Marx and Engels through Klages to Cassirer, Plessner and Gehlen.

[76] Warburg, *Schlangenritual*, pp. 58f.

and because it favors a "dichotomized" approach. It leads straight to the heart of fascist, mythical modernity, shedding further light on the concurrence of magical-mythical and scientific worldviews and situating it in relationship to technology. Furthermore, Ernst Jünger (an author to be considered later in greater detail) chose the image of the serpent when he discussed the issue of technology. "Oh," he addressed it, "you most steely serpent of knowledge [i.e. technology] – you whom we must spellbind if you are not to throttle us."[77] The steel snake could not be mastered by calculation alone and integrated into the human lifeworld.[78] Other means were necessary to put order into the chaos emanating from it – the means of enchantment, as opposed to Weberian "disenchantment."[79] Thus, whereas Warburg sought to control the fear-producing snake/technology by means of logical-scientific objectification, Jünger – and he was by no means alone in this – proposed to charm the steel snake by means of mythical enchantment.

According to Warburg's lecture, the serpent as a symbol of lightning lies at the center of the Pueblo Indians' ritual worship.[80] Their survival depends on the maize harvest, for which the sparse rainfall is in turn indispensable. The serpent is revered as a weather deity, since its shape links it by a magical causality with lightning and therefore with the benefits of a rainstorm. As to the dance ritual that Warburg observed, he interpreted it as a "social means of providing for food through magical practices."[81] The living snakes at the center of the serpent ritual of the Moki Indians in Walpi and Oraibi are "transmuted and sent out as messengers, so that when they come to the souls of the dead they may produce storm in the skies in the shape of lightning."[82]

In his lecture, Warburg traced the symbol of the serpent in antiquity, the Bible and the middle ages, before finally coming to the "technological age." The serpent, his résumé begins, is precisely "an international symbol answering the question of how elemental destruction, death and suffering come into the world. [...] Where human suffering looks helpless for release, the serpent may be found close by as a graphic explanatory cause."[83] Jünger's choice of the snake image is not at all accidental, therefore. He too conceives it as a graphic cause for the "helpless human suffering" experienced in war, which is in quest of salvation. But how,

---

[77] Ernst Jünger, *Das Abenteuerliche Herz. Aufzeichnungen bei Tag und Nacht*, Berlin 1929, p. 224. Immediately after this (pp. 224ff.), Jünger moves on to his experiences at an airfield and describes the fliers there as conquerors of technology.

[78] See Thomas Rohrkrämer, "Die Verzauberung der Schlange. Krieg, Technik und Zivilisationskritik beim frühen Ernst Jünger," in Wolfgang Michalka (ed.), *Der Erste Weltkrieg. Wirkung, Wahrnehmung, Analyse*, Munich 1994, pp. 848–74, p. 865.

[79] Weber, "Science as a Vocation," p. 139.

[80] Warburg, *Schlangenritual*, p. 16.

[81] Ibid., p. 24.

[82] Ibid., p. 44. Snakes are thus the "mysterious incalculable forces" of which Max Weber writes. The ritual in Warburg's lecture is therefore the exact opposite of the disenchantment process described by Weber: "One need no longer have recourse to magical means in order to master or implore the spirits, as did the savage, for whom such mysterious powers existed. Technical means and calculations perform the service" ("Science as a Vocation," p. 139).

[83] Warburg, *Schlangenritual*, p. 55.

Warburg asks, has mankind freed itself "from this inexorable bond with a venomous reptile that it sees as the cause of things?"[84]

The technological age, Warburg argues, is no longer *reliant* on the serpent symbol to understand lightning. City-dwellers are no longer afraid of lightning, and piped water means that they no longer need storms. "Scientific explanation does away with mythological causes."[85] After he had finished the lecture, Warburg showed a slide of an American walking in the street: "Uncle Sam with a top hat," the "human type who overthrew the serpent cult and overcame the fear of lightning." But "electric cabling stretches above his hat. In this copper snake, invented by Edison, he has wrested lightning from nature."[86] Technological mastery of lightning renders magical causality obsolete. The snake dance no longer forces out the storm necessary to life; rather, technology develops new and beneficial sources of energy. However, Warburg also recalls that "the relationship of those seeking salvation to the serpent [moves] in a cycle of cult worship, from crudely sensuous approaches up to the act of overcoming. As we see from the Pueblo Indian cults, it is still today a tangible measure of human development, rising from instinctual-magical proximity up to the spiritual detachment for which the venomous reptile symbolizes the demonic forces of nature that man has to conquer outside and inside himself."[87]

Warburg now assumes that magical-mythical attempts to find order will recur time and time again – and that it is again this very stage that has to be overcome. The precise form of contact with the serpent symbolized the position that a culture occupied within the "cycle" from magic through myth to science. The idea that time's arrow pointed in a progressive direction had anyway yielded to a cyclical ordering of time. The struggle between Athens and Alexandria was bound to repeat itself time and time again.

The key point here is Warburg's (tentative) dissolution of the evolutionary-teleological model of cultural stages. The passage from *mythos* to *logos* was not a one-off process, but had to be fought for repeatedly; enlightenment – in the words of Adorno and Horkheimer – reverted again and again to mythology.[88] This simultaneity and dialectic of *mythos* and *logos*, or rather of the order-generating and the order-dissolving spirit of modernity, is the basis for the placing of fascism within modernity. For fascism was not a relapse into premodernity, but a genuinely modern experience – one that sought to eliminate the "tearing of the world into two" which Heine associated with modernity.[89] It was a response to new phobic causes and an attempt to structure the chaos of the Nameless that had welled up once again. Warburg's theory of culture, on the other hand, should be understood

---

[84] Ibid.
[85] Ibid., p. 56.
[86] Ibid., pp. 58f.
[87] Ibid., p. 57.
[88] Adorno/Horkheimer, *Dialectic of Enlightenment*, p. xvi.
[89] Heinrich Heine, "The Baths of Lucca," in *Travel Pictures*, New York, NY 2008, p. 107.

both as part of the discourse of the period around the First World War and as a condition of possibility for today's perspective on events.

Warburg was not only academically involved in the discourse of his time concerning technology, war and its cultural handling. He was not a detached observer and analyst, remote from any participation in public affairs. On the contrary: his postage stamp design *Idea vincit* represents a link between his cultural theory and his active commitment to the Weimar Republic. The design is a contrasting foil that sets in sharper relief the contributions to aviation discourse considered in the following chapters. For Warburg's encoding of the airplane expresses the attempt on his part to disenchant the steel snake; *Idea vincit* translates his effort at ongoing enlightenment into an image situated in the technological age. The airplane, which epitomizes both man's triumph over nature and the acute threat facing him, is here encoded in the spirit of Enlightenment liberalism and used as a symbol of the triumph of reason and progress. For Warburg, it was reason that made it possible to lead mankind out of the "depths" of magical-mythical thinking into the light of truth. In that consisted progress.

But the fact that Warburg chose an image for his rationalist message may be interpreted as an admission to "visual thinking," to a "soul rooted in poetry and mythology." In other words, Warburg was by then aware of the limits of the Enlightenment.[90] His stamp design is an iconic compression of the liberal world picture, of a scientist's battle cry for the reconquest of Athens and republicanism. At the same time, it is an admission that attempts will continue to capture fear in symbolic images and to bring order into chaos through myth. It was necessary, however, to keep striving to overcome the "eternally unchanging Red Indian in the human soul."[91]

Idea vincit, or the way out of the cave

For a number of years, Warburg wrestled with the idea of writing an art history of the postage stamp. Thus, on August 13, 1927, in his notes for a lecture at the K.B.W. on "The Function of the Stamp Image in Intellectual Intercourse," he recalled: "Felt personal need since 1913 for art history of stamps."[92] And we read in a letter of December 16, 1926 to Ludwig Binswanger:

> I have a huge amount to do. I feel mentally productive, even venturesome, so that my venerable psyche is really beginning to spin out the last extensions of independent thought from the period before the war. To my astonishment, I was seized with a youthful enthusiasm for stamps, an art history of which I was planning to write back in 1913.[93]

---

[90] Warburg, *Schlangenritual*, p. 56.

[91] See Warburg's notes for the draft lecture on serpent ritual, reproduced in Gombrich, *Aby Warburg*, p. 226.

[92] WIA, III, 99.1.1.1, p. 17.

[93] UAT 443/31, Bl. 44, Aby Warburg to Ludwig Binswanger, December 12, 1926.

The stamps of Fascist Italy had attracted his attention, and he recognized the mass psychological potential of these widespread "picture-vehicles."[94] According to Ulrich Raulff, he thought of stamps as "emblems and advertisements of power,"[95] and so his design for a stamp should be seen also in a multidimensional political context. The stamp, as representative and symbol of the state at home and abroad, is a "mirror" reflecting the respective culture.[96] According to a report in the *Hamburger Nachrichten* on an evening lecture at the K.B.W. on "The Stamp as Cultural Document," Warburg had started by assuming that

> numerals, heads, and likenesses of the ruler are all marks of sovereignty that depersonalize the bearer of such signs and turn him into an instrument. He becomes the bearer of an idea. So, a stamp adorned with marks of sovereignty becomes the idea of the body politic and as such is expected to behave with dignity and a sense of detachment. [...] Professor Warburg called stamps the *pictorial language of global communication*, thereby indicating the great significance that attaches to them.[97]

Before Warburg's remarks that evening, there had been a lecture by the Reich Arts Secretary Edwin Redslob, whose office was at the heart of the symbolic politics of the Weimar Republic. The bitter dispute over flags that had led to the resignation of Reich Chancellor Hans Luther on May 12, 1926 lay more than a year in the past, yet the search for tokens of symbolic identification with the Republic had remained rather unsuccessful.[98] Arts Secretary Redslob was also responsible for advising the Post Ministry on artistic or "political-symbolic" matters regarding the design of new stamps, although it was always possible that his proposals would not be accepted.[99]

---

[94] WIA, III, 99.1.1.2, p. 45. On p. 47 we read: "Although the K.B.W. is a cosmic museum for Hist. and Psychology, it belongs to the Bfm [League for Human Rights], since it magically inspires words, detaches them from their bearer, and conveys them to a third party."

[95] See Ulrich Raulff, "Der aufhaltsame Aufstieg einer Idee. 'Idea vincit:' Warburg, Stresemann und die Briefmarke," in Wolfgang Kemp et al. (eds.), *Vorträge aus dem Warburg-Haus*, vol. 6, Berlin 2002, pp. 125–62, 130. Raulff continues: "But nor was he blind to the lesson of the First World War, which had shown that politics alone was no longer the destiny of the contemporary world. Modern technology was connected with it and was dominant all the way through it. This modernity had to be reflected in the most trivial everyday image in which political power presented itself."

[96] Warburg in a letter of February 1927 to Franz Fuchs; quoted in Dorothea McEwan, "IDEA VINCIT – 'Die siegende, fliegende "Idea."' Ein künstlerischer Auftrag von Aby Warburg," in Sabine Flach/Inge Münz-Koenen/Marianne Streisand (eds.), *Der Bilderatlas im Wechsel der Künste und Medien*, Munich 2005, pp. 121–51, 129.

[97] "Die Briefmarke als Kulturdokument. Vorträge von Reichskunstwart Dr. Redslob und Professor Warburg," *Hamburger Nachrichten*, August 15, 1927.

[98] On the flag dispute of 1926, see Bernd Buchner, *Um nationale und republikanische Identität. Die deutsche Sozialdemokratie und der Kampf um die politischen Symbole in der Weimarer Republik*, Bonn 2001, pp. 104–31.

[99] On Redslob's activity as Arts Secretary, especially in connection with the "hassle of stamp design," see Annegret Heffen, *Der Reichskunstwart. Kunstpolitik in den Jahren 1920–1930. Zu*

Warburg's stamp design and his collaboration with Redslob should anyway be seen in the context of disputes over symbols in the Weimar Republic. Warburg considered his *Idea vincit* a suitable way of presenting the Republic at home and abroad. But the *Idea* was symbolic of the peaceful republic, not of martial Germany. Warburg's aim was to spread the image of a non-militarist country within and beyond its frontiers, differentiating Germany from Fascist Italy and its "schizophrenic power mania."[100] The picture on the stamp is meant to convey this *Idea vincit* message to the world. To ensure this, Warburg made contact with the foreign minister, Gustav Stresemann. He would be very happy, he wrote to his brother Max on November 25, to be "able to show my library to Stresemann, whose shrewd intrepidity I have learned to value highly over the years."[101] He thought of showing Stresemann his stamp design at the end of a tour of the library. And Warburg was able to realize his intention when the Foreign Minister visited Hamburg in December 1926 and put in an appearance at the K.B.W.

The story of Warburg's stamp design (Figures 1.1 and 1.2) has been told in detail by Dorothea McEwan and Ulrich Raulff.[102] It may therefore suffice here to recapitulate the background as it is presented in Warburg's K.B.W. journal entry of December 21, 1926:

> On 4/XII I received a call for help from the painter-etcher Alex Liebmann in Munich, which I answered with a commission to draw a new German stamp in landscape format: it should show the sea below and an aeroplane rising steeply, with the inscription "Briand, Chamberlain, Stresemann." Three banal but very hurried specimens were his reply. [...] Since the dinner for Stresemann had been postponed (from the 14th to the 20th) and he [had] meanwhile received the Nobel Prize, it occurred to me that the right artist might be Strohmeyer; I have always (without finding much support) seen his fantastic but real structures as an exceptional instrument. [...] I showed him my measly sketches, for a print now rather than a stamp: soaring flight. Inscription on the underside of the wings, and an iron arch, the top of which is outside the frame. He sketched the aircraft in his tiny notebook.[103]

---

den Bemühungen um eine offizielle Reichskunstpolitik in der Weimarer Republik, Essen 1986, pp. 132–9.

[100] The K.B.W. journal contains the entry: "Antiquity as a stamp (*fasces*) leads in Italy to disclosure of its schizophrenic power mania; Strohmeyer's 'Idea vincit,' on the other hand, is a protest by the 'impractical' idea." See "Eintrag Aby Warburg vom 29.12.1926," in *Tagebuch der Kulturwissenschaftlichen Bibliothek Warburg mit Einträgen von Gertrud Bing und Fritz Saxl*, ed. by Karen Michels and Charlotte Schoell-Glass: Aby Warburg, *Gesammelte Schriften. Studienausgabe*, Siebte Abteilung, vol. VII, Berlin 2001, p. 39.

[101] WIA, FC, Aby Warburg to Max Warburg, November 25, 1926.

[102] See McEwan, IDEA VINCIT; Raulff, "Der aufhaltsame Aufstieg einer Idee;" and Karen Michels/Charlotte Schoell-Glass, "Aby Warburg et les timbres en tant que document culturel," *Protée* 20/2002, pp. 85–94.

[103] "Eintrag Aby Warburg vom 21.12.1926," in *Tagebuch der Kulturwissenschaftlichen Bibliothek*, pp. 23f.

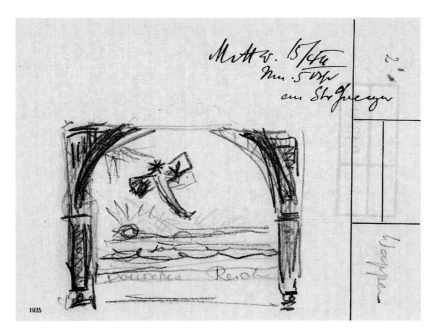

Figure 1.1   One of Warburg's sketches used by Strohmeyer
Source: WIA, IV.87, "Liebmann Correspondence" packet, December 15, 1926.

Figure 1.2   Strohmeyer's linocut that was given to Stresemann
Source: *Idea vincit*, 1926, linocut by Otto Heinrich Strohmeyer, Harvard Art Museum, Fogg Art Museum, gift of Paul J. Sachs, M3027.

So, Warburg first commissioned his friend Alexander Liebmann, who was in financial difficulties, to design a postage stamp. However, when Stresemann's visit was postponed because of the Nobel Prize award, he turned to another graphic designer he knew, Otto Heinrich Strohmeyer, and asked him to develop a linocut on the basis of Warburg's own sketches. On December 20, the big day arrived: the foreign minister was shown round the library. "Stresemann and his entourage were late [...]. I felt pushed by some crude warnings to be brief and could not add any of my planned nuances. [...] Only the blue-leather gift (credit to Saxl) saved the situation."[104] Warburg rushed his remarks and was unhappy with how Stresemann received them. Strohemeyer's print, which he presented as a gift, helped him out of a tight spot. The next day, Warburg wrote to him:

> As I said on the 'phone, your print made a strong impression on Reichsminister Dr. Stresemann: he was really delighted with both the form and the content. If we had not had your artwork he would not have kept such a good memory of the K.B.W., since I had to trim my talk horribly on account of the time pressure.[105]

And in the K.B.W. journal he noted:

> The print did not fail to have an effect on Stresemann; already before midday I saw its impact on Drennig. [...] Counsellor Merck said on the 'phone that Stresemann liked the print a lot. My brother Max made me think this even more. In the late afternoon, Max 'phoned to say that Petersen would ring me to find a way of getting the print to Briand and Chamberlain.[106]

Strohmeyer's linocut was a success, so Warburg ordered another 46 copies and gave around 20 to various friends, acquaintances and family members.[107] He also sent one with New Year's greetings to his psychiatrist Binswanger, "the regular guest in the aircraft of the Idea."[108]

The print shows a simple functional airplane rising from a hangar. Its wings bear the inscription "IDEA VINCIT" and the names Briand, Chamberlain and Stresemann form the lower border. The trio, together with the American Charles Dawes, had been awarded the Nobel Peace Prize for the treaty negotiated in Locarno, where Germany had recognized its borders laid down at Versailles and undertaken to settle any conflicts peacefully under the auspices of the League of Nations. The three statesmen thereby ushered in a brief period of peace and stability in European politics, in the course of which Germany was admitted

---

[104] "Eintrag Aby Warburg vom 20.12.1926," in *Tagebuch der Kulturwissenschaftlichen Bibliothek*, p. 37.

[105] WIA, GC, A. Warburg to Otto Strohmeyer, December 21, 1926.

[106] "Eintrag Aby Warburg vom 21.12.1926," p. 24.

[107] "Eintrag Aby Warburg vom 26.12.1926," in *Tagebuch der Kulturwissenschaftlichen Bibliothek*, p. 26.

[108] UAT 443/31, Sheet 46.

to the League of Nations and Briand began to ponder the idea of a "United States of Europe." Warburg sought to harness this "Locarno spirit" in his gift to Stresemann, a man he greatly admired.[109]

This is apparent also from his statements to the somewhat baffled acquaintances to whom he sent one of Stohmeyer's prints. Thus, in a letter of February 1, 1927 to Cyrus Adler, he writes:

> The "Idea vincit" symbol is meant to express the belief that understanding among the nations of Europe – which Briand, Stresemann, and Chamberlain began to reflect at the Locarno conference – can, will, and must finally lead to a longer respite for reason. Anyone familiar with political relations in Europe will greatly appreciate the apparent naturalness of the statesmen's non-martial sagacity, for the resistance to it is still incredibly powerful. The situation today is more contentious then ever; and that is precisely why, here in Hamburg, we want to strengthen Stresemann's faith in reason by encouraging him with this symbol.[110]

Only the victory of reason would make a peaceful, united Europe possible. "The United States of Europe is in the air," Warburg wrote to Felix von Eckhardt as early as September 1925.[111] During the brief stabilization of the Weimar Republic, this appeared to be a vision worth striving for, and one that might even be achieved in the future. On August 18, 1928 Warburg sent his nephew Erich three copies of the print, with a request to forward one to Frank Billings Kellogg:

> I should be glad if Mr. Kellogg would take the opportunity to publicize this motto as the slogan of the Coming Germany. I would just ask to be placed beneath it: lino engraving by Strohmeyer 1926. You may feel free to tell Mr. Kellogg that the print was presented to Minister Stresemann, when he visited the Kulturwissenschafliche Bibliothek in Hamburg.[112]

That was a few days before Briand and the US Secretary of State Kellogg signed the Treaty for Renunciation of War named after them. On September 9, 1928 Warburg wrote triumphantly in the K.B.W. journal:

> We read in the *Frankfurter Zeitung*: "Reichsminister Stresemann gave the following message to a French paper after the signing of the Kellogg Pact: 'Higher than material force stands the power of the idea that humanity carries with it on its journey'." [...] You see, he did well to receive the "Idea vincit" in 1926.[113]

---

[109] On the Treaty of Locarno and the stabilization of European relations, see Ralph Blessing, *Der mögliche Frieden. Die Modernisierung der Außenpolitik und die deutsch-französischen Beziehungen 1923–1929*, Munich 2008.

[110] WIA, GC, Aby Warburg to Cyrus Adler, February 1, 1927.

[111] WIA, GC, Aby Warburg to Felix von Eckhardt, September 10, 1925.

[112] WIA, GC, Aby Warburg to Erich Warburg, August 18, 1928.

[113] "Eintrag Aby Warburg vom 9.9.1928," in *Tagebuch der Kulturwissenschaftlichen Bibliothek*, p. 340.

But despite this success, very few of the recipients seemed able to make anything of the image. Even the art theorist Oscar Ollendorff wrote to Warburg:

> Many thanks for the fine print! You would like my opinion, and in brief this is roughly what I think. It is emotionally rousing. In that sense, I can feel a wondrous delight at the artistic achievement. But right now, despite my penchant for the symbolic, I find the connection with the 'Idea vincit' motto very difficult.

Be that as it may, another interpretation of the image suggests itself alongside the pacifism and Europeanism proclaimed in the stamp, especially if we consider a passage from the serpent ritual lecture that has already been touched upon.

> Scientific enlightenment puts an end to mythological explanation. [...] Where the technical explanation of cause and effect replaces the mythical imagination, man loses his primitive fears. But we should be loath to decide whether this emancipation from the mythological view really helps mankind to find a fitting answer to the problems of existence. The American government [...] has been admirably active in establishing schools among the Indians. And its intellectual optimism has had this apparent result, that the Indian children now go to school in pretty suits and little aprons and no longer believe in the pagan demons. This is at any rate true in the case of most. And it may denote progress. But I doubt whether it really satisfies the soul of the Indian, who thinks in images and for whom poetic mythology is the true haven. [...] But we do not wish our imagination to fall under the power of the snake image that leads to the primitive subterranean realm. We want to climb onto the roof of the world's house, to direct our head upward, and to think what Goethe too said: Were the eye not sun-like, it could never behold the sun. The whole of mankind joins in worshipping the sun. And it is the right of the savage as well as the learned to take it as the symbol that leads us upward from the depths of night. Children stand before a cave. To bring them into the light is the task not only of American schools but of mankind in general.[114]

For all his skepticism, Warburg had a deep hope in the victory of reason, and this also gave the airplane image its buoyancy. His stamp design should thus be read as an expression not only of republican, pacifist, Europeanist and liberal thinking, but also of the scientific spirit of the Enlightenment outlined earlier in this chapter. Warburg condensed this spirit in the Idea, and it is tempting to read the stamp graphics as an updated version of Plato's cave allegory; the aviator, borne and guided by the Idea, succeeds in rising out of a cave-like hangar. The above quotation from the serpent ritual lecture suggests that Warburg may well have been thinking of the Platonic ascent to the light of reason.[115] The way leads from "the depths of night" (the magical-mythical attitude to the world) up to science

---

[114] Warburg, *Schlangenritual*, pp. 56f.; in part, "A Lecture on the Serprent Ritual," p. 291.
[115] On cave exit and light metaphors, see Hans Blumenberg, *Höhlenausgänge*, Frankfurt/Main 1989.

and the "roof of the world's house;" humanity has the task of "bringing up to the light." It is an Enlightenment *telos*: to lead people out of the shadow-world of the cave up into the light. Warburg gave pictorial form to this normative claim in his *Idea vincit* postage stamp.

There is further evidence of this reading, if any is needed. In the preceding years, both Cassirer and Panofsky had worked intensively in and around the K.B.W. on the Platonic doctrine of Ideas, attempting to position their own theory of symbols or art in relation to the Greek philosopher's "aesthetic."[116] Cassirer gave a lecture at the K.B.W., "Eidos und Eidolon," which appeared in the yearbook of the library,[117] while Panofsky traced the metamorphoses of the Platonic Idea and considered how, despite Plato's supposed hostility to art, it could later become a central concept of art theory. Panofsky's text too was published in the Warburg Library series.[118] Plato was therefore very much in the air during those years in Warburg's milieu.

In their obituaries of Warburg (he died on October 26, 1929), both Panofsky and Cassirer use metaphors suggesting an interpretation of *Idea vincit* in terms of the cave allegory. Warburg's own life and scholarly existence are presented as an ascent from cave to light, or indeed as a soaring flight toward the sun. "It is the curious result of his scholarly work," writes Panofsky, "that all the dark pits into which the spirit of the past had descended and the researcher ventured to follow eventually proved to be adventurous paths to the light."[119]

In his funeral speech, Cassirer presented Warburg as a great seeker after truth. He had been the "living core and focal point of the humanities" at Hamburg University, and his book collection had given the "impression of a great personality and far-reaching destiny in the world of research."[120] Warburg's great theme, according to Cassirer, had been "the opposition and inner tension between freedom and necessity."

> He continually strove to escape all restrictive fetters in the realm of intellectual freedom, the "space of level-headed thinking" that he saw as the highest goal attainable by human knowledge and science. [...] He became master of the night, which threatened to permeate deeper and deeper, because the bright light of the mind glowed inside him – the light of the seeker and researcher, which constantly raised him into the space of level-headed thinking and saved him.

---

[116] Silvia Ferretti, *Cassirer, Panofsky, and Warburg. Symbol, Art, and History*, New Haven, CT 1989, pp. 142–77.

[117] Ernst Cassirer, "Eidos und Eidolon. Das Problem des Schönen und der Kunst in Platons Dialogen," in *Vorträge der Bibliothek Warburg*, vol. 2, Leipzig 1924, pp. 1–27.

[118] Erwin Panofsky, *"Idea." Ein Beitrag zur Begriffsgeschichte der älteren Kunsttheorie*, Leipzig 1924.

[119] Panofsky, "A. Warburg," p. 3.

[120] Ernst Cassirer, "Worte zur Beisetzung von Prof. Dr. Aby M. Warburg," in Anonymous, *Aby M. Warburg zum Gedächtnis*, Darmstadt 1929, unpaginated. The following quotations are also taken from this publication.

Cassirer ended by comparing Warburg to Giordano Bruno, the object of the last studies he made toward the end of his life:

> And there are indeed features that made the thinker seem close and kindred to him. For Giordano Bruno is the first of the Renaissance thinkers who, though originally bound entirely to the sphere of magical thought, consciously wrestled free of it. [...] A certain spiritual-intellectual theme repeatedly makes itself heard in Giordano Bruno's philosophical works and poetry: *the theme of the finite human spirit flying to the sun of an infinite divine truth* [emphases added]. The human spirit knows that it will not and cannot reach the goal, that its flight, like that of Icarus, must finally end in a fall – but it risks the flight nonetheless, because only there can it have its eternal being and eternal destiny assured. A fall from the heights is still better than captivity on the ground, in the depths of existence.

Cassirer too is exercised by "a quite wonderful image," as Schleiermacher puts it in his German translation of Plato's cave allegory in *The Republic* (515a).[121] Like Warburg with his *Idea vincit*, he introduces the Icarus-aviation image to illustrate "ascent and the contemplation of higher things," as in the cave allegory, and the "rising of the soul into the regions of knowledge."[122]

After Warburg's death, living in exile in Britain, Sweden and eventually the USA, Cassirer would continue his wrestling with "the reconquest of Athens from Alexandria." His posthumously published *The Myth of the State* (1946) is testimony of this.[123] Like Warburg, he recognized that myth can never be entirely vanquished. He therefore asked what philosophy could do "to help us in this struggle against the political myths."[124] Cassirer came to the same painful realization as Warburg – the liberal world picture, complete with its idea of progress, had plunged into crisis:

> What we have learned in the hard school of our modern political life is the fact that human culture is by no means the firmly established thing that we once supposed it to be. The great thinkers, the scientists, the poets, and artists who laid the foundations of our Western civilization were often convinced that they had built for eternity. [...] Our science, our poetry, our art, and our religion are only the upper layers of a much older stratum that reaches down to a great depth. [...] The powers of myth were checked and subdued by superior forces. As long as these forces, intellectual, ethical, and artistic, are in full strength, myth is tamed and subdued. But once they begin to lose their strength chaos is come again. Mythical thought then starts to rise anew and to pervade the whole of man's cultural and social life.[125]

---

[121] Platon, *Politeia*. In Friedrich Schleiermacher, *Werke in 8 Bänden*, vol. 4, 2nd edn, ed. Gunther Eigler, Darmstadt 1990.

[122] Ibid., 517b.

[123] On Cassirer and myth, see Tobias Bevc, *Kulturgenese als Dialektik von Mythos und Vernunft. Ernst Cassirer und die Kritische Theorie*, Würzburg 2005, pp. 141–210.

[124] Cassirer, *The Myth of the State*, p. 295.

[125] Ibid., 297f.

The First World War robbed them of their effectiveness and credibility, those "intellectual, ethical, and artistic forces" required for "struggle against the political myths." The crack in the liberal order, already apparent at the turn of the century, widened into a gaping hole. The temporal order in particular, the narrative of progress, suffered damage. In the face of mechanized death, the suffering of millions and the upheavals following the war, an interpretation of history as progress seemed unsustainable.

Liberals such as Warburg and Cassirer clung to the rational teleology underlying progress. Reason remained for them the normative guiding idea that made it possible to find a way out of chaos. Although they themselves crucially contributed to deconstruction of the revolutionary cultural theories reflecting the nineteenth-century narrative of progress, they could not avoid seeing the postwar fascist attempt to bring order into chaos as a relapse into an earlier stage of humanity.[126] Cassirer writes in *The Myth of the State*:

> The sudden rise of the political myths in the twentieth century has shown us that these hopes of Comte and of his pupils and adherents [in an exact social science] were premature. Politics is still far from being a positive science, let alone an exact science. I have no doubt that later generations will look back at many of our political systems with the same feeling as a modern astronomer studies an astrological book or a modern chemist an alchemistic treatise. In politics we have not yet found firm and reliable ground. Here there seems to be no clearly established cosmic order; we are always threatened with a sudden relapse into the old chaos.[127]

Cassirer complained that the scientization of politics and the social realm had not yet succeeded, and that "magical" thinking was still dominant:[128]

> The belief that man by the skillful use of magic formulae and rites can change the course of nature has prevailed for hundreds and thousands of years in human history. In spite of all the inevitable frustrations and disappointments mankind still clung stubbornly, forcibly, and desperately to this belief. It is, therefore, not to be wondered at that in our political actions and our political thoughts magic still holds its ground.[129]

For most liberals, the fascist attempt to order chaos through myth represented the return of a "primitive" stage of humanity that they had thought superseded.

---

[126] On cultural-evolutionary theories and their relationship to the narrative of progress, see Bernd Weiler, *Die Ordnung des Fortschritts. Zum Aufstieg und Fall der Fortschrittsidee in der "jungen" Anthropologie*, Bielefeld 2006. Weiler's considerations on the idea of progress in the critique of cultural anthropology focus especially on the figure of Franz Boas, whom Warburg met at the Smithsonian Institute in Washington before his trip among the Pueblo Indians.

[127] Cassirer, *The Myth of the State*, p. 295.

[128] On the scientization of the social, see Lutz Raphael, "Die Verwissenschaftlichung des Sozialen als methodische und konzeptionelle Herausforderung für eine Sozialgeschichte des 20. Jahrhunderts," *Geschichte und Gesellschaft* 22/1996, pp. 165–93.

[129] Cassirer, *The Myth of the State*, p. 295.

Logos – or what those Enlightenment liberals regarded as the "light of truth" – was always preferable to myth. In their eyes, myth was the Other, since the metaphysical foundations of their own vision of order, its mythical and historical character, occupied a blind spot outside their field of vision, as did the fact that theirs was only one attempt among many to solve the problem of order.

The temporal hierarchy of liberalism and fascism, *logos* and mythos, went back to Enlightenment thought. Not the least of the ways in which the Enlightenment legitimated its vision of an order based on reason was by contrasting it with the Estates-based order of feudal society. According to the modern temporality, the Enlightenment project was new and therefore better, whereas the feudal one was dusty, corroded and antiquated. "Progress and suchlike terms," writes Reinhart Koselleck, "articulated the otherness of the (worse) past in comparison with the novelty of the (better) future."[130] This temporal model, stemming from the Enlightenment but still operational in modernization theory, was one reason for the decades-long tradition of regarding fascism and modernity as incompatible. For if modernity is equated with the "process of Western rationalization,"[131] and the "project of modernity" with the Enlightenment project, then the nexus between fascism and modernity naturally appears paradoxical.[132] But the categorization of fascism as nonmodern or antimodern, which prevailed in research for a long time, goes back to the *normative* equation of modernity with the "progressive" order of Enlightenment liberalism. As we noted in the introduction, however, this study will share Bauman's conception of order as a *task*. The vision of a liberal order then becomes one attempt among others to bring order into chaos.

While Warburg abandoned his chronology in which antecendent myth gave way to posterior *logos*, he stuck to his normative hierarchy that prioritized *logos*. The Idea that lured philosophers from the cave – that is, the light of reason – was supposed to be the lodestar for the politicians and peoples of Europe. There might be repeated backsliding, as the world war and Warburg's own mental illness had shown, but men and women should stick to their goal of advancing under Reason's guidance to their true humanity.

As Warburg already noticed during the First World War, and as Hartmut Böhme has recently shown for the contemporary world, it was a fallacy to apply the Enlightenment vision of complete disenchantment and rationalization to the real world.[133] Freud too clearly realized this. Soberly and laconically, seeming to agree in advance with Bruno Latour's *We Have Never Been Modern*, he wrote:

> These discussions have already afforded us the consolation that our mortification and painful disappointment on account of the uncivilized behavior of our

---

[130] Koselleck, "Fortschritt," p. 389.

[131] On the "process of Western rationalization," see Max Weber, *The Protestant Ethic and the Spirit of Capitalism* [1930], New York 1992.

[132] On the "project of modernity," see Habermas, "Die Moderne – ein unvollendetes Projekt," Reclam, Leipzig 1994, Germany.

[133] See Böhme, *Fetischismus und Kultur*.

fellow world citizens in this war were not justified. They rested upon an illusion to which we had succumbed. In reality they have not sunk as deeply as we feared because they never really rose as high as we had believed.[134]

After the First World War, the fascist vision of modernity emerged in addition to the liberal and Marxist. Although its roots can be traced to Romanticism, its first shoots came through in movements espousing aesthetic and social modernism around the turn of the twentieth century.[135] These vilified science, the kernel of both Marxist and liberal projects, as positivistic and historicist, and therefore inimical to life. They placed art and the Sublime in opposition to it. One representative of these movements was Gabriele D'Annunzio, whose interventionism in May 1915 aroused Warburg's fury. The clash between the two men, which will be described in the following section, may be read as a personalization of two competing projects of order and modernity.[136] The rifts that marked a whole era may already be seen in outline in their respective encodings of aviation. Ulrich Raulff saw the *Idea vincit* as standing in a relationship of "polemical tension" with "D'Annunzio, the aviator and poet of Fascism so hated by Warburg;" it was an "attempt at counter-appropriation" of the aviator *motif*.[137] As we shall see in the next section, D'Annunzio's contributions to aviation discourse were indeed diametrically opposed to Warburg's *Idea vincit*. The polemical tension between the two projects was discharged along a break in the tectonic order of Europe. And whereas Warburg worked tirelessly to close the rift, D'Annunzio did all he could to deepen it. He was successful. From May 1915, Italy too was shaken by the earthquake that brought the liberal European order crashing down.

---

[134] Freud, *Reflections on War and Death*, p. 44. On the "spiritual affinity" between Warburg and Freud, see Georges Didi-Huberman, *L'image survivante. Histoire de l'art et temps des fantômes selon Aby Warburg*, Paris 2002.

[135] On the aesthetic and social modernist movements, see Roger Griffin, *Modernism and Fascism. The Sense of a New Beginning under Mussolini and Hitler*, Basingstoke, UK 2007, pp. 43–69, 130–59.

[136] See Fernando Esposito, "Warburg und D'Annunzio – Antipoden?," in Gottfried Korff (ed.), *Kasten 117*, pp. 301–23.

[137] Raulff, "Der aufhaltsame Aufstieg einer Idee," p. 147.

2

Icarus Rising: D'Annunzio, the Flying Artificer of Myth

Gabriele D'Annunzio, born in 1863 in Pescara, Abruzzo, is the best-known and most controversial Italian writer of the *fin de siècle*. He was not only a renowned lyrical poet, dramatist and novelist, but also a notorious playboy and dandy, as well as a politician and war hero.[138] The British journalist Sisley Huddleston had already singled him out in 1924 as Mussolini's John the Baptist, and consequently his reputation suffered after the Second World War on account of his affinity with Fascism. Yet D'Annunzio remains as a literary point of reference – if only for those who wish to mark a distance from his exalted Symbolist style. Despite the undisputed quality of his work, *Il vate* (the "Bard") – as he is widely known in Italy – commanded so much attention because of his extravagant lifestyle and amorous adventures, which turned him into a media icon during his lifetime. Far from shunning the limelight, D'Annunzio knew well how to attract it. Il Vittoriale, his villa near Gardone on Lake Garda, is one token of this.[139]

On December 22, 1923 the poet donated the villa to the Italian people as a victory shrine,[140] and in 1924 – as a birthday gift to D'Annunzio – Mussolini reassuringly had it declared a national monument. In any event, the "Bard" achieved his own musealization with the villa's furnishings and adorned its buildings and gardens with various war machines he had used for action and agitation. On May 17, 1925 the D'Annunzio collection was enlarged. An S-16 hydroplane *Alcyone* now joined the Italian warship *Puglia*, the SVA-10 on which the Bard made his flight to Vienna, and the torpedo boat he used for the legendary *Beffa di Buccari*

---

[138] John Woodhouse, *Gabriele D'Annunzio. Defiant Archangel*, Oxford 1998. Woodhouse's biography is the most pertinent of the numerous publications on the writer's life, also containing extensive bibliographical references. See also the recently published: Lucy Hughes-Hallett, *The Pike. Gabriele d'Annunzio, Poet, Seducer and Preacher of War*, London 2013.

[139] The villa-museum also houses the D'Annunzio Archive. For an overview, see http://www.vittoriale.it. See also Wolfgang Ernst, "Museale Kristallisation: *Il Vittoriale degli Italiani*," in Hans Ulrich Gumbrecht/Friedrich Kittler/Bernhard Siegert (eds.), *Der Dichter als Kommandant. D'Annunzio erobert Fiume*, Munich 1996, pp. 309–20.

[140] Here and on what follows, see Woodhouse, *Gabriele D'Annunzio*, pp. 353–80.

(the "Bakar joke").[141] However, it is not for his egomania or his collection of memorabilia that D'Annunzio deserves our attention, but because he was the most prominent speaker in Italian aviation discourse and a leading protagonist of mythical modernity. We shall explain here his contextualization of aviation and the extent to which he may be understood as the polar opposite of Aby Warburg and his vision of order. The name *Alcyone* will already put us on the right track.

In 1903 Gabriele D'Annunzio published the second and third books of his *Laudi del cielo, del mare, della terra e degli eroi* [In Praise of the Heavens, the Sea, the Earth and Heroes]. He named the individual books of the series after four of the Pleiades: *Maia*, *Electra*, *Alcyone* and *Merope*. The stars bore the names of the daughters of Atlas and Pleione; they were the nymphs who were said to have flown skyward to escape the pursuit of Orion. In the third book, *Alcyone*, several poems are devoted to the mythical figure of Icarus.[142] One of them, "L'ala sul mare" [The Wing on the Sea], contains the lines: "Who shall gather them? / Who with stronger bonds will know / How to unite the strewn feathers / And try again the mad flight?"[143] A few years later, D'Annunzio himself would unite the "strewn feathers" with mythical bonds and try his hand at "mad flight."

D'Annunzio's use of mythical figures points to one link with Aby Warburg, who saw the nymphs as the Dionysian element in the Renaissance and related the divergence of magic, myth and science to the belief in the stars that still flourished in the age of Luther.[144] But neither nymphs nor astrology were central to Warburg's hostility to D'Annunzio; it was the First World War that inflamed them both. The intellectual paths of Warburg and D'Annunzio crossed in the months surrounding Italy's entry into the war, which the former sought to prevent and the latter to promote. The conflict was rather one-sided in character: Warburg fought his end with great passion, while D'Annunzio probably took scarcely any notice of his efforts. Still, it is worth looking at it more closely, since Warburg's attack on D'Annunzio (and the "Alexandrian spirit" he represented) is paradigmatic of the antagonism that unfolded between the liberal and the newly emerging fascist

---

[141] This was an operation in February 1918, which involved the firing of six torpedoes against ships of the Austro-Hungarian Navy lying at anchor in the Bay of Buccari (Bakar in Croatian). None of them caused any damage, but the three Italian boats did demonstrate the vulnerability of the Habsburg coast and navy. At the end, in a touch recalling his aerial exploits, D'Annunzio placed a sardonic nationalist message in three tricolored bottles and sent them to the enemy shore defenses. See Bojan Budisavliević, "D'Annunzios Torpedowesen. Instrument der Vorsehung und Geschicke des Meeres im Seekrieg um Fiume herum," in Hans Ulrich Gumbrecht et al., *Der Dichter als Kommandant*, pp. 227–59.

[142] On the background and references of the Icarus poems, especially the *Ditirambo IV, di Icaro*, see Pietro Gibellini, "Il volo di Icaro," in Gibellini (ed.), *Logos e Mythos. Studi su Gabriele D'Annunzio*, Florence 1985, pp. 119–32. [All English versions are by the present translator – *Trans. note.*]

[143] Gabriele D'Annunzio, "L'ala sul mare," in Id., *Laudi del cielo, del mare, della terra e degli eroi, Libro Terzo Alcyone*, Milan 1956 [orig. 1903], p. 760.

[144] On Warburg's study of the Florentine nymphs, Ulrich Raulff, *Wilde Energien. Vier Versuche zu Aby Warburg*, Göttingen 2003, pp. 17–47.

order. D'Annunzio is the spiritual father of the fascist political style: its herald, as he himself put it. In a letter to Mussolini, dated January 9, 1923, he wrote:[145]

> Did not my mind produce what is best in the movement called "Fascist?" Did I not herald today's national rising some forty (how that hurts) years ago? [...] Are my concise order dealing with the deserter Misiano and my painful sacrifice during Bloody Christmas not the highest examples of "Fascist" action?[146]

We shall now look more closely at this political style, which had myth at its center. This will put some flesh on the definition of myth offered in the Introduction, and open a vantage point on Fascism that locates the origins of this mythical model of order in the First World War.

The war prophet of May 1915

On June 29, 1915, roughly one month after Italy's declaration of war on Austria-Hungary, Warburg wrote to his assistant Saxl:

> Sometime soon I shall give a lecture on Italian war literature at the Society of Booklovers [...]. I am currently studying the D'Annunzio type, which is actually very interesting. He who wanted to burn up in the 100,000 francs sunset is now, instead of [intervening?] militarily, on his way back to his low Apache dives in Paris and [letting?] Rubinstein earn his keep. Besides he is no Jew; Rappaports are decent people; he will probably have to look for his chain of disasters in a sailors' pub in Pescara: Serapis and Isis are serving there. For me the kid is an Archigallus redivivus, who mimed the ecstasy of the Gutmanns for a fee.[147]

The "Archigallus" supposedly reincarnated in D'Annunzio is the castrated high priest of the goddess Kybele, in the state religion of ancient Rome.[148] Warburg is thus not only denigrating D'Annunzio, but making a connection with an event that took place a few weeks earlier and which led Warburg to occupy himself with the poet: namely, D'Annunzio's opening of the monument to "The Thousand"

---

[145] Renzo De Felice/Emilio Mariano (eds.), *Carteggio D'Annunzio – Mussolini (1918–1938)*, Milan 1971, p. 38.

[146] After a treaty between Italy and Yugoslavia established Fiume/Rijeka as a "Free State," D'Annunzio led a maverick force to occupy the port city and even declared war on Rome. On December 24, 1920, Italian government forces began the operation that led to the eviction of the rebels shortly afterwards. During this period, D'Annunzio passed a death sentence (never carried out) on Francesco Misiano for alleged treason in opposing his "capture" of Fiume. [*Trans. note*]

[147] WIA, GC, Aby Warburg to Fritz Saxl, June 29, 1915.

[148] On the Archigallus and his role in the Kybele cult of antiquity, see Sarolta A. Takács, "Kybele," in Hubert Cancik/Helmuth Schneider (eds.), *Der Neue Pauly*, vol. 6, Stuttgart 1999, cols. 950–56.

(Garibaldi's core fighting force) at the cliffs at Quarto, near Genoa.[149] Kybele, the *Mater Magna*, was worshipped at the sacred rocks, and on May 5 the "high priest" D'Annunzio gave a speech there envisioning Italy's imminent sacrifices. The speech was the opening act in the so-called *maggio radioso* ("radiant May") of 1915, in which the interventionists forced the government to declare war.[150]

D'Annunzio had used this opportunity not only to agitate for Italy's involvement on the side of the Entente, but also to stage-manage his return from his less than glorious exile in France. In the *Taccuini*, his diaries and notebooks, he wrote on March 7:

Is there a more superb occasion? I will go and carry the Garibaldi Legion with me, the red wave. [...] The Red Spirits will hasten there from every part of Italy. A boisterous force will crowd together at the eternal bronze. Where The Thousand weighed anchor, the New Thousand will come and berth. From the Quarto cliff, the Italian army will betake itself to the borders. The movement will be irresistible. [...] Finally the reward for my patient, melancholic wait! A poetic vision is capable of translating itself into military reality. To arrive in Quarto – not as an ordinary speaker, but as leader of the youth, as intermediary between two generations. To cross the Tyrrhenian Sea on a ship, loaded with blood impatient to be spilled. [...] It is impossible that Italy, however blind or deaf it may be, will fail to see the sign, to hear the call, the one as inspired as the other, from this cliff of Quarto, while two thousand armed young men, led by the hero's grandson, surround the celebratory monument and then set off to win and die.[151]

On May 5, 1915 D'Annunzio did indeed begin to translate his poetic vision into military reality. He was able to unite art and politics in a sacral myth, not only giving birth to a new political style that would mark the following decades, but also expanding the religious dimension of politics. By the time of the "Fiume exploit," D'Annunzio would have perfected what had its beginning at Quarto.[152]

---

[149] See Fernando Esposito, "Warburg und D'Annunzio – Antipoden?," in Gottfried Korff (ed.), *Kasten 117. Aby Warburg und der Aberglaube im Ersten Weltkrieg*, Tübingen 2007, pp. 301–23, and Claudia Wedepohl, "'Agitationsmittel für die Bearbeitung der Ungelehrten.' Warburgs Reformationsstudien zwischen Kriegsbeobachtung, historisch-kritischer Forschung und Verfolgungswahn," in Korff (ed.) Kasten 117, pp. 325–68. Cf. Bettina Vogel-Walter, *D'Annunzio – Abenteurer und charismatischer Führer*, Frankfurt/Main 2004, pp. 46–57; and Woodhouse, *Gabriele D'Annunzio*, pp. 283–94.

[150] On the *maggio radioso*, see among others Mario Isnenghi/Giorgio Rochat, *La Grande Guerra*, Milan 2004 [repr.], pp. 133–36. And on D'Annunzio's role and fiery rhetoric, cf. Mario Isnenghi, *Le guerre degli italiani. Parole, immagini, ricordi 1848–1945*, Bologna 2005, pp. 44–52; and Mario Isnenghi, *Il mito della grande guerra*, Bologna 2002⁵, pp. 105–08.

[151] Gabriele D'Annunzio, *Taccuini*, ed. Enrica Bianchetti and Roberto Forcella, Milan 1965, pp. 713–15.

[152] On D'Annunzio in Fiume, see Vogel-Walter, *D'Annunzio*; Claudia Salaris, *Alla festa della rivoluzione. Artisti e libertari con D'Annunzio a Fiume*, Bologna 2002; Gumbrecht et al., *Der Dichter als Kommandant*. (esp. Gumbrecht's "I redentori della vittoria"); and Michael A. Leeden, *The First Duce*, Baltimore, MD 1977.

The future Fascist rhetoric and gestures, the staging of the deed and its related semantics, found their first objectification in D'Annunzio – ranging all the way from his "Roman salute" and the "Eia eia alalá" battle cry to the speaker's total hold on the masses, preferably exercised from a balcony, and the staging of the "March on Fiume" in 1919 from Ronchi. The culmination, however, was his hard-to-rival sacralization of the Nation and its "martyrs." The whole style expanded politics to take in the aesthetic dimension that Walter Benjamin identified as the hallmark of fascism.[153] This aesthetic, but also sacral, dimension of D'Annunzio's politics was already clear in the Quarto spectacle, where the desired aim of entry into the First World War was linked up with the Risorgimento and Garibaldi.[154]

In the hagiography of the Risorgimento, Garibaldi and his thousand volunteers set off for Sicily in 1860 from a cliff just outside the Genoan district of Quarto, with the aim of opening the campaign for Italian unification.[155] It was an obvious place, then, for agitation in favor of the war. The nationalist interventionists sought to place Italy's entry into the war on the side of the Entente within the same Risorgimento tradition. Sober political self-interest dictated that justice should be done to national sentiment and history. For the interventionists wanted the war to be seen as a means to complete Italian unification by integrating the *terre irredente*, the "unredeemed territories" still under Habsburg rule, and hence as a struggle for a *più grande Italia*.

D'Annunzio's appearance on the stage was a powerful act of symbolic politics, which served not only the poet himself and the interventionist cause, but also

---

[153] Walter Benjamin, "The Work of Art in the Age of Mechanical Reproduction," in Benjamin, *Illuminations*, New York, NY 1968, pp. 217–51. On p. 250 we read: "Fascism attempts to organize the newly created proletarian masses without affecting the property structure which the masses strive to eliminate. Fascism sees its salvation in giving these masses not their right, but instead a chance to express themselves. [...] The logical result of Fascism is the introduction of aesthetics into politics." Decisive though this aestheticization was for the new style, the second key factor should not go unmentioned: violence. See Sven Reichardt, *Faschistische Kampfbünde. Gewalt und Gemeinschaft im italienischen Squadrismus und in der deutschen SA*, Cologne 2002.

[154] With regard to the question of a distinctively Fascist antihistoricism, there is an interesting comparison here with Mussolini's official opening of the monument to Anita Garibaldi (Giuseppe's wife) on June 4, 1932, the fiftieth anniversary of Garibaldi's death. Mussolini used the occasion to place his Blackshirts in a direct line of descent from Garibaldi's Redshirts. See Claudio Fogu, "Fascism and Historic Representation. The Garibaldian Celebrations," *Journal of Contemporary History* 31/1996, pp. 317–45. In both cases, history becomes a supply of material for instrumentalization. The linear conception of the course of history is revoked, and with it the liberal idea of progress. A Nietzschean "monumentalist" history appears in its place, which is looked at in terms of its uses and retrieved or represented only if it is useful to "life." On antihistoricist discourse in Germany and its connection with the demise of the liberal category of progress, see Anselm Doering-Manteuffel, "Mensch, Maschine, Zeit. Fortschrittsbewusstsein und Kulturkritik im ersten Drittel des 20. Jahrhunderts," in *Jahrbuch des Historischen Kollegs 2003*, Munich 2004.

[155] On Garibaldi and his "March of the Thousand," see *inter alia* Alfonso Scirocco, *Garibaldi. Battaglie, amori, ideali di un cittadino del mondo*, Rome 2001. And on the hero cult surrounding Garibaldi: Lucy Riall, *Garibaldi. Invention of Hero*, New Haven, CT 2007.

the government of Antonio Salandra. In fact, the signing of the secret London Protocol on April 26 had already clinched Italy's entry into the war,[156] but the government had not yet brought itself to admit it, let alone to inform the Italian public. It believed that the *divino poeta* (another name used to refer to D'Annunzio) could stoke up feelings to such an extent that entry into the war would appear at home and abroad as an expression of the people's will. And in part the calculation paid off. The "high priest" D'Annunzio ended his peroration at Quarto with a very special Sermon on the Mount.

> Blessed are the young who hunger and thirst after fame, for they shall be satisfied. Blessed the merciful, for they shall stem gleaming blood and bandage radiating pain. Blessed the pure of heart, blessed those who return with victories, for they shall behold Rome's rejuvenated face, Dante's newly crowned star, and Italy's triumphant beauty.[157]

The "war prophet of May," wrote Warburg in March 1916, sanctified not so much the monument as the Nation and the longed-for war.[158] This highlights the mythical structure of D'Annunzio's agitation and hence his role in sacralizing and aestheticizing politics and the nationalization of the masses. What he does not do is present rational arguments as to why the war is necessary. Instead, he justifies it by connecting it to the sphere of the holy: he removes the national war from linear time, placing it in a sacred realm of the eternal and transcendental. It is thereby glorified and made to seem natural. Aesthetic transfer into a mythical temporality does not entirely banish existential fears aroused by the war, but it confers a higher, transcendent meaning on death in battle. D'Annunzio sacralizes war, and death in war becomes a sacrifice for the (eternal) life of the Nation. In a speech to students in Genoa on May 7, he said:

> This night, before the break of day (and may it be a daybreak that brandishes in its ruddy fingers the spear of our Roman god), many of you will set off for distant lands, set off for distant hearths. O messengers of faith, o pilgrims of love, may the fire that burned among night's youth on the Quarto cliff be rekindled in your breasts! [...] Tonight, just as in the Homeric night pyres announced victory from mountaintop to mountaintop, we shall see in dream your flares

---

[156] On the negotiations and motivations behind the Protocol, see Gian Enrico Rusconi, "Das Hasardspiel des Jahres 1915. Warum sich Italien für den Eintritt in den Ersten Weltkrieg entschied," in Johannes Hürter/Gian Enrico Rusconi (eds.), *Der Kriegseintritt Italiens im Mai 1915*, Munich 2007, pp. 13–52; and Holger Afflerbach, "Vom Bündnispartner zum Kriegsgegner. Ursachen und Folgen des italienischen Kriegseintritts im Mai 1915," in ibid., pp. 53–69.

[157] Gabriele D'Annunzio, "Orazione per la sagra dei Mille," in D'Annunzio, *Scritti gioralistici 1889–1938*, vol. 2, ed. and introduced by Annamaria Andreoli, Milan 2003 [orig. 1915], pp. 675–85, 684. Further references to this volume of speeches and articles will use the abbreviated form: SG.

[158] WIA, GC, Aby Warburg to Emil Schaeffer, March 7, 1916.

light up all down Italy as far as Marsala, as far as the African sea. "Go forth, arm yourselves, obey your orders": so said the priest of Mars to the hallowed young. "You are the seeds of the new world."[159]

Alongside other prominent interventionists, such as Marinetti or Mussolini, D'Annunzio was a major adversary targeted by Warburg in his *Rivista*, in the struggle for Italian neutrality.[160] When the troops of the Central Powers advanced to the Piave river in October 1917, following the twelfth Battle of the Isonzo, and Italy's defeat in the war seemed imminent, Warburg could not conceal his satisfaction. Italy was finally paying for its switch of alliances. On November 14, barely two weeks after the Caporetto debacle, he wrote to Frau Professor Bulle:

The events in Italy must naturally be very painful for you. But on the other hand, for us who are friends of Italy, it is a great consolation that the land of Dante will not come under a victorious D'Annunzio in Rome. I still hope that Italian mothers who have lost their children will club to death this Orpheus with a French golden tongue. Through Italy's defeat, ethical meaning will for the first time come into the European slaughterhouse.[161]

But D'Annunzio was an opponent who fascinated Warburg as well as arousing his hatred. He was not only "the most heinous of Italy's corrupters of souls"[162] (as Warburg wrote to Jacques Mesnil in 1925) but also an "interesting type." The K.B.W.'s acquisition list bears this out. For between May 10 (five days after Quarto) and July 17, 1915, it purchased no fewer than 19 works by D'Annunzio and another eight about D'Annunzio and the unification of Italy.[163] The work papers for Warburg's lecture at the Hamburg Society of Booklovers, given on July 5, 1915, including typed copies of newspaper cuttings on D'Annunzio's activities,

---

[159] Gabriele D'Annunzio, "Parole dette nell'Ateneo genovese il VII di Maggio, ricevendo in dono dagli studenti una targa d'oro," in D'Annunzio, *Per la più grande Italia*, in *Prose di ricerca*, vol. 1, ed. Annamaria Andreoli and Giorgio Zanetti, Milan 2005 [orig. 1915], pp. 7–157, 27–30; here p. 30. Further citations of this series of *Prose di ricerca* will be referenced as PdR.

[160] On Warburg's *Rivista* project, see McEwan, *Kampf gegen Windmühlen*.

[161] WIA, GC, Aby Warburg to Elise Bulle, November 14, 1917.

[162] WIA, GC, Aby Warburg to Jacques Mesnil, July 3, 1925.

[163] The K.B.W. acquisition book from January 1, 1905, to September 28, 1918, lists the following works: *Contemplazione della morte* and *Il Piacere* (5/14/1915), *Il fuoco, Le vergini delle rocce, Le novelle della Pescara, Il ferro, Pagine disperse, L'innocente* and *La pisanella* (6/4/1915), all by D'Annunzio. On 7/1 Warburg bought Magnin's *D'Annunzio et son rôle actuel*; on 7/10 Blemmerlassed's *Gabriele D'Annunzio*, and D'Annunzio's *Römische Elegien, La Canzone di Garibaldi*, and apparently a collection of his lyrical verse that appeared in the *Neueste italienische Lyrik* series. On 7/15 followed D'Annunzio's *Das Schiff*, on 7/16 *Per la più grande Italia*, and on 7/17 von Puttkammer's *D'Annunzio*, *D'Annunzios Lust*, the *Laudi del cielo, del mare, della terra e degli eroi, Libro 3 Alcione, Laudi del cielo, del mare, della terra e degli eroi, Libro 4, Merope, I Malatesti (1) Francesca da Rimini* and *La Gioconda*. On 5/10/1915 the following were entered in the acquisitions book: Abba, *Von Quarto zum Volturno*, Melene, *Garibaldi*, Meerheimb, *Von Palermo bis Gaëta*, and Rüstow, *Der italienische Krieg 1860*.

testify to the scale of Warburg's interest. The evening, we read in the text of the lecture, would have

> a special timely justification if we consider some of the printed works published or circulated in Italy since the outbreak of the war. For, like real weapons in close combat, the mass dissemination of words and images in print should be seen as a powerful means of struggle in the civil war of the mind that Italy had to wage in Winter 1914/15 for the preservation of its neutrality. And Italy would never have lost that neutrality if the rulers, by means of a mass hypnosis feeding off the press in the most sophisticated way, had not been able for a time to talk the people into believing that they had to wage a war of defense to save threatened ideals, whereas in reality Vittore [sic] Emmanuele appears to us as a condottiere for the English billions, who, on behalf of his financial backers, treacherously attacks his former allies in order to capture the largest possible share of the booty.[164]

D'Annunzio, it can be said, was the strongest of the Italian propaganda weapons. His rousing speeches, military feats and other adventures constantly featured in the landscape of the Italian press up to the Armistice and beyond. He owed this ubiquitous presence not least to his close connections with the publisher of his prose and verse, Emilio (and later Guido) Treves, and with the director of the *Corriere della Sera*, Luigi Albertini.[165] Treves, who also produced the mass-circulation magazine *L'illustrazione italiana*, used pictorial representations of D'Annunzio's adventures as publicity for his books, while Albertini, in addition to poetry by the war prophet, published a quantity of his bellicose speeches in support of the interventionism they both advocated. During the war itself, however, it was D'Annunzio's widely publicized feats of aviation that attracted the greatest attention. In the following section, along with further examples of the juxtaposition of scientific-technological "civilization" and mythical "culture" during the war years, we shall present one of the clearest voices and some of the most striking themes of Italian aviation discourse.

Eia! Eia! Alalà! D'Annunzio the war hero

On May 23, 1915 Italy declared war on Austria-Hungary. But during that "radiant May" of pro-war agitation, D'Annunzio did not stop at his *Per la più grande Italia* – the collection of interventionist speeches, which Treves published in 1915.[166] At dawn on May 25, he gave another address to a private circle:

> O my friends, this war that appears a work of destruction and horror is the most fertile creator of beauty and virtue. [...] The tenth muse, who bears the name

---

[164] WIA, III.86.1 *Italienische Kriegsliteratur*, "Vortrag vor der Gesellschaft der Bücherfreunde zu Hamburg am 5.7.1915."

[165] On D'Annunzio's relations with his publishers, especially Treves and Albertini, see Vito Salierno, *D'Annunzio e i suoi editori*, Milan 1987.

[166] D'Annunzio, *Per la più grande Italia*, Milan 1915.

Energeia, [...] does not like moderate words, but prefers an abundance of blood. Her measures are different, and so too are her yardsticks. She counts the forces, the nerves, the victims, the battles, the wounds, the torments, the corpses; she observes the screaming, the gestures, and the watchwords of the heroic mortal combat. She tots up the hacked flesh, all the nourishment given to the earth, so that she can convert it, once digested, into ideal material, forming an eternal spirit out of it. She takes the horizontal male body as the sole measure of a broad destiny. O friends, this is not the frostiness of daybreak. We are all ashen. Blood is beginning to flow from the body of the fatherland. Can you not feel it? The bloodbath is beginning, the destruction is beginning. [...] The whole people that yesterday bustled on the streets and squares, calling in loud voices for war, is full of veins, full of blood, and that blood is beginning to flow; that blood steams at the base of an invisible grandeur, one greater than this whole people. A sublime mystery, like nothing in the universe. We shudder before it; we blanch before it.[167]

It is open to question whether D'Annunzio's friends shuddered more at his pathos and lurid imagery than at the mystery he evoked. In any event, the "war prophet of May" volunteered for duty with the Fifth Cavalry Regiment, the *Lancieri di Novarra*, and as the War Ministry granted the 52-year-old poet an exemption, he was able to serve under the Duke of Aosta, who provided him with access to the whole front.[168] The authorities gave him a pass, because they did not think he would take part in any action. But D'Annunzio now tried to drop leaflets and other material from the air onto the "unredeemed territories," and when he ran into opposition from the military authorities he turned directly to Prime Minister Salandra.

I beg you, I implore you, my dear great friend. Do something so that the hated veto is lifted. [...] I am not a confused youngster. In this matter [a planned flight over Trieste], as in all others, I have given serious consideration to all the probabilities, and I have studied the most appropriate means. [...] I have not lived, my dear great friend, I have not lived except for this moment.[169]

On August 7, D'Annunzio targeted for the first time the object of irredentist longings, the Habsburg Adriatic port of Trieste. Together with his pilot Giuseppe Miraglia, he flew over the city and dropped little sacks filled with sand and leaflets.[170] The message to its Italian inhabitants was:

Take heart, brothers! Take heart and be steadfast! In Trentino, Cadore, and Carnia, and on the Isonzo, we are fighting with no pause for breath to liberate

---

[167] Gabriele D'Annunzio, "Tacitum Robur. Parole dette in una cena di compagni all'alba del XXV maggio MCMXV," in PdR, vol. 1 [orig. 1915], pp. 61–5.

[168] Woodhouse, *Gabriele D'Annunzio*, pp. 294ff.; and Vogel-Walter, *D'Annunzio*, pp. 58–62.

[169] Letter from D'Annunzio to Salandra, July 30, 1915, quoted from Saverio Laredo de Mendoza, *Gabriele D'Annunzio. Aviatore di guerra*, Milan 1930, pp. 99–104.

[170] On D'Annunzio's attempts to gain official permission for the leaflet drop, and on the historical context of this form of propaganda, see Vogel-Walter, *D'Annunzio*, pp. 72–85.

you more quickly; we are conquering territory every day. [...] Soon the whole of the Karst will be in our hands. I tell you, brothers, I swear to you: our victory is certain. [...] Courage and tenacity! The end of your martyrdom is nigh. The dawn of our joy is nigh. From the height of these Italian wings, flown by my skilful Miraglia, I throw to you this message and my heart as a pledge: I, Gabriele D'Annunzio, in the air of the fatherland, August 7, 1915.[171]

On September 20, the anniversary of the incorporation of the residual Roman ecclesiastical state into Italy (in 1870), D'Annunzio and Captain Ermanno Beltramo flew over the Adige Valley and dropped leaflets on the city of Trento, again in small tricolored sacks; the *Corriere della Sera* reported the event six days later. In fact, although the article in praise of the bold operation spoke of D'Annunzio in the third person, it actually came from the poet's own pen.[172] In its several pages, he again evoked the Risorgimento and the uprising of 1848 against the Habsburgs:

> *Trentini*, our people in love and distress, brothers in the eternal Dante, today is the first Roman festival of the true unification of Italy, today is the festive consecration of greater Italy, the perfect Italy, celebrated by the resolute will of the whole people in arms. The people's will swears, on the bones of its acclaimed martyrs and the fresh blood spilled in strife, that the Trento of our mountains, like the Trieste of our waters, is an Italian city, Italian and sacred like the breast of Narciso Bronzetti, like the tricolours your women sewed as votive offerings, with which that refugee from Strigno wrapped his living flesh and smuggled out to safety to Garibaldi in Peschiera. [...] These messages [the leaflets], enclosed in the cloth of our flag and bedecked with shimmering flames, are twenty-one in number, three times seven, in memory of the twenty-one volunteers who were captured at Santa Massenza by the Austrian soldiery and were shot in the castle moat on April 16, 1848. May one of them [the leaflets] fall on the cemetery, on their burial place, which we shall finally avenge. It is necessary for those forbears to be awoken, to rise again, so that that they may light the way for the liberators.[173]

The *Corriere della Sera* reported that the Habsburg government responded by offering a reward of 20,000 crowns on D'Annunzio's head. But a few months later, on January 17, 1916, the famous poet, transformed into an angel heralding liberation, once more crossed the sky above Trieste. The day before, an Austrian seaplane had forced D'Annunzio and his pilot, Luigi Bologna, to make a forced

---

[171] Quoted from Laredo de Mendoza, *Gabriele D'Annunzio*, p. 105.

[172] This is clear from D'Annunzio's *Taccuini*, whose entries on the event coincide word for word with the newspaper article. See the remarks by the editor of D'Annunzio's journalistic writings: D'Annunzio, SG, vol. 2, p. 1693. For the text of the article, see Gabriele D'Annunzio, "Il volo di D'Annunzio su Trento," in *SG*, vol. 2, pp. 686–89.

[173] From D'Annunzio's Trento leaflet of September 9, 1915, quoted in Laredo de Mendoza, *Gabriele D'Annunzio*, pp. 113–17.

landing near Grado, where D'Annunzio hit his head against his machine-gun with great force.[174] The vain poet completed the Trieste mission the next day, saying nothing at first about his wound to the much younger aviators who surrounded him. But on February 21, the military doctor told him that he was in danger of losing all vision in his right eye. D'Annunzio was ordered to observe strict bed rest, with his eyes bandaged. Yet he returned to Venice in his *Casetta Rossa* and, despite the bandages, managed to start work on his autobiographical and meditative *Nocturne*.[175] Of course, D'Annunzio was not above describing this flight as another heroic deed, both in *Nocturne* itself (which first appeared in 1921) and in the autobiographical essay *Licenza*, which he added to the short story "Leda without Swan" when Treves published it in May 1916. D'Annunzio, like Treves, was adept at using his patriotic activism and wartime popularity to help the sales of his works.[176] D'Annunzio's propaganda activity was thus effective on two fronts.

D'Annunzio's style was certainly capable of turning on the pomp, most notably in *Notturno*, where he proceeds to sanctify the Nation, the War and those who fought in it. On April 21, Good Friday, of the year 1916 he writes:

On what Calvary is the Son of Man sacrificed today? The Son of Man is martyred for us today on the wild mountain named after Saint-Michael the Swordbearer, the mountain of peaks and four wraths, in the shadowy and waterless Carso.[177] And our bile-burnt mouths again breathe spirit and hope. There the infantry find eternal life. [...] They become the teeth of the furious rock. They bite into eternity. I have my four brotherly crosses. Giuseppe Miraglia is crucified on his wings. Luigi Bailo is crucified on his wings. Alfredo Barbieri is crucified on his wings. Luigo Bresciani is crucified on his wings.[178] I do not possess the linens of the disciple from Arimathea, nor the balms of Nicodemus.[179] But I have given them a new monument, "wherein no man was yet laid" [John 19, 41].[180]

---

[174] See Woodhouse, *Gabriele D'Annunzio*, pp. 298–99; and Gianni Turchetta, "Introduzione," in Gabriele D'Annunzio/Turchetta (eds.), *Notturno*, Milan 2003⁴, pp. v–xliii; here pp. vii ff.

[175] D'Annunzio cleverly used numerous individual strips of paper, so that he was able to avoid losing his place among the lines. These slips, "more than ten thousand of them," were then processed and fitted together by his daughter Renata. See Gabriele D'Annunzio, "Notturno," in PdR, vol. 1 [orig. 1921], pp. 161–410; here pp. 161, 395–96.

[176] See Vogel-Walter, *D'Annunzio*, pp. 60–1.

[177] This refers to Monte San Michele, in the vicinity of Gorizia. The "four wraths" are the Battles of the Isonzo (already five by the time of writing, and eventually totaling 12), one focus of which was the capture of this mountain.

[178] These four "crucified" figures all died in the air on wartime missions.

[179] Joseph of Arimathea, a "secret" disciple, asked Pontius Pilate to hand over to him the dead body of Christ. He then wrapped the body in a linen cloth and "laid it in his own new tomb" [Matt. 27, 59]. Nicodemus was a Pharisee who supplied the balm used for the burial of Christ [John 19, 39].

[180] Gabriele D'Annunzio, *Notturno*, English translation by Stephen Sartarelli, New Haven, CT 2011, pp. 272–73.

The aviator as crucified hero and redeemer, the self-sacrificing soldier as suffering Christ: these were the central themes of the mythical order into which the winged dandy inserted the air war and his own active involvement in it.

These mythical tropes are even clearer in "Faith in Italian Aviation," a piece that appeared on March 26, 1918 in the *Corriere della Sera*. In fact, it was the text of a speech that D'Annunzio had given the day before at La Scala in Milan, on the occasion of a banquet in honor of a number of pilots, where an appeal had been made for donations to fund the purchase of more airplanes. Among other things, he says: "Since the war infiltrates one element after another, it rises from the earth and water up into the air. Nike flies as in the myth, only not with two wings but with a thousand, with thousands upon thousands." D'Annunzio continues by referring to the exemplary donations made by a few cities. They were not ordinary gifts, but rather

> a testimony of faith. The most modest contribution is an act of faith in the only weapon that can win the war. [...] These heroes know it. The living know it, and so do the dead. The shadow of the winged machine is akin to the shadow of the wood of sacrifice and redemption. The similarity occurred to me on a day long past from that other war, on the grim field of Gonars so like a levelled Mount Calvary, when I caught sight of the aeroplane piloted by Oreste Salomone, with its deadly load, spattered all over with blood. Its two wings, running transversally between nose and rudder, formed the bloody cross. [...] "O pinion of Italy, thou art my faith," confessed those of ours who were torn and crushed, who had landed as a holocaust to be carried once more skyward by the spirit of fire.[181]

What was the purpose of these metaphors? Why is the stage of industrial warfare described as a Mount Calvary, a soldiers' memorial as a sepulcher of resurrection, the silhouette of the airplane as a cross of redemption and sacrifice, the aviator as Christ and savior? Why is the air force stylized as an item of faith, fallen pilots as burnt offerings? These questions go to the core of our investigation. The case of D'Annunzio and his association of heroism with the theme of sacrifice and redemption permits a first approximation to an answer.

A few years ago, the sociologist Bernhard Giesen pointed to the nexus between the collective identity of a society and its sacrificial rituals.[182] And Sabine Behrenbeck, in a pioneering study, underlined the key role of the "sacramental dimension of sacrifice for the cult of heroes," and recalled that heroism and sacrifice often overlap in the Judeo-Christian context.[183] With Hildegard Cancik-Lindemaier, she explained the function of the sacrificial metaphor in wartime

---

[181] Gabriele D'Annunzio, "La fede nell'aviazione italiana," in SG, vol. 2 [orig. 1918], pp. 736–38.
[182] Bernhard Giesen, *Triumph and Trauma*, Boulder, CO 2004.
[183] Sabine Behrenbeck, *Der Kult um die toten Helden. Nationalsozialistische Mythen, Riten und Symbole*, Vierow bei Greifswald 1996, p. 71.

as, first, "to sanctify killing and being killed," second, "to glorify and honor the fallen," and third, "to relieve the strain on those who kill and their offspring."[184] Nevertheless, the complex relationship between sacrifice and community, or between sacrifice and heroism, is not exhaustively explained by this functional account of the metaphor.

Sacrifice is central to many theories of religion; it is analyzed in detail by, among others, Peter L. Berger, Walter Burkert, Mircea Eliade, and René Girard.[185] Here it will have to suffice, with the help of Peter L. Berger, to underline a key aspect that is reflected in D'Annunzio's rhetoric of redemption and sacrifice.

> Religion [...] maintains the socially defined reality by legitimating marginal situations in terms of an all-encompassing sacred reality. This permits the individual who goes through these situations to continue to exist in the world of his society – not "as if nothing had happened," which is psychologically difficult in the more extreme marginal situations, but in the "knowledge" that even these events or experiences have a place within a universe that makes sense. It is thus even possible to have "a good death," that is, to die while retaining to the end a meaningful relationship with the nomos of one's society – subjectively meaningful to oneself and objectively meaningful in the minds of others.[186]

Berger points out that, especially in times of crisis, such as wars, societies experience such marginal and correspondingly "ecstatic" situations in a collective manner. During the marginal situation, the reality or order previously taken as given is called into question and needs to be stabilized. Attempts are made to develop religious legitimations, especially if a society must push its members to the limits by motivating them to kill or to risk their own lives. "Killing under the auspices of the legitimate authorities has, for this reason, been accompanied from ancient times to today by religious paraphernalia and ritualism."[187] D'Annunzio's sanctifications of war, sacrifice and "heroes" are by no means mere adjuncts. Rather, they discursively associate central threads of the "nomic web" that constitutes society and holds it together – and which a society at war especially requires.

This nomic web, or nomos *qua* order of common meaning, is produced discursively, in interpretation and action.[188] The nomos or order acts as a "shield against

---

[184] Hildegard Cancik-Lindemater, "Opfer. Religionswissenschaftliche Bemerkungen zur Nutzbarkeit eines religiösen Ausdrucks," in Hans Joachim Althaus (ed.), *Der Krieg in den Köpfen. Beiträge zum Tübinger Friedenskongreß "Krieg – Kultur – Wissenschaft,"* Tübingen 1988, pp. 109–20; quoted from Behrenbeck, *Der Kult um die toten Helden*, p. 74.

[185] See Peter L. Berger, *The Sacred Canopy: Elements of a Sociological Theory of Religion*, Garden City, NY 1967; Walter Burkert, *Homo Necans. The Anthropology of Ancient Greek Sacrificial Ritual and Myth*, Berkeley, CA 1986; Mircea Eliade, *The Sacred and The Profane. The Nature of Religion*, Orlando, FL 1959; and René Girard, *Violence and the Sacred*, Baltimore, MD 1979.

[186] Berger, *The Sacred Canopy*, p. 44.

[187] Ibid.

[188] Ibid., pp. 19–20.

terror," against the danger of meaninglessness and disorientation. This is another aspect of the protection from the "absolutism of reality" discussed in the previous chapter.[189] Since meaninglessness and destabilization of the nomic order become an acute threat precisely in time of war, the nomos requires stabilization, updating and adaptation to the circumstances of the (marginal) situation.[190] D'Annunzio's sacralization of war, sacrifice and "heroes" therefore involves "work on nomos." In the extreme marginal situation of the First World War, killing and dying required interpretation and legitimation. On the other hand, it was necessary to create a semantic space for integrating encounters with the "modern world" – a world that seemed alien and unfamiliar to large sections of the population – into the horizon of personal and collective meaning.

The "cult of fallen soldiers" that George L. Mosse has so impressively portrayed did serve such interpretation and legitimation.[191] It offered a semantic possibility to appropriate experience and suffering, and perhaps even to find consolation in the fact that the end of one's own life, and that of relatives or "comrades," might contribute to the life of the nation. The end of the soldier's time on earth was sublated in the eternal, sacred time of the nation. D'Annunzio made a huge contribution to this semantic offer, and the importance of his discursive legacy was similarly enormous for the Italian interpretation of the war. Through its fallen "heroes," Mosse suggests, the nation became associated with the passion of Christ.[192] By analogy, sacrifice purged the nation of its sins and – even more crucial – created the possibility of its eternal life or rebirth.[193]

D'Annunzio too established this analogy. He, whose books were on the Papal Index, needed only to plumb the ever-present reservoir of Catholic symbols and experiential content, and to adapt them to the situation. The elitist poet had recourse to a space of experience that was shared even by the largely illiterate mass of conscripts. On the other hand, D'Annunzio naturally availed himself of the humanist educational tradition and replenished its stock of imagery by blending it with wartime events and charging it with new meaning. The resulting amalgam aestheticized the war, raising it out of the everyday into eternal time, and out of the mundane into the sacred.

Thus D'Annunzio conjured up the following images in his speech at the funeral of Gino Allegri, a fallen aviator and friend.[194] Renaming him after a probably fictitious companion of Francis of Assisi, Frate Ginepro (Brother Juniper), he painted

---

[189] Ibid., p. 22.

[190] Berger calls this danger "the nightmare *par excellence*, in which the individual is submerged in a world of disorder, senselessness, and terror" (p. 22). Cf. p. 23.

[191] George L. Mosse, *Fallen Soldiers. Reshaping the Memory of the World Wars*, New York, NY 1990.

[192] Ibid., p. 76.

[193] See Behrenbeck, *Der Kult um die toten Helden*, pp. 71–6; and Klaus Latzel, *Vom Sterben im Krieg. Wandlungen in der Einstellung zum Soldatentod vom Siebenjährigen Krieg bis zum II. Weltkrieg*, Warendorf 1988, pp. 65ff., 70ff.

[194] On D'Annunzio's funeral speeches for various aviators, see Vogel-Walter, *D'Annunzio*, pp. 89–92.

him as "the clearest of the mystics of this war," the "most guileless of the armed ascetics," and a *"miles Christi."*

> When the first droning of his engine sounded from beneath the protective roof, it was like the matinal ringing that announces the call to the divine service. [...] My guardian appeared in the thundery nimbus of his propeller. And as he did this, I realized once more how great was the spiritual affinity between him and his war machine, and how much it had grown together with him. [...] Like Francesco Baracca, like any really great hero, he was a pinion of the war. The wing had to break with it, to burn with it, to be consumed with it.[195]

Once more D'Annunzio was using the Christian idea of sacrifice and combining it with the Icarus myth. The pilot becomes a hybrid being who has grown together with his machine; his military function is like being in God's service. D'Annunzio's sacralization of death in combat and of the nation, as well as his aestheticization of war, is again apparent in the speech he gave on December 12, 1917, to a group of recruits born in 1899. It was published under the title "Voices of Reconquest," in the Christmas edition of the *Corriere della Sera*.

> O young people, educated or uneducated, still in your first blossom, still warm from your mother's breath, you heard in one moment, as you strode into the firing zone, what years and years of mental effort do not reveal to the fully grown. What Dante thought he understood in the middle of life, rising through the three worlds from one torment to the next and from light to light, you guessed in the blinking of an eye. No power, be it divine or human, can equal the power of sacrifice, which plunges into the darkness of the future to call forth new images and a new order. [...] Well, there is the mother who bore you, who calmed you, who dried your first tears, who taught you to speak, who led your first steps, who advised, forgave, and comforted you; now she gives you to the war, she chases you into the fire, she shouts out to you: "Go and fight. Go and win. Go and die." Why? [...] To regain an Alpine crest? For the sake of a crescent-shaped gulf? For a piece of land hanging like a grape cluster in the sea [Istria]? A garland of islands? A rim of Latin shore studded with precious stones? Yes, of course, for those too. But the real reason has not to do with earth: it is the soul's reason, the reason of immortality.[196]

D'Annunzio creates a horizon of meaning for the hundreds of thousands of Italian deaths in the Alps and the Karst, and on the Isonzo; he promises the soldiers eternal life. At first, nationalist ideology would have produced little effect

---

[195] Gabriele D'Annunzio, "I mistici della guerra. Morte di frate Ginepro," in SG, vol. 2 [orig. 1918], pp. 769–79; here pp. 771, 777.

[196] Gabriele D'Annunzio, "Voci della riscossa. Alle reclute del '99," in SG, vol. 2 [orig. 1917], pp. 709–17; here pp. 709, 711.

among the largely peasant recruits,[197] as the idea of the nation had scarcely any resonance among the rural lower classes from which the great bulk of the infantry was drawn.

The founding of the Kingdom of Italy had not markedly changed the life of a day laborer in Basilicata or Sicily. Since time immemorial, the rural population had equated "the State" with the *signori* at the top, seeing it simply as assigner of work tasks and the power that tore men from the soil and dragooned them into military service. This stereotypical attitude still marks the literary works of writers opposed to Fascism, many of whom were sent into internal exile in the islands or rural areas of the South in the 1920s and 1930s.[198] The idea of the Nation was largely unknown to the *contadini*, and not only them; family, village and land-owner (*il padrone*) marked the boundaries of their world.[199]

The often quoted witticism of Massimo D'Azeglio, prime minister of the Kingdom of Sardinia, had lost none of its force a half-century or more later: after Italy was made, the next task was to make the Italians.[200] Only in the war itself were the masses "Italianized." But it was a faltering process: at first, the ruling liberal elite saw no need to motivate the troops and the home front. Until the military disaster at Caporetto, in late October 1917, the officer corps under its commander-in-chief Luigi Cadorna waged a largely "dynastic" war; it was, as Richard Bosworth put it, the only major European power to be drawn into the

---

[197] The disproportionate number of *contadini* at the front – a group consisting of day labor-ers as well as small farmers – was central to the Fascist memory of the First World War. The Fascists tried to use it to drive a wedge between the rural population and the industrial workers, the so-called *operai imboscati*, who were accused of "draft dodging" for a cushy life behind the lines. In any case, most estimates and statistics clearly confirm the overrepre-sentation of the peasantry at the front and among battlefield losses. One reason for this, of course, was Italy's comparatively low level of industrialization: the last prewar census, in 1911, still showed 58 per cent of the working population in the primary sector, against 23.7 per cent in industry, and 18 per cent in the tertiary sector. For figures on the Italian armed forces and war losses, see Antonio Gibelli, *La grande guerra degli italiani, 1915–1918*, 2nd edn, Milan 2001, pp. 85–92. And on the relationship of ordinary people to the war: Giovanna Procacci, *Dalla rassegnazione alla rivolta. Mentalità e comportamenti popolari nella Grande Guerra*, Rome 1999.

[198] See, for example, Carlo Levi, *Christ Stopped at Eboli* [orig. 1945], New York, NY 2006; Ignazio Silone, *The Abruzzo Trilogy* [orig. 1930–1940], South Royalton, VT 2000; Cesare Pavese, *The Political Prisoner* [orig. 1949], London 1955.

[199] Marco Meriggi has pointed out, however, that *campanilismo*, with its local patriotism and lack of nationalism, was by no means confined to the lower classes. Together with support for the monarchy, it prevailed over nationalism also among local notables (the real political players in the countryside) long after the unification of Italy. See Marco Meriggi, "Soziale Klassen, Institutionen und Nationalisierung im liberalen Italien," *Geschichte und Gesellschaft* 26/2000, pp. 201–18, esp. pp. 214–15.

[200] The original formulation in D'Azeglio's memoirs was *"fatta l'Italia bisogna fare gli italiani,"* but in 1896, following Italy's defeat at the Battle of Adwa in Ethiopia, the education minis-ter Ferdinando Martini popularized the more striking version quoted above. See Simonetta Soldani/Gabriele Turi, "Introduzione," in Soldani/Turi (eds.), *Fare gli italiani. Scuola e cultura nell'Italia contemporanea*, vol. 1 of *La nascita dello Stato nazionale*, Bologna 1993, pp. 9–34; here p. 17.

World War without a *union sacrée*.[201] The heated nationalism of the *maggio radioso*, like that of the "August experience" in Wilhelmine Germany, was largely a phenomenon of the urban middle classes.[202] Above all, however, it gripped a small number of intellectual interventionists, displaying few points of contact with the "Italian" masses.

The lack of Italians noted by D'Azeglio was evident also in the use of the national language.[203] Italian had remained a language mainly of educated people; outside Tuscany and Rome it was taught and spoken only at school, and even there not necessarily. Nor was the dominance of regional dialects the only issue. In 1901, 50 per cent of the Italian population were still unable to read and write.[204] This reflected shortcomings in the educational system, of course, but it also indicated the "weakness" of the state in large parts of the peninsula. Partly because of clerical reservations about secular influence on the population, the state had not managed to enforce compulsory schooling everywhere.[205] Only the war and government propaganda[206] after Caporetto eventually spread language instruction among the lower classes, enabling them to develop and express rudimentary forms of patriotic identity.[207] But what scope was there then for D'Annunzio's propaganda?

D'Annunzio's speeches and texts undoubtedly had an elitist quality, but his tropes were refashioned in such a way that they had a resonance even among

---

[201] Richard J. B. Bosworth, *Mussolini's Italy. Life under the Dictatorship 1915–1945*, London 2006 [repr.], pp. 60ff.

[202] See Isnenghi, *Il mito della grande guerra*; and Isnenghi/Rochat, *La Grande Guerra*, pp. 87–137, esp. pp. 106–11. On the August experience: Sven Oliver Müller, *Die Nation als Waffe und Vorstellung. Nationalismus in Deutschland und Großbritannien im Ersten Weltkrieg*, Göttingen 2002, pp. 56–70; Jeffrey Verhey, *The Spirit of 1914. Militarism, Myth, and Mobilization in Germany*, Cambridge 2000; and Benjamin Ziemann, *Front und Heimat. Ländliche Kriegserfahrungen im südlichen Bayern 1914–1923*, Essen 1997, pp. 39–54.

[203] On the following, cf. Gibelli, *La grande guerra degli italiani*, pp. 92–7.

[204] Giovanni Vigo, "Gli italiani alla conquista dell'alfabeto," in Soldani/Turi, *Fare gli italiani*, pp. 37–66. Apart from the low average rate of literacy by international standards, there were enormous regional differences. In 1901, while 82 per cent of the population in Piedmont and 78 per cent in Lombardy were literate, the figures for Sicily, Basilicata and Calabria were only 29 per cent, 25 per cent and 21 per cent respectively. By comparison, Prussia in the mid-nineteenth century already had a literacy rate over 80 per cent, and in England and Wales only 3 per cent of people were illiterate by the end of the century.

[205] Vigo, "Gli italiani alla conquista dell'alfabeto," pp. 58–9.

[206] On propaganda directed at "semi-illiterates," see Mario Isnenghi, *Giornali di trincea, 1915–1918*, Turin 1977.

[207] See Gibelli, *La grande guerra degli italiani*, p. 151. "The ruling classes," Gibelli concludes, "not only exposed millions of men to slaughter; they also offered and imposed words so as to give a name and a meaning to things that had had none for the lower classes. What happened in the hell of the trenches was ultimately a forcible penetration of patriotic vocabulary, a (superficial and contradictory) process of Italianization that accompanied the bloody and traumatic experience."

the lower middle classes. They crucially marked Italy's political and media landscape. The discourse of "ordinary people" – which in this case often meant illiterates – has often remained a blind spot for historians, and this is especially true of the pre-Fascist period.[208] It cannot be said with certainty whether D'Annunzio's sphere of activity included ordinary soldiers, but we may surmise that his rhetoric – the emotion in his voice and the fury it aroused – did not leave unmoved even the uneducated listeners whom he addressed in the "Voices of Reconquest" speech of December 1917. Giorgio Zanetti, one of the editors of D'Annunzio's journalistic writings, notes that, despite the abundance of cultured references in that and other speeches, he "was not afraid that people would fail to understand him; he trusted more in the power of rhythmic expressions of feeling than in conceptual or cultural modes of understanding."[209] Zanetti points to entry LXXXV, October 17, 1915, in the *Taccuini*, where D'Annunzio reports having come across a chaplain, Semeria,[210] who had to address the Caltanissetta Brigade:

> They [the soldiers] look strong and proud. They belong to the Sicilian brigade: dark, wild, many looking like Arabs or kaffirs. [...] Mass begins [...] "Kneel down!" the General roars. [...] It is a mass attended by animals for slaughter. [...] The Sacrifice is interrupted, so that Father Semeria can speak. [...] He has a simple, commonplace style of speech. There is no *beauty* in his words. He too admits to the misconception that humble souls would not be able to understand a high, noble eloquence.[211]

So D'Annunzio at least thought that the "animals for slaughter" understood him, and that it was mistaken to trim the content of his speeches to "humble souls." The decisive factor was not content but style and appearance, an appeal to the irrational in human beings.[212] The speaker's rapport with his audience was for

---

[208] At least it is when the history relies mainly on textual analysis. In the 1990s, however, Isnenghi and Gibelli began to reduce the area of blindness: Isnenghi, by focusing on the richly illustrated trench newspapers and on the *Case del soldato*, the leisure houses founded by the army chaplain Giovanni Minozzi, where soldiers were often taught to read and write; Gibelli, by including the field post of illiterates. See Isnenghi, *Giornali di trincea*; Gibelli, *La grande guerra degli italiani*, and *L'officina della guerra. La grande guerra e le trasformazioni del mondo mentale*, Turin 1991.

[209] D'Annunzio, SG, vol. 2, p. 1700.

[210] The Barnabite priest Father Giovanni Semeria, who is known for his modernizing theological writings, worked after the war with the aforementioned Don Minozzi to found the Opera Nazionale per il Mezzogiorno d'Italia, which administered to numerous war orphans in its schools and orphanages.

[211] D'Annunzio, *Tancuini*, pp. 793–94.

[212] Vogel-Walter (*D'Annunzio*, pp. 85–97) describes the speeches at the front as a form of postmodern religion. On p. 87 she writes: "D'Annunzio was convinced that the aesthetics of word choice helped to 'elevate' people. This clearly shows that in his view aesthetics – that is, external form – stands above content."

D'Annunzio's a "divine mystery."[213] In his novel *Il fuoco* [Fire], the protagonist Stelio Affrena defines the artist's relationship to the crowd:

> The extraordinary feeling that had filled him with wonder when he addressed the people from the Doge's throne came over him once again. Something mysterious entered into the communion of his soul with the soul of the crowd, something almost divine. [...] In the crowd lay slumbering a hidden beauty, from which only the poet, only the hero, could draw flashes of lightning. When this beauty suddenly manifested itself, whether in a theater, a marketplace, or a prisoner-of-war camp, a stream of joy swelled up in the heart of him who knew how to kindle it through his verse, his speech, or his sword. The poet's word, addressed to the people, was thus a deed like the feat of a hero. It was a deed that all of a sudden brought forth beauty from the dark confines of the soul, as a carver, with one touch of his creative hand, might miraculously shape the statue of a god from a clay block.[214] [...] On the steps of the new theater, he saw the true populace, the huge single-willed crowd, whose scent had previously wafted up to him. [...] Through its mysterious rhythmic power, his art, though not understood, produced a violent commotion in the raw, unknowing souls.[215]

Stelio Effrena, who hopes to bring an Italian Bayreuth into being, is D'Annunzio's alter ego.[216] The quotation not only hints at the unity of art and life (which defined D'Annunzio's existence) but illustrates how the war bard thought of his effect on "raw, unknowing souls." The point to transform the chaotic mass into a crowd imbued with order – and through this order to bring forth beauty, the ultimate aim of the aesthete.[217] This needed the right setting, first of all to create the necessary space for emotions. What counted was a "shared atmosphere" generated by natural surroundings, by the rhythms of "music, dancing, and lyric poetry."[218]

These were significant images for Gustave Le Bon, the French sociologist and father of mass psychology, but also evidently for D'Annunzio. Indeed, it appears that D'Annunzio had internalized Le Bon's theories. His imagery and his "high and noble eloquence" were supposed to captivate the masses. For Le Bon, moreover:

> Handled with skill, they possess in sober truth the mysterious power formerly attributed to them by the tenets of magic. They cause the birth in the minds of crowds of the most formidable tempests, which in turn they are capable of

---

[213] See George L. Mosse, "The Poet and the Exercise of Political Power: Gabriele D'Annunzio," in Mosse, *Masses and Man. Nationalist and Fascist Perceptions of Reality*, New York, NY 1980, pp. 87–103; here 93ff.

[214] Mussolini used this image of forming a "god's statue" from a shapeless block when, in view of Italy's frequent defeats in battle, he complained of the "material" he had available to produce the New Fascist Man. See Chapter III.2 below.

[215] Gabriele D'Annunzio, *Il fuoco*, in *Prose di romanzi*, vol. 2, pp. 195–518; here pp. 161–62.

[216] Woodhouse, *Gabriele D'Annunzio*, p. 188.

[217] Mosse, "The Poet and the Exercise of Political Power," p. 94.

[218] Ibid., p. 93.

stilling. [...] The power of words is bound up with the images they evoke, and is quite independent of their real meaning. Words whose sense is the most ill-defined are sometimes those that possess the most influence.[219]

"It is always," Le Bon writes, "the mysterious and legendary side of events that especially strikes crowds." As to the question of how "the imagination of crowds is to be impressed," he replies that it is never "by attempting to work upon the intelligence or reasoning faculty."[220] It is necessary to address the "beliefs" of the masses, not their reason. And these beliefs have a peculiar form

> that I cannot better define than by giving it the name of a religious sentiment. This sentiment has very simple characteristics, such as worship of a being supposed superior, fear of the power with which the being is credited, blind submission to its commands, inability to discuss its dogmas, the desire to spread them, and a tendency to consider as enemies all those who do not accept the sentiment. Whether such a sentiment applies to an invisible god, to a wooden or stone idol, to a hero or to a political cause, provided that it presents the preceding characteristics, its essence always remains religious. The supernatural and the miraculous are present to the same extent. [...] Sentiment has never been vanquished in its eternal conflict with reason.[221]

D'Annunzio spoke precisely to this "religious sentiment" and constantly summoned it up. But we should note here that another part of D'Annunzio's impact strengthened his words: that is, his credibility and authenticity resulting from direct participation in the war. As we have seen, he went off at the age of 50 to the front, attempted to strike at the Habsburg navy with torpedo boats, and carried out several aerial missions. D'Annunzio was no *imboscato*, no shirker calling for sacrifice to the nation from a cushy position far from the front. Such behavior would have contradicted his own idea that, as a poet of genius and angelic herald, he had been called to set an example of heroism. He, the "artist of his life," saw himself as the prototypical Nietzschean *Übermensch*, who aimed to rise ever higher. In D'Annunzio, Nietzsche's "Excelsior!" became an Icarian *più alto e più oltre* (higher and farther).[222]

D'Annunzio's elevation to superman status was due not least to his flight to Vienna on August 9, 1918 – certainly the most spectacular, and most distant, of his aerial leafleting missions. He knew how to make of it an *avant garde* work of

---

[219] Gustave Le Bon, *Psychology of* Crowds, Southampton 2012, p. 95.

[220] Ibid., pp. 60, 62.

[221] Ibid., pp. 65, 67.

[222] Friedrich Nietzsche, *The Gay Science*, translated by Walter Kaufman, New York, NY 1974, pp. 229–30. (§285). The motto *più alto e più oltre*, which D'Annunzio used both during and after the war, appeared in an article entitled "Il saluto di D'Annunzio agli aviatori prima della battaglia" (D'Annunzio's greeting to the aviators before the battle), *Corriere della Sera*, May 27, 1917: reprinted in SG, vol. 2, p.693. See also Chapter III.1 below.

art, assuring him of all the more attention and reverence. "People of Vienna!" his leaflet began. "Learn to know the Italians! If we wished, we could drop tons of bombs on your city, but we are sending you only a tricolor greeting – the tricolor of liberty!" Once it was realized, following the Caporetto debacle and the appointment of a new military leadership under Armando Diaz, that propaganda had become a decisive factor in warfare, there was a greater appreciation of the value of D'Annunzio's actions. The poet-hero saw the mobilizing potential of the "vehicle of pictures and words," but above all he understood that the airplane was an "index fossil" of the era, a token of power, and a key symbol of modernity that should be instrumentalized for the national cause.[223]

Having flown back safely from Vienna, D'Annunzio gave a speech to his men that was published on August 18 in the *Corriere della Sera*. The Nietzschean "Excelsior" is here combined with active belligerence and false modesty. And, of course, the *comandante* calls for further sacrifices:

> What has been achieved no longer matters to the warrior. Nothing matters but what has still to be achieved. How happy we were in enemy skies! But why was each of us sad when we returned to the aerodrome? We had tasted a new joy, a complete joy, on the long flight. We were already impatient to leave again. [...] We felt that the real joy of our war was finally beginning, because our victory was really beginning. We felt there was no longer a life for us except in the newness of the deed, in the continuing newness of an energy that, instead of pausing or slowing, grew stronger and faster. [...] As we approached our objective, as we braved fate and victory with each heartbeat, that visage [of beautiful Italy] brightened and shone forth. It was the visage of our love, to which we had all surrendered ourselves in sacrifice and more. And love and death, fame and fatherland, were and are for us a unique beauty that we shall carry each time in our dilated breast, between wing and wing, where everything is purity and hope. [...] O friends, we defy tomorrow and the unknown with our battle cry: Eia! Eia! Alalà![224]

The *comandante* bathes himself and his men in the beauty of an aestheticized war that he has himself created. Even more striking in the quotation, however, is the cult of the deed, which would later become one of the hallmarks of fascism.

D'Annunzio had soared over the whole population of Vienna and made it dependent on his mercy; the words he dropped instead of bombs had demonstrated a plethora of power. And with his flight D'Annunzio had clearly inscribed this power over man and nature in his aircraft. Heroism and "living dangerously,"

---

[223] On Aby Warburg's use of the term "index fossil" (*Leitfossil*), see Ernst H. Gombrich, *Aby Warburg. Eine intellektuelle Biographie*, Frankfurt/Main 1984, p. 349.

[224] Gabriele D'Annunzio, "Parole di G. D'Annunzio dopo il vuolo su Vienna," in SG, vol. 2, pp. 757–62; here pp. 759, 762.

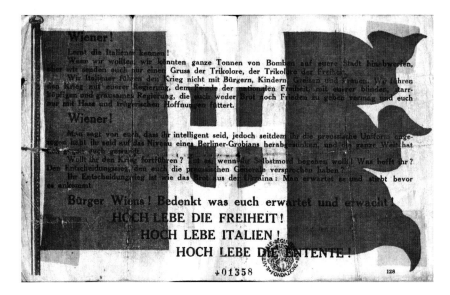

Figure 2.1 Words instead of bombs – D'Annunzio's Vienna leaflet
Source: D'Annunzio's Vienna leaflet, reproduced with the kind permission of the Archivi del Vittoriale, Archivo Iconografico

virility and avant-gardism, were now among the central messages of the airplane; it testified to the essentialization of war as a central category of life. Such were the tropes that the Fascists were to take over from D'Annunzio and the Futurists, when they too used the airplane as a symbol (Figure 2.1).

D'Annunzio's cynical dropping of "words instead of bombs" has been described as "rhetorical Douhetism."[225] "Poetic Douhetism" seems equally apposite, however, since D'Annunzio presents bombing as a poetic means for the catharsis of the Viennese population; the aeronautic spectacle is supposed to bring about

---

[225] Bernhard Siegert, "L'Ombra della macchina alata. Gabriele D'Annunzios *renovatio imperii* im Licht der Luftkriegsgeschichte 1909–1940," in Gumbrecht et al. (eds.), *Der Dichter als Kommandant*, pp. 261–306; here p. 291. The theory of aerial warfare developed by the Italian general Giulio Douhet (1869–1930) is known as Douhetism; see his *Command of the Air*, Tuscaloosa, AL 2009. In the preface to the German edition of this book, first published in Italian in 1921, Hilmer von Bülow writes: "Future wars will be total wars. Douhet's guiding idea is: defense on land and sea, massive attack in the air! Victory on land or sea no longer forces a decision; only aerial supremacy creates the possibility of imposing your will on the enemy with the help of an effective offensive, an aerial offensive. Therefore, according to Douhet, the air armada must gain command of the air and win victory by annihilating the enemy air force and its resources, by means of violent destruction of material and spiritual centers, and massive attacks on the enemy's whole living space." Hilmer Freiherr von Bülow, "Vorwort," in Giulio Douhet (ed.), *Luftherrschaft*, Berlin 1935, pp. 5–9; here p. 7.

a purification, which the Viennese themselves could initiate by breaking with their government and the Prussian generals. But D'Annunzio did not stop at poetry: he also envisaged Douhetism proper, with plans for area bombing of the cities of Vienna and Pula. As he put it in a memorandum of May 11, 1917 to Italian chief of staff Cadorna, "On the Use of Bomb Squadrons in the Coming Operations":

> Equipped with everything necessary for an eight-hour flight, it [the Caproni triplane] can carry a bomb load of a ton or more. It will therefore carry a thousand kilos of explosives to Vienna, but over Pula it will drop two thousand kilos. The first of these aircraft will be ready in the early part of June. We will soon have thirty of them. [...] On the port of Pula (where the face of Rome can be seen in the amphitheater) and on the large Austrian ships (which, carefully armored in their monkish seclusion, offer Lissa [Vis] a little fame), we could rain down a bomb load of more than sixty tons in one go, and then reload to repeat the roaring storm with true Roman persistence.[226]

The firestorm did not happen for the time being. But this did not prevent D'Annunzio from using to excess the theme of cathartic fire and burnt sacrifices. The Futurist variant of purification by fire – war as "the world's sole hygiene" – will be examined more closely in a later chapter.[227] Here we should simply underline that the idea of fire and its sacrificial victims, bound up with rebirth of the nation, was not "mere rhetoric."[228] It should be taken seriously as an act that imparted meaning, with major consequences also in the practice of warfare.

Anomie and political religion

By 1918, Italy had more than 650,000 war dead – roughly 7.5 per cent of all males between the ages of 15 and 49.[229] As it was inconceivable that these myriads had perished in vain, their death was embedded in a horizon of meaning, or indeed eschatologically transfigured. And it was the Christian semantics of sacrifice for the nation that made it possible to transcend, by conferring meaning upon, the

---

[226] USAM, R.A., cart. no. 77, "D'Annunzio G." Quoted [and translated] from Gumbrecht et al. (eds.), *Der Dichter als Kommandant*, pp. 27–36; here p. 28.

[227] See Chapter III.1 below. The idea of war as "the world's sole hygiene" made its debut in the first Futurist manifesto, published in *Le Figaro*, February 20, 1909. See Filippo T. Marinetti, "Fondazione e Manifesto del Futurismo," in Marinetti, *Teoria e invenzione futurista*, ed. by Luciano De Maria, Milan 1983 [repr.], pp. 7–14; here p. 11.

[228] See D'Annunzio's previously quoted speech of May 7, 1915 to Italian students, as well as "La fede nell'aviazione italiana." Cf. Mosse, "The Poet and the Exercise of Political Power," pp. 91–2.

[229] Isnenghi/Rochat, *La Grande Guerra*, pp. 471–72.

often militarily pointless, "unheroic" deaths on the field of battle.[230] Their death became a prerequisite for the life and rebirth of the collective. The ending of the individual's life became meaningful as it was indispensable for the eternal life of the nation.

Following on from George L. Mosse, Emilio Gentile adjudged the cult of fallen soldiers during and after the First World War to be "the first universal, liturgical manifestation of the sacralization of politics in the twentieth century."[231] It gave a new impetus to the sanctity of the nation. And D'Annunzio was the driving force behind the construction of an Italian national religion in the war period and during his adventurous occupation of Fiume.[232] The sacralization of politics was the context within which the aeronautic tropes received their comprehensive significance. Fiume was not just an epilogue to the war, but a prologue to the construction of the Fascist dictatorship in Italy. But before we turn to that small Istrian port (today known, in Croatian, as Rijeka), we need to explain more clearly how we shall be using the terms "sacralization of politics" and "political religion."

The term "political religion" has not only been discussed in numerous publications; it has even, since 2000, had a journal devoted to it: *Totalitarian Movements and Political Religion*.[233] The term was already in use during the Fascist period, and although the political theorist Eric Voegelin is often credited with its invention[234] it was actually first coined by Condorcet during the French Revolution.[235] Whereas the concept of civil religion (frequently confused with political religion) originated with Jean-Jacques Rousseau, Raymond Aron introduced the term "secular religion" in 1944.[236] Suffice it to recall here that in the 1960s George L. Mosse investigated the religious, or pseudo-religious, *völkisch* roots of National Socialism, and that Voegelin's disciple Klaus Vondung explicitly reintroduced the term into Fascism studies with his work "Magic and Manipulation."[237]

With the collapse of the Soviet system, debate flared up again over political religions and the uses and disadvantages of the concept of totalitarianism. Key contributions came from the circle in Munich around the political theorist Hans

---

[230] On the cult of the fallen soldier in Italy, see Oliver Janz, "Zwischen privater Trauer und öffentlichem Gedenken. Der bürgerliche Gefallenenkult in Italien während des Ersten Weltkrieges," *Geschichte und Gesellschaft* 28/2002, pp. 554–73, and "Monumenti di carta. Le pubblicazioni in memoria dei caduti della Prima Guerra Mondiale," in Fabrizio Dolci/Oliver Janz (eds.), *Non Omnis Moriar. Gli opuscoli di necrologio per i caduti italiani nella Grande Guerra*, Rome 2003, pp. 1–44.

[231] Emilio Gentile, *The Sacralization of Politics in Fascist Italy*, Cambridge, MA 1996, p. 17.

[232] Ibid.

[233] For an overview, see Emilio Gentile, "Political Religion. A Concept and its Critics – A Critical Survey," *Totalitarian Movements and Political Religion* 6/2005, pp. 19–32.

[234] See Eric Voegelin, *The Political Religions*, in *Voegelin, Modernity without Restraint. The Collected Works of Eric Voegelin*, Vol. 5 ed. by Manfred Henningsen, Columbia, MO 2000.

[235] Emilio Gentile, *Politics as Religion*, Princeton, NJ 2006, p. 2.

[236] Gentile, *"Politics as Religion,"* p. 21.

[237] See George L. Mosse, *The Crisis of German Ideology: Intellectual Origins of the Third Reich*, London 1966 [repr.]; and Klaus Vondung, *Magie und Manipulation. Ideologischer Kult und Politische Religion des Nationalsozialismus*, Göttingen 1971.

Maier.[238] But it was above all Emilio Gentile who, with his book on the Lictor cult in Fascist Italy, contributed to a heuristically fruitful revival of the concept of political religion in Fascism studies.[239] Shortly after the turn of the century, Michael Burleigh then successfully adopted the concept in his account of Nazi Germany.[240]

Gentile's definition of political religion provides the foundation for the present study. A political movement brings about a sacralization of politics when it

(a) consecrates the primacy of a secular collective entity by placing it at the center of a system of beliefs and myths that define the ultimate purpose of the social existence and prescribe the principles for discriminating between good and evil; (b) formalizes this concept in an ethical social code of commandments that binds the individual to the sacralized entity and imposes loyalty, devotion, and even the willingness to lay down one's life; (c) considers its followers to be the community of the elect and interprets its political action as a messianic function to fulfill a mission; and (d) develops a political liturgy for the adoration of the sacralized collective entity through the cult of the institutions and images that embody it, and through the mythical and symbolic representation of its sacred history – a regular ritual evocation of events and deeds performed over a period of time by the community of the elect.[241]

As to political religion, Gentile defines it as

a form of sacralization of politics that has an exclusive and fundamentalist nature. It does not accept the coexistence of other political ideologies and movements, it denies the autonomy of the individual in relation to the collectivity, it demands compliance with its commandments and participation in its political cult, and it sanctifies violence as a legitimate weapon in the fight against its enemies and as an instrument of regeneration. In relation to traditional religious institutions, it either adopts a hostile attitude and aims to eliminate them, or it attempts to establish a rapport of symbiotic coexistence

---

[238] See Hans Maier, *Politische Religionen. Die totalitären Regime und das Christentum*, Freiburg 1995, but especially Hans Maier (ed.), *Totalitarianism and Political Religions*, vol. 1, New York, NY 2004; Hans Maier/Michael Schäfer (eds.), *Totalitarianism and Political Religions*, vol. 2, *Concepts for the Comparison of Dictatorships*, New York, NY 2007; Hans Maier (ed.), *Wege in die Gewalt. Die modernen politischen Religionen*, Frankfurt/Main 2000; and Hans Maier (ed.), *Totalitarianism and Political Religions*, vol. 3, *Concepts for the Comparison of Dictatorships. Theory and History of Interpretation*, New York, NY 2007.

[239] See Gentile, *The Sacralization of Politics in Fascist Italy*.

[240] See Michael Burleigh, "National Socialism as a Political Religion," *Totalitarian Movements and Political Religions* 1/2000, pp. 1–26; *The Third Reich. A New History*, London 2000; and, on the development of political religion in the long 19th century, *Earthly Powers. Religion and Politics in Europe from the Enlightenment to the Great War*, London 2005.

[241] Emilio Gentile, "Die Sakralisierung der Politik," in Hans Maier (ed.), *Wege in die Gewalt. Die modernen politischen Religionen*, Frankfurt/Main 2000, pp. 166–82; p. 169; [cf. the slightly variant Gentile, *Politics as Religion*, p. 139].

by incorporating the traditional religion into its own system of beliefs and myths while reducing it to a subordinate and auxiliary role.[242]

But what does Gentile understand by religion? Well, religion is the result of that process of sacralization, whether the entity in question is transcendent or immanent. This often criticized extension of the concept of religion may already be found in Voegelin.[243] Distinguishing between otherworldly and worldly religions, he attributed "a religious dimension to any *absolutization* of reality that creates meaning and is ultimately decisive for a person's thinking, feeling, wishing, and acting."[244] Although theorists of religion may not regard such absolutization as sufficient to characterize a belief system as a religion, it is adequate for an analysis of totalitarian ideologies.

This brings us to another central concept of fascism studies: totalitarianism. Since the appearance of Hannah Arendt's *The Origins of Totalitarianism* in 1951, it too has been the object of heated debate.[245] To avoid a long and tedious exegesis, it again seems to make most sense to select a plausible and heuristically fruitful definition,[246] drawing on the work of the Roman historian Emilio Gentile.[247] For our purposes, the following points are most pertinent. First, totalitarianism is a "political experiment": that is, its aspirations are independent of the degree of totalitarian penetration of society already achieved by the regime. The aim is the "conquest of society," involving palingenesis or an "anthropological revolution" that produces a new man.[248] The concepts of political religion and totalitarianism therefore both point to an essential aspect of fascist ideology and practice. A political order is not all there is to it. The goals of fascism concern man as a totality, not just as *zoon politikon*. Man and society must be created anew; a new time must begin. With such a comprehensive aspiration, the concept of political

---

[242] Gentile, "Die Sakralisierung der Politik"; *Politics as Religion*, p. 140.

[243] "The spiritual religions, which find the *realissimum* in the Ground of the world, should be called trans-worldly religions, and all others, that is, those that find the divine in subcontents of the world, should be called inner-worldly religions." Voegelin, *The Political Religions*, pp. 32–3.

[244] Mathias Behrens, "'Politische Religion' – eine Religion? Bemerkungen zum Religionsbegriff," in Hans Maier/Michael Schäfer (ed.), *Totalitarismus und Politische Religion*, vol. 2, Paderborn 1997, pp. 249–69; here p. 254 [emphases added]. See also Juan J. Linz, "The Religious Use of Politics and/or The Political Use of Religion. Ersatz Ideology versus Ersatz Religion," in Hans Maier (ed.), *Totalitarianism and Political Religion*, vol. 1, New York, NY 2004, pp. 102–19. Gentile starts from the same concept of absolutization: see Emilio Gentile, "Fascism, Totalitarianism, and Political Religion. Definitions and Critical Reflections on Criticism of an Interpretation," *Totalitarian Movements and Political Religions* 5/2004, pp. 326–75; here p. 364.

[245] Hannah Arendt, *The Origins of Totalitarianism*, New York, NY 1951.

[246] For an overview of the history of the concept, see Abbott Gleason, *Totalitarianism: The Inner History of the Cold War*, New York, 1995; Alfons Söllner/Ralf Walkenhaus/Karin Wieland (eds.), *Totalitarismus. Eine Ideengeschichte des 20. Jahrhunderts*, Berlin 1997; and Eckhard Jesse (ed.), *Totalitarismusim 20. Jahrhundert. Eine Bilanz der internationalen Forschung*, 2nd edn, Baden-Baden 1999.

[247] See Gentile, *Politics as Religion*, pp. 45ff.

[248] On the concept of palingenesis in relation to fascism, see Roger Griffin, *The Nature of Fascism*, London 1993 [repr.], pp. 26, 32–6; and Roger Griffin, *Modernism and Fascism. The Sense of a New Beginning under Mussolini and Hitler*, Basingstoke, UK 2007.

religion and also that of myth are well suited to render essential aspects of fascism intelligible. For we are talking of a "worldview," in the full sense of the word.

Fascist ideology breaks down the Enlightenment barriers between particular values and spheres of society; the political sphere is religious, and the religious sphere is political, not a private matter concerning the individual alone. This has its historical roots in Romanticism, but, as Peter L. Berger has shown, rests upon "anthropological constants."[249] The First World War did not merely destabilize the prevailing nomos, but led to its dissolution. The result was a crisis of socially constructed reality – which means, also, of the reputedly stable pillars that offer a sense of orientation and make action possible by giving it legitimacy.[250] Thus, only a worldview equivalent to a religion could offer a way out from the ubiquitous relativism. For religion is an "instrumentality of legitimation" that "maintains socially defined reality": it grounds the "tenuous realities of the social world" on a "sacred *realissimum*, which by definition is beyond the contingencies of human meanings and human activity";[251] it thereby transforms "human products into supra- or non-human facticities," so that "the human nomos becomes a divine cosmos, or at any rate a reality that derives its meanings from beyond the human sphere."[252] Fascism will be seen here as such an attempt to conceive of a human nomos as a "supra- or non-human facticity," and hence to escape the charge of relativism that arose in the wake of the crisis of historicism. Fascists opposed relativism by absolutizing something immanent: the Nation or *Volk*. The means of doing this was myth.

In the following chapters, an interpretation of various aspects of aviation discourse will shed further light on this nexus of political religion, myth, fascism and modernity. We shall begin by looking more closely at the occupation of Fiume/Rijeka, which took place in the postwar context of a radical dissolution of the existing nomos. The old liberal order had already broken up during the war, but there was a growing awareness of this in its aftermath, not least because of the collapse of empires, revolutions and the recasting of the European system of states. The collapse went together with a longing for order and a strengthening of the kind of mythical thought that generates order.

Fiume and the sacralization of politics

The military and "moral" catastrophe of Caporetto, which enabled the Central Powers to break through the Southwest Front in October 1917 and to drive the Italians to the Piave, some 40 kilometers from Venice, is followed in Italian histories by the glorious victory at Vittorio Veneto on November 4, 1918.[253] Only

---

[249] See Manfred Frank, *Der kommende Gott. Vorlesungen über die Neue Mythologie*, Frankfurt/Main 1982, pp. 188–211.

[250] See Otto Gerhard Oexle (ed.), *Krise des Historismus – Krise der Wirklichkeit: Wissenschaft, Kunst und Literatur 1880–1932*, Göttingen 2007, pp. 11–116.

[251] Berger, *The Sacred Canopy*, p. 32.

[252] Ibid., p. 89.

[253] On the course and consequences of the two battles, see Isnenghi/Rochat, *La Grande Guerra*, pp. 373–406 and 466–71.

in the year between the two battles did the patriotic arousal of Italians, especially of larger sections of the middle classes, begin in earnest.[254] The mobilization of hearts and minds was a result of the alarming situation after Caporetto. Pressure to defend against the external, and by now also internal, enemy led to greater realization of the need to manufacture national cohesion. And the presence of the "Germanic barbarians" on Italian soil provided grounds for the war that seemed plausible even to non-interventionist Socialists. It mobilized people for a war that had previously been justified only in terms of the *sacro egoismo* of the latest and smallest of the European great powers.

Although D'Annunzio had taken up the task of mobilization for war as early as 1915, with his speeches and spectacular actions designed for media effect, only the specter of total defeat induced the "liberal" ruling strata to address the phenomenon of "the masses."[255] When Italy concluded an armistice with the Habsburg Empire on November 4, 1918 at the Villa Giusti, the war ended only on the battlefield. It soon became obvious that, apart from demobilization of the army, there was an urgent need to convert the economy to peacetime purposes, to stabilize the shaky government finances,[256] and to defuse social tensions intensified by the war. And on top of these Herculean tasks, the popular wartime mobilization had made it necessary to cash in the promises made to Italy under the Treaty of London of 1915.[257]

America's entry into the war in 1917, together with Wilson's 14 Points, had nevertheless placed new obstacles on the road to the *più grande Italia*. Wilson did not feel tied by the Treaty of London and supported both the ending of secret diplomacy and the right of nations to self-determination. Moreover, some of the Entente powers, especially France, did not think highly of the efforts that Italy had made during the war. The Italian delegation at the Paris peace talks therefore faced daunting foreign policy conflicts, made all the more acute by the country's domestic crisis.[258] The situation was spinning out of control for Italy, the sorcerer's apprentice of European nationalism.

---

[254] See Gibelli, *La grande guerra degli italiani*, pp. 308–13.

[255] Ibid., pp. 310–11.

[256] See Douglas Forsyth, *The Crisis of Liberal Italy. Monetary and Financial Policy 1914–1922*, Cambridge 1993; and Charles S. Maier, *Recasting Bourgeois Europe. Stabilization in France, Germany, and Italy in the Decade after World War I*, Princeton, NJ 1975.

[257] To make it more attractive for Italy to join the war on their side, the Entente powers had offered it the prospect of territorial booty in the shape of the Trentino and South Tyrol, the Brenner frontier area, Trieste and Istria as far as the Kvarner Gulf (Fiume was not part of the deal), Dalmatia, a kind of protectorate over Albania, sovereignty over the Dodecanese islands and parts of the Ottoman bankruptcy estate, and a clarification of border issues in the cases of Eritrea, Somalia and Libya. See Nicola Tranfaglia, *La prima guerra mondiale e il fascismo*, Turin 1995, p. 46.

[258] On the talks in Paris, the "Adriatic question," and the deepening political crisis inside the country, see ibid., pp. 131–46, and Vogel-Walter, *D'Annunzio*, pp. 99–119. For a general survey of the situation in Italy after the First World War, see also Hans Woller, *Geschichte Italiens im 20. Jahrhundert*, Munich 2010.

The crisis shook the nineteenth-century liberal order to its foundations. The stratum of notables, who had ruled the country after unification, were divided among themselves and incapable of rising to the tasks. The liberal era, identified with the name of the five-time prime minister, Giovanni Giolitti, appeared to be at the end of its tether, since it was no longer possible to ignore the masses in the work of government. The wartime mobilization and the integration of opponents of the war had not been achieved through nationalist agitation alone; hopes had been raised that, once the war was over, the masses would have a larger stake in both the political process and the means of production, especially agricultural land. As expectations grew, the rulers became less and less able or willing to fulfill them. But there were alternatives to the established liberal stratum, and new ones were also appearing on the scene.

The war had enlarged the power base of the Socialists and their labor unions, as well as strengthening political Catholicism.[259] In 1919, Don Luigi Sturzo founded the Partito Popolare Italiano, the Catholic "People's Party."[260] Relations between the Vatican and the secular state had thawed and the Pope's *Non expedit* of 1874 (forbidding active Catholic involvement in politics) had been quietly dropped. And, not least, the growth of nationalism – whose various forces and currents had still to be knitted together – threatened to undermine any stabilization of the critical state of affairs.[261]

When victory came in sight, D'Annunzio began to amalgamate the theme of sacrifice more and more directly and frequently with the question of war booty.[262] Now calling his speeches "sermons," he launched in October 1918 at Sernaglia the nationalist phrase that would mark the coming years: *Vittoria nostra, non sarai mutilata* ("Our victory, you shall not be mutilated").[263] And a few days later, in his "sermon" at Aquilea, he uttered an Our Father of a most peculiar kind:

> O ye dead, who are in the earth and in heaven, blessed be your names, your kingdom come, your will be done on earth. Give daily bread to our faith and

---

[259] On the history of Italian Socialism and the labor movement, see Gaetano Arfè, *Storia del socialismo italiano 1892–1926*, Turin 1992 [repr.]; Zeffiro Ciuffoletti, *Storia del PSI*, vol. 1, *Le origini e l'età giolittiana*, Rome 1992. On the *biennio rosso*, the wave of strikes, occupations and fear of revolution in 1919 and 1920, see Nolte, Ernst Nolte, *Three Faces of Fascism: Action Française, Italian Fascism, National Socialism*, New York 1969, pp. 248ff.; Nicola Tranfaglia, *Fascismi e modernizzazione in Europa*, Turin 2001, pp. 180–92; and Elio Giovannini, *L'Italia massimalista. Socialismo e lotta sociale e politica nel primo dopoguerra italiano*, Rome 2001.

[260] For an overview of political Catholicism in Italy and the founding of the PPI, see Guido Formigoni, *L'Italia dei cattolici. Fede e nazione dal Risorgimento alla Repubblica*, Bologna 1998; Tranfaglia, *La prima guerra mondiale*, pp. 165–71; and Sergio Zoppi, *Dalla rerum novarum alla democrazia cristiana di murri*, Bologna 1991.

[261] On the development of nationalism in Italy and its relationship with Fascism, see Stefan Breuer, *Nationalismus und Faschismus. Frankreich, Italien und Deutschland im Vergleich*, Darmstadt 2005, pp. 125–44; Alexander J. De Grand, *The Italian Nationalist Association and the Rise of Fascism in Italy*, Lincoln, NE 1978; Emilio Gentile, *La Grande Italia. The Rise and Fall of the Myth of the Nation in the Twentieth Century*, Madison, WI 2008; and Tranfaglia, *La prima guerra mondiale*, pp. 147–65.

[262] On the following, see also Woodhouse, *Gabriele D'Annunzio*, pp. 311ff.

[263] Gabriele D'Annunzio, "La preghiera di Sernaglia," in PdR, vol. 1, pp. 593–99; here p. 599.

keep alive in us our sacred hate, as we shall never forswear your love. [...] If need be, we shall fight until the Coming of the Lord to direct the living and the dead. So be it.[264]

In his sacralization of politics, the demagogue knew neither shame nor limits. At the center of his rhetoric was a nation whose renaissance was supposed to be achieved by the sacrificial victims of the war. He was thus a crucial figure linking the nationalization of the masses described by George L. Mosse with the cult of fallen soldiers and the establishment of a Fascist political religion.

The sacrifices had been offered up "on the altar of the fatherland" for the renewal and expansion of the nation. D'Annunzio's rhetoric was thus by no means simply a manipulative device; it was needed to give meaning to the myriad deaths resulting from the war – a meaning that fastened onto familiar patterns of interpretation and adapted them to the new context. This happened in such a way – see the discourse model presented in the Introduction – that links were created between previously unconnected concepts, an existing semantic affinity was increased or minimized, and relations were suspended. In Italy and Germany, but also in other European countries, death on the battlefield turned into the "seed" of a new life and a new man. The sacrifice became the basis for national rebirth and renewal. And as we can see from the following quote, taken from a speech "The Rebirth of Heroic Man" given by the nationalist Hans Schwarz in 1928, neither D'Annunzio's tropes nor his emotive style were peculiar to him:

> But the dead of Langemarck are like a new aristocracy promised to us again. Their traits can be sculpted, and we must avow before them: they did not die for Germany, to save the idea of Germany; they died for a new world, to save Germany! Here they planted the marks of revolt among the nations and will become emissaries of rejuvenation! Their memory wrenches them from the darkness of the grave into the light of resurrection: they make the meaning of the divine cross into a genuine truth again, in a way that is human. While in the West a specter is adorned with the pale crimson of victory, these young men arise from the new earth and affirm the will of the Father, who was none other than the war![265]

In the case of Italy, now fearful for the fruits of its victory, the religious aura surrounding the nation and its fallen was intensified by talk of its *unredeemed* territories. As in the passion of Christ, this emphasized a causal link between sacrifice and redemption; only the sacrifice granted redemption, and the fact of redemption conferred meaning upon the sacrifice. Given the course of the peace talks in 1919, it was now feared that despite the numerous sacrifices there would be no "redemption" of the *terre irredente*. But then the sacrifice would lose its

---

[264] Gabriele D'Annunzio, "La preghiera di Aquilea," in PdR, vol. 1, pp. 600–603; here p. 600.
[265] Hans Schwarz, *Die Wiedergeburt des heroischen Menschen. Eine Langemarck-Rede vor der Greifswalder Studentenschaft am 11.11.1928*, Berlin 1930, p. 13.

meaning. The attempt to preserve this horizon of meaning led to radicalization of the nationalist "theologeme of salvation."[266]

This radicalization becomes apparent in the intensity of a further religious theme: purification. If sacrifice was not sufficient to trigger redemption, probably atonement was necessary too; redemption had failed to occur because the nation was unclean and tainted with sin. This logic of sacrifice, purification and redemption, this whole discursive mechanism, may be further illustrated with a recurrent theme in D'Annunzio: the cathartic fire and holocaust.[267] In a funeral speech in 1919 for the aviators Bini and Zeppegno, who had been brought down over Fiume and cremated in the city, D'Annunzio said:

> At Whitsun, when the martyrdom of Fiume had reached the limits of human suffering, it was said that Fiume appeared the only living city, the only glowing city, the only animated city in this bleak world, all breath and fire, all pain and rage, all purification and wasting [*consunzione*]: a holocaust, the most beautiful holocaust offered for centuries on a desolate field. And it was said that the real name of the city is not Fiume but Olocausta: completely consumed by fire. [...] In the middle of Olocausta these two young Italians kindled their holocaust – in the middle of the land of embers, these two young Italians kindled their sacrificial pyre. [...] Glory to him who added fire to fire. [...] Glory to the two celestial messengers, who, in the course of their brief lives, revealed to our mind how this life we live can be an eternal life! My pilots, cover the two coffins! Perform the rite in the sign of the cross that the winged machine makes with its two cross wings between the nose and the rudder.[268]

Only fiery sacrifice and purifying embers, it is suggested, are capable of bringing new, eternal life and national rebirth. The nation was in need of catharsis, to be saved and reborn. The nexus propagated in Fiume between violence and sacrifice, purification and redemption, eternal life and rebirth, was decisive for the Fascist political religion. But it is not only D'Annunzio's metaphors of sacrifice and redemption that make his Fiume adventure paradigmatic of the sacralization of politics and present us with the Fascist political religion *in statu nascendi*. Other

---

[266] On the "theologeme of salvation," discussed in greater detail below, see Gumbrecht, "I redentori della vittoria." On the nexus of sacrifice, purification, violence, nation and political religion, see Emilio Gentile, "Der Liktorenkult," in Christof Dipper/Rainer Hudemann/Jens Petersen (eds.), *Faschismus und Faschismen im Vergleich. Wofgang Schieder zum 60. Geburtstag*, Cologne 1998, pp. 247–61, esp. p. 251.

[267] "Holocaust" is used here in the original Greek sense of "burnt offering"; it first came to be applied to the Shoah in the 1950s and, despite its religious connotations, remained unchallenged in this use until the 1980s. See "'Wohin die Sprache nicht reicht ...' Sprache und Sprachbilder zwischen Bilderverbot und Schweigegebot," in Bettina Bannasch/Almuth Hammer (eds.), *Verbot der Bilder – Gebot der Erinnerung. Mediale Repräsentationen der Schoah*, Frankfurt/Main 2004, pp. 147–66.

[268] Gabriele D'Annunzio, "Il primo olocausto," in PdR, vol. 1, pp. 984–86; Gabriele D'Annunzio, L'Ala d'Italia è liberata, in: PdR Bd. 1, [Or. 1919], S. 879–894, S. 880.

central tropes of Fascism are also established here. It is therefore worth dwelling for a moment on D'Annunzio's trajectory after the Armistice.

On July 9, 1919, *Il Vate* invoked the aviators gathered at Centocelle airfield. After the Armistice a ban on flying had been decreed, but now this was lifted:

> We are not sated. We are not returning sated from the war. Of all the fighters, we are the privileged ones: we can still win victories, we can still die; we want to win more victories and we still want to die. Our fields will not turn into exercise grounds, but will remain fields of combat and battle. [...] Just as we gave our heroes to the war, just as we have given them to the armistice, so will we give them to the peace. We shall fly! We shall fly. Italy's pinion is freed.[269]

D'Annunzio here invokes *combattentismo*, the spirit and community of frontline fighters, albeit in a variant specific to the world of aviation. He then refers to Guido Keller,[270] another aviator who would accompany him to Fiume, and speaks of his caged eagle and the pity or scorn felt for it by a donkey:

> "It is the image of our flying. Until recently they played at being our masters. Poorly disguised enemies of aviation and aviators: tired old men or ambitious retards, unused to the new ways and opposed to the divine instinct, incapable of understanding the spirit of the race or of supporting and encouraging it."[271]

Remaining on the theme of the two Italies, he notes that the contemptible well-fed old political class bows and scrapes to the other Entente powers, but that

> there is also an Italy that looks upward and into the distance, [...] and loves faraway places without refuge. This bold, energetic, venturesome Italy does exist. One need only look into your eyes. [...] Part of that good material will remain tied to the earth; but another part will set out for flight, adventure, and conquest; a part will put on wings and enter the thousands upon thousands of blue roads, [...] will head for the trackless paths and waterless lanes to the Orient. Let us free ourselves of the Occident, which neither loves us nor wants us. [...] Let us split away from the degenerate West, [...] which has become a giant Jewish bank in the service of the pitiless transatlantic plutocracy. [...] Leave us our privilege as fighters who can still win and still want to win, who can still die and know how to die. Once the black flames, green flames, crimson

---

[269] Gabriele D'Annunzio, "L'Ala d'Italia è liberata," in PdR, vol. 1, pp. 879–94; here p. 880.

[270] The artist, esoteric, vegetarian and aviator Guido Keller founded a yoga group in Fiume and practiced nudism there. On November 14, 1920, in the best manner of D'Annunzio and the Futurists, he flew to Rome and dropped roses on the Vatican and the Quirinale Palace, as well as a chamber pot on the Chamber of Deputies. See Bettina Vogel, "Guido Keller – Mystiker des Futurismus," in Hans Ulrich Gumbrecht et al., *Der Dichter als Kommandant*, pp. 117–32; and Salaris, *Alla festa della rivoluzione*.

[271] D'Annunzio, "L'Ala d'Italia è liberata," p. 881.

flames[272] have dispersed, the blue flames will gather and rise "higher and farther" into the unknown.[273]

The "Jewish-transatlantic plutocracy," the "inert established political class of yesterday": this brings together many fascist enemy-images. But D'Annunzio is mainly depicting the opposite: he himself is the dynamic, martial, risk-loving aviator who flies ever higher. This human type, undeterred by new and unfamiliar paths, would never betray his comrades who died for the salvation of the *terre irredente*; he would prevent Italy's victory from being mutilated. And so the struggle for Fiume came to symbolize these two Italies. It was a struggle that the hapless, overstretched government of Vittorio Emanuele Orlando and his foreign minister Sydney Sonnino could win neither at home nor in its dealings with Woodrow Wilson, however much it conjured up the danger of revolution in the event of a "mutilated peace."[274]

The situation in Italy was indeed becoming ever tenser. Nationalists, agitating against both the government and the Socialists, raised the specter of a Bolshevik takeover. And Benito Mussolini, formerly chief editor of the Socialist *Avanti!*, now in charge at *Il Popolo d'Italia*, called a *Fascio* into being on March 23, 1919 on the Piazza San Sepolcro in Milan, acting in league with Futurists, syndicalists, Arditi and other veterans. A few weeks later, the storming of the *Avanti* building ushered in a wave of anti-Socialist violence by the *squadristi*, which reached a climax in 1921 and 1922.[275] It only involved a small minority, although the *Fasci di combattimento* were active also in other towns of North and Central Italy. The highly disparate bunch of groups associated with nationalist "actionism" shared a tendency to violence rooted in the war, an antibourgeois and anti-Socialist stance, and a contempt for the traditional political caste and its modes of operation.[276] Yet, most of all it was their aura of a "third way" and of national renewal that began to fascinate Italians. At the time in question, the *Fasci* were not yet the center of attention. The spotlight was on D'Annunzio, who was girding himself for the occupation of Fiume after it had been surrendered by the Nitti government.[277] Mussolini, the future Duce, was then no more than D'Annunzio's sidekick.[278]

---

[272] The elite Arditi shock troops wore a variously colored flame on their uniforms: black for the infantry, green for Alpine units and crimson for the Bersaglieri. Many Arditi accompanied D'Annunzio to Fiume. They were among the founding members of the Fascist squads, and the black fez and black flame, which eventually became insignia of the MVSN Fascist militia, had their origins in them. On the Arditi, see Vogel-Walter, *D'Annunzio*, pp. 189–202; Friedrich Kittler, "Il fiore delle truppe scelte," in Gumbrecht et al., *Der Dichter als Kommandant*, pp. 205–25; and Giorgio Rochat, *Gli arditi della grande guerra. Origini, battaglie e miti*, Milan 1981.

[273] D'Annunzio, "L'Ala d'Italia è liberata," pp. 883, 887, 890.

[274] The government too adopted D'Annunzio's image: see Tranfaglia, *La prima guerra mondiale*, p. 147.

[275] See Reichardt, *Faschistische Kampfbünde*, pp. 60ff.

[276] On San Sepolcro and early Fascism, see Gentile, *The Origins of Fascist Ideology*, pp. 104–54.

[277] On the Italian negotiations and the events that led to the occupation of Fiume, see Tranfaglia, *La prima guerra mondiale*, pp. 132–46, 191–200.

[278] According to Ernst Nolte (*Three Faces of Fascism*, p. 245), this was Mussolini's "darkest year," when he was "little else than D'Annunzio's personal journalist in Italy."

On the night of September 11–12, 1919, D'Annunzio set out on his march from Ronchi, a village near Trieste, where the Sardinian grenadiers – the initial core of D'Annunzio's "legionaries," along with the Arditi – had taken up residence after their expulsion from Fiume. With the march from Ronchi, he provided a further matrix for things to come; he even executed it by car, as Mussolini marched his "March on Rome" three years later by train. D'Annunzio began with more than 2,000 volunteers and regular troops, but after a few days the force had swollen to more than 10,000, as the Comandante occupied the city of Fiume and proclaimed its annexation the same evening from the balcony of the governor's palace.[279] It was not only in the eyes of nationalists and Fascists that the Poet had saved the city and rekindled the beacon of the Risorgimento.

This identification of the march with the Risorgimento, the cult of the deed, the Roman salute, the balcony speeches, the characterization of the soldiers as legionaries, the cult surrounding the Comandante himself: all these were first successfully tried out on the masses in Fiume. From there, D'Annunzio sent many celestial bodies of the Fascist semantic-symbolic cosmos into orbit, and anticipated large parts of its ritual and liturgical practice. Yet, as De Felice and Leeden warned as early as the 1970s, it would be overhasty to think of D'Annunzio principally as Mussolini's "John the Baptist" and Fiume simply as the birthplace of Fascism.[280] Alceste De Ambris's Charter of Carnaro, the constitution of D'Annunzio's regime in Fiume, bore the marks of Mazzinian syndicalism and had little in common with Fascist-style corporatism. Left-wing revolutionaries too sympathized for a time with the adventure, and, as Claudia Salaris has shown, it attracted a variety of Bohemians, Dadaists and Futurists, as well as esoterics and Italian "lifestyle reformers," while a mixed bag of cosmopolitan oddballs tried to establish a different social order there in a kind of Hippie commune *avant la lettre*.[281] D'Annunzio's own political position, according to Leeden, wavered between militarist-imperialist hypernationalism, enthusiasm for the "liberation" of oppressed peoples, and radical syndicalism. But does this not contradict the proto-Fascist character of his Fiume adventure?

If Fascism is understood as a modern movement in search of a "new beginning,"[282] rather than a "reactionary" attempt to turn the clock back, Fiume appears precisely as a proto-Fascist phenomenon. Neither its political polyphony nor (as the German case confirms) its collection of "lifestyle reformers," artists and Bohemians therefore seems discrepant. On the contrary, such circles sought to convert the confusion, fragility and decline of the prewar world into the kind of new dawn which they had been yearning for in the years around the turn of the century.[283] Fiume was an experiment in a "third way" or a different modernity, initially at the level of a single polis.

---

[279] See Vogel-Walter, *D'Annunzio*, pp. 131–54. On p. 149 she offers precise estimates of the numbers of soldiers in Fiume.
[280] Leeden, *The First Duce*; De Felice, *D'Annunzio politico*.
[281] Salaris, *Alla festa della rivoluzione*.
[282] See Griffin, *Modernism and Fascism*.
[283] Here, and on the following, see ibid., pp. 130–46.

The "sense of a new beginning," as Roger Griffin called it, appears to have been essential: it aimed at national renewal or rebirth, with the aviator as its key symbol. D'Annunzio and Mussolini agreed about the necessity of palingenesis or *renovatio*.

> I think a meeting between us would be useful for our common cause. After all, our ideas overlap on these fundamental points: 1) Italy's victory must not be mutilated on account of pretexts coming from Croat democrats or Wilson. 2) On the foundation of victory, it is necessary to begin a profound renewal [*rinnovazione*] of our national life. 3) The way must be barred to those who sabotaged the war – secular priests, Giolitti supporters, and the social-*boches*.[284]

---

[284] Letter from Mussolini to D'Annunzio, January 1, 1919. See De Felice/Mariano, *Carteggio D'Annunzio*, p. 3.

3
Longing for Order – Summary

The last two chapters have centered on two attempts, embodied by Warburg and D'Annunzio, to bring order into a world that the war had thrown out of joint. This perception of the world as chaos in need of order, and especially the conception of order as a task, has been identified by Zygmunt Bauman as characteristic of modernity.[285] Warburg's and D'Annunzio's statements in the discourse of aviation, their linguistic order and its contextualization, served to mark the opposition between the liberal and fascist ideal types of order. Warburg's detailed account of liberal order forms a contrasting backdrop to the fascist mythical modernity that is the focus of our investigation.

The art theorist and cultural historian Aby Warburg introduced the dichotomy that not only marked the age of the world wars but subsequently also led to a decades-long interpretation of fascism as reactionary. On both sides of the Iron Curtain, historical analyses of fascism based themselves on the normative premise that modernity equals "disenchantment of the world" (understood as progress) and the victory of "rationalism."

Warburg's postage stamp design is an objectification of this vision of modernity; the airplane depicted on it symbolizes the Enlightenment project and its orientation to progress. The liberal Warburg strove for a "scientific" ordering of the world that was "in accordance with reason," reflecting the Enlightenment ideal of human emancipation from the hold of magic, superstition and myth. A rational world order was the condition of possibility for future progress. But the ordering of time as progress, or the chronopolitical narrative of Enlightenment liberalism that tried to speed up the ageing of experiential space and to integrate it into one's own life, was profoundly shaken by the First World War.[286] This exposed the idea of human advance guided by the light of reason as no more than a belief system.

The First World War, together with his own mental illness, led Warburg to revise his progressive, cultural-evolutionist and historicist worldview.[287] He now

---

[285] Zygmunt Bauman, *Modernity and Ambivalence*, Cambridge 1991.

[286] On "chronopolitics," see Peter Osborne, *The Politics of Time: Modernity and the Avant-garde*, London 1995.

[287] See especially Anselm Doering-Manteuffel, "Mensch, Maschine, Zeit. Fortschrittsbewusstsein und Kulturkritik im ersten Drittel des 20. Jahrhunderts," in *Jahrbuch des Historischen Kollegs*

felt compelled to make distinctions in this ordering of time. On the one hand, the "progress" from myth to *logos* no longer seemed a historical fact but became a normative program. The Enlightenment conquest of "Alexandria" by "Athens" had to be repeated again and again, since the "disenchantment of the world" was not a one-off event in which a "primitive" order was overcome by a higher one. Rather, the conceptual space developed by the order of science had to be repeatedly wrested from the causes of fear. Warburg therefore saw himself facing (as Horkheimer and Adorno might have said) a reversion from the Enlightenment to mythology.[288]

On the other hand, Warburg realized that the replacement of myth with scientific explanation had made the world a less frightening place, since "in principle" it was possible to "master all things by calculation."[289] But it was not self-evident that this "helps mankind to find a fitting answer to the problems of existence."[290] With the passing of time, those who had lived through the war, whether at the front or at the rear, were confronted with the massive problem of the meaning of existence, and especially of death. And they could not impose a meaning by rationally ordering events and interpreting the flow of time as progress.

It would seem to have become increasingly clear to Warburg that, as Claude Lévi-Strauss put it,

> we may be able to show that the same logical processes operate in myth and in science, and that man has always been thinking equally well; the improvement lies, not in an alleged progress of man's mind, but in the discovery of new areas to which it may apply its unchanged and unchanging powers.[291]

Nevertheless, Warburg continued to uphold the normative claim that man could be free only by wresting the space for thought from mythical order.

Gabriele D'Annunzio, the poet, "war prophet," and "war hero," has been presented as the polar opposite of Warburg. As we have seen, the one-sided feud between the two paradigmatic representatives of a vision of order was sparked off in May 1915, when D'Annunzio was agitating for Italy to join the war on the side of the Entente, and Warburg was doing his best to avert this. But the dichotomy between the two men and their respective conceptions of order becomes clearest if we compare their different codings of the airplane and the aviator. Ulrich Raulff's suspicion that Warburg's "choice of the aviator motif" may well have involved a "polemical tension (an alternative appropriation) with the hated aviator and poet

---

2003, Munich 2004, pp. 91–119; Wolfgang Hardtwig, "Die Krise des Geschichtsbewusstseins in Kaiserreich und Weimarer Republik und der Aufstieg des Nationalsozialismus," in *Jahrbuch des Historischen Kollegs 2001*, Munich 2002, pp. 47–75; and Otto Gerhard Oexle, "Krise des Historismus – Krise der Wirklichkeit. Eine Problemgeschichte der Moderne," in Oexle (ed.), *Krise des Historismus – Krise der Wirklichkeit. Wissenschaft, Kunst und Literatur 1880–1932* [Veröffentlichungendes Max-Planck-Instituts für Geschichte, vol. 228], Göttingen 2007, pp. 11–116.

[288] Theodor Adorno/Max Horkheimer, *Dialectic of Enlightenment* (orig. 1944), New York, 1972.
[289] Max Weber, "Science as a Vocation," in H. H. Gerth/C. Wright Mills (eds.), *From Max Weber. Essays in Sociology*, Boston, MA 1948, p. 139.
[290] "A Lecture on Serpent Ritual," *Journal of the Warburg Institute*, vol. 2, no. 4 (April 1939), p. 291.
[291] Claude Lévi-Strauss, "The Structure of Myth," in *Structural Anthropology*, New York, NY 1963, p. 230.

of Fascism D'Annunzio" has been substantiated in so far as their codings of aviation symbolism open up diametrically opposed horizons of meaning.[292] D'Annunzio enchanted the "technology serpent" disenchanted by Warburg. He tried to solve the mysterious "problems of existence" not through the penetration of rational thinking, but through the aesthetic transfiguration and sacralization of technology.

Not least by transferring the Icarus trope and Christian interpretive patterns to the airplane and its pilot, D'Annunzio played a considerable role in constructing the Italian "myth of the war experience" – a myth that, according to George L. Mosse, gave the war a new dimension of national and personal renewal.[293] The famous poet was able to embed the war into a horizon of meaning, and as D'Annunzio himself took part in the war his myth gained in both scope and effectiveness. The brave deeds of the ageing poet, as well as the rousing narrative that accompanied them, became Italy's most potent propaganda weapon. D'Annunzio sanctified technological warfare and the nation in whose name it was fought, interpreting death in battle as a heroic act of sacrifice that made national survival and renewal possible. In D'Annunzio's texts, the aviator turns into a superhuman new man but also a *miles Christi*; he acquires the features of a savior, while his aircraft morphs into the cross of redemption. These "theologemes of salvation," which D'Annunzio transferred to the aviator, formed the basic material for the Fascist political religion and its underlying palingenetic myth.[294]

By means of "mythological explanation," D'Annunzio sought to confer "suprahistorical" meaning on the war, and thereby to justify it. In keeping with the definition given earlier, we may say that he created a myth by overcoming chaos through an order based on the sacred nation.[295] Reference to an "intersubjectively undisputed" holy of holies justified the war in general, and killing and dying in particular. Myth served both to ward off the "absolutism of reality," chaos and contingency and to authenticate the community and its values.

D'Annunzio – Michael Leeden calls him "the first Duce" – did not anticipate only the political "style" of fascism.[296] His "lyrical order" formed the basis for the fascist mythical order, which was sustained by longing for a "sacred time."[297] The "profane" time that liberals construed as progress was thought of by fascists as

---

[292] Ulrich Raulff, "Der aufhaltsame Aufstieg einer Idee. 'Idea vincit': Warburg, Stresemann und die Briefmarke," in Wolfgang Kemp et al. (eds), *Vorträge aus dem Warburg-Haus*, vol. 6, Berlin 2002, pp. 125–62, 147

[293] See George L. Mosse, *Fallen Soldiers: Reshaping the Memory of the World Wars*, New York, 1990, p. 159.

[294] On "theologemes of salvation," see Hans Ulrich Gumbrecht, "I redentori della vittoria. Über Fiumes Ort in der Genealogie des Faschismus," in Hans Ulrich Gumbrecht/Friedrich Kittler/Bernhard Siegert (eds.), *Der Dichter als Kommandant. D'Annunzio erobert Fiume*, Munich 1996, pp. 83–115.

[295] Here and on the following, see the discussion of myth in the Introduction; cf. Manfred Frank, *Gott im Exil. Vorlesungen uber die Neue Mythologie, Frankfurt/Main 1988*, p. 16.

[296] Michael A. Leeden, *The First Duce*, Baltimore, MD 1977. Cf. Emilio Gentile, *The Origins of Fascist Ideology, 1918–1925*, New York, NY 2005 [repr.].

[297] On D'Annunzio's "lyrical order," as distinct from Mussolini's "political order," see Gentile, *The Origins of Fascist Ideology*, pp. 134–54.

decadence.[298] And they sought to transcend that profane time – defined by Eliade as "ordinary temporal duration, in which acts without religious meaning have their setting" – and to replace it with sacred time.[299] The sacralized nation at the center of the fascist mythical order was the bearer of that time, which is "by its very nature reversible" and "indefinitely repeatable."[300] The association of death in battle with the sacralized nation helped to conquer profane, linear, finite time, since sacrifice for the nation allowed the individual to participate in its eternal time and to overcome death in its eternal life.

Thus, it was not only because of intoxication, the breakdown of order, and the reversal of values that war turned into the "festival" extolled by the Futurist Marinetti.[301] In the mythical horizon of meaning into which it was integrated, the war became a "festival" in the sense that it offered a means of renewing time and re-entering the mythical, sacred time of the nation.[302] Killing and dying became rituals on the altar of the fatherland, where the war victim was the *sacrificium*. This analogy between war and religious festival will be further illustrated below in relation to the German myth of the "spirit of 1914."[303] Eliade – who was himself not immune from the temptations of fascism and, as professor of religious studies at Chicago, pleaded for the spiritual *renewal* of modern man[304] – described as follows the nexus of festival, elimination of profane time and renewal of sacred time:

> The participants in the festival become contemporaries of the mythical event. On other words, they emerge from their historical time – that is, from the time constituted by the sum total of profane personal and interpersonal events – and recover primordial time, which is always the same, which belongs to eternity. Religious man periodically finds his way into mythical and sacred time,

[298] See Roger Griffin, *Modernism and Fascism. The Sense of a Beginning Under Mussolini and Hitler*, Basingstoke, UK 2007, p. 221.

[299] Mircea Eliade, *The Sacred and the Profane*, p. 68.

[300] Ibid., pp. 68–9. Cf. Peter L. Berger, *The Sacred Canopy: Elements of a Sociological Theory of Religion*, Garden City, NY 1967, pp. 35–6.

[301] On Marinetti's concept of the *guerra festa*, see Chapter III.1 below. See also especially Mario Isnenghi, *Il mito della grande Guerra*, Bologna 2002, pp. 179–83; and George L. Mosse, "The Political Culture of Italian Futurism. A General Perspective," *Journal of Contemporary History* 25/2–3, 2000, pp. 253–68.

[302] See Eliade, *The Sacred and the Profane*, pp. 68–9. For more on the distinction between profane and sacred time, see his *Patterns in Comparative Religion*, Chapter 11: "Sacred Time and the Myth of Eternal Renewal," Lincoln, NE 1996.

[303] On the myth that the outbreak of war welded the nation together into a community, see Jeffrey Verhey, *The Spirit of 1914: Militarism, Myth, and Mobilization in Germany*, Cambridge, 2000. Cf. Stefan Bruendel, *Volksgemeinschaft oder Volksstaat. Die "Ideen von 1914" und die Neuordnung Deutschlands im Ersten Weltkrieg*, Berlin 2003; Peter Fritzsche, *Germans into Nazis*, Cambridge, MA 1998; and Michael Wildt, *Hitler's Volksgemeinschaft and the Dynamics of Racial Exclusion. Violence against Jews in Provincial Germany, 1919–1939*, New York, NY 2012, pp. 26–68.

[304] See Douglas Allen, *Myth and Religion in Mircea Eliade*, New York, NY 2002, pp. 291–331; and Robert S. Ellwood, *The Politics of Myth. A Study of C. G. Jung, Mircea Eliade and Joseph Campbell*, Albany, NY 1999, pp. 79–126.

reenters the "time of origin," the time that "floweth not" because it does not participate in profane temporal duration, because it is composed of an eternal present, which is indefinitely recoverable.[305]

The war had shaken the volatile order of reality. It confronted people with many "marginal situations" that they thought "progress" had banished in the "golden age of security."[306] Death – which, according to Peter L. Berger, "radically challenges *all* socially objectivated definitions of reality"[307] – became a ubiquitous presence again and required an overarching nomos to order it. Like Warburg's attempt to ward off chaos through an order anchored in "reason," D'Annunzio's religious or quasi-religious construction of reality and nomos served to assuage the widespread longing for order radicalized by the war itself. The distinction between the two orders was not phylogenetic: the fascist order was not more "primitive," "archaic" or "reactionary" than the liberal one. But the mythical order grounded on the sacral, absolutized nation would prove to be altogether more violent.

---

[305] Eliade, *The Sacred and the Profane*, p. 88. Cf. Eliade, *Patterns in Comparative Religion*, Chapter 11: "Sacred Time and the Myth of Eternal Renewal."

[306] On the "golden age of security," see Stefan Zweig, *The World of Yesterday. An Autobiography*, New York 1943, pp. 1ff. On p. 3 he writes: "Earlier eras, with their wars, famines, and revolts, were deprecated as times when mankind was still immature and unenlightened. But now it was merely a matter of decades until the last vestige of evil and violence would finally be conquered."

[307] Berger, *The Sacred Canopy*, p. 43.

Part II
Fractured Order

1

Don Quixote of the Air

> In fact, now that he had utterly wrecked his reason he fell
> into the strangest fancy that ever a madman had in the whole
> world. He thought it fit and proper, both in order to increase his
> renown and to serve the state, to turn knight errant and travel
> through the world with horse and armor in search of adventures,
> following in every way the practice of the knights errant he had
> read of, redressing all manner of wrongs, and exposing himself
> to chances and dangers, by the overcoming of which he might
> win eternal honour and renown.[1]

Such is Cervantes' portrait of the impoverished nobleman who, as Don Quixote,
rides forth into the world on his old nag Rocinante. From countless books of chivalry, the "knight of sad countenance" takes the paradigms and ways of thinking
that he will use for his encounter with reality.

> He filled his mind with all he read in them, with enchantments, quarrels, battles,
> challenges, wounds, wooings, loves, torments and other impossible nonsense;
> and so deeply did he steep his imagination in the belief that all the fanciful stuff
> he read was true, that to his mind no history in the world was more authentic.[2]

To the "madman" Don Quixote, his books are a model for the reading of reality.
Through them, not only does reality become book, but book becomes reality. As
Foucault puts it in *The Order of Things*, the book is Don Quixote's task:

> He is constantly obliged to consult it in order to know what to do or say, and
> what signs he should give himself and others in order to show that he really is
> of the same nature as the text from which he springs. The chivalric romances
> have provided once and for all a written prescription for his adventures. And

---

[1] Miguel de Cervantes Saavedra, *The Adventures of Don Quixote* [Penguin edn], trans. by
J.M. Cohen, Harmondsworth 1950, Part One, Chapter 1, p. 33.
[2] Ibid., p. 32.

every episode, every decision, every exploit will be yet another sign that Don Quixote is a true likeness of all the signs that he has traced from his book. [...] Flocks, serving girls, and inns become once more the language of books to the imperceptible degree to which they resemble castles, ladies, and armies.[3]

Don Quixote, which is often described as the archetypal novel of "modernity,"[4] is an expression of the profound change that Spanish society underwent between the sixteenth and seventeenth centuries. With his "knight of sad countenance," Cervantes created a symbol for reactions to the threat of disorientation in times of radical upheaval and rapid transformation. The tragicomedy was that his hero borrowed his paradigms from books that idealized a bygone world.

In the context of aviation discourse and the development of a mythical modernity, *Don Quixote* may serve as a pointer to the reflexivity of "literature" and "reality."[5] The "effect" of the texts analyzed, which is neither quantifiable nor strictly "verifiable," will be seen precisely in the circular reflexive relationship between literature or text and reality. Texts – that is, complex sign-systems with a communicative function – are ways of interpreting and ordering reality that also enable the reader to construct ordered realities.[6] The aesthetic dimension of fascist myth, its tangibility to the senses, was based not least on the fact that "texts" spread the mythical order within society and molded its perception and shaping of reality.[7] Texts diffused paradigms and concepts of order that helped to structure reality and to handle crises of orientation.

In his *The Great War and Modern Memory*, a study of British war experiences on the Western Front first published more than 30 years ago, the literary theorist Paul Fussell pointed like Cervantes to the circular reflexive relationship between fiction and reality. The task he set himself was to examine the "simultaneous and reciprocal" process whereby "life feeds materials to literature, while literature returns the favor by conferring forms upon life."[8] If he had given his book a subtitle, he tells us, it would have been "An Inquiry into the Curious Literariness of Real Life."[9]

This "curious literariness of real life" can be "verified" only in first-person documents, and even then seldom as clearly as in Raleigh Trevelyan's reminiscences

---

[3] Michel Foucault, *The Order of Things: An Archeology of the Human Sciences*, New York 1994, pp. 46–7.

[4] See Raymond Geuss, "The Actual and Another Modernity. Order and Imagination in *Don Quixote*," in Geuss, *Politics and the Imagination*, Princeton, NJ 2010, pp. 41–60.

[5] In what follows, it is essential to maintain the separation between literature and real life, or between fictional and nonfictional texts – a "reality difference" apparent, for example, in the contrast between reportage and fictional narrative. The assumption here, of course, is that even in reportage we are dealing with a discourse-dependent construct, whose reference to "physical" events is determined by text-structuring rules different from those to be found in a novel. Both kinds of text may be "true" in their way, but only the nonfictional must be "factually accurate."

[6] The term "text" is here used in a broad sense: it includes not only the written word but also speeches, images, films and much else besides.

[7] The aesthetic dimension of myth will be discussed at greater length in Chapter III.2.

[8] Paul Fussell, *The Great War and Modern Memory*, New York, 2000 [repr.], p. ix.

[9] Ibid.

of the Battle of Anzio: "We were jammed head to toe, completely immobile, with volleys of tracer like whiplashes a matter of inches overhead. It was a complete *All Quiet on the Western Front* film set once more."[10] All the same, communicative discourse rests upon this mostly unconscious adoption of perceptual and interpretive models from texts that provide participants in the discourse with common (language-)pictures or ideas.

The "curious literariness of real life" forms the backdrop to this chapter. On the one hand, texts of aviation discourse refer us to the reality from which they derived, and which they seek to organize. On the other hand, the effect of the texts lies in their unquantifiable or "unverifiable," yet clearly present, shaping of the reader's models of perception and interpretation. The migration of certain tropes between different discourses and texts can only rarely be traced, since the paths they take are neither obvious nor clearly marked. But this study will nonetheless assume that books have an "effect" and that the "curious literariness of real life" is quite common.[11]

The next section will present the initial embedding of aviation texts in the contexts of mass entertainment, sport and international competition, and identify the media in which aviator images spread among the population. By means of the Brescia air show and D'Annunzio's novel of 1910 *Forse che si forse che no* [Maybe Yes, Maybe No], we shall see just how porous were the boundaries between literature and reality. For in his novel, which drew on experiences at Brescia, D'Annunzio establishes one of the central themes of aviation discourse: the trope of the superman. He already puts forward an interpretation of the flier as the heroic new man who has left the old life behind. The flier has triumphed over gravity, and his upward striving has become a highly versatile sign. The aviator and his flying machine as symbols of ascent might refer, in the general context of progress, to man's scientific-technological mastery of nature; or they might be deployed, as in Warburg, within an Enlightenment narrative of reason and its conquest of magic, superstition and irrationality. D'Annunzio, however, stylizes the aviator – in the context of the air show – as a victor over self, others and the masses, over the bourgeois lifeworld and human nature as such. The superman image found a crystallization point here, so that D'Annunzio became a herald transferring this elevation of the flier to the context of war. But this transfer

---

[10] Ibid., pp. 221–22.

[11] In a polemical article for the *Frankfurter Allgemeine Zeitung*, Jörg Baberowski wrote: "Description of causal relations between events is always associated with the idea that events have causes. But how can we know that events have causes? [...] If we trace the outbreak of the Russian Revolution back to miserable living conditions, we claim that miserable conditions cause revolutions. But those who lived in such conditions had no idea that their life was a cause of a future event. Only historians had that knowledge – and only historians used to associating poverty with revolutions. We might say with as much reason that bad temper or bad weather causes revolutions, because in the years before the outbreak of the Russian Revolution many people were bad-tempered and it always rained in the fall." See Jörg Baberowski, "Über die schöne Schwierigkeit, Geschichte zu schreiben," *Frankfurter Allgemeine Zeitung*, July 29, 2009, p. N3.

did more than overstep the genre boundaries between novel and reportage, or between speech and its claims to authenticity; it conjured away the boundaries between reality and fiction, war and game. The "curious literariness of real life" brought about actual consequences.

This chapter will also look at the dissemination of myth in society through the popular media. We shall see how events and cultural products may be read as originators and bearers of a newly developing narrative of relevance to society, whose effectiveness lay not least in the normative sphere: of myth, in other words.[12] And we shall see how popular culture diffused and intensified certain tropes or ideas, inscribing them into the discourse. A central focus will be the reception and production of images of the aviator and aerial warfare in the popular media – in particular, their development in Germany through the literary mobilization that took place during the First World War. It will be shown how the paratexts of such books made possible the transition from paradigms originating in sports and mass culture or fictional high-cultural contexts to paradigms rooted in war.[13] As in *Don Quixote*, interpretations were transferred to a reality that had little in common with the reality of the books.

Despite the gulf separating the First World War from conflicts in the nineteenth century, paradigms from earlier wars were transferred to mechanized mass warfare. Indeed, it would appear that the existential threat posed by the war and the resulting break with the old order actually "forced" this continuity. An increasingly disoriented society was on the lookout for precedents and models, and it found them, again like Don Quixote, partly in images of heroes still in circulation. The already unclear, porous boundaries between "fiction" and "reality" became more blurred and more plentiful in crossings.

Aerial acrobats or supermen – the Brescia air show

In September 1909, Franz Kafka went on a trip to Lake Garda with the Brod brothers, Max and Otto.[14] On September 10, after a few days in Riva, the tourists from Prague continued to Brescia, the Lombard city on the edge of the Po valley. They had learned of an air show there from a report in the *Sentinella Bresciana*:

> In Brescia we have a concourse of people such as never before, not even at the time of the great motor-car races; visitors from Venetia, Liguria, Piedmont, Tuscany, Rome, indeed even from as far as Naples; distinguished persons from

---

[12] See the definition of myth proposed in the Introduction, as well as Manfred Frank, *GottimExil. Vorlesungen über die Neue Mythologie*, Frankfurt/Main 1988, pp. 15ff.

[13] On the concept of a paratext, see Gérard Genette, *Paratexts. Thresholds of Interpretation*, New York, NY 1997. The term will be explained further in the section below entitled "'The War for the Price of One Mark': the promise of authenticity".

[14] See Peter Demetz, *The Air Show at Brescia, 1909*, New York, NY 2002; Felix Philipp Ingold, *Literatur und Aviatik. Europäische Flugdichtung 1909–1927*, Frankfurt/Main 1980, esp. pp. 19–27; and Robert Wohl, *A Passion for Wings. Aviation and the Western Imagination 1908–1918*, New Haven, CT 1994, pp. 97–123. There is an illustrated account of this vacation in Hartmut Binder, *Mit Kafka in den Süden. Eine historische Bilderreise in die Schweiz und zu den oberitalienischen Seen*, Prague 2007, esp. pp. 39–84.

France, England, America; all are jostling in our squares, in our hotels, in every spare corner of our private houses; all the prices are rising splendidly; the means of transport are inadequate to bring the crowds to the *circuito aereo*; the restaurants on the airfield could serve two thousand people admirably, but so many thousands are bound to defeat them; troops would be needed to protect the buffets; in the cheap areas of the field there are 50,000 spectators standing all day long.[15]

The threesome board a "miserable train," which struggles amid this mass of people to cover the route to the *circuito*, the airfield at Montichiari. There – Kafka wrote later that month in the *Bohemia* magazine – they face the following picture: "An artificial wasteland has been created here in an almost tropical region, and the high nobility of Italy, glittering ladies from Paris and all the other thousands are here assembled, to look for hours on end with narrowed eyes into this sunny waste."[16] But Kafka too is fascinated by the bustle in the aerodrome and looks expectantly skyward. "Thank heavens, no one is flying yet!"[17] The three friends have been hoping to take a look behind the scenes, and perhaps to catch sight of the protagonists: "We come past the hangars, which stand there with their curtains drawn like the closed stages of traveling players. On their pediments stand the names of the aviators whose machines they conceal, and above them the flags of their homelands."[18]

A few years later, the hangars like the "closed stages of travelling players" would become known as the "flying circus," and the air shows already demonstrated numerous analogies with those itinerant spectacles. For with the appearance of bicycle and automobile races, and finally of the air show, the popular culture of mass entertainment had entered the machine age.[19] The fliers assembled in Brescia, as a few weeks earlier in Rheims, were up-to-date performers, acrobats and sportsmen who offered the masses a rare spectacle. At first, though, this seems to have aroused little enthusiasm in Kafka. There was nothing on the airfield "of the kind that lends variety to sports fields otherwise."[20]

What was so exciting that it drew people there by the tens of thousands? First there was the event itself: for many, the parade of distinguished figures, the influx of "high society," was attraction enough.[21] But the central focus was the fliers and their machines. One year earlier, the Wright brothers had begun to perform

---

[15] Quoted in Franz Kafka, "The Aeroplanes at Brescia," in Kafka, *The Transformation and Other Stories*, trans. by Malcolm Pasley, New York, NY 1992, p. 1.

[16] Kafka, "The Aeroplanes at Brescia," p. 5.

[17] Ibid., p. 3.

[18] Ibid., p. 5.

[19] On the changing forms of popular culture, see Kaspar Maase, *Grenzenloses Vergnügen. Der Aufstieg der Massenkultur 1850–1970*, Frankfurt/Main 1997.

[20] Kafka, "The Aeroplanes at Brescia," p. 5.

[21] On p. 7 we read: "People point out to one another the Princess Laetitia Savoia Bonaparte, the Princess Borghese, an elderly lady whose face has the color of dark yellow grapes. [...] Looking out over the rail of the stand is the strong face of Puccini, with a nose that might be called the nose of a drinker."

exhibition flights in the USA and France, and as recently as July 25 Louis Blériot had made the first flight across the English Channel. The "dream of flying" had come true; the airplane, the tangible symbol of progress, had made it possible to see the triumph of technology over nature with one's own eyes.[22] And, of course, everyone wanted a glimpse of the men who controlled the machines:

> And *Blériot?* we ask. Blériot, of whom we have been thinking all the time, where is Blériot? [...] And there on the field quite close, just fancy, stands a little aeroplane, with real yellow paint, that is being got ready to fly. [...] Leaning against one of the two wings of the machine stands, instantly recognizable, Blériot, and with his head set firmly on his shoulders he is keeping a close eye on his mechanics as they work on the engine.[23]

People gazed in wonder at the man who just weeks before, with his daredevil feat of crossing 38 kilometers of foggy sky, had put Britain's island status in a new perspective. Blériot was already a well-known hero for the crowds in Brescia. Yet there is a note of skepticism in Kafka's report:

> A long pause and Blériot is in the air, one can see the upper part of his body held very straight above the wings, his legs are hidden deep down as part of the machinery. [...] Everyone gazes up at him enraptured, in no one's heart is there room for anyone else. He flies a small circuit and shows himself then almost vertically above us. And all of us can see, craning our necks, how the monoplane rocks, how it is seized hold of by Blériot and even climbs. What is happening here? Over our heads, twenty meters above the earth, is a man entangled in a wooden frame, defending himself against an invisible danger that he has freely taken on. But we stand down below, quite left behind and insignificant, and we watch this man.[24]

The art theorist Felix Philipp Ingold sees in Kafka's account a "manifestation of the everyday mythology of the time," no doubt in Barthes's sense of mythology, and an anticipation of coming sports spectacles for the masses.[25] And it is true that it contains all the elements that would define such events in the future: large crowds, popular icons they look up to, competition among nations and staging for the media.[26]

The media – which at the time basically meant daily newspapers, periodicals and illustrated magazines – did more than just report the events; they called them into being in the first place. Alongside the ubiquitous aeroclubs and societies organized

---

[22] On man's "dream of flying," see Wolfgang Behringer/Constance Ott-Koptschalijski, *Der Traum vom Fliegen. Zwischen Mythos und Technik*, Frankfurt/Main 1991.

[23] Kafka, "The Aeroplanes at Brescia," pp. 5–6.

[24] Ibid., pp. 7–8.

[25] Ingold, *Literatur und Aviatik*, p. 23.

[26] See Maase, *Grenzenloses Vergnügen*, pp. 94–5.

on a national basis, as well as the newly developing aviation "industry," it was the press barons who took the initiative in organizing flying competitions.[27] Lord Northcliffe's *Daily Mail*, for example, put up a prize of £1,000 for the first aviator to cross the Channel, and the publisher of the *Corriere della Sera*, Luigi Albertini, was a member of the organizing committee for the Brescia air show.[28] Concerned as ever to boost sales and build a loyal readership, newspaper publishers made it a major priority to offer exciting and, if possible, exclusive stories related to aviation; these helped to satisfy the public thirst for adventure and danger, centered on the struggle of "real men" and powerful nations against the forces of nature. Since air travel involved accidents and mishaps as well as triumphs, those who were skeptical about progress and critical of civilization found themselves taking as much interest as did techno-optimists and utopian dreamers. In any case, more than 60 international dailies and periodicals carried reports on the Brescia show.[29]

The commercial interests of press barons were not, however, the only factor behind the numerous air shows and competitions held between 1909 and 1914. A host of up-and-coming little flight companies, together with their suppliers and related businesses, also eagerly grasped the opportunity to make their products and services better known to the public. Take, for instance, the prospectus of the Aviation Week organized in Berlin in late May 1912 by the Flug- und Sport-Platz Berlin-Johannisthal GmbH; it contains publicity not only for the airplane producers Luft-Verkehrs-Gesellschaft AG and Rumpler GmbH, but also for the Berlin-based Deuta Werke (rotation and speed indicators) and Continental (rubber tires). The menswear retailer S. Adam vaunts its clothing for "fliers," "sportsmen" and "boys," as well as its "fashionable items for gentlemen."[30] Thus, at this event held "under the patronage" of the Berlin Airship Travel Association, the Imperial Aero Club, the Imperial Automobile Club and the Reich Aviation Association, diverse protagonists pursued primarily economic objectives. But the national interest also played a role: the connection between "merchants and heroes" was much closer than Werner Sombart would claim a few years later.[31]

At the latest by the time of Blériot's Channel crossing, governments and military leaders were also waking up to aviation. Admittedly the main focus of German "air policy" was the Zeppelin,[32] but along with the "Prize of His Majesty the King and Emperor" (a trophy adorned with the Kaiser's face), the Royal War Ministry offered an award of 30,000 marks to the Johannisthal aviation week.[33] The interlinking of commercial, national and sporting objectives explains the program and rules of the German Test Flight on the Upper Rhine – an event that took place in May 1911. The "purpose of the undertaking," held "under the patronage

---

[27] Here and in the following, see Wohl, *A Passion for Wings*, pp. 99ff.

[28] Demetz, *The Air Show of Brescia, 1909*, p. 45.

[29] Ibid., p. 58.

[30] ADM, LR 000728–01, Prospekt Flugwoche Berlin-Johannisthal 1912.

[31] Werner Sombart, *Händler und Helden. Patriotische Besinnungen*, Munich 1915.

[32] See Fritzsche, *A Nation of Fliers*, pp. 9–58.

[33] ADM, LR 000728–01, Prospekt Flugwoche Berlin-Johannisthal 1912, pp. 26–7.

of His Royal Highness Prince Heinrich of Prussia" and organized by the Kartell der südwestdeutschen Luftschiffer-Vereine, was multiple:

> But the main emphasis now is not on speed records or the reaching of great heights, but on the need to keep increasing the *reliability* of machine and flier. Germany is currently far behind other countries, especially France, in its construction of flying machines, its aeronautic technology, and its numbers of aviators. We can close this lead. [...] The event should result in the cultivation of an aeroplane that can be kept in service for several days in succession; fliers should become practiced [...] in the difficult art of orientation. [...] The flight should be conducted through one of the most beautiful regions of Germany, and one that is climatically favourable to such events; it should take in a wide arc of as many cities as possible, promoting interest among the widest circles and a willingness to make sacrifices for the development of our German aeronautics.[34]

The commercial concerns of media, organizers, aviation industry and related businesses were thus thoroughly consistent with the military and national interest, as well as with the ambition to set new records. All sides profited from the further development of "German aeronautics." And this mutual gain by no means conflicted with the spirit of sporting competition that inspired the events. The rivalry among nations that would soon be acted out on the battlefield expressed itself in 1909 not only at the level of the economy and the race for colonies and spheres of influence, but also in shows of strength in the sporting arena.

Ten years after the war, when the Brescia air show was already two decades in the past and Lindbergh and others were flying across the Atlantic, Kurt Tucholsky published under his pseudonym Theobald Tiger a poem entitled *Meine Flieger – deine Flieger* [My fliers – your fliers]. It gave expression to the nexus of national rivalry, the media, aeronautical competition and aviation industry that has just been mentioned. Of course, the phenomenon was more marked by 1928, but it was already present at Brescia and the poem is therefore worth quoting in full:

> Our fliers have performed over the ocean – / German energy! German energy! / Our fliers had a terrible night – / there's never been one so bad! / Here their biography! / Cock-a-doodle-do – ! / And we yelled so much it rang through the streets: / "Our fliers are number one in the world!" / Your fliers are perfectly OK, / just fine for the small print on page two. / Our fliers are the country's pride! / Vive la France! Quelle rumeur! / Our fliers are top of their class – / Réception et la Légion d'Honneur! / And behind them stand industries, / and they smirk in Paris as in Berlin ... / In the end your fliers are just runners-up. / Our fliers are flying to Mexico today! / *God's own country – our America !* / Our fliers keep up standards – / *For the colonel: / Hip, Hip, Hurra!* / Every paper told us: / if someone there risks a flight, / even the dullest reader of all / rises to the highest firmament. / "Because you, fellow-countryman, are a chip off the same block, /

---

[34] ADM, LR 000728–01, Prospekt Deutscher Zuverlässigkeitsflug am Oberrhein 1911, pp. 3–4.

I too am a hero who hoots as he dances." / Tell me what you are proud of, / and I'll tell you what you can do for me. / Our fliers! Our fliers! / They are victors! They are victors! / Your fliers, no comparison, / can't hold a candle to ours. / Any group of regulars / in any bar in the world / need their little bit of pleasure – / can't do without movies, church, and patriotism.[35]

Tucholsky's satirical gaze should not, however, obscure the deadly seriousness of the rivalries or lead us to underestimate the financial muscle and power over people's minds that media corporations already possessed, and that the aviation industry was beginning to concentrate, in the interwar period.[36] Already before the war, various economic interests stood behind the growing popularity of aviation, but that does not mean there was no genuine enthusiasm for it in society. The point, rather, is that such enthusiasm was directed along certain channels, and that particular interests made it into a potent force.

As Peter Fritzsche has shown in relation to the Zeppelin cult in Germany, a combination of nationalism and technological fervor led not only to the identification of broad strata of the population with the nation, but also to the survival of the Zeppelin itself.[37] On Thursday, August 4, 1908, the LZ 4 airship set off from Friedrichshafen over Lake Constance, with the aim of fulfilling the conditions set by the Reich war ministry for purchases of the model. After an emergency landing in Oppenheim, the "Luftlohengrin," as it was called, pressed on the next day, but was forced to make another break in Echterdingen, near Stuttgart.[38] There a gust of wind carried the airship aloft, whereupon it tilted forward, fell to the ground and burst into flames. Since the construction had not satisfied the war ministry conditions, Count Zeppelin was at risk of running short of funds. The *Schwäbische Kronik* reported: "The airship is lost, but the idea lives on! And its Count Zeppelin. The German people will build a new ship for him. Of that we are sure!"[39] And indeed, by the end of the day after the disaster, 5,359 marks had been donated for the Count and his idea. Six weeks later, the total had risen to five million marks, allowing the foundation of the Zeppelin airship construction company in Friedrichshafen.[40]

---

[35] Kurt Tucholsky, "Meine Flieger – deine Flieger," *Weltbühne* 24/18, May 1, 1928, I, p. 686; quoted (and translated) from *Gesamtausgabe*, vol. 10, *Texte 1928*, Reinbek bei Hamburg 2001, pp. 191–92.

[36] On the aviation industry in the interwar period, see Lutz Budraß, *Flugzeugindustrie und Luftrüstung in Deutschland 1918–1945*, Düsseldorf 1998.

[37] Here and on what follows, see Fritzsche, *A Nation of Fliers*, pp. 9–22. On the relationship between technology and nation in Germany and Britain, see Bernhard Rieger, *Technology and the Culture of Modernity in Britain and Germany, 1890–1945*, Cambridge 2005. On the Zeppelin craze in particular, see also Helmut Reinicke, "Zeppelin, Karl May und die deutschen Auffahrten nach Dschinnistan," in Dieter R. Bauer/Wolfgang Behringer (eds.), *Fliegen und Schweben. Annäherung an eine menschliche Sensation*, Munich 1997, pp. 317–43.

[38] On the use of the name "Luftlohengrin" for the Zeppelin LZ 4, see Fritzsche, *A Nation of Fliers*, p. 29.

[39] Anonymous, "Andere Berichte," *Schwäbische Kronik*, August 6, 1908, No. 362.

[40] On the history of the Zeppelins, see Guillaume de Syon, *Zeppelin! Germany and the Airship, 1900–1939*, Baltimore, MD 2002.

The "Miracle of Echterdingen," as the fundraising operation came to be known, further illustrates the symbiotic alliance among the budding aviation industry, the media, the event organizers, the "national interest" represented by politicians and military leaders and the technologically enthusiastic as well as sensationalist masses.[41] The "grassroots nationalism" expressed in donations for the "flying cigar" was also a factor uniting different social strata at the air shows. Through a mechanism of patriotic identification, the victory of an aviator made his compatriots in the audience feel that they too had conquered the air. Even the mere onlookers were able to bathe in the "modernity" of the whole nation that the pilot's triumph supposedly manifested. As Tucholsky put it: "Because you, fellow-countryman, are a chip off the same block, / I too am a hero who hoots as he dances!"

The trope of the Superman, established in the wake of the progress euphoria, at first displayed an affinity with its typical narrative, but later, after the war, was directed against it. In an aviation context, the theme of progress mainly involved the conquest of nature or (in Warburg's case) of spontaneous, "superstitious" and mythical thought. In the fascist mythology that took shape after the war, however, the theme of conquest turned against the world in which it had originated. What had once been seen as progress came to be interpreted as decay and disintegration, the aim now being to conquer the "decadent" world that had once drawn legitimacy from its overcoming of the Ancien Régime.

This metamorphosis and recoding of the trope of progress or ascent, as well as the divergent instrumentalizations of the aviator narrative, were possible because the aviator qua cipher exhibited a basic structure that was attractive to a wide range of forces and recurred in a number of mutually exclusive contexts. The investigation of aviation discourse reveals a metalevel in which these contradictions and intersections first become visible. At this metalevel, the aviator and his airplane themselves turn out to be polyvalent signs, whose position has shifted in the network of meanings. Hence, the perception and articulation of one and the same event can simultaneously resemble and diverge from each other. When Kafka gazed in wonder at "a man entangled in a wooden frame," who was "defending himself against an invisible danger he ha[d] freely taken on," and when he felt surprise at the crowds of people "standing insignificant down below" and looking skyward,[42] Gabriele D'Annunzio, also present at Brescia, was beginning to establish a different way of seeing things and to form an alternative network among concepts.

One major reason for this was that – as Kafka said – D'Annunzio "danced about in an apparently bashful way in front of Count Oldofredi" (an important member of the organizing committee).[43] The *divino poeta* was thus allowed to fly in an airplane together with Curtis and then with Calderara. Luigi Barzini reported in the *Corriere della Sera*: "He sat on a skimpy board, feet resting on a bamboo brace,

---

[41] Fritzsche, *A Nation of Fliers*, p. 15.
[42] Kafka, The Aeroplanes at Brescia," p. 8.
[43] Ibid., p. 7.

hands on the vertical axes of the frame, in an instinctual and justifiable search for support, locked in a cage of criss-crossing steel wire. D'Annunzio was smiling, lost in expectation." (Figure 1.1)[44]

People waited excitedly to hear D'Annunzio's impressions of the flight. The questions therefore began as soon as he returned to earth:

> He glowed with enthusiasm. [...] It's really divine! – he said at once. Divine and indescribable. The moment you leave earth is infinitely charming. At that moment you feel the birth of a new feeling. My heart is still filled with it. [...] It is a new need, a new passion.

A new passion that D'Annunzio again indulged when he took off shortly afterwards with Calderara. "As you cast off from the earth, it seems as if you are leaving matter behind. You feel light, ethereal, transfigured."

The Poet, who, despite his statements to the contrary, was by no means lost for words at the time, would later incorporate the experience into a novel first published in 1910, *Forse che si forse che no*:

> It [the machine] soon gained height. [...] Then it swung into the wind, with oscillating movements, like a red kite when it spirals upward, like an acrobat on the high wire. [...] [The machine] sped just like an arrow down Ghedi's green, poplar-lined avenue, [...] floated in the pure white reflection of the clouds, as beautiful as the image of the sun god of Edfu, all wings like the emblem above the Egyptian temples.[45]

D'Annunzio certainly did not stop at his earlier description of the experience as "divine!" In *Forse che si forse che no*, which appeared the same year in German translation, the intoxication with speed is foregrounded alongside an almost futuristic account of a mechanical-organic hybrid of man and machine:

> Giulio Cambiaso had never felt as much as now the concordance between the machine and his body, between his will and this tamed force, between his instinctual movements and that mechanical motion. From the blades of the propeller to the edge of the rudder, he felt the whole floating structure as an organic prolongation and enlargement of his own being. When he bent over the control to maneuver against a gust of wind, when he leaned into a circling movement so that the pressure of his hip effected the torsion of the wing tips,

---

[44] Luigi Barzini, "L'ebbrezza del volo," quoted (and translated) from Laredo de Mendoza, *Gabriele D'Annunzio*, pp. 44–7; the next two quotations are taken from the same source.

[45] Gabriele D'Annunzio, *Forse che si forse che no*, Milan 1910, p. 98. Aviation technology made its world literary debut in D'Annunzio's novel. In a way, it was the founding act of European air poetry – D'Annunzio boasted that it introduced two thousand neologisms – and has therefore often been written about. Both Ingold and Wohl discuss it in detail. See Ingold, *Literatur und Aviatik*, pp. 28–49; and Wohl, *A Passion for Wings*, pp. 114–22. See also Woodhouse, *Gabriele D'Annunzio. Defiant Archangel*, pp. 242–49.

Figure 1.1 D'Annunzio blissfully entangled in Curtiss's flying *Gestell*
Source: Reproduced with kind permission of the Archivi del Vittoriale, Archivo Iconografico.

> when he unfailingly restored equilibrium on hauling to the wind and peri-
> odically shifted the flight line, he had the sensation that his bodily tissue was
> interwoven with his two white trapezoid wings.[46]

D'Annunzio continues to describe the flight of the two bosom friends, Cambiaso
and Paolo Tarsis. Their friendship "was sealed on the deck of a battleship" and
"consolidated in the cavernous depths of submarines."[47] In both, "the pride of the
small aristocracy forming among the random masses of eager fliers had already
begun to mount" when they gathered for the competition in Montichiari.[48]
There, on the Brescia airfield, "the crowd flocked to the show as to a transfiguration of the human species."[49]

The novel already marshals important tropes of the aerial warfare narrative that
would enter into the fascist myth: from technical dexterity and fusion with the
machine, through the intoxication with speed, to virile camaraderie, misogyny,
aristocratic conduct and contempt for the masses. In addition, the figures of

---

[46] D'Annunzio, *Forse che si forse che no*, pp. 98–9.

[47] Ibid., pp. 71–2.

[48] Ibid., p. 79.

[49] Ibid., p. 69.

Cambiaso and Tarsis (another alter ego of D'Annunzio) disseminate the writer's vision of the Nietzschean superman.

> The new machine seemed to exalt man above his fate, endowing him with not only a new domain but also a sixth sense. As the darting structure of iron and fire had devoured time and space, the Daedalean device triumphed over both and over gravity itself. Nature lowered one after another of its barriers.[50]

With the airplane, man celebrated a triumph over space, time, and nature:

> The soul of mankind had trespassed the millennium, speeded up time, deepened man's gaze into the future, ushered in the new age. The sky had become his third kingdom, conquered not by rolling huge blocks of stone in the manner of Titans but by capturing lightning and turning it into a slave."[51]

In principle, D'Annunzio's depiction of the Promethean aviator is here still consistent with the utopian-technological narrative of progress. But it has already been divested of its emancipatory component. D'Annunzio is not describing the universalist ascent of "humanity"; he is looking ahead to the birth of a new nobility – one to which not all aviators will belong:

> Nevertheless – Giulio Cambiaso said – I'm rather fonder of these Icaruses than of those paid mercenaries. [...] He was referring to the steering-wheel professionals, who considered the new machine as nothing more than a simplified racing car. [...] Hired by the artificial bird-manufacturers, they sniffed out the popular liking for the new circus game and made a packet out of their bones and their daring.[52]

The real supermen were Tarsis and Cambiaso, who did not compete for the money but to play with death; for "death is there in every game worth playing."[53] With *Àrdea* ("Heron," as both their machines were called), Tarsis rose above the masses, who called out the name *Àrdea* "in thousands upon thousands of voices."[54] Tarsis and Cambiaso had two "identical" machines, their shadows like that of the heron – so called also because they had built them in the Lazio town of Ardea, a location in Virgil's Aeneid, which bears "the name of the high-flying bird."[55] The crowd, participating aesthetically in this "elegant and dangerous game," was "intoxicated" by it.

---

[50] Ibid., p. 68.
[51] Ibid., pp. 101–2.
[52] Ibid., p. 81.
[53] Ibid., p. 88.
[54] Ibid., p. 100.
[55] Ibid., p. 77.

On the one hand, we see here the themes that would reappear during the war in modified form: the shadow of the aircraft changed from heron to cross of salvation; *vivere pericolosamente* outgrew simple rejection of the world of security and became the warrior's contempt for death. On the other hand, the birth of D'Annunzio's superman needs to be put in context. Tarsis soars above the crowd that admires him, but also metaphorically away from everyday life on earth:

> "We are alone, brother, free and a long way from the torments on earth! [...] How virile the sky is today!" He left everything behind: the turbulence of his passion, Isabella's exciting laughter, the feverish, hostile gaze of her younger brother, the vanity of girlfriends, the stupidity of the men accompanying them, all that intrusive crowd that had assailed and oppressed him. He rediscovered his silence, his solitude, his duty.[56]

Tarsis is weary of a self-satisfied bourgeois existence. Through his characters, D'Annunzio expresses a feeling that, as Robert Wohl has shown, the generation of 1914 would take up and radicalize.[57] And it is the flier who symbolically breaks loose, exchanging security for a dynamic life filled with adventure and dangers.[58] This soaring up and away does, however, demand sacrifices on the part of "the multitude."[59]

One of the aviators who has fallen to earth lies "soundless" in "a circle of horror, a mute wreck on its still warm and smoking heart. Dumbfounded, the eager crowd caught a sniff of the corpse."[60] Another aviator crashes burning to the ground, and again the crowd lets out "a visceral cry, not out of pity for the man about to die but in a rush of excitement at the deadly game. [...] The circus bloodlust rose again in every breast – a sudden surge of life before the imminence of death."[61]

Tarsis, "a herald of life at its highest," remains in the sky, victorious. Meanwhile his friend Cambiaso climbs again to set a new record, but then, as he returns from the "incalculable height" where he was "totally detached from his fellow-humans," the propeller of his machine fails and the man-machine plunges to the ground. The despised crowd "push and shove as they hurry to the spot, eager to catch a glimpse of the grim spectacle"; "the back of Cambiaso's head is stuck to

---

[56] Ibid., pp. 99–100.

[57] See Robert Wohl, *The Generation of 1914*, Cambridge, MA 1979.

[58] Felix Ingold writes in *Literatur und Aviatik* (p. 35): "D'Annunzio's heroes in *Forse che si, forse che no* act out, so to speak, the author's ideas in dramatic form; they are people without a social context, exemplary and – in a Nietzschean sense – pioneering prototypes. [...] This poetic space is vertically-dialectically aligned with D'Annunzio's central metaphor (motorized flight as an image of suprahuman, 'transfiguring' elevation); it is thus associated also with ancient religious conceptions of space."

[59] D'Annunzio, *Forse che si forse che no*, p. 102.

[60] Ibid., p. 103.

[61] Ibid., pp. 103, 105.

the engine in such a way that the seven rough cylinders of the cooling fans form a kind of corona heavy with blood-soaked earth and grass."[62]

The saint's corona glorifying the dead aviator is a further sacral motif that D'Annunzio used as early as 1910, and it would often recur in his speeches and writings on aviation during and after the war. But the transformation of the Icarus theme appears crucial here.[63] The heaven-storming superman Cambiaso crashes to earth, as Icarus did when he went beyond his limits. Unlike Icarus, however, who out of hubris left the middle course between sea and sun recommended to him by Daedalus, Cambiaso is hallowed precisely because of his striving to cross boundaries and rise ever higher. He is not punished with death but ennobled by it. His fearless contempt for life testifies to his superhuman qualities.

This is confirmed at the end of the book, where Tarsis sets off again on a daredevil flight over the Tyrrhenian waters "of Odysseus and Aeneas," content so long as he "crashes into the sea at the maximum distance from shore." But that is not what happens, and Tarsis begins to wonder: "Could death become life? the day of immolation become the day of transfiguration?" He reaches Sardinia, "and his heart pounded because the will to live was reborn in it, the will to live for the sake of victory."[64] We have already referred to the Christian and palingenetic themes in D'Annunzio. Here it needs to be clarified that he makes gambling with death the precondition of the new life; only a willingness to sacrifice one's life can give it a superhuman quality. Tarsis pulls off the Icarean flight to the vicinity of the sun; he does not come crashing down. Yet Tarsis does burn himself – not from the sun, but from the hot exhaust gases of the engine that carries him there. When he lands again, "there is no shouting, no triumphal tone, no pale-faced crowd with outstretched hands"; only "boorish silence" and "solitary glory." The burns mean that he can walk only with unbearable pain. His ties to the earth have been cut; the superman has lost his terrestrial grip.

Aside from the numerous allusions, the overabundant symbolism, and the mythological and sacred themes that D'Annunzio weaves into *Forse che si forse che no*, it must be said that he constructs a straightforward dichotomy between those destined for superhuman status and the "primitive" masses hungry for sensation. The hero is a solitary figure and D'Annunzio's elitism is individualist. The individualism of the aviation narrative was planed down in fascist mythmaking, whereas the elitism remained intact.

The elitist message of the novel is in keeping with its style. *Forse che si forse che no* was by no means written with an eye on those without a literary culture, nor did it become a book for the masses. But it did play an essential role for the media channels most at issue here, offering the first detailed artistic-literary treatment of the new phenomenon of aviation. The author succeeded in integrating this into a context of meaning, and in recasting, for the interpretation of aviation, a number

---

[62] Ibid., pp. 115–16.

[63] For a survey of the Icarus myth, see Behringer/Ott-Koptschalijski, *Der Traum vom Fliegen*, pp. 121–24. The main literary source for the myth is Book Eight of Ovid's *Metamorphoses*.

[64] D'Annunzio, *Forse che si forse che no*, pp. 515, 509, 519, 520.

of existing tropes that would continually reappear in the future. This does not mean that the establishment of this semantic context or perceptual framework was the achievement of D'Annunzio alone. Naturally he was also embedded in various discourses, and he drew upon both a pan-European educational canon and set of imagery out of the national experiential space. It was not least because of his prominence, especially in the media, that these tropes enlisted for aviation became so commonplace. Some of D'Annunzio's expressions stand paradigmatically at the beginning of a long process of repetition and incorporation that culminated in a mythical order.

On the one hand, the air show theme and D'Annunzio's related novel have enabled us to present the broader framework that initially existed for the interpretation of aviation. On the other hand, the various forms of participation in the aviation narrative need to be illustrated. The case of Kafka provided an example of the polyphonous interpretations that the phenomenon generated. The First World War subsequently limited the interpretative options, but there remained multiple forms of involvement in the process of giving meaning and order to a phenomenon – all the way from publicity for an air show prospectus through poster design and photographic depictions of the event to press reports and the publication of a "high culture" novel. The key form of involvement, however, lay in the individual reception of the event in question: that is, the transfer and reproduction of its meaning within the individual's store of experience and particular lifeworld. Not only aerial acrobats and composers of text, but also ostensibly passive spectators at air shows or readers of written material contributed to the meaning of the aviation narrative. Independently of the role played by individuals, aviation discourse was being constantly updated and reproduced, in a process that marked how individuals perceived and spoke about the phenomenon. Even the most passive onlooker became an "author" who produced meaning.[65]

An overemphasis on high-cultural products by prominent authors, or on their constructions of meaning, can scarcely be avoided in the reconstruction of discourse. The reason for this lies not least in the established methodologies of history, but also, of course, in the nature of the sources that have been handed down. Nevertheless, a concentration on particularly well-known figures also seems justified from the point of view of discourse theory, since their constructions of meaning cannot be understood as the accomplishments of autonomous, monadic subjects. D'Annunzio and other prominent representatives should rather be seen as disseminators of meaning: they constitute nodal points in the meaning production network who foster particular insertions into the discourse: that is, the repetition of certain statements and the establishment of certain tropes.

By contrast, the production or reproduction of meaning by "the man in the street" can be investigated only in rare cases. In the following sections, a focus on mass reading material connected with aviation during the First World War will at least allow an attempt to be made in this direction. Even such trivial literature is

---

[65] On the reader as "creator" of "meaning," see Roland Barthes, "The Death of the Author," in idem (ed.), *Image – Music – Text*, New York, NY 1977, p. 148.

not an accurate reflection of the meanings on offer to the lower strata in society. In Italy, as we have already seen, large parts of these strata were illiterate;[66] and even in Germany, where official propaganda boasted of high education spending, low rates of illiteracy, and sizeable book production,[67] it seems unlikely that the war books in question found great resonance among Ruhr miners or Pomeranian day laborers.[68] Their perception of aviation, their constructions of meaning and their attempts to create order therefore mostly remain blind spots for historical research. But at least the analysis of mass reading material gets us closer to the (re)production of meaning and the participatory opportunities of those who did not belong to the cultural elites.

As the analyses in the next chapter will show, although mass reading material differed in form and style from the products of high culture, the wartime context had a unifying effect on the interpretation of aviation. Certain tropes and constructions of meaning grew in importance, while others receded into the background or disappeared altogether if they proved incapable of structuring the chaos of war and giving meaning to the experience of suffering. The next section will attempt, as far as possible, to reconstruct the framework within which the consumption and reception of the "everyday myth" of war developed. It will present those paratexts that fostered a Don Quixote-style transfer of literary interpretations to reality.

"The War for the Price of One Mark": The promise of authenticity

Our first task will be to contextualize some 50 popular German war books from the years 1914–1918 that represent the flier and the war in the air.[69] The mythical aviation narrative that took shape in these books reached large parts of society. In the next chapter we shall look more closely at the actual imagery and interpretive patterns, but here the focus will be on the place of such literature in the media landscape of the German Empire at the time of the First World War, the framework of its reception, and the production of the books themselves. In particular, it will be shown how the books actually promoted the "curious literariness of real life."

The war books fit into a tendency that ran parallel to the growth of technological mass warfare. As the disorienting world war mobilized the home front, it generated a

---

[66] See Antonio Gibelli, "'Letteratura degli illetterati' nella Grande Guerra. Lineamenti di un percorso storiografico," in Massimo Bacigalupo/Roberto De Pol (eds.), *Grande Guerra e letteratura*, Genoa 1997, pp. 37–50.

[67] See the First World War poster designed by Louis Oppenheim and printed by Dr. Selle und Co. GmbH: "Sind wir die Barbaren?," http://www.dhm.de/lemo/objekte/pict/pl002758/index.html, accessed on April 3, 2008.

[68] On the reading habits of "men and women in the street," see *inter alia*: Jost Schneider, *Sozialgeschichte des Lesens. Zur historischen Entwicklung und sozialen Differenzierung der literarischen Kommunikation in Deutschland*, Berlin 2004.

[69] On "the war for the price of one mark," see Thomas F. Schneider, "Zur deutschen Kriegsliteratur im Ersten Weltkrieg," in idem (ed.), *Kriegserlebnis und Legendenbildung. Das Bild des modernen Krieges in Literatur, Theater, Photographie und Film*, 3 vols., Osnabrück 1999, pp. 101–14, 105.

greater need for information and orientation to which publishing houses responded with their new technologies and products. The transformation of literature into a commercial mass product had gathered pace in the nineteenth century, and especially since the turn of the 20th,[70] and the war books should be seen as a result of this process and of the rise of consumerist entertainment.[71] The book was a commodity produced by the "culture industry," which tried to satisfy people's needs and demands for everything from pulp fiction to the classics, from glossy magazines to serious dailies.[72] However, instead of top down manipulation of readers, it seems more sensible to think in terms of a symbiotic relationship involving the demand of the home front for "authentic" war reports, the need of soldiers and direct observers to process their experience through written material, and the financial and patriotic interest of publishers in marketing the corresponding goods.[73]

Thus, in the early August days of 1914, it was by no means only the armed forces that were mobilized; publishers, editors, house readers, authors, typesetters and printers also performed their service to the fatherland. In a war involving an advanced division of labor, words and media workers turned into weapons and soldiers,[74] so that Marieluise Christadler could speak appositely of a "literary call-up."[75] The media landscape was flooded with gung-ho verse, prose and reportage that self-styled poets, thinkers and correspondents immediately began to churn out.[76] As in the case of the "August experience," however, the way in which the media hyped up the war should not necessarily be seen as an expression of enthusiastic support.[77] As Helmut Fries has shown, the "massive production of prowar literature by civilians" was "an act of psychological unburdening, an attempt to find positive meaning in an event that, for all the hurrahs, was also deeply unsettling."[78] The war did not only disturb the daily round of people's lives; it

---

[70] On the so-called reading revolution, see Rolf Engelsing, "Die Perioden der Lesergeschichte in der Neuzeit," *Archiv für Geschichte des Buchwesens* 10/1969, pp. 945–1002.

[71] On the development and commercialization of the book market, see Helmut Fries, *Die große Katharsis. Der Erste Weltkrieg in der Sicht deutscher Dichter und Gelehrter*, vol. 1, *Die Kriegsbegeisterung von 1914. Ursprünge – Denkweisen – Auflösung*, Konstanz 1994, pp. 35–60; and Harro Segeberg, *Literatur im technischen Zeitalter. Von der Frühzeit der deutschen Aufklärung bis zum Beginn des Ersten Weltkrieges*, Darmstadt 1997, pp. 227–33.

[72] On the rise of popular literature and "reading matter for ordinary people," see Kaspar Maase/Wolfgang Kaschuba (eds.), *Schund und Schönheit. Populäre Kultur um 1900*, Cologne 2001; Peter Nusser, *Trivialliteratur*, Stuttgart 1991; and Rudolf Schenda, *Die Lesestoffe der kleinen Leute. Studien zur populären Literatur im 19. und 20. Jahrhundert*, Munich 1976.

[73] Thomas F. Schneider, "Endlich die 'Wahrheit' über den Krieg. Zu deutscher Kriegsliteratur," *Text und Kritik* 124/1994, pp. 38–51, here p. 47.

[74] Joachim S. Heise, "Sprache im Dienste des Völkerringens. Linguistische Perspektiven zum Ersten Weltkrieg," *Krieg und Literatur – War and Literature*, V/1999, pp. 37–54.

[75] Marieluise Christadler, *Kriegserziehung im Jugendbuch. Literarische Mobilmachung in Deutschland und Frankreich vor 1914*, Frankfurt/Main 1978.

[76] See, Helmut Fries, "Deutsche Schriftsteller im Ersten Weltkrieg," in Wolfgang Michalka (ed.), *Der Erste Weltkrieg. Wirkung, Wahrnehmung, Analyse*, Munich 1994, pp. 825–48, here p. 827; Thomas F. Schneider/Hans Wagner, "Einleitung," in idem (ed.), *Von Richthofen bis Remarque. Deutschsprachige Prosa zum I. Weltkrieg*, Amsterdam 2003, pp. 11–16, here p. 12.

[77] Verhey, *The Spirit of 1914*; and Ziemann, *Front und Heimat*.

[78] Fries, *Deutsche Schriftsteller im Ersten Weltkrieg*, p. 827.

brought the whole "age of security" to an abrupt end.[79] The media therefore sought to inform the public and to erect a barrier against the advancing chaos.[80] Since neither army reports nor newspaper articles could satisfy this need for information and orientation, book publishers too responded to the patriotic mood and sought to fill the lucrative gap in the market.[81] Soon after the outbreak of war, two nonfictional literary genres that had not existed before established themselves: the letter from the front and the war report.[82] Alongside the quantities of memoirs, anthologies, chronicles, picture books, regimental histories and battle accounts, publishers pumped out collections of letters, journals and first-hand reports that ostensibly brought home the experience of war to readers. Thomas F. Schneider reports in meticulous detail on a total of 7,973 literary works dealing with the First World War that appeared during the years from 1914 to 1939;[83] this only included texts "whose main focus was on the experience of individuals or groups of individuals at the front, behind the lines, or on the 'home front,'"[84] but even so there was a total of 3,585 books for the years between 1914 and 1918.[85] In what follows, we shall be concerned mainly with ostensible journals kept by air force men, although we shall also look at reports contained in war anthologies

---

[79] On the "golden age of security," see Zweig, *The World of Yesterday*, pp. 1ff.

[80] Much has been written in recent years on this theme of war and the media. For an overview of the research, see Jörg Becker, "Bibliographie zum Thema 'Krieg und Medien'," in Ulrich Albrecht/Jörg Becker (eds.), *Medien zwischen Krieg und Frieden*, Baden-Baden 2002, pp. 267–79; and Christian Filk, *Im Bann der Live-Bilder. Krisenkommunikation, Kriegsberichterstattung und Mediensprache im Informationszeitalter. Studien nach dem Ende der Ost/West-Konfrontation*, Siegen 2006.

[81] In the case of the German-Jewish publishers Mosse und Ullstein, for example, it is likely that the "huge loyalty pressure on Jews as an exposed minority," as well as "the possibility of [finally] proving their ties with the fatherland," ratcheted up their commitment to the war. See Sieg, *Jüdische Intellektuelle im Ersten Weltkrieg*, p. 61.

[82] In addition to the previously mentioned works on literature and the First World War, see Helmut Fries, *Die große Katharsis. Der Erste Weltkrieg in der Sicht deutscher Dichter und Gelehrter*, vol. 2., *Euphorie – Entsetzen – Widerspruch. die Schriftsteller 1914–1918*, Konstanz 1995; Bernd Hüppauf (ed.), *Ansichten vom Krieg. Vergleichende Studien zum Ersten Weltkrieg in Literatur und Gesellschaft*, Königstein i. Ts. 1984; Hans-Harald Müller, *Der Krieg und die Schriftsteller. Der Kriegsroman der Weimarer Republik*, Stuttgart 1986; Thomas F. Schneider, "Zwischen Wahrheitsanspruch und Fiktion. Zur deutschen Kriegsliteratur im Ersten Weltkrieg," in Rolf Spilker/Bernd Ulrich (eds.), *Der Tod als Maschinist. Der industrialisierte Krieg 1914–1918, Eine Ausstellung des Museums für Industriekultur Osnabrück im Rahmen des Jubiläums "350 Jahre Westfälischer Friede," 17.5.–23.8.1998*, catalog, Bramsche 1998, pp. 142–53; Uwe Schneider (ed.), *Krieg der Geister. Erster Weltkrieg und literarische Moderne*, Würzburg 2000; Harro Segeberg, *Literatur im Medienzeitalter. Literatur, Technik und Medien seit 1914*, Darmstadt 2003, esp. pp. 15–34; and Jörg Friedrich Vollmer, *Imaginäre Schlachtfelder. Kriegsliteratur in der Weimarer Republik. Eine literatursoziologische Untersuchung*, http://www.diss.fuberlin.de/diss/receive/FUDISS_thesis_000000001060, accessed on August 6, 2008.

[83] See Thomas F. Schneider/Julia Heinemann/Frank Hischer/Johanna Kuhlmann/Peter Puls, *Die Autoren und Bücher der deutschsprachigen Literatur zum 1. Weltkrieg 1914–1939. Ein bio-bibliographisches Handbuch*, Göttingen 2008.

[84] Ibid., p. 8.

[85] Ibid., p. 9.

and histories.[86] Apart from the theme of the air war, these various kinds of text have another thing in common: they claim to describe what really happened. And it was precisely this truth postulate that speeded up the amalgamation of fictional and nonfictional interpretations.

Books on the war tried to strike a balance between diary and journalism, literature and propaganda.[87] They stood out by their claim to authentic, first-hand reportage, and this also differentiated them in style and content from writings of the prewar period. The reporters were characterized by a historical/historicist attitude: they wanted to bear witness to a "great age" and to leave important source material for posterity; to observe and describe "naturalistically" how people experienced the war. Georg Wegener's preface to his *Der Wall von Eisen und Feuer* [The Wall of Iron and Fire] provides a good example of this:

"I am supposed to go out into the theatre of war – one of the few, apart from the soldiers, allowed to do this – and to report on what happens. How am I to do this? How can I possibly convey the immensity of it? I think the most truthful course will be simply to tell what I see, experience, and feel." I wrote these words when I started out last August as a field reporter for the *Kölnische Zeitung*. [...] A war book cannot offer a comprehensive account, calmly weighing everything up on the basis of historical sources; it can only be an original source itself. The following is therefore a mere reproduction of what I saw as a correspondent in the Western theater of war.[88]

Since the aim is "simply to tell," the literary yardsticks of the prewar period increasingly lost their validity. The demand for "artistic quality" was supplanted by a craving for authenticity.[89] Publishers were willing to go along with this, since the risk was small enough. Even if the authors were literary dilettantes, their names and "glorious deeds" were already known from army reports and the press, and so readers would want to learn more about them. Ullstein Verlag, in particular, but also Scherl Verlag were capable of rushing out a book in which participants or eye-witness observers reported on current developments at the front. A news-hungry public had to wait only a couple of weeks to read a more detailed account than newspapers, magazines and official statements gave them of (say) the naval Battle of Jutland or Germany's unrestricted submarine warfare.[90]

---

[86] To achieve a complete picture of the tropes in use, we shall also consider published novels and collections of letters from the front (in so far as they contain letters from aviators or speak of the air war).

[87] For an overview of war reporting in general, see Ute Daniel (ed.), *Augenzeugen. Kriegsberichterstattung vom 18. zum 21. Jahrhundert*, Göttingen 2006; and Manuel Köppen, *Das Entsetzen des Beobachters. Krieg und Medien im 19. und 20. Jahrhundert*, Heidelberg 2005.

[88] Georg Wegener, *Der Wall von Eisen und Feuer. Ein Jahr an der Westfront*, Leipzig 1915, unpaginated.

[89] See Müller, *Der Krieg und die Schriftsteller*, pp. 13–14.

[90] See Schneider, *Zwischen Wahrheitsanspruch und Fiktion*, pp. 146–47.

The war books aimed at realistic observation and depiction of reality, taking further a general trend in German literature around the turn of the century that might be described as "journal-writing" or "the age of testimony."[91] This led in the 1920s to the new style of objective reportage, which, according to the publicist Siegfried Kracauer, was driven by a "hunger for directness." In 1929 Kracauer could write:

> For a number of years now, reportage has enjoyed in Germany the highest favor among all types of representation, since it alone is said to be able to capture life unposed. Writers scarcely know any higher ambition than to report; the reproduction of observed reality is the order of the day.[92]

The success of war books also fed off this enthusiasm for supposedly genuine "reproduction of observed reality." It may therefore be assumed that few of Kracauer's contemporaries shared his critical attitude to their reality content:

> But existence is not captured by being at best duplicated in reportage. [...] Reality is a construction. Certainly life must be observed for it to appear. Yet it is by no means contained in the more or less random observational results of reportage; rather, it is to be found solely in the mosaic that is assembled from single observations on the basis of comprehension of their meaning.[93]

From a constructivist point of view, the claim to "reproduce reality" seems as illusory as the readers' expectation that war books would tell them "the truth" about the war. But on no account should this lead to the anachronistic idea that the claim and the expectation were not to be taken seriously. On the contrary, as we shall see, the war books in question were characterized precisely by the claims they made to reality and truth.

The different attitudes of readers to books and their content may again be clarified by a passage from Cervantes's *Don Quixote*.

> "See here, brother," began the priest, "there never were such people in the world as Felixmarte of Hyrcania, or Don Cirongilio of Thrace, or any of the other knights in those books of chivalry. They are all fictions, invented by idle brains. [...] For I swear to you that really such knights never existed in the world, and all these feats and follies never happened."

---

[91] On the genre of "journal-writing" in German literature between 1871 and 1914, see Fries, *Die große Katharsis*, vol. 1, pp. 4–9. And on the "age of testimony," see Ingold, *Literatur und Aviatik*, p. 232, where we read: "The 'age of testimony' ushered in by the First World War helped [...] to promote subliterary genres such as diaries, correspondence, reportage, and memoirs."

[92] Siegfried Kracauer, *The Salaried Masses. Duty and Distraction in Weimar Germany*, London/New York, NY 1998, p. 32.

[93] Ibid.

"Try that bone on another dog!" replied the innkeeper. [...] "It's a nice thing for you to try and persuade me that all these fine books say is only nonsense and lies, when they're printed by license of the Lords of the Privy Council – as if they were people who would allow a pack of lies to be published, and enough battles and enchantments to drive you out of your wits!"[94]

Were we to apply the priest's disbelieving attitude directly to the kind of books we have been discussing, finding in them nothing but "nonsense and lies," we would certainly not do justice to their mass production and success with readers. We shall therefore have to consider in greater detail why "the innkeeper" in *Don Quixote* could not imagine that nonsense might be printed "by license of the Lords of the Privy Council," or that such "fine books" might contain lies.

The credibility of the war literature partly rested upon the authority of the printed word; the very materiality of a book lends persuasiveness to its content. But an additional factor was the semi-official aura, bound up with authoritative eye-witness reporting, that emanated from these popular books. Their explicit claim to truth was bolstered by the author's frontline credentials and the military context of the action; his experiences, and usually also his military rank, vouched for the authenticity of the reports and the truth of his statements. This aura of the books, as well as its institutional reinforcement, will become clearer if we consider some valuable points made by Roger Chartier[95] and, above all, Gérard Genette. An account of the reception framework originating with the books themselves will further serve to contextualize them.

The concept of "paratext" that the French structuralist Genette developed in the 1980s makes it possible to outline this reception framework.[96] A text, he argues, "is rarely presented in an unadorned state, unreinforced and unaccompanied by a certain number of verbal or other productions."[97] These accompaniments first make the text present and guide its consumption and reception. In the words of Philippe Lejeune, who had a similar concept in mind, we are talking of "a fringe of the printed text which in reality controls one's whole reading of the text."[98] Since

---

[94] Cervantes, *Don Quixote*, Part I, Chapter 32, p. 280.

[95] See *inter alia*: Roger Chartier, *Forms and Meanings. Texts, Performances, and Audiences from Codex to Computers*, Philadelphia, PA 1995; "Laborers and Voyagers. From the Text to the Reader," *Diacritics* 22/1992, pp. 49–61; and "Texts, Prints, Readings," in Lynn Hunt (ed.), *The New Cultural History*, Berkeley, CA 1989, pp. 154–75.

[96] Genette, *Paratexts*. See also his *Palimpsests. Literature in the Second Degree*, Lincoln, NE 1997. Here (on p. 3) he explains the concept of paratext for the first time: "a title, a subtitle, intertitles; prefaces, postfaces, notices, forewords, etc.; marginal, infrapaginal, terminal notes; epigraphs; illustrations; blurbs, book covers, dust jackets, and many other kinds of secondary signals, whether allographic or autographic. These provide the text with a (variable) setting and sometimes a commentary, official or not, which even the purists among readers, those least inclined to external erudition, cannot always disregard as easily as they would like and as they claim to do."

[97] Genette, *Paratexts*, p. 1.

[98] Philippe Lejeune, *Le pacte autobiographique*, Paris 1975, p. 45. Quoted in Genette, *Paratexts*, p. 2.

most readers first approached the war books through such "fringes," it seems useful to present the corpus by reference to them. We shall begin by explaining how the "fringes" (which Genette calls "paratexts") guided such reading, and then focus on the actual production of the text.

Genette subdivides paratexts into "peritexts" and "epitexts." Peritexts are all the paratexts found "around the text and within the same volume."[99] An epitext, on the other hand, is "any paratextual element not materially appended to the text within the same volume but circulating, as it were, freely";[100] Genette gives as examples press reviews and interviews with the author. What is central is that the reception of the text is influenced by the "illocutionary force" and "functional character" of the paratext.[101] As we shall show with reference to paratexts appended to the most diverse war books, their "illocutionary force" consisted precisely in their making a claim to truth and underpinning the authenticity of the books. The paratexts signaled that here was the true face of war; but they also announced that the reader could expect to find adventure, heroism and entertainment.

The reader might come across war books in a number of ways – through excerpts in the press, for instance, or publicity, or magazine articles about the author. Advertisements were paradigmatic "epitexts," placed by the publisher outside the actual book; they positioned it in the market, addressing possible readers and shaping their first impression of the book. We have already referred to the many reports on D'Annunzio's adventures in *Illustrazione Italiana*, which allowed his publisher, Treves, to draw attention to his literary works; for example, magazine articles about his prowar agitation steered the public perception of *Per la più grande Italia*.

Ullstein Verlag operated in a similar way. Alongside a highly successful series of war books, the *Berliner Illustrierter Zeitung* (*BIZ*) also appeared under its auspices. Of course, the Berlin publishing house took the opportunity to advertise other publications as well: the *BIZ* of March 18, 1917 printed an excerpt with accompanying photographs from one of the chapters of *Die Abenteuer des Ostseefliegers* [Adventures of the Baltic Airman] by Lieutenant Erich Killinger. It was presented as follows:

> Second Lieutenant Killinger and Lieutenant von Gorissen were shot down on a flying mission. Both men fell into the Baltic Sea and were captured by the Russians. They were treated in a barbaric manner, being taken first to the underground dungeons of the Peter-Paul Fortress and then to Siberia. *En route* Lieutenant Killinger jumped from a moving train, made his way with indescribable difficulty to China and Japan, following this with a mad dash through America and finally risking the passage via England to Norway as an able seaman. Like *Kapitänleutnant* Plüschow, Lieutenant Killinger has recorded his

---

[99] Genette, *Paratexts*, p. 4.
[100] Ibid., p. 344.
[101] Ibid., p. 12.

exciting adventures in a recently published book (price 1 mark, Verlag Ullstein & Co.). Here we reproduce part of the chapter entitled "British Manhunt on the High Seas."[102]

The illustrated sample told "how Lieutenant Killinger travelled as Able Seaman Jean Pâhu on board a Norwegian steamer and pulled the wool over the eyes of the British investigators."[103] In effect, the *BIZ* epitext advertised a "cops and robbers" or "spy" story. But the fact that it involved a German officer and navy pilot gave a serious touch to the round-the-world adventure – from the skies of the Baltic through Russia's dark dungeons to China, the United States and back to Europe. If Killinger's escapade seemed to border on pulp fiction, his military rank and the book's inclusion in the Ullstein war book series attested to its dignity and historical accuracy.

Nevertheless, the story drew less on Killinger's heroic deeds as a military aviator – which lasted only a few pages, until he was taken prisoner – than on the exotic nature of his travels. The reference to *Kapitänleutnant* Plüschow further underlined that it was an airman's tale from remote foreign lands, since a year earlier the latter's *Abenteuer des Fliegers von Tsingtau. Meine Erlebnisse in drei Erdteilen* [Adventures of the Flier from Tsingtao. My Experiences in Three Parts of the World] had been published as No. 23 in the Ullstein war books series.[104]

Plüschow's bestseller from the Middle Kingdom had been printed in more than 500,000 copies by the end of the war, and reached a total of 700,000 by 1939.[105] Killinger's report was supposed to build on this success, and that is indeed what happened: his *Die Abenteuer des Ostseefliegers* had been printed in 420,000 copies by 1939.[106] But such sales of war books were by no means attributable to their exoticism alone. What was fairly marginal in Plüschow and Killinger was the central theme in most of the other books: war and the soldier's life, together with the fascination of technology. This did not exclude exoticism, of course, as the

---

[102] Erich Killinger, "Die Abenteuer des Ostseefliegers," *Berliner Illustrierte Zeitung* No. 11, March 18, 1917, Year 26, pp. 149–51; here p. 149.

[103] Ibid.

[104] Gunther Plüschow, "Die Abenteuer des Fliegers von Tsingtau. Meine Erlebnisse in drei Erdteilen," *Berliner Illustrierte Zeitung* No. 50, December 10, 1916, Year 25, pp. 757–60. The presentation (p. 757) states: "*Kapitänleutnant* Plüschow was the only flier in Tsingtao. Before the stronghold fell, he flew with important documents to China. From there he travelled as an 'American millionaire' via Japan to America, and from America as a 'Swiss journeyman' to Europe. In Gibraltar, after a test in 'Swiss German' went badly for him, he was arrested by the British and shipped to England. There he escaped from a concentration camp and, in a miraculous journey, finally got back to Germany. The reader will follow with breathless excitement the story of his adventures, which have just been published in book form (price 1 mark, Verlag Ullstein and Co.). Here we reproduce parts of the chapter entitled 'Escape from British Imprisonment'."

[105] See Schneider, *Zur deutschen Kriegsliteratur*, p. 106, and Schneider et al., *Die Autoren und Bücher der deutschsprachigen Literatur zum 1. Weltkrieg 1914–1939*, p. 10.

[106] Schneider et al., *Die Autoren und Bücher der deutschsprachigen Literatur zum 1. Weltkrieg 1914–1939*, p. 11.

following presentation of Lieutenant Hans Henkelburg's *Als Kampfflieger am Suez-Kanal* (published by the August Scherl Verlag) demonstrates:

> The book shows our German airmen in the alien world of the Orient. One understands the excitement that the first German aeroplane must have aroused in Jerusalem, as well as the feelings of the aviator as he soars over the venerable old sites of Palestine in the most modern of all means of transportation. Hard is the struggle against the British, and German airmen have to face many a tussle with the enemy, who is tough and capable also in the air. The vivid, enthralling accounts of these combat missions excite our interest all the more because work in the desert sands and blazing heat of the sun, far from all the aids of civilization, brings huge difficulties and places great demands on the aviator's capacity for adjustment.[107]

Technology as such, however, played a secondary role. Its "modern wonders" seem to have had a kind of "fetishistic" appeal: that is, its actual technical side does not seem to have interested readers much at all.[108] In any event, some of the most successful books – apart from those on aviation – were ones dealing with warships or submarines.[109]

C.F. Amelangs Verlag in Leipzig, the publisher of Adolf Viktor von Koerber's *Luftkreuzer im Kampf* [Airships at War], placed an advertisement for another work by this author in the endpages. It mainly consisted of snippets from reviews in various newspapers. The epitext for the second edition of *Feldflieger an der Front* said:

> Adolf Viktor von Koerber, also known as a stagewriter under the name of Dolf von Korb, depicts the airman's soldierly life in his brilliant little book that bears the title of the first sketch, doing so with a finesse and penetration that only someone who puts their heart and soul into it can achieve. **K. Langenbachin** *Tägliche Rundschau*, **Berlin**. This work is warmly recommended to those who wish to have an idea of what "flying" means, of what the field aviator accomplishes out there. The layman is told the facts in a thorough and captivating way, while the expert, noting with satisfaction that it was a "colleague" who revealed these delightful pictures, will learn a lot from them or find confirmation of his own experience. A. V. von Koerber – himself decorated with the Iron Cross – has created a lasting memorial of incomparable beauty to those who are out there in the field every minute, every hour of their life, by day and by night. **Dr. Dietrich Helfenberg, in the** *Dresdner Anzeiger*. The little book is so splendidly written, in such a fresh and enthralling manner, that one will gladly pick it up again and again. *Berliner Lokal-Anzeiger*.[110]

---

[107] *Börsenblatt für den deutschen Buchhandel*, No. 71, March 26, 1917, Year 84, p. 2141.

[108] On the "modern wonders," see Bernhard Rieger, *Technology and the Culture of Modernity in Britain and Germany, 1890–1945*, Cambridge 2005.

[109] See the tables in Schneider et al., *Die Autoren und Bücher der deutschsprachigen Literatur zum 1. Weltkrieg 1914–1939*, pp. 10ff.

[110] Adolf-Victor von Koerber, *Luftkreuzer im Kampf*, Leipzig 1916, unpaginated.

This is a different way of steering the reader. Instead of playing upon the appeal of the foreign and exotic, as in the preceding epitexts, the publisher highlights the proximity to the field aviator. The epitext also delineates the possible circle of readers; the hint that the author is also active pseudonymously as a stagewriter, yet his "heart and soul" is with the soldier, suggests that the book will thoroughly satisfy literary demands without descending into *belles lettres*. The praise of Dietrich Helfenberg, on the other hand, addresses the reader with a more special-ist interest. The book informs lay people "thoroughly," and even the "expert" will find "delightful" confirmation of his views. The *Berliner Lokal-Anzeiger*, for its part, refers to the cost-benefit ratio: the book, whether in the "lightly wrapped" two-mark edition or the "clothbound" three-mark edition, is so "captivating" that one will read it several times. Von Koerber's book was thus eminently affordable and satisfied both factual and "literary" requirements. Above all, however, it informs the reader "in a thorough and captivating way" about the field aviator, a guaran-tee of this being that the author himself has been "decorated" with the Iron Cross. The publisher's recipe for success was based not least on this military credential. The authors might be literary dilettantes – although this is disputed in the case of the "stagewriter" Adolf Viktor von Koerber – but they were highly decorated officers reporting straight from the front.

As the peritexts in the actual books show, the tension in them between serious-ness and triviality, military-official reportage and popular literature, was the key to their success. They satisfied the reader's thirst for adventure and entertainment, while also giving them the information they required and making them feel that they were part of the war. The publishers' offerings matched the public demand for "sensation" and military "starlets," as well as for serious models, soldierly icons and national recognition. "Pulp fiction" was by no means incompatible with national-military heroism. Rather, it seems possible to speak of an updat-ing or "modernization" of nationalism in the course of the First World War – a phenomenon that spread also through innovative channels of popular culture.

A sweeping categorization of these books as "propaganda" would therefore be misleading, since propaganda was too tainted with Americanism and democracy for the traditional political and military elites to warm to it. They did not feel they needed popular consent. Even after the Army High Command began to see propa-ganda as an important means to secure the loyalty of the masses and to discipline them, "official communications policy and public information" and "state-guided opinion formation" were still a long way from constituting an apparatus in the style of Goebbels's later Ministry of Public Enlightenment and Propaganda.[111] As will become clear, the air force press department and other bodies wielded consid-

---

[111] On German propaganda in the First World War, see Anne Lipp, *Meinungslenkung im Krieg. Kriegserfahrungen deutscher Soldaten und ihre Deutung 1914–1918*, Göttingen 2003; Ulrike Oppelt, *Film und Propaganda im Ersten Weltkrieg. Propaganda als Medienrealität im Aktualitäten- und Dokumentarfilm*, Stuttgart 2002, esp. pp. 99–118; Anne Schmidt, *Belehrung – Propaganda – Vertrauensarbeit. Zum Wandel amtlicher Kommunikationspolitik in Deutschland 1914–1918*, Essen 2006; David Welch, *Germany, Propaganda, and Total War, 1914–1918. The Sins of Omission*, New Brunswick, NJ 2000.

erable influence over a number of publications, and all books were subject to censorship. But the propaganda category does not take into account the spontaneous and autonomous emergence of these war books; they were not the product of an information policy decreed from above, but arose out of (and updated) a "grassroots nationalism" and a military discourse radicalized by the war.[112] To underline what we have said before: they had their origin in the consumption-oriented strategy of publishers and in the patriotic mood of dilettante writers, war observers, and soldiers.

The tighter links with military institutions and dignitaries did, however, enhance the official character of the books and underpin their claim to authenticity. Institutional reinforcement of their truth postulate – which went far beyond highlighting the author's military rank – can also be found in the peritexts. Take, for example, the dedication in Lieutenant Heydemark's *Doppeldecker "C 666"*: "Respectfully dedicated to my old section leader Captain Mohr."[113] The book *Der Luftkrieg 1914–1915. Unter Verwendung von Feldpostbriefen und Berichten von Augenzeugen* [The Air War 1914–1915; with letters from the front and eye-witness reports], published in 1915 in Leipzig by Hesse & Becker, was not only "presented by an aerial technician" but printed "by permission of the Royal Prussian War Ministry and the Imperial Navy Office."[114]

The 1917 edition of *Unsere Luftwaffe*, the yearbook of the officially registered Airmen's Relief Association, was edited by a Captain Funk. On the last page we are told that it is the task of the society to "supplement state welfare" by supporting "fliers and aeronauts involved in an accident, as well as those they leave behind." The book underlines the significance of this task and – the peritext suggests – will serve as an appeal for donations and perhaps contribute to

the great national duty of gratitude to our airmen. [...] Hence it is the duty of every single German to lend a hand and to help ensure that some of our huge debt of gratitude is discharged to the fliers and airshipmen who, in this war, have kept protective watch over us and our beloved fatherland from the heights of heroic valor.[115]

---

[112] For an overview of German militarism, see Thomas Rohkrämer, *Der Militarismus der "kleinen Leute." Die Kriegervereine im Deutschen Kaiserreich, 1871–1914*, Munich 1990; Wolfram Wette (ed.), *Schule der Gewalt. Militarismus in Deutschland 1871–1945*, Berlin 2005. On the alternative concept of "bellification" see: Frank Reichherzer, *Alles ist Front. Wehrwissenschaften in Deutschland und die Bellifizierung der Gesellschaft vom Ersten Weltkrieg bis in den Kalten Krieg*, Paderborn 2012. On the radicalization of nationalism in the Wilhelmine Empire, see Peter Walkenhorst, *Nation – Volk – Rasse. Radikaler Nationalismus im Deutschen Kaiserreich 1890–1914*, Göttingen 2007. And on the development of nationalism during the First World War, see *inter alia* Sven Oliver Müller, "Die umkämpfte Nation. Legitimationsprobleme im kriegführenden Kaiserreich," in Jörg Echternkamp/Sven Oliver Müller (eds.), *Die Politik der Nation. Deutscher Nationalismus in Krieg und Krisen*, Munich 2002, pp. 149–72.

[113] Georg Heydemarck, *Doppeldecker "C 666." Als Flieger im Westen*, Berlin 1916, unpaginated.

[114] Anonymous, *Der Luftkrieg 1914–1915. Unter Verwendung von Feldpostbriefen und Berichten von Augenzeugen*, Leipzig 1915, unpaginated.

[115] (no forename) Funk, *Unsere Luftwaffe. Jahrbuch des Luftfahrerdank e.V. 1917*, Leipzig n. d., unpaginated.

In lending a solemn official aura to a book meant as entertainment, with numerous illustrations and photographs, this paratext supplemented the noble statement of aims in the preface: "While many efforts today are directed at making and keeping our youth proficient with weapons for future times, the content of this book will also contribute to that end; how better could young people find new strength for the distant struggle for survival!"[116]

The war books delivered models that young people were supposed to emulate. This is suggested by the epigraphs for the two-volume *Heldentum im Weltkriege. Berichte von Heldentaten* [Heroism in the World War. Reports of Heroic Deeds], which, like the other peritexts mentioned here, served as instructions, as it were, on how to use the war books.[117] Dr Johann Nieden, who collected the reports, gave his readers a quotation from Schiller "by way of presentation": "A great example stirs to emulation / And teaches higher laws that we should judge by."[118] Whether drawing on *Wallenstein* or on ancient stores of quotation, Nieden and other authors thought it clear which interests and purposes their books should serve: "Words instruct, but examples enrapture." Heroes served to define how young people should act in the future. We find this confirmed in the preface to *Vom Jäger zum Flieger. Tagebuchblätter und Briefe von Leutnant Schäfer* [From Hunter to Flier. Diary Pages and Letters of Lieutenant Schäfer]: "May these pages keep alive in the German people's heart the memory of those who fell so early, and spur our young people to follow in their footsteps."[119]

As in the case of Carl Emil Schäfer, the father or another close relative often took responsibility for the preface to written material left by the fallen aviator. It was evidently thought good form to emphasize that the hero had only reluctantly described his experiences. But the authentic nature of the published material was also highlighted:

> When my son was on leave and spoke of his experiences in the field and his air battles, I often asked him to put them down in writing. The answer I always received was: "At war I have no time to write books." But when I was going through his things after his death, I noticed that my words had stuck in his mind after all. I found sketches from which I could tell that he had thought of collecting his experiences and eventually publishing them, perhaps only after the war. I have now loosely arranged his letters and diary pages, in merely temporal succession.[120]

[116] Ibid.

[117] Johann Nieden, *Heldentum im Weltkriege. Berichte von Heldentaten*, Strasbourg 1914, unpaginated.

[118] Ibid. [Translation quoted from Friedrich Schiller, "Prologue," in Schiller *Wallenstein*, translated by F. J. Lamport, London 1979, p. 165.]

[119] Emil Schäfer, *Vom Jäger zum Flieger. Tagebuchblätter und Briefe*, Berlin 1918, pp. 7–8.

[120] Ibid., p. 7.

In a publisher's preface to Max Immelmann's *Meine Kampfflüge. Selbsterlebt und selbsterzählt* [My Flying Missions, as lived and told by the author], August Scherl Verlag pointed out that, by publishing the "letters that Lieutenant Max Immelmann sent to his mother during the war," it was fulfilling "one of the last wishes of the memorable flying officer." And it quoted a letter in which he had finally agreed to describe his "flying experiences," since "what is usually presented to readers as 'air battles and airmen's letters' is often irresponsible" and does not have an "educative" effect.[121]

The market was well supplied, and so an emphasis on the authenticity of one's own book supposedly made it stand out from the many others. Whether readers at the time accepted such "proofs" and protestations at face value remains unclear. In any event, there was something compulsive about the way in which the publishers kept harping on authenticity – not only in the peritexts, but also in the texts themselves, through devices such as the insertion of photographs. Especially popular were pictures of destroyed enemy aircraft, with captions such as "the fifteenth aeroplane shot down by Immelmann."[122] Ernst Udet added a slightly more personal touch – "my third victory"[123] – while Manfred von Richthofen baldly recorded "the thirtieth!" or "the fortieth!"[124] A further factor made the books seem more trustworthy: the names of combatants and superiors were abbreviated, and locations were sometimes disguised. This gave the reader the flattering impression of being privy to military secrets, though only to the extent that left individuals and sites protected in case the book "fell into enemy hands."

Everything was done to give the books an official aura of veracity, but at the same time any suggestion that they were dry presentations of facts was scrupulously avoided. Rather, publishers tried to create the impression that the books were exciting tales of war and adventure. The format of a book – which Genette includes among the publisher's peritexts designed to steer the public – and its external appearance in general naturally tend to shape how it is received. Many war books, and particularly the Ullstein Verlag series, appeared in a small octavo format, with a spine approximately 16 cm high. In other words, they were pocket books. But in addition to the format, the book jackets were suggestive of what the Americans call "pulp fiction," if not downright trashy literature. The jackets or dust covers of many war books were garish and sensationalist (Figure 1.2).

The jacket illustrations of the Ullstein war book series, like those of the August Scherl Verlag in Berlin, certainly stood out. Iconographically, they already seem to point ahead to Pop Art; the cover of Plüschow's *Abenteuer des Fliegers von Tsingtau*, in particular, testifies not only to an inhuman, imperialist sense of superiority,

---

[121] Max Immelmann, *Meine Kampfflüge. Selbsterlebt und selbsterzählt*, Berlin 1916, pp. 5–6.

[122] Ibid., p. 112.

[123] Ernst Friedrich Eichler (ed.), *Kreuz wider Kokarde. Jagdflüge des Leutnants Ernst Udet*, Berlin 1918, p. 128.

[124] Manfred von Richthofen, *Der rote Kampfflieger*, Berlin 1917, pp. 96–7; *The Red Fighter Pilot. The Autobiography of the Red Baron*, St. Petersburg, FL 1997.

Figure 1.2 Covers for war books published by Ullstein and Scherl

but also to a comic strip aesthetic. The Killinger cover, like the illustration in Plüschow's book, fuels the expectation of exoticism. But apart from that, it is the aircraft that are meant to awaken the reader's interest. And whereas Richthofen's red "double-decker" climbs again after shooting down an enemy plane, the other aircraft swoop down like raptors on their prey (an allusion to Immelmann's nickname as the Eagle of Lille).

The covers promised exciting adventures, fascinating technology and expert daredevilry in a series of heroic battles: in short, entertainment. In the case of the Ullstein series, the price was also printed on the cover; the war was on offer for the price of one mark. The war books of August Scherl Verlag cost one mark "stapled" and two marks "elegantly bound." *Der Krieg in der Luft* [War in the Air], part of the "war book series for young people" published by Julius Beltz Verlag, could even be bought for 30 pfennigs.

We cannot know exactly who read these books. But it is likely that the readers were mostly young, male and middle-class. At the price of one mark, the war books were also theoretically available to "lower strata of the population" – even if they had to pay the full cost upfront.[125] It is doubtful, however, whether the urban and rural lower classes, faced with rising prices, shortages and undernourishment, spent the little money they had on books. But we may assume that lower middle-class and working-class people often accessed such reading matter through read-and-exchange schemes, lending libraries and shops that ran a library as a sideline.[126] As we shall see, despite their price and presentation, these war books were not thought of as "trashy literature," the opponents of which were by no means inactive during the war.[127]

The trade journal, *Börsenblatt für den Deutschen Buchhandel*, referred in early 1917 to a couple of books that were leading the charge against "trashy war literature." Precisely because young people were "often deprived of paternal authority" during the war years, the "question of trashy and smutty literature" presented itself with even greater urgency.[128] However, it was far from clear what did and did not count as trash. The author of the book *Die Kriegsschundliteratur und ihre Bekämpfung* [The Fight against Trashy War Literature], Wilhelm Tessendorf, therefore took his lead from the vague

declaration of 1914 by Youth Literature Watch: "All trash books should be denied any literary or artistic value. They appear as mass commodities, and their diffusion is pure business. They do not seek to ennoble the mind, but are directed at the reader's base instincts." The last sentence should be stressed in particular, since enough items appear on the book market that, without having any literary or artistic value, are sold as mass commodities for purely commercial reasons, yet should not be considered trash literature. Once again we see how difficult it is to apply the term "trash" correctly and accurately.[129]

---

[125] By comparison, the price of wheat bread in Berlin stood at 54 pfennigs per kilo in 1910, and rye bread cost 30 pfennigs per kilo in 1913. See Ashok Valji Desai, *Real Wages in Germany 1871–1913*, Oxford 1968, pp. 128–29. The Ullstein war books, priced at one mark each, were less expensive than other books. According to a record of shop prices in 1908, art and music theory books averaged 9.22 marks, books for young people 1.48 marks, religious works 2.06 marks, and literary works 2.22 marks: see Reinhard Wittmann, *Geschichte des deutschen Buchhandels. Ein Überblick*, Munich 1991, p. 297.

[126] "There are many statistics and contemporary statements to suggest that even in the countryside it was fairly common to read a book without owning it": Wittmann, *Geschichte des deutschen Buchhandels*, p. 324.

[127] Kasper Maase, "'Schundliteratur' und Jugendschutz im Ersten Weltkrieg – Eine Fallstudie zur Kommunikationskontrolle in Deutschland," kommunikation@gesellschaft3/2002, http://www.uni-frankfurt.de/fb03/K.G/B3_2002_Maase.pdf, accessed on August 5, 2014; Schenda, *Die Lesestoffe der kleinen Leute*, pp. 78–104.

[128] Kurt Loele's reviews of Elisabeth Süersen, *Die Stellung der Militär- und Zivilbehörden zur Schundliteratur*, Berlin 1916 and Wilhelm Tessendorf, *Die Kriegsschundliteratur und ihre Bekämpfung. Mit einem Verzeichnis empfehlenswerter Schriften*, Halle 1916, in *Börsenblatt für den Deutschen Buchhandel*, No. 1, January 2, 1917, Year 84, pp. 3–4.

[129] Wilhelm Tessendorf, *Die Kriegsschundliteratur und ihre Bekämpfung. Mit einem Verzeichnis empfehlenswerter Schriften*, Halle 1916, p. 4. See also Anonymous, review of Paul Samuleit, *Kriegsschundliteratur. Vortrag, gehalten in der öffentlichen Versammlung der Zentralstelle zur*

In any event, most of the war books in question here were not considered trashy literature. Evidence of this is the way they were handled at the Royal University Library in Tübingen, which began to collect war literature extensively from 1914 on. It was supported in this by Louis Laiblin, a man of independent means from a family of paper manufacturers, who was known as a patron in and around his home town of Pfullingen.[130] On January 20, 1915 Dr Carl Geiger, the head librarian in Tübingen, reported to the rectorate: "Further to his July donation (a large part of which I have already used with his agreement for the acquisition of war literature), Herr Laiblin in Pfullingen has transferred another 500 marks especially for the purchase of German and foreign literature on the war."[131]

It appears that, on a number of occasions between 1914 and 1919, Laiblin donated a sum of between 500 and 2,000 marks for the purchase of war literature. The inside cover of the books bore an *Ex libris* inscription, which noted, in addition to the year of acquisition, that they were "war literature from the Louis Laiblin donation." The *Ex libris* should also be seen as a peritext, which enhanced the "semi-official" character of the books.

Correspondence among the head librarian, the donor, and the principal's office indicates the importance attached to these war books, as well as evoking the mentality reminiscent of Heinrich Mann's *Untertan*[132] that prevailed in provincial Swabia. This, for example, is what Laiblin wrote on September 28, 1916 to Head Librarian Geiger:

> With reference to our conversation earlier today, I should like on the occasion of His Majesty the King's jubilee to transfer a further donation for your commendable collection of war literature that is of such value to the general public. For this purpose, I therefore take the liberty of sending you herewith a further check to the value of 2,000 marks.[133]

The library collection of war literature was an important matter, overseen by the royal (and, after the Revolution, the Württemberg) ministry of churches and schools:

> Acceptance of the 1,000-mark donation from Privy Councilor Laiblin in Pfullingen for acquisitions of war literature by the university library is hereby approved. The academic principal's office would also like to convey to the

---

Bekämpfung der Schundliteratur zu Berlin am 25.3.1916, Berlin 1916, in *Börsenblatt für den Deutschen Buchhandel*, No. 156, July 8, 1916, Year 83, p. 900.

[130] Hermann Taigel, *Louis Laiblin, Privatier. Ein schwäbischer Mäzen*, Pfullingen 2005. Between April 18, 1905 and March 27, 1919, Laiblin made donations totaling 19,000 marks (of which 13,000 marks for war literature alone) to the university library in Tübingen: see ibid., pp. 113 and 120. On the Laiblin collection see: Wilfried Lagler, "Karl Geiger und sein 'Liebeskriegswerk'. Die Sammlung 'Kriegsliteratur Laiblin' in der Universitätsbibliothek Tübingen," in Julia Freifrau Hiller von Gaertringen (ed.), *Kriegssammlungen 1914–1918*, Frankfurt/Main 2014, pp. 423–33.

[131] UAT167/172, Bl. 9, 1.20.1915.

[132] For an English translation, see Heinrich Mann, *Man of Straw*, London/New York, NY 1984.

[133] UAT 167/172, Bl. 17, 9.28.1916.

donor the warmest thanks of the ministry for this new contribution and for the active support that he has given to the war collection through his contributions.[134]

In fact, due thanks for his generosity was bestowed upon Laiblin only five years after the event. On December 23, 1924 he was appointed honorary senator of the University of Tübingen.[135] The importance that Head Librarian Geiger attached to the war literature collection is clear from his letter of May 12, 1915 to the donor:

What I and our whole people experienced in those August days, when the greatest event in our German history made such a huge impact on me, soon stood in my mind as an official duty for my profession and my institute. The records of what we lived through should, perhaps from the beginning, also be collected methodically & as extensively & completely as possible for our librar- ies & universities, for people's use & benefit in the present & future. The little exhibition I put on in fall 1913 with the resources of our library, to commemo- rate the Battle of the Nations & the year 1813, showed me almost palpably how valuable the smallest pamphlet, the tiniest image, may later prove to be as testimony & record of a great age![136]

The head librarian would be proved correct, at least in his view that "the smallest pamphlet" had value as a source. But the war books were not collected for their value to the history of ideas, but rather because they were considered *depictions* of war or even "testimony & record of a great age." Anyway, as a letter from April 12, 1919 shows, the head librarian was extremely grateful to the donor and firmly believed that the war collection would be very useful:

Just in the last few days, I have to my great joy learned once again how the col- lection is already being put to good use & how surprised people are in academia at the abundance of material that is available. We estimate that our war collec- tion comprises roughly 10,000 volumes & each volume bears the name of the donor. So the great war collection with its thousands of volumes [...] will keep your name alive as that of a great benefactor of our library, & convey to future generations the memory of the world war and of its heroic deeds and tragic outcome. A later age – I am absolutely convinced of this – will fervently desire to be informed about what things were really like in 1914–18.[137]

Even the shelf marks – peritexts in their own right – suggest that the books were regarded as sources of great historical value, and that it was by no means feared

---

[134] UAT 167/172, Bl. 27, 4.10.1919.

[135] UAT 117/291a, Bl. 30.

[136] Geiger's letter to Laiblin of May 12, 1915 is quoted from Taigel, *Louis Laiblin, Privatier*, p. 118.

[137] Geiger's letter to Laiblin of April 12, 1919 is quoted from Taigel, *Louis Laiblin, Privatier*, p. 121.

that they would serve to arouse "base instincts." The firsthand reports of Boelcke, Plüschow and von Richthofen – together with many other examples of their kind – bear the shelf mark "Fo XIIa...," which, in the classification system used at the Royal University Library of Tübingen, designated "history and geography of particular countries and peoples" (Fo), and in this case Germany and Austria (XII) and General (a).[138] The mark "Kg," on the other hand, given to such books as those on Boelcke by Anton Luebke and Rudolf Oskar Gottschalk, or to Ernst Udet's firsthand report edited by Ernst Friedrich Eisler, was reserved for "biographies and autograph correspondence." Other war books, such as Heydemarck's *Doppeldecker "C 666,"* Killinger's *Abenteuer des Ostseefliegers,* or Julius Schoenthal's *Flieger über London. Eine Erzählung aus den Spätherbsttagen 1915,* bear the mark "Dk XI," which designated "fine literature" (Dk), and in this case "Germanic – German."

It should be clear by now that people at the time held these books in high regard. The library classification "Fo XIIa" was an institutional reinforcement of the books' own claim to truth and authenticity; marks on the spine or inside cover did not involve a deliberate guiding of the reader's interest, but such paratexts certainly did influence the reception of the books. And although Head Librarian Geiger may just have been flattering the donor Laiblin, it would seem that the collection was "put to good use." Even the hint that Tübingen students were reading the books cannot, however, give us a precise answer as to the actual readership. The recipients and reception of works or ideas remain a blind spot for historical research. Still, a closer look at the *Börsenblatt für den deutschen Buchhandel* will further help to identify the hypothetical readership.

The *Börsenblatt,* the magazine for the book trade, devoted several articles to the question of "what soldiers at the front are reading." A piece with this title appeared on February 16, 1917, written by one K. Imwolde out "in the field." Having apparently been given the opportunity to study statistics on what recruits at one depot were borrowing from "libraries and reading rooms," he wanted "to make a few suggestions and to highlight certain shortcomings in the way mobile libraries have been constituted."[139] The statistics were admittedly of limited significance, since they were based on only a few libraries for a six-week period in late 1916, but it is still worth examining them in some detail.

Of the 1,393 books lent out, "1,048 came under the heading of stories and novels." Of the remaining 345 books, 275 belonged to the category "humor, poetry, art, and science." "Biographical works" were issued "18 times, war literature 43 times, and religious writings only 7 times."[140] The author stressed that the soldiers did not demand books that were "intellectual heavy going." Future acquisitions should therefore ensure that

---

[138] The classification of the old stocks in the university library can be determined from the registers of books received.

[139] K. Imwolde, "Was liest der Frontsoldat?," *Börsenblatt für den Deutschen Buchhandel,* Year 84, No. 39, February 16, 1917, pp. 149–50; here p. 149.

[140] Ibid., p. 149.

the soldier [has to make] little effort, that his reading cheers him up and reflects real life as he actually experiences it. Some of the Ullstein books do this, and that is why the soldiers ask for them so much. Even a so-called robbers' story does no harm, since the dangers of trash literature, which used to warn us away from such books, mainly cease to apply here.[141]

It cannot be said whether Imwolde was referring to the Ullstein war series; probably he had titles from their novels list in mind. Dr Faaß, the manager of another mobile library, confirms Imwolde's view. Soldiers look for "distraction at all events, as well as humor and in a few cases nature."[142] An officer on the Eastern front, who "had the opportunity to examine field bookstores and what they sold," reported the following: "Strikingly little can be done with pamphlets, even those of a so-called topical nature, especially if they refer to the war in some way. People don't want to read that kind of stuff; they have enough of the war in their own lives."[143]

On the other hand, a member of the national book trade society who was put in charge of the "battalion books" played "an important role [...] in the reader's selection from what was on offer."[144] "Books dealing with the present war" were at first "rejected out of hand," but

when the man handling the books managed to overcome this apparent prejudice, it turned out that the soldiers continued to reject anything written about the land war, whereas the reader of a book about German feats at sea (Emden, the SMS Möwe,[145] submarines, etc.) was grateful and enthusiastic, borrowing one after another all other works in the library on the same or related topics.[146]

Soldiers were also part of the readership for aviation books, since these too were entertaining, "easy going" tales of adventure that had little to do with the land

---

[141] Ibid., p. 150.

[142] See (no forename) Faaß, "Lesestoff fürs Feld," *Börsenblatt für den Deutschen Buchhandel*, No. 189, August 15, 1917, Year 84, pp. 969–70; here p. 969. He continues: "Adventure and crime stories that keep a simple soul in a state of tension and excitement and do not allow homesickness or sad thoughts to surface; cheerful and amusing stuff that eases the worst troubles of the heart; tales of strong heartfelt love and powerful fates that show a vicarious way forward for souls torn between contradictory feelings; and now and then a serious book, as easy to understand as possible, that refreshes and enlarges the knowledge required in a civilian profession: these are what a private asks for most often and most insistently."

[143] Anonymus "Was verlangen die Feldgrauen zum Lesen?," *Börsenblatt für den Deutschen Buchhandel*, No. 144, June 23, 1917, Year 84, p. 748.

[144] Anonymous, "Unsere Feldgrauen beim Lesen. Erfährungen und Erlebnisse eines Batallions-Bücherwarts," *Börsenblatt für den Deutschen Buchhandel*, No. 190, August 17, 1916, Year 83, pp. 1086–87; here p. 1086.

[145] SMS Möwe, a merchant raider of the Imperial German Navy, which carried out many successful operations against Allied shipping in the First World War.

[146] Anonymous, "Unsere Feldgrauen beim Lesen," p. 1087.

war. Like the books about "German feats at sea," they fed off the fascination with technology and conjured up an image of the war that corresponded to its legitimating pretexts. Their Don Quixote quality consisted precisely in the fact that they presented deeds of heroism and a possibility for the individual to prove his mettle. Ground troops, it was suggested, turned to such books when they needed an escape from the war "as it really was" in the trenches.

The main readers of such books, however, were the droves of schoolboys and youngsters for whom the world war was the setting for their early experiences of life. War books seem to have best satisfied their hunger for information, their need for models and their longing for "tales of chivalry." Sebastian Haffner gives a good example of this in his widely read memoirs, conveying something of the mood that fueled such books and accounted for their success:

> The analogy with the soccer fan can be carried further. In those childhood days, I was a war fan just as one is a soccer fan. [...] What counted was the fascination of the game of war, in which, according to certain mysterious rules, the numbers of prisoners taken, miles advanced, fortifications seized, and ships sunk played almost the same role as goals in soccer and points in boxing. [...] It was a dark, mysterious game [...] My friends and I played it all through the war: four long years, unpunished and undisturbed.[147]

War was a game for many young people, and the war books provided them with body counts and glamorous individual characters complete with faces. The "real men" they found there served as models, which played a role in the formation of an identity of their own that would bear strange fruit in the years to come:

> From 1914 to 1918 a generation of German schoolboys experienced war daily as a great, thrilling, enthralling game between nations, which provided far more excitement and emotional satisfaction than anything peace could offer; and that has now become the underlying vision of Nazism. [...] Many things later bolstered Nazism and modified its character, but its roots lie here: in the experience of war – not by German soldiers at the front, but by German schoolboys at home. [...] The truly Nazi generation was formed by those born in the decade from 1900 to 1910, who experienced war as a great game and were untouched by its realities.[148]

Further evidence that the "war youth generation" saw the war as an "exciting, inspirational game" comes from the diaries of Heinrich Himmler (b. 1900):[149] "Played at sword-fighting with Falk. This time with 40 army corps and Russia, France and Belgium against Germany and Austria. The game is very interesting.

---

[147] Sebastian Haffner, *Defying Hitler. A Memoir*, New York, NY 2003, pp. 14–15.
[148] Ibid., p. 17.
[149] On the war youth generation, see Ulrich Herbert, *Best. Biographische Studien über Radikalismus, Weltanschauung und Vernunft, 1903–1989*, 3rd edn, Bonn 1996.

Victory over the Russians in East Prussia (50,000 prisoners)."[150] For the young Himmler, according to his biographer Peter Longerich, the boundaries between game and reality became blurred, as they probably did for many of his contemporaries. War books played a major role in this, by presenting war as a competitive game, and it seems undeniable that they affected the view of the war and the world with which the next generation grew up.

Firsthand reports by war pilots underpinned the illusion of heroism that schoolchildren already knew from prewar texts. For according to the books in question, the air war offered opportunities for fame and adventure in which one could prove one's worth and masculinity. This was the meaning that such literature purported to create, by making war seem a glorious enterprise and killing a virile sport. It took no account of the technologized slaughter that had stripped war of its enchantment, so that in its pages war was re-enchanted and the attendant suffering and sacrifices were legitimized in the name of the Nation. Chaos was overcome by means of mythical thinking. Before we turn to the kind of meaning that this thinking generated, we must first clarify who the writers were and how the war books were produced.

Media workers and the everyday myths of warfare

On June 26, 1917 the author Willi Hackenberger, then employed as an engineer at the Albatros aircraft manufacturer, send the following request to Dr Proskauer at the chief censor's office of the general in command of the air force:

> I would ask you personally to use your influence to ensure that my book is released. I have no great wish to have sacrificed in vain all my time and effort, as well as significant sums of money, for this purpose. In fact, the interests of the fatherland do not militate against publication of this book. There may be differences of opinion about its success. So much of inferior quality has appeared in this field of war literature that this book should be judged by a different measure. If some of the reports are not so interesting, God knows that is not my fault.[151]

Hackenberger's manuscript *Die Helden der Lüfte* [Heroes of the Air], which unfortunately exists only in a 1935 version in the Freiburg military archives, evidently did not meet with the approval of the general in command of the air force. His press department replied to Hackenberger in July 1917:

> Publication of the planned book is authorized as a *purely private matter*. Neither in the book nor in the requisite appreciations and reviews should any mention be made of promotion or support of the work by the general in command of

---

[150] Diary entry for August 29, 1914, quoted from Peter Longerich, *Heinrich Himmler*, Oxford/New York, NY 2012, p. 22.

[151] BA-MA, MS g 170/10–6, letter from Willi Hackenberger to Dr Proskauer, June 26, 1917.

the air force. The foreword under consideration until now must be dropped. Before publication, the print-ready manuscript must be resubmitted here to the censorship.[152]

The fact that the general in command of the air force withdrew his patronage of the book, together with Hackenberger's attempts to prevent this and to clear up "differences of opinion," explains the importance of the paratexts. But Hackenberger's letter deserves attention for a further reason: he suspected that he would miss out on "promotion or support" because the book was "not so interesting," by which he simply meant that the "gentlemen" – the said "Heroes of the Air" – had either "had no interesting experiences" or had not known how to speak of them.[153] In any case, Hackenberger refused to attribute to the aviators "heroic deeds that they might not have performed." Was that perhaps common practice among the authors of war books? Their narratives were not simply "made up": the persons described in them had really existed; the aerial battles, shoot-downs, and bombing raids had taken place. But were the "gentlemen's" accounts not exaggerated? Were their experiences not given a literary flourish and, where need be, embroidered a little?

These questions cannot be definitely answered without a comparison of the airmen's own contributions, that is, their diary entries, correspondence and firsthand reports, with their manuscripts as originally written and edited by the media, as well as their published books. The materials collected in the archives do not permit this to be carried out. But the correspondence between Hackenberger and the air force press department, together with material discovered in the archives of Ullstein Verlag and the German Museum of Technology, do allow us to piece together at least in part how the aviation war books came to see the light of day. One thing that is clear is that, despite – or because of – their many protestations of authenticity, they can by no means be treated as "ego-documents." Even if media workers did not sprinkle in some "heroic deeds," the diary or suchlike was certainly "worked up" into a manuscript fit for publication.[154]

Even an original ego-document cannot capture the lived experience of a frontline combatant, since in the end current models of perception and interpretation, as well as the linguistic conventions of the genre, insinuate themselves between the experience and the written text. A document on which a ghostwriter has been at work, and that a copy-editor has then prepared for printing, is a long way from the "truth" about war. And yet, even if the experience of war remains a historiographic chimera, war books are still significant sources concerning the "shaping of war by the media" and the "history of what is sayable."

---

[152] BA-MA, MS g 170/10–6, letter from the air force press department to Willi Hackenberger, July 3, 1917.

[153] BA-MA, MS g 170/10–6, letter from Willi Hackenberger to Dr Proskauer, June 26, 1917.

[154] A communication from the Military Aviation Inspectorate also testifies to such procedures. See BayHStA-KA, Iluft 211, draft communication of the Military Aviation Inspectorate to all commanders of aerial combat troops and sections including Bavarian airmen, April 1918.

In any event, the aeronautics enthusiast Hackenberger had published as early as 1915 the picture book *Deutschlands Eroberung der Luft* [Germany's Conquest of the Air], which he "respectfully dedicated to His Excellency Count Ferdinand von Zeppelin."[155] Under the surtitle "Über alles in der Welt!" he wrote:

> Germans are covering distances by air from one part of the earth to another, as no other nation has done before; Germans fly over the highest mountains and the widest bodies of water. Go and fight, German eagle, shake your feathers in anger, stretch your wings upward to the sun, to the light, to freedom! Now is the time! Take the sword in your talons, hold out against storm and foul weather, and yours shall be the victory in the air![156]

Probably no "interests of the fatherland" militated against the sponsorship of Hackenberger's new manuscript. In September 1916, at an inspection of airmen, he had apparently asked for assistance with the publication of another aviation book, and the air force chief of staff, Hermann von der Lieth Thomsen, had offered his support. According to Thomsen himself, he had already "thought of a number of officers and NCOs in the aircraft and airship combat forces who, by virtue of their outstanding achievements, deserve[d] to be publicly known."[157]

Furthermore, Thomsen proposed that "the gentlemen's contributions" should be presented to him in advance for correction, together with Hackenberger's own "individual requests to the gentlemen." Indeed, Thomsen sent him a blank form to be completed for this purpose: "I suggest you send me the text of a printed letter in 100 copies, in which the gentlemen – roughly in line with the enclosed model – are asked for their participation. I will then fill in the addresses and get the contributions submitted directly to me." Hackenberger evidently planned to accompany the potted life histories of Germany's "heroes of the air" with a number of illustrative pictures. But Thomsen had a different idea – not least because "the number of colleagues involved [would have to be] limited for important reasons," so that "a work of the scale you intend to publish [would not be] achievable." Consequently, Thomsen advised "including brief descriptions that [he would] select from the experiences of his colleagues." He even thought that this would "comply with the wishes of the readership."

The staff officer seemed to have thought of everything, since he also spoke of the financial side of the venture: "I assume that the aim is *not* to give associates a share in the profits of the book. But I am willing to feed part of the profits into the fund run by the aircraft and airship troop inspectorates for the surviving dependants of aircraft and airship combatants."[158] Evidently Hackenberger agreed to

---

[155] See Willi Hackenberger, *Deutschlands Eroberung der Luft. Die Entwicklung deutschen Flugwesens an Hand von 315 Wirklichkeitsaufnahmen*, with a preface by Hellmuth Hirth, Siegen 1915, p. 22.

[156] Ibid.

[157] BA-MA, MS g 170/10–6, letter from Thomsen to Hackenberger, October 17, 1916.

[158] Ibid.

everything, for – as a communication from the press department informs us – the manuscript was ready by April 1917 for headquarters to look through. Before this could happen, however, a wish was expressed that "the pictures of His Excellency Höppner [the general in command of the air force] and Lieutenant-General Thomsen should be omitted from the book [...], since they [were] not among the 'heroes of the air'."[159]

The interventions by the press department or the air force commander were considerable. It is doubtful, however, whether the "press" (i.e. propaganda) department was involved to the same extent in the production of all aviation books. Hackenberger was a special case; the air force commander should be mentioned as the sponsor of the picture book and was indeed approached by Hackenberger directly. In other cases, the process of writing a "heroes' report" or something similar had to be authorized, and then the war book manuscript would be submitted for pre-publication censorship. This means that the press department probably prevented the publication of undesirable material (against which the "interests of the fatherland" militated), and also that it deleted certain references to time and place, the stationing of part of a particular regiment, or possibly some inglorious chapter or another.

A letter of June 1917 from Hackenberger to Höppner, in which he explains previous "misunderstandings," provides us with further details of the book project, as well as indicating the events that led to its discontinuation:

> When I applied through the air combat forces inspectorate for approval of a work "Biographies of Successful Fliers" as the second volume of my Montanus book "Germany's Conquest of the Air," it was done in an effort to make readers more familiar with our heroes of the air. At Your Excellency's request, a circular was sent to all successful aviators, but it did not produce the hoped-for results, since only a limited number of the gentlemen in question submitted interesting accounts of their flights. [...] I would mention here that I received all reports after they had been *censored*, and I originally assumed that nothing would be changed in the editing of the reports to safeguard the privacy of the texts.[160]

Hackenberger was mistaken in this belief, however, since others saw it as their duty to do precisely that: "At the prompting of the chief censor's office, I first handed over the available material unaltered for checking, but I did not claim to regard it as the finished manuscript ready for censorship." The "unfinished and unarranged material" was then forwarded to headquarters, which naturally turned it down. Thereupon, "at the prompting of the chief censor's office," Hackenberger got down to "*free editing of the material*," but "the book was returned a second time with a flat refusal." This seemed completely incomprehensible to Hackenberger. Despite the adverse circumstances, the book was "so interesting

---

[159] BA-MA, MS g 170/10–6, letter from Proskauer to Hackenberger, April 10, 1917.
[160] BA-MA, MS g 170/10–6, letter from Hackenberger to Höppner, June 21, 1917.

[...] that it would probably have sold well, since the Luftfahrerdank [airmen's relief association] wanted to take on the distribution. Ullstein & Co. committed me to a certain delivery schedule, which I was unable to fulfil because the material came in sparsely and slowly."[161]

It was all to no avail. Hackenberger ought to have sent back the "material": the (neither particularly interesting nor well written) airmen's reports. But he could no longer do that, since in the editing process he had "cut up the original reports [...] and, after making a copy, partly destroyed them." Although he stressed that he did not think of "the publication of the book as a business," but "simply took pleasure in the thing itself," the rejection continued to stand. For want of embellishments and literary assistance, the book was probably not suitable for public consumption.

A similar fate, though for different reasons, befell Dr Emil Leimdörfer's book on the Austro-Hungarian airman Gottfried Alois Freiherr von Banfield. On October 31, 1918 Ullstein Verlag was forced by "circumstances" to inform the author of the surviving typescript that the project was cancelled: "We have received the Banfield manuscript, which is really interesting and written in an exciting manner. You have worked diligently to extract everything possible from the material. Under other circumstances, the book would have been a valuable addition to our war books series, but we cannot go up against world history."[162]

Evidently the air force press department was not alone in wanting the "original reports" to be dealt with freely. Flexibility was the order of the day at Ullstein Verlag too, where it seems to have been not only permissible but desirable – or anyway common practice – to do a little "touching up" of war reports, diaries, correspondence and so on. This is confirmed by contracts in the Ullstein Verlag archives.

In 1917, *300 000 Tonnen versenkt! Meine U-Boots-Fahrten* [300,000 Tons Sunk! My Submarine Missions] was published as No. 29 in the Ullstein war books series; its author, Max Valentiner, was the highly decorated captain of the U38 submarine, who wrote in the foreword:

U38 alone had sunk some 300,000 tons of shipping when I got down to writing. [...] For military and other reasons, I have had to take a few liberties with people's names and with times and places. Otherwise I have kept strictly to the truth and reported only my own experiences.[163]

In fact, however, someone else actually wrote the book: the prolific writer and former naval officer Reinhard Roehle.[164]

---

[161] Ibid.

[162] DTM I.4.045 NL Willy Stiasny [re: Banfield typescript], letter to Leindörfer, October 31, 1918.

[163] Max Valentiner, *300,000 Tonnen versenkt! Meine U-Boots-Fahrten*, Berlin 1915, pp. 9–10.

[164] Thomas F. Schneider's biographical-bibliographical handbook mentions six books that Reinhard Roehle published under his own name between 1915 und 1917. Apparently

The Ullstein archives also contain a copy of the contract signed on March 21, 1917 between Max Valentiner and Roehle. This states:

> Herr Roehle undertakes to prepare a book to the length of the well-known Scherl war books, basing himself on diaries about the U38 missions made available to him. The print-ready manuscript shall be delivered to Kapitänleutnant Valentiner, who has the right to publish it under his own name. [...] The work shall be commenced at once and, immediately after delivery, offered to a suitable publisher. Herr Kapitänleutnant Valentiner reserves the right to alter the manuscript as he sees fit.[165]

So, the manuscript was written by a ghostwriter on the basis of diaries. When Valentiner's Ullstein war book appeared, it was 155 pages in length. Evidently Roehle had "commenced it at once," since a "note on the Reinhard Roehle contract" was already filed at Ullstein Verlag on May 25, 1917: "Herr Roehle has written the Valentiner manuscript and is authorized by the author to arrange for publication."[166]

Valentiner was the "author," it is true, but Roehle had written the manuscript and received 4,000 of the 20,000 marks that Valentiner was paid for the book. The writing of war books was lucrative, and it was also bargained over. The note on the contract reveals that, in addition to the Ullstein war book, Valentiner and Roehle planned another "book on the U-boat experiences of Herr Valentiner. Accordingly Roehle had offered us one part of the manuscript and Scherl another part. It therefore seemed appropriate to guarantee Roehle 6,000 marks in advance in order to keep all the material."[167]

The idea seems to have been to sell the manuscript to the highest bidder, and in this case it was Ullstein's rival, Scherl, that lost out. Other contractual documents, concerning the most famous of all German aviator heroes, Manfred Freiherr von Richthofen, and his *The Red Fighter Pilot*, suggest that what happened with Valentiner's book was not exceptional.

The Berlin publishing house agreed a royalty of 8 per cent with the "red baron" and guaranteed him 10,000 marks.[168] The book, No. 30 in the Ullstein war series,

---

he specialized in naval war books. His titles include: *Emden – Ayesha. Heldenfahrten und Abenteuer deutscher Seeleute im Weltkrieg. Nach Berichten von Teilnehmern erzählt*, Stuttgart 1915; *Über Anden und Meer ins deutsche Heer. Ernste und heitere Abenteuer deutscher Reservisten auf ihrer Fahrt von Bolivien in die Heimat. Der reiferen Jugend erzählt*, Stuttgart 1916; *Kriegsfahrten unseres Kreuzergeschwaders*, Stuttgart 1916; *Graf Dohnas Heldenfahrt auf S.M.S. "Möwe,"* Stuttgart 1916 (with a second edition in 1917); and *Als Flüchtling um den halben Erdball. Die abenteuerlichen Erlebnisse des Prisenoffiziers S.M.S. "Emden" Kapitänleutnant der Reserve Julius Lauterbach*, Stuttgart 1917. See Schneider et al., *Die Autoren und Bücher der deutschsprachigen Literatur zum 1. Weltkrieg 1914–1939*, p. 528. It emerges from a contract between Ullstein Verlag and Roehle, dated June 5, 1917, that he also wrote a book for young people: *Auf verschlungenen Pfaden*. See UVA, Akte Valentiner.

[165] UVA, Akte Valentiner, Abschrift des Vertrages zwischen Valentiner und Roehle vom 21.3.1917.
[166] UVA, Akte Valentiner, Notiz zum Vertrag Reinhard Roehle vom 25.5.1917.
[167] Ibid.
[168] UVA, Akte von Richthofen, Vertrag zwischen Ullstein und Manfred von Richthofen, 6. 29.1917.

would earn far more for him and (since he died on April 21, 1918) his family: 400,000 copies were sold by the end of the war,[169] in little over a year, and all editions together added up to a total of 1,226,000 by 1939.[170] It is not clear who actually wrote the bestseller, however, since von Richthofen gave the following powers to Captain Erich von Salmann, the author of *Über die Weltmeere zur deutschen Front in Flandern*, in dealings with Ullstein Verlag:

[T]o make any changes or improvements, without consulting me in advance, that may seem necessary to him concerning the book "Der rote Kampfflieger" that the company is scheduled to publish, and to represent my interests vis-à-vis Ullstein & Co. or such persons as are involved in the production of the book. Herr von Salzmann is authorized to settle any matters regarding the book independently, in my name.[171]

The contract between the publisher and von Richthofen states in §3: "Captain von Salzmann is empowered to agree the definitive text with the publisher."[172]

On September 20, 1918, Ullstein Verlag struck a deal for a further book with Lothar von Richthofen, the brother of the deceased and himself a fighter pilot. Its title was to be *Manfred Freiherr von Richthofen. Ein Heldenleben* [A Hero's Life], and despite Germany's defeat, or rather because of it, the book eventually appeared in 1920. The contract states: "The editing of the material and the putting together of the book shall be carried out by Captain Erich von Salzmann."[173]

Von Salzmann's contribution to the two books cannot be established. But it is clear that they are not ego-documents, since the censorship, the publisher and media professionals all worked on the texts. For the public that eventually read these war books, however, they reflected "the reality of the war" as participants had "themselves experienced and narrated it," to quote the prefatory remarks. An inhuman war became the theme of popular literature marketed for consumption by the "entertainment industry." An indescribable war was made tangible with the narrative schemas of pulp fiction and anachronistic interpretations handed down from earlier conflicts. A "curious literariness," understood to be the actual historical truth, thus began to overarch a reality that seemed to elude other ways of arranging it. Readers then converted this into a "curious literariness of real life," as they began in turn to update the received interpretative models.

---

[169] Schneider, *Zur deutschen Kriegsliteratur*, p. 106.

[170] Schneider et al., *Die Autoren und Bücher der deutschsprachigen Literatur zum 1. Weltkrieg 1914–1939*, p. 10.

[171] UVA, Akte von Richthofen, Vollmacht für von Salzmann.

[172] UVA, Akte von Richthofen, Vertrag zwischen Ullstein & Co. und Manfred von Richthofen, 6.29.1917.

[173] UVA, Akte von Richthofen, Vertrag zwischen Ullstein & Co. und Lothar von Richthofen, 9.20.1917.

2
Flying Swords and Mechanized Warfare

The great battles of Verdun and the Somme were long since over, unrestricted submarine warfare had been resumed, and the Russian February Revolution and the US declaration of war on Germany had already taken place, when in 1917 a book entitled *Das fliegende Schwert. Wesen, Bedeutung und Taten der deutschen Luftflotte in Wort und Bild* [The Flying Sword. Nature, Significance and Actions of the German Air Fleet in Word and Pictures] appeared in print. The editor of the book was the Deutsche Luftflottenverein e.V., founded in 1908 as a typical civilian association of the time combining nationalist politics with vested interests.[174] To the question: "What do we want?," the program of the Verein tersely replied: "Our future lies in the air!"[175]

Aviation advanced during and after the First World War to become the bearer of nationalist dreams and technological visions of the future. The airplane gradually supplanted the ship in this role. In any event, the Luftflottenverein was comparable in its structure and aims to the (maritime) Deutscher Flottenverein founded in 1898. The future of "laggard nations" – the kind of language used in and around the Luftflottenverein – lay in the air rather than on the seas. The nineteenth century had found in the ship an image that covered a range of themes, from the

---

[174] The lobbying and interest-guided politics of the Luftflottenverein supposedly benefited its first chairman and president (until Summer 1914), the Mannheim manufacturer of agricultural machinery Dr Karl Lanz, who in 1909 had founded the Schütte-Lanz airship company. As with the authors and publishers discussed in the last chapter, financial interests and nationalist ideology tended to complement, and perhaps reinforce, each other in the case of the Deutsche Luftflottenverein chairman. On the phenomenon of the civil association (*bürgerlicher Verein*), see Otto Dann (ed.), *Vereinswesen und bürgerliche Gesellschaft in Deutschland*, Munich 1984. For a survey of organized nationalist agitation, see Wehler, *Deutsche Gesellschaftsgeschichte*, vol. 3, pp. 1071–81, and for a review of the current state of research: Walkenhorst, *Nation – Volk – Rasse*, pp. 15–24.

[175] The program of the Verein was published in *Illustrierte Aeronautische Mitteilungen* 12/1908, No. 19, pp. 579–81, quoted here from Sabine Höhler, *Luftfahrtforschung und Luftfahrtmythos. Wissenschaftliche Ballonfahrt in Deutschland 1880–1910*, Frankfurt/Main 2001, p. 179. On the context of the *Luftfahrtvereine* see ibid., pp 144–206 and Helmuth Trischler, *Luft- und Raumfahrtforschung in Deutschland 1900–1970. Politische Geschichte einer Wissenschaft*, Frankfurt/Main 1992, pp. 34–48.

Enlightenment collection of knowledge through the occupation of global space to imperialist geopolitics and power play.[176] But the days of the steel battleship were numbered, and "Britannia Rules the Waves!" would also soon be history. The key task now was to conquer airspace, as the aircraft and the aviator became images of the new world order and mass society; the roots of this change, and of the thinking that underlay it, were to be found in the wartime environment. In the aftermath of that "great seminal catastrophe of this [twentieth] century,"[177] it seemed as if the future had already begun to show its face in the pivotal years of 1916 and 1917, with the industrialized killing in gigantic battles, with the steel helmet, the tank and the aircraft. In historical perspective too, the world war appears as a period of transition, when the future announced its coming, but often still did so in older guises.

The Deutscher Luftflottenverein was not the least example of the future appearing in "venerable disguise."[178] Among the tasks of the association, whose membership soared during the war from 10,000 to 50,000, were

> to promote a strong air fleet, to foster love and understanding of air travel, to maintain its own pilot schools [...], to advance peaceful air transport after the war, to involve its members in air trips, to support exigent families of pilots and aviators, and to find places in suitable spas for members in need of rest and convalescence.[179]

The book also served these aims; its cover featured the flying sword emblem of the Verein, which represented the dialectic of persistence and change, continuity and innovation, break-making and bridge-building, in an age of transition. The book collected together numerous "glorious actions of our flying sword" (Figure 2.1).[180] One account, originally published in the *Osnabrücker Zeitung*, described a flight over Verdun:

> It was a special feeling for me to fly like a king, laden with bombs, over the very terrain where my father had fought and won the Iron Cross 46 years

---

[176] On the British navy as a tool for the appropriation of the world through knowledge, see Julia Angster, *Erdbeeren und Piraten. Die Royal Navy und die Ordnung der Welt 1770–1860*, Göttingen 2012. On geopolitics and power politics, as well as the antagonisms resulting from the naval arms race, see Rolf Hobson, *Imperialism at Sea. Naval Strategic Thought, the Ideology of Sea Power, and the Tirpitz Plan, 1875–1914*, Boston, MA 2002; Paul Kennedy, *The Rise and Fall of British Naval Mastery*, London 2004; Michael Salewski (ed.), *Die Deutschen und die See. Studien zur deutschen Marinegeschichte des 19. und 20. Jahrhunderts*, Stuttgart 2002.

[177] On this much quoted formulation, see George F. Kennan, *The Decline of Bismarck's European Order. Franco-Russian Relations, 1875–1890*, Princeton, NJ 1981, p. 12; and on the new order of mass society, see Chapter II.3 below.

[178] Karl Marx, "The Eighteenth Brumaire of Louis Bonaparte," in Marx *Political Writings*, vol. 2, *Surveys from Exile*, ed. by David Fernbach Harmondsworth 1973, pp. 143–249, here p. 146.

[179] Deutscher Luftflottenverein (ed.), *Das fliegende Schwert. Wesen, Bedeutung und Taten der deutschen Luftflotte in Wort und Bild*, Oldenburg i. Gr. 1917, p. 15.

[180] Ibid., p. 15.

Figure 2.1   The flying sword as image of the dialectic of persistence and change

before. [...] I dropped my bombs on target and saw them crash with a bang down below! Then I counted again the bridges over the Meuse and flew happily home. Never in my life have I experienced anything so wonderful! Raised above everything terrestrial, flying safe and sound, you imagine you are a god! Far below on earth, a ring of smoke lay around the town: nothing but exploding grenades. The fires blazed skyward, the whole earth was churned up and torn apart – a gruesome spectacle! Apart from that, the earth looks like a plaything: green meadows and forest alternate with brown fields, and in among them lie villages like white and red dots. Hole after hole in the earth, columns of smoke in the villages; flashing explosions immediately followed the glow

and roar of the heavy artillery, and everywhere were steam, smoke, and fires –
Hell itself! And then you think of the soldiers fighting down there, who have
to capture every meter with blood, and of the losses! And I? You float like a god
above all this horror and hurl thunderbolts down on the enemy![181]

The future announces itself in this text: the airman sees himself as "a king laden
with bombs," to be revered like a god raining thunderbolts down on the enemy!
After the war, it was amid the ruins of the old world order that a search began for
the conditions under which other, new orders might come to pass.[182] In the East,
the Marxist-Bolshevik order was already taking shape against the bourgeois-liberal
system that had collapsed in the war and emerged from it as a mere shadow of its
former self. But it was still trapped, antithetically, in the bourgeois world and its
categories. The central focus of the present work, however, is the different, mythical
order that, in a break with "yesterday's world," was adumbrated in the trenches and
craters along the Isonzo or the Aisne, and in the air above Champagne and the Karst.

In this chapter, we shall illustrate this gradually emerging "new world" by reference to German hero-narratives and their characteristic interpretive models. The
next section defines the figure of the hero and explains its relationship with society. Then follows a brief sketch of the air forces operating in the First World War,
which will serve to situate the protagonists of popular aviation literature in the
actual context of the war. We shall see that the meaning of the aviation hero narrative derived from a semantic counterposition to the ground war. It was trench
warfare that produced the aviator-hero. The dynamic hero with technical skills
made up for the stalemate on the ground and its sacrificial heroism, creating an
up-to-date image of war more in line with the space of experience and the horizon
of expectation of people at the time.

The social construction of the hero

"It is a great subject, and a most grave and wide one, this which, not to be grave about
it, I have named *Hero-worship*. It enters deeply, as I think, into the secret of Mankind's
ways and vitalest interests in the world, and is well worth explaining at present."[183]
These words, which stand at the end of the famous lectures *On Heroes, Hero-Worship,
and the Heroic in History*, given in 1840 by the Scottish historian Thomas Carlyle
(1795–1881), will serve as a guiding thread for our account of the heroic.

Precisely because for Carlyle the "heroic" constituted a "great subject," he
might help to clarify the figure of the hero. It is explicitly the historicist conception of a "history" made by "great men" that underlay the cult of heroes in the
First World War, and as it was such a powerful social phenomenon such heroism

---

[181] "Ein Flug über Verdun," in ibid., pp. 69–70.

[182] See esp. Modris Eksteins, *Rites of Spring: The Great War and the Birth of the Modern Age*, New
York, 2000 [repr.]

[183] Thomas Carlyle, *On Heroes, Hero-Worship, and the Heroic in History*. Notes and Introduction
by Michael K. Goldberg, Berkeley, CA 1993 [orig.1841], p. 208.

requires definition. Thus, before we turn to actual hero-images in the air war literature, we need to clarify the social function of heroes and their relationship with those who worship them. Only when the social construction of heroism and the force of the hero-narrative have been elucidated will it be possible to grasp the significance of the popular books on wartime aviators.

For Carlyle, "great men" were as a matter of fact the ones who made history and moved in his eyes the masses;[184] in his eyes the "leaders of men" were true heroes.[185] But his insights into the function of hero worship remain valid even if one thinks of the hero as a media phenomenon and of heroism as a social construction. Indeed, that makes it all the more apparent how the hero acquired the role of "creator [...] of whatsoever the general mass of men contrived to do or to attain."[186]

The "great men" described by Carlyle were classical leaders, whose heroic character was "exclusive" since it was acquired "from a privileged political or social position" that, by definition, was not available to everyone.[187] The self-sacrificing hero, by contrast, who came to the fore with the French Revolutionary armies and general conscription in the wars of liberation, and who compensated for the cult of the leader-hero during the First World War, should be understood – to quote René Schilling –as an expression of the "longterm democratization of the cult of the dead hero." In the long nineteenth century, the self-sacrificing hero thus became a complementary figure to the leader-hero.[188] Although there are large differences between these two types, they both function as "leaders of the masses" and offer them a model to follow. Heroes, or rather the protagonists of media-transmitted narratives, may be said to personify a certain order. Hero-worshipers internalize this order by taking the hero as a model. According to Carlyle, a "divine relationship" joins the hero to the common man. But what does this actually mean?

For the Puritan Carlyle, the king represented the summit of heroism. The hero-king – he gave Cromwell and Napoleon as examples – were those

---

[184] Ibid., p. 3. See also Ute Frevert, "Herren und Helden. Vom Aufstieg und Niedergang des Heroismus im 19. Und 20. Jahrhundert," in Richard van Dülmen (ed.), *Erfindung des Menschen. Schöpfungsträume und Körperbilder 1500–2000*, Vienna 1998, pp. 323–44.

[185] Using Carlyle as an example, historians since Hayden White have raised the question of the literary nature of historiography and the affinity between myth and history or the heroes of epics and the "great men" of history. See *inter alia* Hayden White, *Metahistory. The Historical Imagination in Nineteenth-Century Europe*, Baltimore, MD 1973; and Hayden White, "The Value of Narrativity in the Representation of Reality," *Critical Inquiry* 7/1980, pp. 5–27.

[186] Carlyle, *On Heroes*, p. 3.

[187] Schilling, "Kriegshelden," p. 25. On the distinction between leader-hero and self-sacrificing hero, see ibid., pp. 25–7, and Hans Günther, *Der sozialistische Übermensch. M. Gorkij und der sowjetische Heldenmythos*, Stuttgart 1993. On the self-sacrificing hero, see Behrenbeck, *Der Kult um die toten Helden*, esp. pp. 71–6. Behrenbeck (p. 65) also provides a very useful general definition of the hero.

[188] Schilling, "Kriegshelden," p. 26. See also Michael Naumann, *Strukturwandel des Heroismus. Vom sakralen zum revolutionären Heldentum*, Königstein i.Ts. 1984, pp. 41–107.

to whose will our wills are to be subordinated and loyally surrender themselves, and find their welfare in doing so [...]. [W]hatsoever of earthly or spiritual dignity we can fancy to reside in a man, embodies itself here, to command over us, to furnish us with constant practical teaching, to tell us for the day and hour what we are to do.[189]

One subjects one's will to that of the hero and thereby takes over the values and goals that the hero embodies: such was the social function that Carlyle recognized in heroism, despite, or precisely because of, his own worship of heroes as a historian. The hero offers practical teachings to the society that worships him; he "leads" the hero-worshipper, in so far as he embodies a history to be imitated. The hero narrative encourages people to transfer his values and norms to their own life story. The subordination of one's own will to that of the hero, so dramatically portrayed by Carlyle, may be understood as a process of inscribing the hero into the worshipping subject. The history of the hero becomes the model for one's own life, providing the patterns and imagery to be used in it. This receptive activity of the hero worshipper may be described, in Foucauldian terms, as a kind of *assujetissement*, a "subjectification" that includes the sense of subjection.[190]

Like the stories of Christian saints, the hero narrative serves as the matrix into which one's own life is inserted, as the yardstick by which one's own life should be measured. The approximation to the hero that the worshiper strives to achieve becomes the means of "disciplining" him; his *imitatio heroica* "trains" him, so to speak. Thus, as soon as the hero worshiper attempts to grasp his own life in accordance with the models placed at his disposal, hero worship becomes practically relevant. The hero then forms a porous border between real fiction and fictional reality, between media and life. The "curious literariness of real life," highlighted by Paul Fussell in his portrayals of war participants, is based not least upon the force of the hero narratives circulating at the time, from which are taken

---

[189] Carlyle, *On Heroes*, p. 169. We can already see why fascists appreciated Carlyle so highly. Kershaw tells us that Goebbels, who was "gripped" by Carlyle's biography of Frederick the Great, gave it to Hitler as a gift in March 1945: "He read out to him the passages relating the King's reward for his unbending resolution in circumstances of mounting despair during the Seven Years War by the sudden and dramatic upturn in his fortunes. Hitler's eyes filled with tears" (Ian Kershaw, *Hitler 1936–1945. Nemesis*, New York, NY 2000, p. 783). The American historian Jacob Salwyn Schapiro even called Carlyle a prophet of fascism: see his "Thomas Carlyle, Prophet of Fascism," *Journal of Modern History* XVII/1945, pp. 97–115. Ernst Cassirer too asserted a nexus between Carlyle and Nazi ideology, and his characterization of Carlyle is particularly significant in this connection: "He wished to stabilize the social and political order and he was convinced that for such a stabilization he could recommend no better means than hero worship. [...] To him hero worship was the oldest and firmest element in man's social and cultural life." Ernest Cassirer, *The Myth of the State*, pp. 189ff.

[190] See Michel Foucault, *Discipline and Punish* [1975], New York, NY 1995. Jochen Hellbeck, the historian of Eastern Europe, has shown with reference to Foucault's concept of *assujetissement* how Soviet diary writers inscribed the ideology of the new Soviet man in their own lives, thereby realizing the self-transformations they were expected to carry out. The procedure was similar to the one through which hero narratives provided the model for self-transformation. See Jochen Hellbeck, *Revolution on My Mind. Writing a Diary under Stalin*, Cambridge, MA 2006.

the models of perception and interpretation, as well as the gender roles, images of war and maxims of conduct.[191] The hero is the medium through which "real life" is given a literary orientation, and literature itself is converted into real life.

But the hero does not only have effects on society and its characteristic forms of behavior; he is also a social construct. He is a factor, but also an indicator, since the hero is always a screen onto which the values and norms – as well as the wishes, hopes, and fears – of the society that worships him are projected. As a personified ideal of virtue, he refers back to the society from which he stems. Deconstruction of the hero therefore permits conclusions to be drawn about the collective that produced him. But how should we understand the relationship between hero and society, or between hero and hero worshipper? What is the nature of Carlyle's "divine relationship" between the two?

The phenomenon that Carlyle highlighted in his lectures may be understood with the help of the concept of "charisma," which Max Weber applied three-quarters of a century later, under the impact of the Great War, to a form of rule.[192] For Weber the hero displays "a certain quality [...] by virtue of which he is considered extraordinary, and treated as endowed with supernatural, superhuman, or at least specifically exceptional powers or qualities," which are "regarded as of divine origin or as exemplary, and on the basis of [which] the individual concerned is treated as a 'leader'."[193]

The concept of charisma makes clear the foundations of the relationship between the figure of the hero and his worshipers. Whereas in Carlyle "great men" seem to be really endowed with "divine" gifts, so that the hero has a "transcendent" foundation, charisma is here understood as a powerful (discursively produced) social ascription and function that helps to identify the relationship between heroes and hero worshipers. The fact that the hero can develop his charismatic effect by virtue of what hero worshipers ascribe to him does not mean, however, that the hero has no "exceptional" gifts. The war heroes of the narratives under consideration stood out on account of their special military achievements. But their "gift" became the basis of hero worship only because of its historical context and the projections of admirers that gave the hero's feats a special meaning.

The liminal experience resulting from the constant presence of death in war is favorable to the development of charismatic relationships. As Weber writes:[194]

> As the consummated threat of violence among modern polities, war creates a pathos and a sentiment of community. War thereby makes for an unconditionally

---

[191] Fussell, *The Great War and Modern Memory*, p. ix.

[192] The concept of "charismatic rule" or "domination" was developed in *Economy and Society* (1922). But according to Joachim Radkau, Weber first used it with regard to Stefan George in a letter of June 9, 1910 to Dora Jellinek, and the "theory of charisma" was present from around 1913. See Joachim Radkau, *Max Weber. A Biography*, Cambridge 2009, p. 394.

[193] Max Weber, *Economy and Society*, Berkeley, CA 1968, vol. 1, p. 241 [Chapter III.10]. See also Stefan Breuer, "Das Charisma der Nation," in Stefan Breuer (ed.), *Bürokratie und Charisma. Zur politischen Soziologie Max Webers*, Darmstadt 1994, pp. 110–43, and "Das Charisma des Führers," in ibid., pp. 144–75.

[194] Cf. Breuer, "Das Charisma der Nation," p. 142.

devoted and sacrificial community among the combatants[;...] religions can show comparable achievements only in heroic communities professing an ethic of brotherliness. Moreover, war does something to the warrior which, in its concrete meaning, is unique: it makes him experience a consecrated meaning of death which is characteristic only of death in war. The community of the army standing in the field today feels itself – as in the times of the warlord's "following" – to be a community unto death, and the greatest of its kind. Death on the field of battle differs from death that is only man's common lot. [...] Death on the field of battle differs from this merely unavoidable dying in that war, and in this massiveness *only* in war, the individual can *believe* that he knows he is dying "for" something. [...] The very extraordinary quality of brotherliness of war, and of death in war, is shared with sacred charisma and the experience of communion with God, and this fact raises the competition between the brotherliness of religion and of the warrior community to its extreme height.[195]

This passage originates from the year 1916. It may thus be read as a primary source in its own right, not just as sociological secondary literature. Here, however, it is not possible to go further into exegetical detail by considering how Weber's thought is related to the First World War; nor is it necessary to elaborate the nexus of charisma, religious communities and political domination that Weber set out to analyze.[196] For an analysis of the popular war literature, what we need to establish is that the hero served as the embodiment and medium of that belief in "dying for something" to which Weber referred. The aviator-hero, who is usually both leader-hero and self-sacrificing hero, saint and martyr, embodies that belief and gives a meaning to death and suffering in war.

In any event, an "existential crisis breaking up all traditional ideas"[197] is of central importance in the establishment of a charismatic relationship between hero-figure and hero worshiper. The existential crisis, which should also be understood as an orientation crisis, increases the disposition of society to hero worship. The First World War was such a crisis, which also operated as a "transcendence generator" *par excellence*.[198] As we saw in Chapter I.1, it was not only the ubiquitousness of death but also the discrediting of the rationalist worldview that tended to rein-

---

[195] Hans Heinrich Gerth/C. Wright Mills (eds.), *From Max Weber. Essays in Sociology*, Boston, MA 1948, pp. 335–36.

[196] On the link between hero worship (our main focus here) and charismatic rule (Weber's central preoccupation), suffice it to say that the former is a precondition for the establishment of the latter. On charisma, the success of the Mussolini and Hitler cults, and the establishment of charismatic rule, see *inter alia*: Gentile, *The Origins of Fascist Ideology, 1918–1925*, New York, NY 2005; Ian Kershaw, *The "Hitler Myth". Image and Reality in the Third Reich*, New York, NY 2001; Franz Neumann, *Behemoth. The Structure and Practice of National Socialism*, New York, NY 1942; Hans-Ulrich Wehler, *Deutsche Gesellschaftsgeschichte*, vol. 4, *Vom Beginn des Ersten Weltkrieges bis zur Gründung der beiden deutschen Staaten 1914–1949*, Munich 2003.

[197] Wehler, *Deutsche Gesellschaftsgeschichte*, vol. 4, p. 553.

[198] On the concept of "transcendence generator," see Justin Stagl, "Immanenz und Transzendenz – ethnologisch," in Jan Assmann/Rolf Trauzettel (eds.), *Tod, Jenseits und Identität. Perspektiven einer kulturwissenschaftlichen Thanatologie*, Freiburg 2002, pp. 562–74.

force supposedly atavistic dispositions to magic, supernaturalism and transcendentalism only superficially covered over by the faith in reason. The charismatic hero, like the Christian saint, mediates between the divine and the earthly.[199] And he must, as Weber puts it,

> work miracles, if he wants to be a prophet. He must perform heroic deeds, if he wants to be a warlord. [...] The mere fact of recognizing the personal mission of a charismatic master establishes his power. Whether it is more active or passive, this recognition derives from the surrender of the faithful to the extraordinary and unheard-of, to what is alien to all regulation and tradition and therefore is viewed as divine – surrender which arises from distress or enthusiasm.[200]

Regardless of the supernatural origins ascribed to charisma, there is a "real" foundation to the "surrender" and the hero worship. As we mentioned before, the aviator-heroes performed martial feats that people admired. But, according to Sabine Behrenbeck,

> the hero's greatest and most admirable feat was to risk his life for the cause (the common good), and if necessary to sacrifice it. The value of the cause (existence of the group, preservation of its order) as something higher than personal well-being or survival can be gauged only by his preparedness for this ultimate, unconditional commitment.[201]

Here we should recall once more our underlying definition of myth, which may throw further light on the socially significant function of the hero. The achievement of myth, whose protagonist is often a hero-figure, is of a normative character.[202] By staking his life, the hero grounds the prevailing norms and values, legitimizes going to war, and sacralizes death in battle for the fatherland.[203] His sacrifice attests to the supreme value or "divinity" to which it is made, and which in turn legitimizes customs and virtues. The staking of his own life in war is thereby justified, and so too is the cause for which he fights. The hero validates death for the community and Horace's famous saying *dulce et decorum est pro patria mori*. In so far as his example conveys the meaningfulness of "dying for something," the hero serves to provide an orientation and a guide to action; he legitimizes both deeds on the battlefield and passive endurance of events. In the words of Mircea Eliade:

> Human life becomes meaningful by imitating the paradigmatic models revealed by supernatural beings. The imitation of transhuman models constitutes one

---

[199] On the hero-figure in Christianity, see Schilling, "Kriegshelden," pp. 186ff. and 283ff.
[200] Weber, *Economy and Society*, vol. 2, pp. 1114–115.
[201] Behrenbeck, *Der Kult um die toten Helden*, p. 66.
[202] Frank, *Gott im Exil*, pp. 16–17.
[203] Here and on the following, cf. Behrenbeck, *Der Kult um die toten Helden*, pp. 65–6. and 71–2.; and Latzel, *Vom Sterben im Krieg*.

of the primary characteristics of "religious" life, a structural characteristic [...]. From the most archaic religious documents that are available to Christianity and Islam, *imitatio dei* as a norm and guideline of human existence was never interrupted.[204]

Imitatio heroica served the same end and helped to integrate ubiquitous death into one's own life. For in the hero narratives, it is not only that death does not appear purposeless; the hero, through his death, transcends ordinary, mundane time and enters into sacred time. This temporal aspect again refers back to the mythical dimension of the hero-narrative:

> The victorious hero receives the thanks and reverence of those he has saved. If he gave his life in the action, contributing to the victory against disaster or enemy forces (or sealing the validity of the moral code with his blood), the community will accord him a lasting place in its hall of fame. Therein lies the immortality of the hero. [...] In this way, the hero becomes a savior who still functions after death and transcends profane time: he has overcome death or has risen again from the dead. [...] Myth interprets the hero's death as a change in his mode of existence; the transformation makes it bearable.[205]

It should be noted that the central function of the hero-narrative is to give a purpose to action and a meaning to suffering, war, and death that transcends profane time. The hero's death wins him eternal life in the memory of the nation or people, as well as ensuring actual life for that nation or people. Thus, in the previously quoted Langemarck speech *Die Wiedergeburt des heroischen Menschen*, which the conservative revolutionary Hans Schwarz delivered in 1928, we read:

> A young men's dance of death then passed over from the chronicler's time-bound account to the eternal youth that belongs to the mythical estate of all peoples. [...] The end of the war has embittered people's hearts, and many think of it as the last word of history. [...] One can turn the past into a coffin, because it has taken all life away from us – but one can also move rocks with it and tear the present apart! It depends on our attitude to death. When the dead no longer have any rights, the living lose them too – and in the end the best of presents sinks down among the fish. But when life embraces the dead in our midst, the fire that creates myths begins to believe. They point into the future and appeal to freedom, and no history can be so full of attack! They adorn

---

[204] Mircea Eliade, "Preface," in Eliade, *The Quest. History and Meaning in Religion*, Chicago, IL 1969, unpaginated.

[205] Behrenbeck, *Der Kult um die toten Helden*, p. 67. Cf. Cassirer (*The Myth of the State*, p. 49): "Death, [myth] taught, means no extinction of man's life; it means only a change in the form of life. [...] In mythical thought the mystery of death is 'turned into an image' – and by this transformation, death ceases being a hard, unbearable physical fact; it becomes understandable and supportable."

their ancestors' faces with features that only their grandchildren will bear, and they rejuvenate the peoples with that elan that can carry away the defeats of history as on wings.[206]

The heroes stepped out of history into an eternal time. When the war was lost, it seemed as if their sacrifice had been in vain. It was therefore necessary to destroy the past (which had led to defeat) as well as the present, and to revive the eternal ideals for which the heroes had died. The dead joined together the past, present and future, anticipating the latter through their actions. The heroes of the aviation narratives ensured the "elan" that "can carry away the defeats of history as on wings." They popularized a mythical order based on the nation, and they inscribed it in people's everyday lives.

Before we turn to the hero-narratives themselves, the next section will outline the factual context in which they had their roots.

The eyes of the army: a brief sketch of the air war 1914–1918

And along with the giant airships, the almost birdlike aeroplanes developed into a weapon more important than anyone could have imagined five years earlier. [...] How to use the air fleet, was for a long time an unsolved puzzle. [...] The war has shown us that we have in the aircraft an extraordinary means of scouting the terrain. [...] But it can also be used in battle. Almost every day we read that our airmen appear over enemy cities or defenses and drop bombs on them. But we also read how battles take place in the air between aviators who chase and shoot at each other, though mostly with negative results.[207]

This accurate picture of the development of the German air force until 1915 appeared in the seventh war issue of *Die Luftflotte*, a periodical published by the Deutscher Luftflottenverein e.V. In this section, we shall briefly sketch the course of the air war and the development of aerial combat forces.[208] The German air

---

[206] Schwarz, *Die Wiedergeburt des heroischen Menschen*, pp. 6–7.

[207] Anonymous, "Über unsere Luftflotte," *Die Luftflotte*, VII/1915, No. 2, pp. 10–11.

[208] Although there is no shortage of popular accounts of the air war in 1914–1918, critical studies are harder to find. Special mention should be made of: Lee Kennett, *The First Air War. 1914–1918*, New York, NY 1991; three works by John H. Morrow Jr: *German Air Power in World War I*, Lincoln, NE 1982, *The Great War in the Air. Military Aviation from 1909 to 1921*, Washington, DC 1993, and *Knights of the Sky*; and Harald Potempa, *Die Königlich-Bayerische Fliegertruppe 1914–1918*, Frankfurt/Main 1997. Despite its ideological slant, Olaf Groehler, *Geschichte des Luftkrieges 1910 bis 1970*, Berlin 1970 is also worth consulting. If read with a critical eye, the books by Ernst von Hoeppner (former commander of the air force) and especially Georg Paul Neumann (director of the Berlin-Adlershof pilots' school and former adjutant of the air corps inspectorate), which appeared shortly after the war, are informative about the factual course of the air war. See Ernst von Hoeppner, *Deutschlands Krieg in der Luft. Ein Rückblick auf die Entwicklung und die Leistungen unserer Heeres-Luftstreitkräfte im Weltkriege*, Leipzig 1921; and Georg Paul Neumann, *Die deutschen Luftstreitkräfte im Weltkriege. Unter Mitwirkung von 20 Offizieren und Beamten der Heeres- und*

force, and especially its fighter pilots on the Western Front, are our main focus here, because they were at the center of the popular war literature analyzed in this chapter. This now needs to be further contextualized.[209]

The air force (*Luftstreitkräfte*) was an integral part of the division of labor in an increasingly technological[210] and industrialized "total" war.[211] In 1914, however, it was "still an unknown concept" (according to Georg Paul Neumann, writing in 1920).[212] Yet the skeptically viewed air corps, which had grown out of "the meager beginnings of German air sports in 1909,"[213] would become a military branch that revolutionized warfare in the twentieth century. At first, in Wilhelmine Germany, the airship stood in the way of a straightforward emphasis on the airplane as a military factor.[214] The eventual decision in favor of the airplane came only during the war itself, so that in 1914 the German Reich had at its disposal 13 airships

---

Marine-Luftfahrt, Berlin 1920 (for a partial English translation, see *The German Air Force in the Great War, Comp. by Major Georg Paul Neumann, from the Records and with the Assistance of 29 Officers and Officials of the Naval and Military Air Services*, London 1921). Both would find a place in many later works on aerial warfare.

[209] The air war took place on all fronts in the First World War, but, as in the case of the ground war, the Western Front shaped how it was perceived. It was there that the great bulk of troops were deployed and most of the aerial battles took place. On the Eastern front too, the main activity of the air force consisted of aerial observation – although it was considerably more difficult because of the large spaces involved. Aerial battles were much less frequent there. According to Kennett (*The First Air War*, pp. 175–206), of the 7,425 German air victories, only 358 were won in the East. German air forces were also active, especially in observation, in the Balkans, Turkey and the Middle East, as well as in the sea war.

[210] On the nexus of war and technology in the First World War, see: Mary R. Habeck, "Technology in the First World War: The View from Below," in Jay Winter/Geoffrey Parker/Mary R. Habeck (eds.), *The Great War and the Twentieth Century*, New Haven, CT 2000, pp. 99–131; Kehrt, *Moderne Krieger. Die Technikerfahrungen deutscher Militärpiloten 1910–1945*, Paderborn 2010; Dennis E. Showalter, "Mass Warfare and the Impact of Technology," in Roger Chickering/Stig Förster (eds.), *Great War, Total War. Combat and Mobilization on the Western Front, 1914–1918*, New York, NY 2000, pp. 73–93; Rolf Spilker/Bernd Ulrich (eds.), *Der Tod als Maschinist*, catalog, Bramsche 1998; Peter Wilding, "Krieg – Technik – Moderne. Die Eskalation der Gewalt im 'Ingenieur-Krieg.' Zur Technisierung des Ersten Weltkrieges," in Petra Ernst/Sabine A. Haring/Werner Suppanz (eds.), *Aggression und Katharsis. Der Erste Weltkrieg im Diskurs der Moderne*, Vienna 2004, pp. 163–86. On "technological heroes" in the First World War, see Schilling, "Kriegshelden," pp. 252–88.

[211] See Chickering/Förster (eds.), *Great War, Total War*, and *The Shadows of Total War. Europe, East Asia, and the United States, 1919–1939*, New York, NY 2003. Also, Stig Förster, "Das Zeitalter des totalen Krieges, 1861–1945. Konzeptionelle Überlegungen für einen historischen Strukturvergleich," *Mittelweg 36/1999*, 8, pp. 12–29.

[212] Neumann, *Die deutschen Luftstreitkräfte im Weltkriege*, p. 3.

[213] Ibid., p. 58.

[214] On the airship, see Wolfgang König, *Wilhelm II. und die Moderne. Der Kaiser und die technisch-industrielle Welt*, Paderborn 2007, pp. 68–83; Syon, *Zeppelin!*; and Fritzsche, *A Nation of Fliers*, pp. 9–58. On the prewar building of the air force and the priorities of the war ministry, see Morrow, *The Great War in the Air*, pp. 17ff., and *German Air Power in World War I*, pp. 3–13.

alongside its 218 to 250 combat-ready aircraft.[215] Austria-Hungary had a further 48 flying machines. The Entente, on the other hand, already had more than 500 airplanes, including 244 Russian (although not all of these were suitable for use at the front).[216] The numerical inferiority of the Central Powers persisted all through the war.[217]

Whereas in August 1914 the flying personnel of the German armed forces was 500 strong, there were around 5,000 on active service in 1918. These figures testify to the huge expansion of the air force, but they also confirm that, in contrast to its high media profile, this branch of the armed forces remained numerically insignificant: the Kingdom of Italy, for example, ended the war with 2,000 airmen and a ground army of 3,500,000.[218] The casualty figures tell a similar story. In the years from 1914 to 1918, approximately 6,840 German airmen died, and a further 1,371 were reported missing.[219] If we look more closely at the losses, another aspect becomes clear: 1,962 of the dead German airmen lost their lives on service in training flights or accidents, and another 1,859 were wounded in training flights or on planes returning to base.[220] The accident statistics show how undeveloped the technology still was, and how great the dangers of flying were even in the absence of enemy fire.

"The aircraft types available to us at the beginning of the war," wrote Lieutenant-Colonel and Air Corps Inspector Wilhelm Siegert, "were like the archaeopteryx in the dinosaur age or, to use a modern comparison, flying wire fences."[221] But

---

[215] Neumann, *Die deutschen Luftstreitkräfte im Weltkriege*, p. 19.

[216] Morrow, *The Great War in the Air*, pp. 81ff.

[217] By January 1, 1919 a total of 47,637 airplanes and 40,449 aircraft engines had been delivered to the German air force. During the same period, the French aviation industry built 67,982 aircraft; the British roughly 50,000; the Italian 12,031; and the American between 11,227 and 13,840. The figures indicate how greatly the various air forces expanded in the course of the war; all the totals, including those for air losses, come from Neumann, *Die deutschen Luftstreitkräfte im Weltkriege*, p. 596. Groehler, for his part, puts total aircraft production at 47,831 for Germany and 138,685 for the Entente Powers, while Morrow estimates that the German air force had between 45,704 and 47,931 aircraft at its disposal. See Olaf Groehler, *Geschichte des Luftkrieges 1910 bis 1970*, (East) Berlin 1970², pp. 51 and 58; Morrow, *German Air Power in World War I*, S. 202. See also Sönke Neitzel, "Zum strategischen Mißerfolg verdammt? Die deutschen Luftstreitkräfte in beiden Weltkriegen," in Bruno Thoß/Hans-Erich Volkmann (eds.), *Erster Weltkrieg – Zweiter Weltkrieg. Ein Vergleich. Krieg, Kriegserlebnis, Kriegserfahrung in Deutschland*, Paderborn 2002, pp. 167–92. On the aviation industry, see Martin Bach, *Luftfahrtindustrie im Ersten Weltkrieg. Mobilisierung und Demobilisierung der britischen und deutschen Luftfahrtindustrie im Ersten Weltkrieg*, Allershausen 2003.

[218] Kennett, *The First Air War*, pp. 83ff.

[219] All in all, the German Reich lost 8 per cent of all male combatants, 2,037,000 soldiers, and a further 700,000 civilians. In Italy, a similar number of civilians died, together with a total of roughly 460,000 soldiers. See Rüdiger Overmans, "Kriegsverluste," in Gerhard Hirschfeld/Gerd Krumeich/Irina Renz (eds.), *Enzyklopädie Erster Weltkrieg*, Paderborn 2003, pp. 663–66.

[220] Neumann, *Die deutschen Luftstreitkräfte im Weltkriege*, p. 586.

[221] Ibid., p. 114.

in the course of the war, new kinds of deployment and a resulting increase in specification standards produced technological breakthroughs: "Engine power tripled in a normal observation aircraft and increased fivefold in large aeroplanes [...]. Airspeed rose [from 70 to 80 kms] to 150–160 kms and in high-performance models to 200 kms an hour."[222] Flying altitude, which originally averaged 800 meters, increased to around 7,000 meters. There was a similar spurt in industrial capacity: whereas the German aviation industry was delivering 50–60 aircraft a month in summer and fall 1914, it managed 2,195 aircraft and 1,915 engines in October 1918.[223] Its Italian counterpart produced a total of approximately 12,000 aircraft, and in 1918 was achieving some 541 machines a month.[224]

At the beginning of the war, however, it was still unclear what the air force would be used for; in fact, the military leadership distrusted "the detail [...] of the [air reconnaissance] reports and was happy to await confirmation by other means."[225] When the front became more or less stationary in fall 1914, following the Battle of the Marne and the "race to the sea," the generals turned to reconnaissance aircraft and began to place new demands on the men and machines of the air force; close-range tactical observation complemented operational and strategic purposes. The aim was to scout the frontline for artillery emplacements, ammunition depots and troop concentrations, and to monitor the highways and railroads for enemy movements. The front and its hinterland were photographed from the air, and growing importance was attached to the spotters who helped the artillery to find its range.

Many innovations that would revolutionize aerial warfare were introduced as early as 1915.[226] Whereas airmen had previously been armed at most with pistols, carbines or automatic rifles, French aircraft appeared at the front in spring 1915 with fixed machine-guns capable of firing in parallel to their longitudinal axis. This innovation brought into being the fighter ace,[227] whose success on the German side increased considerably when the Dutch manufacturer Anthony Fokker managed to synchronize the machine-gun with the engine. In Spring 1915, the Germans had already achieved a more efficient coordination of their air force through the appointment of a head of battlefield aviation. Now, in the second half of 1915, the Fokker single-seater and the equipment of aviation divisions with armed C-aircraft made it possible to overcome their technological inferiority.[228]

---

[222] Ibid., pp. 114–15.

[223] Morrow, *German Air Power in World War I*, pp. 138 and 204–05.

[224] Andrea Curami, "I primi passi dell'industria aeronautica italiana," in Paolo Ferrari (ed.), *La grande guerra aerea 1915–1918. Battaglie – industrie – bombardamenti – assi – aeroporti*, Valdagno 1994, pp. 97–139; here p. 98.

[225] Hoeppner, *Deutschlands Krieg in der Luft*, p. 8.

[226] On the new deployments and tactics in 1915, see also BayHSt-KA, Iluft 34, Entwurf zu einer Anleitung über den Kampf mit Flugzeugen vom 1.10.1915.

[227] See Wohl, *A Passion for Wings*, pp. 203–10.

[228] Hoeppner, *Deutschlands Krieg in der Luft*, p. 42.

When the Kingdom of Italy declared war on Austria-Hungary on May 23, 1915 – the declaration against the German Reich followed only in August 1916 – it had just 58 aircraft.[229] The Italian aviation industry was a negligible quantity in comparison with the German or French, mainly producing – apart from its own Caproni bombers – French models under license. Yet as early as 1911 Italy had deployed an aircraft formation in its war with Turkey, gaining experience both in photography and in bombing with darts. In 1915 the Italian air force, like its counterparts, was used mainly for observation purposes. Aside from the exploits of Gabriele D'Annunzio, the Italian theatre was marked by the characteristics of mountain warfare and the special importance given to Gianni Caproni's bombers.[230] Despite the four battles of the Isonzo initiated by Chief of Staff Cadorna, the Italian annual report for 1915 mentioned only seven air battles, 41 aerial photography missions and 28 raids with aerial darts.[231] In the following year, however, the air war intensified over the Alps and the Karst as well as on other fronts.

The early phase of improvization gave way to greater specialization of aircraft and pilots and a generally more methodical use of air forces. Distant and close observation were organized separately, and a distinction was made between bombers and the use of aircraft for aerial combat, artillery direction or infantry support, and battlefield operations; special principles and tactics were developed for the different types of aircraft. In addition, it became ever clearer that the state had to organize the further industrialization of aircraft production in order to achieve the necessary quantitative targets.

While reconnaissance became more extensive and remained the central func-tion of the air force, the emergence of the fighter-pilot was the second most important development. He was the principal figure in the war literature we have been analyzing, since it was "in the nature of things" – as Georg Paul Neumann put it – that individual achievements

> in the many other areas of responsibility did not have such a high profile [...].
> These lacked the distinctive romanticism that came down to us from solitary
> heights like an echo from the long-lost chivalrous age of heroic battles and,
> perhaps unconsciously, was and still is received by the German soul with

---

[229] On the construction of the Italian air force in the First World War, and the course of the air war on the Italian front, see Alessandro Massignani, "La guerra aerea sul fronte italiano," in Ferrari (ed.), *La grande guerra aerea 1915–1918*, pp. 17–55; Alessandro Fraschetti, *La prima organizzazione dell'aeronautica militare in Italia dal 1884 al 1925*, Rome 1986, pp. 41–84; Alessandro Massignani, "La grande guerra. un bilancio complessivo," in Paolo *Ferrari* (ed.), *L'aeronautica italiana. Una storia del Novecento*, Milan 2004, pp. 267–99; and Morrow, *The Great War in the Air*. For a detailed bibliography on Italian military aviation, see Andrea Curami, "L'aeronautica italiana dalle origini ai giorni nostri," in Piero Del Negro (ed.), *Guida alla storia militare italiana*, Naples 1997, pp. 191–209.

[230] Caproni owed this not least to its promotion by the air war theorist Giulio Douhet.

[231] Massignani, *La guerra aerea sul fronte italiano*, p. 28.

double gratitude, in its refreshing contrast to our present-day experience that has desperately little in common with romanticism and chivalry.[232]

In fact it was positional warfare – conducted also in the air – and the great technological battles that enhanced the popularity of the fighter-pilot. For the killing fields of Verdun and the Somme brought together in huge concentrations not only artillery, munitions and troops, but also, for the first time, military aircraft.[233] At Verdun, where the Fifth Army began the slaughter on February 21, 1916 with fire from 1,220 artillery pieces, "twelve aircraft detachments, four fighter squadrons and 30–40 single-seaters" were in action.[234] Their main function was still reconnaissance, as well as the creation of an air barrier to conceal preparations for an attack. At first the Germans did succeed in gaining air superiority, but by May the balance of forces had shifted in favor of the French, who not only had a new aircraft, the Nieuport, superior in many respects to the German machines, but had developed a promising new tactic. Fighters were no longer used only to protect observation aircraft, but were formed into *groups de combat* to take on the enemy air force. At the Somme, the Germans responded in kind by introducing fighter squadrons under the command of Oswald Boelcke.[235] The Battle of the Somme became, as Wilhelm Siegert put it, "the university of aviation, decisively influencing its organizational and technological development as well as the training of flying crews until the end of the war."[236]

At first, though, the German air force was anything but successful; whatever the military command might say, no amount of "guts" and "heroism" could make up for its technological and numerical inferiority. The air force commander could still write in 1921:

The enemy's huge numerical advantage, limitless input of pilots and technical personnel, and abundant raw materials available from around the world for his highly developed industry gave him a considerable superiority in the air. [...] Our adversaries relied on plentiful supplies; we constantly had to devise new makeshifts. [...] If the Entente still failed to crush us in the air, we owe it to the bright, conquering spirit of our airmen. [...] What outsiders may sometimes have condemned as youthful levity, was nothing other than the result

---

[232] Neumann, *Die deutschen Luftstreitkräfte im Weltkriege*, p. 452.

[233] On the battles of Verdun and the Somme, see John Keegan, *The First World War*, New York, NY 1999, pp. 278–86 and 286–99; and David Stevenson, *1914–1918. The History of the First World War*, pp. 200ff. On the Somme, see also Gerhard Hirschfeld, "The Battle of the Somme 1916," in Gerhard Hirschfeld/Gerd Krumeich/Irina Renz (eds.), *Scorched Earth. The Germans on the Somme 1914–1918*, Barnsley 2009.

[234] Hoeppner, *Deutschlands Krieg in der Luft*, p. 51.

[235] On the command structure and other organizational aspects, see Karl Köhler, "Organisationsgeschichte der Luftwaffe von den Anfängen bis 1918," in Militärgeschichtliches Forschungsamt (ed.), *Handbuch zur deutschen Militärgeschichte 1648–1939*, vol. 3, Part V, Munich 1979, pp. 283–311; and Potempa, *Die Königlich-Bayerische Fliegertruppe*, pp. 40–61.

[236] Neumann, *Die deutschen Luftstreitkräfte im Weltkriege*, p. 469.

of activity based upon optimism, daring, and disregard for one's own life. Given the length of the war and the enemy's growing numerical advantage, the efforts of all superior officers in the force went into keeping alive this fresh and open spirit.[237]

More and more often, the German forces had to counter the Entente's strength in personnel and industry – which grew considerably after the USA joined the war in April 1917 – with a reliance on their own "moral superiority."[238] It is true that, after the slaughter on the Somme, tactical and organizational modifications, the input of fresh forces, and the supply of new and better aircraft (the so-called *D-Flugzeuge*) brought about a temporary improvement in the situation. But in the longer term, the Germans could do little to counter the superior industrial potential of the Entente and the quantitative superiority of its air forces.

The massing of forces in the sky over the next two years was also reflected in the emergence of the wing squadron, an attack group of 30–40 aircraft. The memorandum of the most successful German fighter-pilot, Freiherr Manfred von Richthofen, recalls: "The aim of such a strong squadron is to annihilate enemy squadrons. In this case it is inappropriate for the commander to attack individual pilots. [...] A close attack formation is the secret of success."[239] Sometimes, of course, a dogfight would break up the squadron, but the individual pilot would be fighting a losing battle if he did not subsequently manage to rejoin the formation. "To avoid unnecessary losses, if individual members do not connect up again, they must fly home and not remain isolated at the front."[240] Whereas popular war books emphasized the daredevilry of youthful lone rangers, the "man to man" struggle of "knightly duels," the reality was clearly different; it was not the isolated fighter but the tightly integrated attack formation that led to success. Even so, the "tactics of aerial combat" were pretty rudimentary. Von Richthofen, who notched up 80 kills, had this to say:

> You could sum up the "tactics of aerial combat" in one sentence: "I go as close as fifty meters behind the enemy and take aim, then my adversary falls." Those are the words Boelcke fobbed me off with when I asked him for his special trick. Now I know that that is the whole secret of shooting a plane down. You don't need to be a flying acrobat or trick shooter, only have the courage to fly right up close to the enemy.[241]

---

[237] Hoeppner, *Deutschlands Krieg in der Luft*, pp. 77–8.

[238] Ibid., p. 78.

[239] BayHStA-KA, Iluft 38, Abschrift eines von von Richthofen verfassten Erfahrungsberichts, pp. 3–4. In 1938 a slightly edited version of this first-hand report was published by Mittler & Sohn as von Richthofen's "testament": *Manfred von Richthofen. Sein militärisches Vermächtnis*, ed. the Kriegswissenschaftliche Abteilung der Luftwaffe, Berlin 1938.

[240] BayHStA-KA, Iluft 38, Abschrift eines von von Richthofen verfassten Erfahrungsberichts, p. 5.

[241] Ibid., p. 12.

These "aerial combat tactics" found expression in the fighter squadrons formed after the Battle of the Somme and consolidated in the air force reorganization of October 1916. "In line with the growing importance of aerial warfare, all the resources of aerial combat and defense, at the front and in the rear," were brought together in a "single service" under the commander of the air force, who thus "took charge of the unified construction, supply, and deployment of this means of warfare."[242] In 1917, the arms drive initiated by Hindenburg and Ludendorff (who had taken over the Supreme Army Command in August 1916) began to show results. The target of tripling aircraft production was not achieved, but it did increase to twice its previous level.[243]

The changes in 1916–1917, involving the formation of fighter squadrons and groups, and the restructuring and expansion of artillery observation, reconnaissance and protective air cover, were the prerequisite for successful operational deployment of the air force. To ensure the hoped-for breakthrough at the front, the military on both sides thought it necessary to hide its offensive preparations from enemy scouts and to carry out thorough reconnaissance of its own. The key objectives were to locate the weak points in the enemy lines, to attack well-chosen targets, and to prevent the strengthening of weak points by destroying the flow of information, munitions and troops.[244] But the Entente breakthrough attempts in 1917 near Arras, in Champagne and in Flanders ended in failure. True, the Central Powers did not achieve a breakthrough either on the Western Front, but they successfully beat back the Kerensky offensive in the east and forced revolutionary Russia to withdraw from the war, eventually sealing this with the Treaty of Brest-Litovsk in March 1918.

In October 1917 German and Austro-Hungarian troops also broke through on the Southern front, near Caporetto, and pushed Italy to the brink of military collapse. Nevertheless, its air force, considerably expanded in the aftermath of 1916, had won its first victories in the air.[245] And on November 4, 1918 – after the victorious Battle of Vittorio Veneto, which sealed the collapse of the Habsburg Empire also militarily – the Italian army's air force had more than 504 aircraft available to it at the front. This was little in comparison with the German force, but the two had undergone the same kind of growth. What is central to our investigation, however, is neither the size of the new air forces nor their development into a decisive military weapon, but rather the symbolic effect of the airplane and pilot. The figure of the aviator-hero sank deep roots in the Italian collective consciousness,

---

[242] Decision of the Imperial Cabinet, October 8, 1916, quoted from Hoeppner, *Deutschlands Krieg in der Luft*, p. 82.

[243] This was also the reason for the creation of the air corps inspectorate under Major Wilhelm Siegert: to ensure the further mobilization and rationalization of the aircraft industry, and to allocate the limited resources most efficiently. See Morrow, *German Air Power*, pp. 73–94.

[244] On the concentration of air power prior to ground operations, see BayHSt-KA, DV Preuß. Geh. XV 29, Weisungen für den Einsatz und die Verwendung von Fliegerverbänden innerhalb einer Armee, May 1917.

[245] On the following, see Morrow, *The Great War in the Air*, pp. 193ff., 262ff. and 332ff.; and Massignani, *La grande guerra*, pp. 275–91.

even though the country's "literary mobilization" cannot be compared to Germany's and really came into its own only during the two Fascist decades. Apart from D'Annunzio, the "flying ace" Francesco Baracca – who notched up a record of 34 victories in the air – loomed the largest in the wartime propaganda.[246]

Less spectacular, though with greater international impact, was the theoretical work on aerial warfare that Giulio Douhet published in 1921 under the title *Il dominio dell'aria*.[247] In between, the German air force too had carried the bombing war behind enemy lines. But although it replaced the vulnerable airships with more effective large aircraft, its bombing raids failed to achieve a strategic effect.[248] The same was true of the new battle squadrons: they did not decide the outcome of battles in the German offensive of Spring 1918, but they were harbingers of the future shape of war.

Germany's defeat put an end to such aspirations for the time being. The Versailles treaty of June 28, 1919 prohibited the Weimar Republic from developing an air force,[249] and the protagonists of the air war recruited by Georg Paul Neumann to work on his history adopted seemingly conciliatory tones: "The fighting birds that used to carry destruction beneath their wings turned into express shuttles, called upon to spin the placatory threads of human happiness from country to country, continent to continent."[250] The peaceful intent did not last long: the Rapallo Treaty of April 1922 meant that by 1924 the Reichswehr and the Red Army were jointly practicing aerial warfare.[251]

---

[246] On the Italian "flying aces," see Maurizio Longoni, "Gli 'assi' sul fronte italiano," in Ferrari (eds.), *La grande guerra aerea 1915–1918*, pp. 291–322. On Baracca, see Irene Guerrini/ Marco Pluviano, *Francesco Baracca una vita al volo. Guerra e privato di un mito dell'aviazione*, Udine 2000.

[247] The first English translation, *Command of the Air*, appeared in London in 1943, and was republished in New York, NY in 1972.

[248] Between May and August 1917, the German air force flew eight sorties against England, killing 401 and wounding a further 983. In the period up to September 1918, the Gotha and the so-called Giant Airplane (*Riesenflugzeug*) carried out many more raids on England (killing 1,414 and wounding 3,416) and France (237 and 539 respectively). Germany too was a target for bombing from the earliest days of the war, suffering 729 dead and 1,754 wounded. On the 1914–1918 bombing war, see in addition to previously cited works: Ralf Blank, "Strategischer Luftkrieg gegen Deutschland 1914–1918," at http://www.erster-weltkrieg.clio-online.de/_Rainbow/documents/einzelne/Luftkrieg14_18-1.pdf, accessed on 11.4.2008; Christian Geinitz, "The First Air War against Noncombatants. Strategic Bombing of German Cities in World War I," in Chickering/Förster (eds.), *Great War, Total War*, pp. 207–25; and Achille Rastelli, "I bombardamenti sulle città," in Ferrari (ed.), *La grande guerra aerea 1915–1918*, pp. 183–249.

[249] On the Treaty of Versailles, see Gerd Krumeich (ed.), *Versailles 1919. Ziele, Wirkung, Wahrnehmung*, Essen 2001.

[250] Neumann, *Die deutschen Luftstreitkräfte im Weltkriege*, p. 599.

[251] On the military cooperation between Germany and the Soviet Union, see Manfred Zeidler, *Reichswehr und Rote Armee 1920–1933. Wege und Stationen einer ungewöhnlichen Zusammenarbeit*, Munich 1994.

Between 1914 and 1918, the airplane developed into an indispensable weapon that, according to military historians, influenced the course of *battles* but not of the *war*.[252] In the size of their personnel, the air forces were irrelevant. The ubiquitous media presence of the "flying sword" during the First World War was not based on its military significance, but rather on the fact that the war in the air could be represented and understood with age-old perceptual and interpretive schemas. Since the pilot upheld the traditional "matrix of the male warrior," which had been destroyed in the trenches, the media built him up into a powerful weapon in the struggle to give the war a meaning.[253] As we shall see, news coverage and popular literature on the air war compensated for the horrors of industrial-technological warfare on the ground. The significance of the popular hero-narratives also lay in their "bridge function": they helped to overcome the gulf that appeared between space of experience and horizon of expectation, thereby integrating events into a suprahistorical perspective.

Birth of the aviator-hero from the "battle of materiel"

The aviator-hero narratives stood in a dialectical relationship with what was happening on the ground. The most striking condensation of the positional warfare of the first two years was the so-called "Langemarck myth," in which "regiments of young men" went singing to their death in a spirit of heroic self-sacrifice.[254] This epic construction really caught on only in the Nazi period, when it found an echo in the attitudes and political guidelines of the time. During the First World War, it still contradicted too sharply all the ideas that had sucked German society into the conflict.

The mass self-sacrifice of the Langemarck myth cut across bourgeois images and expectations of the war. The bourgeoisie yearned for heroes to display what it took as the key virtues: assertiveness, energy, proficiency and will to achieve. The Langemarck narrative, like most other visions of the positional war, foregrounded very different virtues. Even in the Social-Democrat paper *Vorwärts*, there was a longing for individual, knightly heroes:

> Just when it was thought that war is no more than mutual bloodletting by faceless weapons that leave no trace of individual action or achievement, the thoroughly personal heroism that people thought to have been buried for

---

[252] Isnenghi/Rochat, *La grande guerra 1914–1918*, p. 220; Neitzel, "Zum strategischen Mißerfolg verdammt?," p. 191; Kennett, *The First Air War*, p. 220.

[253] On the "male matrix" of war, see Lutz Klinkhammer, "Der Partisanenkrieg der Wehrmacht 1941–1944," in Rolf-Dieter Müller/Hans-Erich Volkmann (eds.), *Die Wehrmacht. Mythos und Realität*, Munich 1999, pp. 815–36; here p. 834.

[254] Here and on the following, see Bernd Hüppauf, "Schlachtenmythen und die Konstruktion des 'Neuen Menschen'," in Gerhard Hirschfeld/Gerd Krumeich (eds.), *Keiner fühlt sich hier mehr als Mensch. Erlebnis und Wirkung des Ersten Weltkrieges*, Essen 1993, pp. 43–84.

good with the world of the knight developed once again in certain areas of warfare.[255]

The meaninglessness of the ground war, culminating in the great "battles of materiel" in 1916, called forth a semantic reaction that led to a part-new, part-old narrative. This helped to overcome "discursive blocks," making it possible to speak about and theorize the war.[256] The aviator-hero narratives were part of this counter-tendency, so that the aviator-hero himself may be seen as a product of the "battle of materiel"; he served to compensate discursively for the horrors of the ground war, as Eric J. Leed noted back in 1979.[257] This compensatory function involved keeping open the horizon of meaning and expectation that had drawn men into the war in the first place, and offering a means of imaginative flight from the mud of the trenches and the deadly "storm of steel." It conjured up a different, glorious war of movement, in which the individual and his deeds still had value, and the soldier was not the victim of all-powerful technology but had it under his control.

Thus, if we want to understand the meaning that the aviator-hero narratives generated, not only for soldiers but for Germany's and Italy's wartime societies as a whole, we need to relate the accounts of the air war to the prevailing interpretations of positional warfare. We shall therefore compare accounts of aerial warfare with "canonical" descriptions of the ground war, most of them written sometime after 1918, when a language had been found to express the novelty and massiveness of the "battles of materiel." This look ahead to the postwar period will also anticipate the change in the compensatory function of the aviator-hero. For the popular hero-narratives created a disposition and responsiveness to the palingenetic visions of the postwar years, when the pilot became the symbol of a new age, a new man, a new order, and therefore of a different modernity.

The meaning generated by popular accounts of aviator-heroes becomes accessible only if it is seen in the context of the war as a whole. Whereas stasis characterized the ground war, movement was the defining feature of the war in the air. Whereas men in the trenches felt threatened by the autonomous power of technology, the aviator-hero epitomized human control over the machine. Whereas the "battles of materiel" were mass phenomena, the air war was fought by individuals. And whereas hardship, sacrifice and anonymous death typified warfare among the trenches and shell craters, the "heroes of the air" had fame, medals and photos in magazines to show for themselves.

These dichotomies that marked the narratives can be spelled out more precisely. Four features, in particular, attract our attention. *First*, the ways in which man

---

[255] "'Flieger Richthofen gefallen'," *Vorwärts*, April 24, 1918, p. 2. Quoted from Schilling, "Kriegshelden," p. 253.

[256] See Aribert Reimann, *Der große Krieg der Sprachen. Untersuchungen zur historischen Semantik in Deutschland und England zur Zeit des Ersten Weltkriegs*, Essen 2000.

[257] Eric J. Leed, *No Man's Land. Combat and Identity in World War I*, New York, NY 1979, p. 134. On the following, see ibid., pp. 115–62; and Aribert Reimann, *Der große Krieg der Sprachen. Untersuchungen zur historischen Semantik in Deutschland und England zur Zeit des Ersten Weltkrieges*, Essen 2000.

and space, warrior type and landscape, were associated with each other point to a central opposition not only between movement and stasis, but also between above and below, constriction and expanse, blindness and sight. *Second*, in the nexus of man and machine, the trench combatant appears as the victim of technology, whereas the aviator uses it as a means to power. After the war, this instrumental relationship to technology would make the pilot the prototype for the creation of a new man or "worker," in Ernst Jünger's sense of the word. *Third*, images of individual, supposedly chivalrous, aerial combat are set against those of mass warfare on the ground. *Fourth*, the living conditions of trench soldiers differ sharply from those of aviators. While the infantryman lived in filth and suffered deprivation, the pilot led a luxurious and hedonistic existence. Furthermore, if a machine-gunner mowed down hundreds of enemy troops in a single action not one word was lost about it, but every plane that an airman shot down won him honor and recognition in the press.

Overcoming stasis

The German "Schlieffen Plan" envisaged rapid encirclement of the French armies, a "modern Cannae," following which the bulk of troops would be deployed against the supposedly more cumbersome enemy in the East. This would avoid an exhausting, long-drawn-out war on two fronts. With the creation of a vast railroad network, rapid movement had become an axiomatic military principle.[258] In the Southeastern and Eastern theaters of the war, the fronts repeatedly became mobile, but the Western Front seized up after the Battle of the Marne and the race for the sea in fall 1914;[259] since it was there that the greatest numbers of troops were deployed, it came to shape how the First World War was seen.

As Aribert Reimann has shown, the static front and the dominance of defensive warfare (achieved *inter alia* by the machine-gun) led to "the 'metallization' of discursive conceptions" and the development of an "armor-plated consciousness."[260] The more terrible the living conditions, and the more impenetrable the 860 kilometer barrier stretching from the North Sea to the Alps, the greater was the longing to break the blockade and to become mobile. "The only way out of the dispositions of positional warfare," wrote Reimann, "seemed to be in the air."[261]

---

[258] The mobility paradigm did not define only military strategy; it was "a central idea of the [long] nineteenth century," epitomized by the locomotive. In this sense, positional warfare may be read as a blocking of the movements that characterized the 19th century. See Wolfgang Schivelbusch, *The Culture of Defeat. On National Trauma, Mourning, and Recovery*, New York, NY 2003, p. 255. On the locomotive as epitome of the nineteenth century, see also Wolfgang Schivelbusch, *The Railway Journey. The Industrialization of Time and Space in the 19th Century*, Berkeley, CA 1986.

[259] The fact that war of movement was possible in the Eastern theater was partly due to the lower troop/space ratio prevailing there. See Stevenson, *1941–1918*, pp. 219–42 and p. 240: "In the Winter of 1915–16 the Western allies were deploying 2,124 men per kilometer of front, but Russia only 1,200."

[260] Reimann, *Der große Krieg der Sprachen*, pp. 48–9. The dominance of metal and armor found its most striking expressions in the steel helmet, the "ironclad Roland," and the tank.

[261] Ibid., p. 69.

In flying over no-man's-land, the aviator broke through the barrier and carried movement into the discursive stasis.

To clarify the compensatory function of the aviation narrative, we should first recall the report *Ein Flug über Verdun* that we quoted at the beginning of this chapter. This made it seem effortless to overcome the space where "the magic of mechanical death held sway."[262] The bomb-laden flying king had never "experienced anything so wonderful" in all his life![263] The zone of death was, it is true, "a gruesome sight," but apart from that "the earth looked like a plaything." The gulf between aviator and ground soldier could not have been greater, and it becomes particularly clear in the descriptions of the battlefield and the landscape of war.[264] The trench warrior found himself in a chaotic landscape, a battered earth pockmarked with moon craters. Henri Barbusse gave an impression of this apocalyptic space in his novel *Le Feu* (1916):

> Up there, on high, far away, a flight of fearsome birds, panting powerfully and with broken breath, which can be heard but not seen, spirals upwards to look down upon the earth. The earth! The desert has started to appear, vast and full of water, beneath the long and desolate light of dawn. Ponds and craters, their waters pinched and shivering under the sharp blast of early morning; tracks left by last night's troops and convoys in these fields of sterility, streaked with ruts which shine like steel rails in the weak light; and piles of mud with here and there broken stakes emerging from them, broken trestles in the shape of a cross, packages of rusted wire, twisted, like bushes. With its puddles and its banks of slime it looks like an oversized gray canvas sheet floating on the sea, submerged in places.[265]

Even before Apollinaire brought surrealism to life, the Western Front had been a surreal landscape that defied the imagination.[266] Arnold Zweig, initially an enthusiastic supporter of the war, underwent a "reeducation" before Verdun, and later, in 1935, published a novel that depicted the battlefield as a new Armageddon:

> "That was once a trench," observed Süssman, as they changed direction and made for a dark smudge upon the landscape that was called Douaumont village, once a group of neat houses and a little church, and now an indistinguishable heap of ravaged earth. Moreover, that earth was beginning to smell; first the sickly odor of decay, and then the scorched, sulfurous reek of poisoned earth.[267]

---

[262] Ernst Jünger, *Der Arbeiter. Herrschaft und Gestalt*, Hamburg 1932³ [orig. 1932], p. 104.

[263] Deutscher Luftflottenverein (ed.), *Das fliegende Schwert*, p. 70.

[264] On the construction of the war landscape, see Bernd Hüppauf, "Räume der Destruktion und Konstruktion von Raum. Landschaft, Sehen, Raum und der Erste Weltkrieg," *Krieg und Literatur* III/1991, pp. 105–23; and "Das Schlachtfeld als Raum im Kopf," in Steffen Martus/ Martina Münkler/Werner Röcke (eds.), *Schlachtfelder. Codierung von Gewalt im medialen Wandel*, Berlin 2003, pp. 207–34.

[265] Henri Barbusse, *Under Fire*, trans. by Robin Buss, New York, NY 2003, p. 7.

[266] See Modris Eksteins, *Rites of Spring*, p. 146.

[267] Arnold Zweig, *Education Before Verdun*, New York, NY 1963, p. 175.

The aviator, on the other hand, did not have to endure cramped trenches or their mud and foul odors. As we can read in *Artillerieflieger*, published in 1918 by the Dresden publisher Das Größere Deutschland [Greater Germany]: "How free, young, and strong one feels up there in the exhilarating sunshine, in the pure and frosty air! How often I cut through the monotonous hum of the engine with a high-pitched cheer: you feel like hugging the whole of Mother Earth, and all that goes with it."[268]

Positional warfare drove infantrymen to seek cover, to press themselves into the ground. As an American soldier put it, in a letter quoted by Eric J. Leed, it was a "cowering war – pygmy man huddles in little holes and caves praying to escape the blows of the giant who pounds the earth with blind hammers."[269] The trench warrior was thus constantly on the lookout for dangers, seeking protection in the earth from a death approaching from afar. In *All Quiet on the Western Front* (1928), Erich Maria Remarque impressively conveys this bond with the earth:

> To no man does the earth mean so much as to the soldier. When he presses himself down on her long and powerfully, when he buries his face and his limbs deep in her from the fear of death by shell-fire, then she is his only friend, his brother, his mother; he stifles his terror and his cries in her silence and her security; she shelters him and releases him for ten seconds to live, to run, ten seconds of life; receives him again and often for ever. – Earth! Earth! Earth! – Earth with thy folds, and hollows, and holes, into which a man may fling himself and crouch down![270]

The earth was the element of the trench warrior: it surrounded and protected him, limited and restricted him. The flier's element was the air, whose connotations and properties were also transferred to him. In the air, von Richthofen wrote, one is free: "The glorious thing in the flying service is that one feels that one is a perfectly free man and one's own master as soon as one is up in the air."[271]

The flier displayed other qualities that the trench soldier lacked. In the maze of trenches and craters, the infantryman was always in danger of getting lost; he had no overview beyond his immediate section of the front. He felt this as a lack, however, as Ernst Jünger explained in *Das Wäldchen 125*:

> If there were a great creature who could effortlessly take in the whole view from the Alps to the ocean, these goings-on would seem to him a dainty battle of ants, a single labor of delicate hammering. But for us who see no more than a tiny detail, our petty fate weighs us down, and the figure of death is most terrible.[272]

---

[268] F. W. Eddelbüttel, *Artillerieflieger*, Dresden 1918, p. 18.

[269] Edward F. Graham, in *War Letters of Rochester's Veterans*, quoted from Leed, *No Man's Land*, p. 133.

[270] Erich Maria Remarque, *All Quiet on the Western Front*, New York, NY 1987, p. 55.

[271] Manfred von Richthofen, *The Red Fighter Pilot*, pp. 62–3.

[272] Ernst Jünger, *Das Wäldchen 125. Eine Chronik aus den Grabenkämpfen 1918*, Berlin 1925, pp. 67–8.

For the troglodytes (as Paul Fussell calls them) who were forced into and under the earth, the flier was just such a creature. The air force was the "eyes of the army" that seemed to provide those on the ground with an overview of everything: "My observer wanted to show me the area. We flew up to our positions, and then along them at a height of 2,000 meters. It is amazing how precisely you can see every trench, indeed every crater."[273] The flier even enjoyed the "panorama":

> Since I was flying at a height of 1,000 meters, the whole Alpine chain lay like a panorama behind me, with the Rhine below, the immense Black Forest to the right, which, seen from a plane, lives up to its name as dark and menacing, and the rolling hills of the Vosges to the left, with their isolated castles and fortresses.[274]

The "blindness" of those on the ground was unknown to the aviator, as Captain Boelcke's field report makes clear:

> Yesterday I went with the foot artillery and saw the battlefield from there. Actually that means you see next to nothing. There was no sight of troops – only a few riders and people here and there. All you could see was the places where artillery shells had exploded and burning villages on all sides; but you hear all the more as a result.[275]

Both the fliers' view and their aloofness gave them an aura of the grand and super-human, intensified not least by the semantics of sky and ascent.[276] The differences between the flier's and the foot-soldier's experiential spaces could hardly have been greater. Moreover, the possibilities of escape and compensatory achievement evident in the aviation books were not just a question of greater mobility, freedom and elevation. The man-machine relationship was fundamental to the dichotomies of movement and stasis, above and below, constriction and openness, blindness and vision, impotence and strength. Mechanized warfare was responsible for all of the oppositions we have been discussing.

The machine as instrument of power

The perceived omnipotence of the aviator was based not least on his being presented as the master of technology, whereas the soldier in the trenches was at the mercy of mechanized warfare in which the enemy remained faceless, hidden and out of reach.[277] Since he could not simply flee the zone of death and soar

---

[273] Immelmann, *Meine Kampfflüge*, p. 35.

[274] Eddelbüttel, *Artillerieflieger*, pp. 17–18.

[275] Oswald Boeckle, *Hauptmann Bölcke's Feldberichte*, Gotha 1916, pp. 22–3.

[276] Leed, *No Man's Land*, pp. 136–37.; Mosse, *Fallen Soldiers*, pp. 125–26.; and Richthofen, *The Red Fighter Pilot*, pp. 57–8.

[277] On images of technology in and after the First World War, see *inter alia* Habeck, *Die Technik im Ersten Weltkrieg*; Jeffrey Herf, *Reactionary Modernism. Technology, Culture, and Politics in Weimar and the Third Reich*, Cambridge, MA 1984; Alexander Meschnig, *Der Wille*

above the trenches, the "cowering" soldier saw the aviator-hero as a kind of *deus ex machina*. Robert Musil conjures up this impression in a short story that harks back to his wartime experiences: *The Blackbird*. He describes a flier's approach to a battlefield in South Tyrol:

> Once during that time an enemy plane appeared in the sky over our quiet encampment. [...] All of a sudden a machine-gun barrage spotted the sky with little white clouds of shrapnel, like a nimble powder puff. It was a cheerful sight, almost endearing. And to top it off the sun shone through the tricolored wings of the plane as it flew high overhead, as though through a stained-glass church window, or through colored crepe paper. The only missing ingredient was some music by Mozart.[278]

In Arnold Zweig's *Education before Verdun*, the airman appears as a higher stage of evolution: "An airman, after all, knew everything. He was superior to his enemies, or rather he was set above them, a being of a higher order, a step ahead in the sluggish development of the vertebrate called man."[279] In Zweig, but also in other interwar literary treatments of the First World War, the flier is the new man who, thanks to technology, has progressed beyond the "lower life form," the trench soldier. For Zweig's sapper, Lieutenant Eberhard Kroysing, the dream was to ascend high above his earthbound existence:

> [...] the earth and not the sky was what damned his efforts, the clay on which man is born and condemned to crawl about till he dies and is reabsorbed in it once more. [...] How fortunate that we had discovered the combustion engine – we, the lords of fire and of detonation! And in that moment, there flashed into his mind the unshakeable resolve that he would become an airman. [...] To scratch about in muck was good enough for men like Süssmann and Bertin, men without the fighting instinct, without weight in their striking arms, men worn out too soon. But he, with a frail craft beneath him, two broad wings, and a whirling propeller – he would be off over the cloud-sea like the lark on Sunday; not, indeed, trilling songs, but hurling bombs that would scatter gas and bullets among the crawling multitudes below, and seeking out the duel from which only one returns.[280]

---

zur Bewegung. Militärischer Traum und totalitäres Programm. Eine Mentalitätsgeschichte vom Ersten Weltkrieg zum Nationalsozialismus, Bielefeld 2008, pp. 161–208; Radkau, *Technik in Deutschland*, pp. 239–53; Aribert Reimann, *Der große Krieg der Sprachen. Untersuchungen zur historischen Semantik in Deutschland und England zur Zeit des Ersten Weltkrieges*, Essen 2000, esp. pp. 257–78; Bernhard Rieger, *Technology and the Culture of Modernity*; Thomas Rohkrämer, *Eine andere Moderne? Zivilisationskritik, Natur und Technik in Deutschland 1880–1933*, Paderborn 1999; Rohkrämer, "Die Verzauberung der Schlange"; Showalter, "Mass Warfare and the Impact of Technology"; and Spilker/Ulrich (eds.), *Der Tod als Maschinist*.

[278] Robert Musil, "The Blackbird," in Musil, *Posthumous Papers of a Living Author*, New York, NY 2006, p. 157. See Ingold, *Literatur und Aviatik*, pp. 248ff.

[279] Zweig, *Education before Verdun*, p. 217.

[280] Ibid., pp. 217–18.

Aronld Zweig's figure Eberhard Kroysing is a product of postwar discourse and already reflects the construction of a new type of man and fighter. The airman here becomes superhuman, because he is not a victim of technology but makes it his instrument. By contrast, the ground soldier experiences mechanized warfare as a loss of power, since he himself becomes an impotent object of anonymous machinery. In Ernst Jünger's *Feuer und Blut* (1925), we read:

> Here the age from which we come reveals its cards. The rule of machines over men, of slave over master, becomes apparent, and a deep split that already began to shake the social and economic order in peacetime makes its deadly appearance on the battlefield too. Here the ways of a materialist race are unveiled, and technology celebrates a bloody triumph.[281]

Technology appeared to have rendered obsolete the "heroic struggle" for which people were drawn into the war,

> regardless of the will to power that inspires individuals, or the moral and spiritual values that make them distinctive. Free will, education, enthusiasm, and an intoxicating contempt for death are not enough to overcome the force of gravity on the few hundred meters where the magic of mechanical death holds sway.[282]

Soldiers felt powerless in the trenches: indeed, according to George Soldan, the military journalist and editor of the *Schlachten* series of the Reich Archive, they had become mere materiel:[283] "Matter triumphs, man himself is valued only as materiel. We experience the mechanical annihilation performed by our own creation, the machine; we stand there powerless, finding no way to change it."[284] The dichotomy between soldier and pilot in relation to the machine was a theme in Ernst Jünger's *Der Kampf als inneres Erlebnis* [Combat as Inner Experience], first published in 1922. He writes:

> Machine combat is so prodigious that man almost disappears in the face of it. [...] Expressing itself as a gigantic dead mechanism, it spread an icy, impersonal wave of annihilation across the area. [...] And yet, man is there behind everything. He gives direction and meaning to the machines. He drives projectiles, explosives, and poison out of them. He crouches in their belly when they heave and spew fire over the battlefield. He is the most dangerous, bloodthirsty, and purposeful creature that the earth has to carry.[285]

---

[281] Ernst Jünger, *Feuer und Blut. Ein kleiner Ausschnitt aus einer großen Schlacht*, Berlin 1929 [orig. 1925], p. 31.

[282] Jünger, *Der Arbeiter*, p. 104.

[283] Leed, *No Man's Land*, pp. 33–4.

[284] George Soldan, *Der Mensch und die Schlacht der Zukunft*, Oldenburg 1925, p. 35, quoted from Meschnig, *Der Wille zur Bewegung*, p. 179.

[285] Ernst Jünger, *Der Kampf als inneres Erlebnis*, Berlin 1922, p. 114.

In the pilots' reports, it was man who gave "direction and meaning" to machines. While men in the trenches felt threatened by an automaton, a mechanical-organic hybrid was coming into being in the skies above them. On the ground, largely because the enemy was invisible, "technology" often seemed to be what brought death. But the flier gave a face to anonymous machinery.

In the now almost iconic publicity issued by the Fokker works in 1916, the flier fuses with his machine and its built-in machine-gun. He has put technology back in the service of man. For this reason among others, the figure of the pilot was so successful with the public after the war, when the lack of acceptance of technology was recognized as one of the causes of defeat (Figure 2.2).[286]

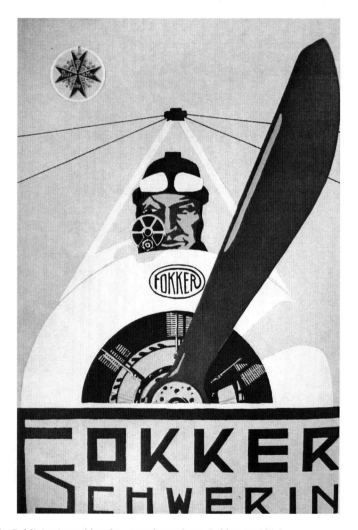

Figure 2.2   Publicity issued by the aircraft producer Fokker in 1916

---

[286] See Radkau, *Technik in Deutschland*, pp. 239ff.

The airplane became the flier's prosthesis, extending his body and his capacities. In *Kreuz wider Kokarde* [Cross versus Cockade], an account of Lieutenant Udet's fighter missions, we read of all those "for whom the joystick grew into their fist."[287] And in his war memoirs, Hans Joachim Buddecke writes: "The pilot must grow together with his machine in the sky, so that he makes all his turns and movements automatic and mechanical."[288] Here the relationship between man and machine is symbiotic. The pilot brings it to life, but he himself is dependent on his machine:

> Alone for the first time at that height [...].Left entirely to your own resources. [...] Basing yourself only on the machine! You get to have a special kind of love for your machine, which has carried you so far aloft. More than the rider for his horse. But you also have a sense that you are standing with a pair of scissors in front of the thread by which your life hangs. One jerk of your hand and you have cut the thread that would not otherwise have been broken. Up there you really have your life in your hands.[289]

The fighter pilot of the popular narratives also epitomizes D'Annunzio's *vivere pericolosamente*, the "living dangerously" that airmen appear to enjoy:

> The moment when he shoots down an enemy aircraft is the most precious in a fighter pilot's life. [...] The moment came when you had searched carefully all around, calculating the best possible position for this or that Nieuport to be in, when your little bird turned downward and a physical effort suppressed the dizziness of the movement, when you then swooped vertically several hundred meters on the enemy to slice through him – that was perhaps a more splendid moment than the victory that followed, and the uncertainty as to whether your wings would hold only increased the excitement.[290]

Though seldom as explicit as in this quotation, the aviator-hero narratives in general testify to the cult of speed and danger, and to a prototypical enrichment of life by the machine. But the popular literature in question did not treat technology only as a means to power and a more intense experience of life. It also presented the aircraft more straightforwardly as a fascinating "modern wonder."[291] Here, for example, is Max Immelmann: "The atmosphere or mood surrounding

---

[287] Eichler (ed.), *Kreuz wider Kokarde*, p. 115.

[288] Hans Joachim Buddecke, *El Schahin (Der Jagdfalke). Aus meinem Fliegerleben*, Berlin 1918, p. 99.

[289] Leonhard Müller (ed.), *Fliegerleutnant Heinrich Gontermann*, Barmen n.d. [1918 or 1919], p. 64.

[290] Buddecke, *El Schahin*, pp. 100–1.

[291] On "modern wonders," see Bernhard Rieger, *Technology and the Culture of Modernity in Britain and Germany, 1890–1945*, Cambridge 2005.

me fills me with satisfaction. The hum of the engines is incessant: wherever you look, there are only engines, automobiles, motorcycles, aircraft of all kinds, airships – in short, you are truly up and running."[292] Here the machinery of war still arouses enthusiasm in the target groups; it consists of coveted technological toys and fulfils the dreams of youth. "My material side here comes fully into its own. I exist in the midst of engines, as it were. I get to ride either a motorbike (we have three of them) or a car. Or else I fly. [...] I need hardly say how much this all appeals to me."[293]

This enthusiasm was still part of a basic social consensus about technological development and its association with nationalist visions.[294] The premise for this was a (partial) acceptance of the industrial world, as expressed in *Fliegerleutnant Heinrich Gontermann*:

> This is the atmosphere into which our Heinrich Gontermann, Jr. was born – amid the smokestacks, amid the twitching flames of the vaulted furnaces, amid the whistling of the power plant and the screeching of the slowly turning barrels that exposed their hulls for so long to the sharp chisel until they had precisely the right form and curve. It was in such surroundings that the boy and young man later grew up. [...] It is a school for any industrially minded spirit, who consciously or unconsciously wishes to gain mastery over matter, whether by chopping wood, sawing a block of iron, or swimming safely in the air ocean.[295]

The fliers' testimonies reproduced the basic acceptance of technological progress that characterized Imperial Germany, in opposition to reactionary (and often overemphasized) tendencies to agrarian romanticism.[296] As Jeffrey Herf has shown, however, a pro-technology position by no means necessarily implied support for liberal Enlightenment modernism.[297] The horrors of mechanized warfare delegitimized the bourgeois-liberal vision of modernity, as its faith in progress gave way, amid the shell bursts, to a muffled cultural pessimism that had its origins not least in the "steel snake of knowledge." The snake of technology had become a seemingly omnipotent creature outside human control, which terrified even people with a tendency to techno-enthusiasm.[298]

---

[292] Immelmann, *Meine Kampfflüge*, p. 10.

[293] Ibid., pp. 21–2.

[294] On this basic social consensus about technology that existed despite the many enemies of progress, see Rohkrämer, *Eine andere Moderne?*, pp. 38–56. And on the association of technology with nationalist visions, see Fritzsche, *A Nation of Fliers*.

[295] Müller (ed.), *Fliegerleutnant Heinrich Gontermann*, pp. 5–6.

[296] See Rohkrämer, *Eine andere Moderne?*, p. 71. On the critique of progress, see *inter alia* Gunther Mai, "Agrarische Transition und industrielle Krise," *Journal of Modern European History* 4/2006, pp. 5–37.

[297] See Herf, *Reactionary Modernism*.

[298] See Leed, *No Man's Land*, pp. 30ff.

But the figure of the aviator-hero was reconciled to the powerful demon of technology, since in his case it was applied not in the name of universal human progress, but in that of the nation fighting for survival. In the hands of the aviator-hero, technology appeared to be a means for the salvation and advancement of the nation. Thus, after the war these hero-narratives would offer many links for a "heroic" reinterpretation of technology in the direction of radical nationalism; they endorsed and celebrated a technology that stood in the service of the nation and was enwrapped in national values and norms. The media promotion of aviator-heroes was a precursor of postwar visions of a new man who would use technology to bring about the renewal and revival of the nation.

A knightly duel of man against man

The compensatory function of the popular aviator-hero narratives did not concern only the spatial-technological dimension of the First World War. Rather, the aviation literature supplemented the accounts of anonymous death in the trenches by laying stress on one-to-one "knightly" combat.[299] A letter written on the Somme front in September 1916 reports laconically: "Let's not speak of that anymore; I prefer a heroic battle to having to fight against weapons we cannot see, and that we cannot fight against heroically. It is mass combat, with no reward for the individual in the mass, etc."[300]

In the popular accounts of the air war, the individual stood out against a visible enemy, and it was his courage, vigor and skill that decided the question of life and death, victory and defeat. In this way, media representations of the aviator-hero updated the bourgeois virtues and conceptions of order that defined Imperial Germany. The war correspondent Georg Wegener summarized as follows this compensatory function of the aviation literature:

> This war, with its huge increase in mass phenomena, has a more impersonal quality than any other before it. We are aware that it produces heroes at the front in greater abundance than at virtually any other time in human history, but the fame of the individual merges into the whole; it is curious how few names have been loved and glamorized at a popular level during the last two years. Fighter pilots are an exception in this respect. People know the names of the most successful among them, love them by name, follow their feats, and acclaim them when their victories mount up. More clearly than in any other of the forces, it is realized what they achieve as individuals, how thoroughly personal their actions are. The flier's boldness and cunning, decisiveness and

---

[299] See Omer Bartov, "Man and the Mass. Reality and the Heroic Image in War," *History and Memory* 2/1989, pp. 99–122; John H. Morrow Jr., "Knights of the Sky. The Rise of Military Aviation," in Frans Coetzee/Marylin Shevin-Coetzee (eds.), *Authority, Identity, and the Social History of the Great War*, Providence, RI 1995; and George L. Mosse, "The Knights of the Sky and the Myth of the War Experience," in Robert A. Hinde/Helen E. Watson (eds.), *War: A Cruel Necessity? The Bases of Institutionalized Violence*, London 1995, p. 133.

[300] Letter from Maximilian Jackowski to Kurt Böhning, September 6, 1916; see Hirschfeld/Krumeich/Renz (eds.), *Scorched Earth*, pp. 109–10.

presence of mind, aerial mastery and shooting skills, confirm that he has made a success of his life.[301]

While the ground war was seen as an anonymous mass phenomenon, "personal feats" and successes were distinguishable in the air war. Up there it seemed possible to act out the ideas of heroism that had drawn men into the war but had proved obsolete in positional warfare.[302] The trenches were the realm of the mechanized death that the writer Wilhelm Lamszus had predicted in 1912 in his *Das Menschenschlachthaus* [The Human Slaughterhouse]:

> 240 rounds or more a minute! What a miracle of technology is such a machine-gun! You let it whirr, and it sprays bullets thicker than rainfall. And the automat bares its teeth from left to right. It aims at the middle of bodies and covers the whole firing line. It is as if death has thrown the scythe onto the scrap heap, as if it has now become a machine operator. [...] Once it was a horseman's death, an honorable soldier's death. Now it is a mechanical death.[303]

Although death in the air was also a "machine operator," it came to be represented as an "honorable soldier's death," a "horseman's death." The popularity of the aviation books was based on this updating of the traditional ways of seeing and interpreting combat. By contrast, the military capacities of ground troops became irrelevant to the raging fire spewed out from cylindrical machines.[304] While the infantryman could do nothing but hide in the earth, the flier could take action in the face of danger or at least swerve to avoid it: "When will the first shrapnel come up from the English defenses, and where will it hit? When we see the first gunfire, the tension will be over. Then the trained frontline pilot can take countermeasures, even if they are sometimes ineffective."[305]

---

[301] Georg Wegener, *Der Wall von Eisen und Feuer*, vol. 2, *Champagne – Verdun – Somme*, Leipzig 1918, p. 332.

[302] See Schilling, *"Kriegshelden,"* pp. 252–53.

[303] Wilhelm Lamszus, *Das Menschenschlachthaus. Bilder vom kommenden Krieg*, Hamburg 1912, pp. 19–20., quoted from Peter Borscheid, *Das Tempo-Virus. Eine Kulturgeschichte der Beschleunigung*, Frankfurt/Main 2004, pp. 250–51.

[304] See Ernst Jünger, *In Stahlgewittern. Aus dem Tagebuch eines Stoßtruppführers*, Berlin 1930 [orig. 1920], p. 162: "You crouch alone in your hole in the earth and feel given over to a blind, merciless will to destroy. With disgust you suspect that all your intellect, your abilities, your mental and physical assets, have become ridiculously insignificant. It may be that, while you are thinking this, the chunk of iron that will smash you to a shapeless void has already begun its hurtling journey."

[305] Eddelbüttel, *Artillerieflieger*, p. 47.

Aerial combat seemed to give scope for daredevilry, risk-taking and "genuinely heroic" manhood.[306] The flier was in a position to execute "bold deeds of youth," in the tradition of the heroes of Germany's anti-Napoleonic Wars of Liberation:[307]

Quickly making up my mind, I climbed to a greater height and, flying alone, attacked a squadron of eighteen enemy planes over Mulhouse. [...] I swooped down into the pack with my engine running full blast, lined the thickest one up in my sights, and opened fire from a distance of roughly 50 meters. The result was that the enemy immediately caught fire and fell brightly glowing into the city of Mulhouse. I cannot describe the feeling I had: I could have shouted with pride and joy.[308]

The accounts of aerial combat maintain the prewar expectations that it would enable fliers to prove their worth and satisfy their longing for adventure. As the vision of "male," "knightly" combat sank deeper into the mud of the trenches, as the casualty figures rose and the enemy superiority became clearer, the desire for "individual hero-myths" that had originally drawn men into the war gradually became more insistent. Up in the air, the popular narratives suggested, there was still man-to-man combat, akin to a sporting contest or knightly duel.[309]

---

[306] On gender conventions, ideals of manhood, and constructions of military masculinity, see *inter alia*: Ute Frevert, "Das Militär als 'Schule' der Männlichkeit. Erwartungen, Angebote, Erfahrungen im 19. Jahrhundert," in Ute Frevert (ed.), *Militär und Gesellschaft im 19. und 20. Jahrhundert*, Stuttgart 1997, pp. 145–73; Marcus Funck, "Vom Höfling zum soldatischen Mann. Varianten und Umwandlungen adeliger Männlichkeit zwischen Kaiserreich und Nationalsozialismus," in Eckart Conze/Monika Wienfort (eds.), *Adel und Moderne. Deutschland im europäischen Vergleich im 19. und 20. Jahrhundert*, Cologne 2004, pp. 205–35; Marcus Funck, "Bereit zum Krieg? Entwurf und Praxis militärischer Männlichkeit im preußisch-deutschen Offizierskorps vor dem Ersten Weltkrieg," in Karen Hagemann/Stefanie Schüler-Springorum (eds.), *Heimat-Front. Militär und Geschlechterverhältnisse im Zeitalter der Weltkriege*, Frankfurt/Main 2002, pp. 69–90; Thomas Kühne (ed.), *Männergeschichte – Geschlechtergeschichte. Männlichkeit im Wandel der Moderne*, Frankfurt/Main 1996; Sonja Levsen, *Elite, Männlichkeit und Krieg. Tübinger und Cambridger Studenten 1900–1929*, Göttingen 2006; George L. Mosse, *The Image of Man. The Creation of Modern Masculinity*, Oxford 1998 [repr.], esp. pp. 107–32; Ute Planert, "Kulturkritik und Geschlechterverhältnis. Zur Krise der Geschlechterordnung zwischen Jahrhundertwende und 'Drittem Reich'," in Wolfgang Hardtwig (ed.), *Ordnungen in der Krise. Zur politischen Kulturgeschichte Deutschlands 1900–1933*, Munich 2007, pp. 191–214; Schilling,"*Kriegshelden*"; Stefanie Schüler-Springorum, "Vom Fliegen und Töten. Militärische Männlichkeit in der deutschen Fliegerliteratur, 1914–1939," in Karen Hagemann/Stefanie Schüler-Springorum (eds.), *Heimat-Front*, pp. 208–33.

[307] Schilling,"*Kriegshelden*," p. 260.

[308] Eichler (ed.), *Kreuz wider Kokarde*, p. 38.

[309] Laurence Goldstein has established that British accounts of aerial combat have a great structural affinity with Sir Thomas Malory's tales of Arthur: "the catalogs of warriors, the repeated trials by combat, the attention to gestures and the set pieces of dialogue, the relishing of all mechanical description, and the avowal of moral purpose throughout. [...] The duellings between British and German aces resemble the battles of champions like Aeneas und Turnus." Goldstein, *The Flying Machine and Modern Literature*, p. 89. See also Stefan Goebel, *The Great War and Medieval Memory. War, Remembrance and Medievalism in*

According to Peter Fritzsche, this image of "knights of the sky" defined the "folklore" of fighter pilots.[310] In *Briefe eines deutschen Kampffliegers an ein junges Mädchen* [Letters from a German Fighter Pilot to a Young Girl], the "letters of heroism and love" by the fallen Lieutenant Erwin Böhme published in 1930, the editor of the book, Johannes Werner, remarked that the "fighter pilot" was the only combatant still left "in modern mass warfare" who could "wage a knightly, man-to-man duel."[311]

In one of these "letters" to "Dear Miss Annemarie," Erwin Böhme writes:

> Every fight for us is a tournament, a knightly or, in more modern terms, a sporting contest. I have nothing at all against the individual I take on: all I want is to put him and his plane *hors de combat*, so that they can no longer be of any harm to us. If my adversary were then to make an emergency landing on our territory, I would gladly shake his hand – assuming he had fought honorably and bravely – and hold him in respect. But I would still notch him up as one in a number of "kills." As you see, that does not involve any disdain for the enemy. It is understandable to record your victories: every action is personal, you enter successes on your own credit side, and you inevitably keep score as their number increases.[312]

Whereas success on the ground often came down to sheer survival, it was easier to quantify achievement in the air.[313] Individual success – in terms of the number of shoot-downs – became the measure of heroic qualities; it told people how many "knightly" contests you had come out of victorious.

Although in reality the air war at the front by no means consisted of knightly duels, use was made of this trope (and of the "sporting contest") in descriptions of combat. It evoked a human war, in which the enemy, instead of being an object of hatred, was a worthy adversary with whom fraternization was not only possible but normal: "Scarcely had I landed when he [the shot-down enemy] came over to me. He shook my hand and said I was a 'real sportsman.'[314] We smoked a good English perfumed cigarette."[315] Behind the stress on fair, chivalrous conduct, an exoneration strategy seems to be at work here: "On March 25, I shot down a two-seater over the British lines. It was reduced to

---

Britain and Germany, 1914–1940, Cambridge 2007, pp. 223–30. On p. 227 he writes: "In sum, the tale of chivalrous air warfare fulfilled a compensatory function. It pictured the kind of battle the war as a whole should have been but was not: a fair and straightforward man-to-man-fight."

[310] Fritzsche, *A Nation of Fliers*, pp. 59–101.

[311] Johannes Werner (ed.), *Briefe eines deutschen Kampffliegers an ein junges Mädchen*, Leipzig 1930, p. 7.

[312] Ibid., p. 67.

[313] On the birth of the flying ace and his counting of "kills," see Wohl, *A Passion for Wings*, pp. 203–51.

[314] In English in the original.

[315] Eichler (ed.), *Kreuz wider Kokarde*, pp. 76–7.

dust in mid-air. It looked horrific. It's a grim business, but we are only doing our duty."[316] Violent killing is thus morally encoded as an act of duty. On the other hand, the "grim business" is covered up by the trope of a knightly or sporting contest:

> I kept on his tail most of the time and then opened fire. [...] After three minutes or so, the pilot seemed to be wounded. His machine descended in large regular curves. [...] I hoped he would land smoothly. But at around 1,000 meters the machine tilted over and somersaulted down to earth. [...] Both the occupants were dead. You always feel sorry about that, but there was nothing else I could have done. He fought back too well.[317]

Gontermann's grandfather and editor, Pastor Leonhard Müller, comments: "Gontermann was sorry that he could not save the enemy's life, but we are glad that, even in the heat of a life-and-death battle, his sense of chivalry did not shed feelings of humanity and compassion, so that he sought to preserve the enemy's life whenever possible."[318] This "sense of chivalry" and these "feelings of humanity" were not always features of aerial combat, as the popular hero-narratives occasionally demonstrate. In *Kreuz wider Kokarde*, for instance, we read of Manfred von Richthofen:

> During many flights I made with him at the front, I only once saw an enemy get away with his life. Otherwise, all the Englishmen he attacked bit the dust. He set all the enemy aircraft on fire. He said that in his view that was the only real shoot-down. Once I shot down a Sopwith, whose pilot had been killed with a bullet in the head. His plane did not catch fire but crashed into the ground. When I reported the kill to Richthofen, I was nearly ashamed that I had to answer "no" to his question: "Burning?"[319]

The life-and-death struggle was often described as a hunt. In Richthofen's memoirs, for example, he writes:

> My father discriminates between a hunter and someone who just enjoys shooting. When I have shot down an Englishman, my passion for hunting is satisfied for a quarter of an hour. If one of them comes down I have a feeling of complete satisfaction. Only much, much later have I overcome my instinct and become a shooter.[320]

---

[316] Müller (ed.), *Fliegerleutnant Heinrich Gontermann*, p. 83.
[317] Ibid.
[318] Ibid., p. 88.
[319] Eichler (ed.), *Kreuz wider Kokarde*, p. 132.
[320] Richthofen, *The Red Fighter Pilot*, p. 163; translation modified.

In the next section, entitled "I Kill a Bison," the author has been invited to a hunting party by Prince von Pless. Here we read:

> At any rate, at the moment when the bull came near, I had the same feeling, the same feverishness which seizes me when I am sitting in my aeroplane and notice an Englishman at so great a distance that I have to fly perhaps five minutes to get close to him. The only difference is that the Englishman defends himself.[321]

The hunting trope, like the knight metaphor, fulfilled a social function. And even if its connotations did less than those of chivalry or sport to cover up the exercise of violence, man-to-man aerial combat appeared qualitatively less atavistic, and the act of killing essentially more detached, than when two common soldiers fought it out in the trenches. Ernst Jünger writes this in *Storm of Steel*:

> In a mixture of feelings due to blood lust, rage, and alcohol, we set off with difficulty but unstoppably to the enemy lines. [...] I was boiling with a fury I could no longer comprehend. An overpowering desire to kill winged my steps. Rage squeezed bitter tears from me. [...] Then I saw the first enemy. [...] A drama without spectators was in the offing; it was a relief finally to see the enemy as a tangible figure. Gritting my teeth, I lay the muzzle against the temple of the fear-frozen man and dug my other fist into his tunic; with a cry of pain he reached into his pocket and held a card in front of my eyes. It was a picture of him surrounded by a large family... I overcame my mad rage and walked on.

Then, a few pages later:

> He [the combatant] does not want to take prisoners; he wants to kill. He has lost sight of any objective and is under the spell of violent primal instincts. Only when blood has flowed does the fog clear from his brain; he looks around as if waking from a bad dream. Only then is he once more a modern soldier, capable of solving another tactical problem.[322]

Jünger's description of a primal encounter between two enemies is already saturated with the bloodthirsty cult of violence that would develop in the postwar years.[323] But the rift is becoming palpable between life-and-death combat in the trenches and similar encounters in the air:

> After we had flown around nine or ten times, vainly trying to line each other up in our sights, I managed to get a clearer view of my enemy. He wore a

---

[321] Ibid., p. 165.

[322] Jünger, *In Stahlgewittern*, pp. 227ff and 234–35.

[323] See Thomas Rohrkrämer, "Kult der Gewalt und Sehnsucht nach Ordnung – Ernst Jünger und der soldatische Nationalismus in der Weimarer Republik," *Sociologus* 51/2001, pp. 28–48.

fluttering scarf and black headgear, and I could see he was clean shaven. He looked at me for quite a long time, then raised his right hand and began to wave. I don't know why, but the man in the machine had become immensely likeable. I waved back involuntarily, and this mutual greeting lasted for five or six full banking turns. All of a sudden I had the strange feeling that there was no enemy in front of me; it was as if I was competing at practice turns with my comrade.[324]

The technology distancing killer from killed was one of the main reasons why the air war seemed more refined and civilized. The flier always remained a "modern soldier," in Jünger's sense. He killed cleanly and not at close quarters. Yet aerial warfare was "modern" also because it involved a division of labor and a high degree of rationalization, being fought to an increasing degree with massed forces. After the great "battles of materiel" on the ground, the air war too began to pit whole groups of fliers – not just individuals – against one another. But once the "enemy group" was torn apart, as Richthofen put it,

the shooting down of an adversary is only an individual battle. This creates a danger that individual fighters will get in each other's way, giving many Englishmen an opportunity to escape in the tumult of battle. [...] It would be quite wrong – and therefore to be avoided – that several should descend with one enemy aircraft. I have seen pictures where 10–15 machines are all pitching in and following an Englishman down to the ground, while up above the enemy squadron flies on unmolested.[325]

Ten or fifteen to one did indeed have little in common with "chivalrous combat." More important, however, was the fact that most of the shot-down aircraft were outclassed in weapons, maneuverability and speed. As Richthofen admits in his report, the main focus of the air war was to destroy such aircraft designed for another purpose:

I lie in wait for *Artillerieflieger* [artillery direction aircraft]. The best moment to attack such aircraft is when the enemy is flying up from beyond the front. Then I pounce, factoring in the wind conditions (East-West) and diving on him from out of the sun. He who gets to the enemy first has a prior claim to shoot. The whole squadron goes down together. [...] If the first plane jams up, the second has a go, then the third, etc.[326]

It remains undisputed that the capacities and achievements of the individual, his courage and competence in the face of victory or defeat, life or death, were

---

[324] Eichler (ed.), *Kreuz wider Kokarde*, pp. 150–51.
[325] BayHStA-KA, lluft 38, Abschrift eines von von Richthofen verfassten Erfahrungsberichts, pp. 4–5.
[326] Ibid., pp. 8–9.

more decisive in aerial warfare than under heavy shellfire. But the "honorable" soldier's death, if ever there had been such a thing, was an anachronism also in the air. There too better technology and cleverer tactics were decisive, not to speak of industrial production capacities. Above all, it was coordinated action and a kind of Taylorist division of labor, rather than individual boldness, that offered the promise of success: "The squadron must fly itself in – that is, not get used to one place or anything like that. Each individual must get used to all the others, so that he already knows from the movement of an aircraft what the man at the controls wants to do."[327]

The reproduction of traditional conceptions of man-to-man or "knightly" contest was thus based less on the actual shape of aerial warfare or dogfights than on a longing for "fair," "chivalrous" combat at an individual rather than a mass level. Nevertheless, the air war did offer sufficient support for perceptual, interpretive, and representational models that had been shattered by the reality of "battles of materiel." Whereas the ground war had upset the traditional matrix, gradually supplanting it with a radical type of sacrificial heroism centered on comradeship, life at the front, and a "steel-hard" warrior temperament, the descriptions of aerial warfare preserved the older interpretive models in an updated form. Since the media and official propaganda recognized in it the "individual hero-myths" that corresponded to public expectations and to prevailing literary conventions, the air war formed a central part of the media mobilization.

Here it was still the individual, boosted by technology yet largely self-driven, who fought in the name of Kaiser and Fatherland, engaging in mobile combat that could produce heroic deeds and tangible results. Whereas, in many respects, the anonymous trench soldier contradicted the bourgeois ideals of the age, the aviator-hero actually embodied them by virtue of his individualism, achievement, mobility, overcoming of space-time limitations, mastery of nature and technological know-how. The popular hero-narratives showed that, despite all the gunpowder, there was still plenty of room for the *Iliad* scenarios that had drawn so many men into the war and inspired young people waiting at home.[328] As we have seen, however, the attractiveness of the aviator-hero was also fuelled by other than strictly military contexts. If the flier became a model for emulation and a focus of longing, it was not only on account of his mastery of space and machines, nor exclusively because he counterposed the individual to the masses. His popularity also rested upon his embodiment of images that were not specific to war.

"Elegant like at a dinner": the hero's glamorous life

In the context of an expanding culture industry, aviator-heroes played a role comparable to that of today's stars. The *bon viveur* image aroused by the figure of the aviator-hero, as well as his elite habitus, also contributed to his popularity. The

---

[327] Ibid., p. 7.
[328] In his hundredth birthday address, Ernst Jünger remarked: "Karl Marx summed it up: 'Is an Iliad possible with gunpowder?' That is my problem." Quoted from Helmuth Kiesel, *Ernst Jünger. Die Biographie*, Munich 2007, p. 175.

wartime function of this figure should not obscure the fact that it was a product of the "dream factory" then in the process of crystallization. The aviator-hero was a media icon, who satisfied the hopes and longings both of young male "war enthusiasts" and of "little shopgirls."[329]

Richard Euringer, a flier and then flying school director during the First World War, who later developed into one of the best-known Nazi writers, packed a number of non-military tropes into a section at the beginning of his *Fliegerschule 4* (1929):[330]

> We do not know the tedium of the trenches or the crushing devilry of an artillery barrage. [...] We fliers sleep in silk beds, dine in castles, and drink champagne – though not every day, of course. But every day our airbase snuggles up to us and places its gently flattering arms around our neck. They don't even have to be the arms of a little Frenchwoman. We shoot hares and make cocoa, we harness a horse for riding, we wallpaper our huts, install electric lighting, have gramophone records sent to us, carry luxuries around with us in our trucks, keep dogs, read books, wash and rub ourselves down, stick little flags on special maps, whiz into town by car, buy top-quality Valenciennes lace, commandeer hard liquor, make ourselves popular with the assistant in charge of fine soaps, get around and about, save stocks from liquidation, become the favorites at the general staff, blue-eyed boys for the infantry, regular guests at the battery, highly appreciated photographers, elegant like at a dinner, fresh and zappy and well rested.[331]

The fascist reinterpretation of the aviator-hero would strip away this taint of luxury. Indeed, the author himself concluded: "And we knights of dubious sportsmanship will do our squadron drill, perform about-turns when ordered to do so, and become disciplined soldiers as in the time of Frederick the Great."[332] The individualism and willfulness that characterized the figure of the aviator-hero had to be eliminated; the flier had to be inserted into the faceless, obedient community of ordinary soldiers, whose image was the man in a steel helmet visible on propaganda posters.[333]

During the war, however, it was precisely the aviator's individualism, exclusive habitus and relative independence that made him such an attractive figure, by

---

[329] On the comparison between "football enthusiasts" and "war enthusiasts," see Haffner, *Defying Hitler*, pp. 14–15. And on "little shopgirls," see Siegfried Kracauer, "The Little Shopgirls Go to the Movies," (1927) in Kracauer, *The Mass Ornament. Weimar Essays*, Cambridge, MA 1975, pp. 291–306.

[330] On Richard Euringer, see Ernst Klee, *Das Kulturlexikon zum Dritten Reich. Wer war was vor und nach 1945*, Frankfurt/Main 2007, p. 143.

[331] Richard Euringer, *Fliegerschule 4. Buch der Mannschaft*, Hamburg 1929, pp. 17–18.

[332] Ibid., p. 21.

[333] See, for example, the poster designed by Fritz Erler in 1917 to advertise the sixth war loan: accessed on August 6, 2014.

no means just an object of envy and resentment. The quotation from Euringer's *Fliegerschule 4* may well be "over the top," but it illustrates the aura that surrounded the aviator even after the war, in no small measure as a result of the popular hero-narratives we have been investigating.

At the same time, unlike the "shirkers" behind the front, fliers were not accused of having a "cushy number." Since they were mostly officers, they anyway enjoyed more privileges than common soldiers, although the lifestyle depicted in the war books was more dissolute than in other branches of the armed forces:

> When the orderly came to wake me up, I had a pretty bad headache. He drew the curtains. [...] Then he picked up the military clothes scattered around the room and went back out. I turned over to doze a little longer. After five minutes, my orderly returned and claimed I'd slept three more hours! So, I slowly began to remember: the group of comrades, mmmm, Franz the pupil... I was supposed to fly for the Prince of Brunswick... Really, he and Prinz August Wilhelm wanted to come. Ah, and my head was buzzing so! The sun broke through. Good flying weather – how dreadful![334]

Since the heroes risked their lives every day, some loose living was tolerated. Festivities, heavy drinking and lax discipline seemed to be part of the male airman's make-up.[335] Indeed, such excesses confirmed the gender-specific conventions and ideal models of masculinity, especially since, when it came to it, the heroes "fought like men."

The second sex hardly figures in the popular hero-narratives – or at most it does so in a maternal role.[336] But, as Immelmann suggests, it is also part of the hero's aura to be popular with women:

> So, am I supposed to wear this leaf on me all the time? If I did that with every lucky flower, cloverleaf, and so on, I'd have a little vegetable garden on me all the time. And then, in order to be fair, I'd also have to carry around the rosaries, crucifixes, and other talismans I am given. There are too many girls who get such clever ideas – all of them very sensitive young ladies, no doubt.[337]

Immelmann's *Meine Kampfflüge*, in particular, conveys the impression of aviator-heroes as media icons or stars *avant la lettre*.[338] This image related primarily to their military feats and qualities, but the rather private details, as well as the ordinary, everyday decorations that came with the job, increased the attractiveness of the

---

[334] Buddecke, *El Schahin*, p. 51.

[335] See also Eichler (ed.), *Kreuz wider Kokarde*, pp. 141–42.

[336] See Schüler-Springorum, *Vom Fliegen und Töten*, p. 215.

[337] Immelmann, *Meine Kampfflüge*, p. 127.

[338] For a survey of the "star" concept in media studies, see Stephen Lowry, "Star," in Hans-Otto Hügel (ed.), *Handbuch Populäre Kultur. Begriffe, Theorien und Diskussionen*, Stuttgart 2003, pp. 441–45.

type they embodied.[339] The honor and fame accorded to them made their military achievements seem even more glamorous. Thus, in the context of an expanding culture industry, war heroes become idols.[340] They personified a model of masculinity that deserved to be imitated and honored, but also to be coveted. Immelmann for one had trouble answering the flood of fan mail:

> The letters pile up monstrously. [...] It is completely impossible for me to answer them in any kind of detail. [...] At the most I write a card with a picture of myself. I would not have thought that fame is bound up with such inconvenience. I cannot even tell you of the little honors that others do me every day and every hour.

And a few pages later: "I'm really snowed under, hardly have any time. I don't think I need assure you that the number of letters I owe has increased immeasurably. Every day 30–40 letters and postcards. Dreadful."[341]

Aviator-heroes were popular in a twofold sense: because of the aesthetic impact of their accounts and because they were "loved by the people." Their narratives were escapist products of the culture industry: that is, they enabled readers to flee from the unbearable and incomprehensible reality of mechanized mass killing into a structure of warfare more in line with their expectations.[342] Although the state apparatus of opinion formation steered and promoted them, the popular hero-narratives were not manipulated by forces behind the scenes. Rather, bourgeois publishers and authors met the needs of a largely bourgeois public by reproducing its ideal visions of war as a field of honor, fame, self-vindication and adventure.

Although the idolization of aviator-heroes is not altogether comparable to today's worship of film, sports, or music stars, it involved an analogous popular phenomenon. The fact that the idols in question were soldiers is telling evidence of the prevailing "militarism of ordinary people," but also of the incipient tendency to put the whole of society on a war footing.[343] The following quotation again illustrates the affinity with the pop icons of today:

> Two photographers immediately pounced on me and requested the honor of taking my picture. I must have been photographed twenty times, and also filmed with a movie camera. Some three weeks later the pictures were appearing in every cinema in Germany. So you should now make sure you get to a movie theater, so that you don't miss them. They took me first with the fourth

---

[339] See *inter alia* Werner Faulstich (ed.), *Image, Imageanalyse, Imagegestaltung. 2. Lüneburger Kolloquium zur Medienwissenschaft*, Bardowick 1992.

[340] See Walter Uka, "Idol/Ikone," in Otto Hügel (ed.), *Handbuch Populäre Kultur*, pp. 255–59.

[341] Immelmann, *Meine Kampfflüge*, pp. 87 and 96.

[342] See Kracauer, "The Little Shopgirls Go to the Movies," pp. 291ff.

[343] On the militarism of "ordinary people," see Rohkrämer, *Der Militarismus der "kleinen Leute."* And on the "bellification" of society in general: Frank Reichherzer, *"Alles ist Front!."*

Englishman I shot down, then with Captain Rosenmüller, and finally in conversation with the king. [...] The king showed up around 10:30 in the morning.
He came straight up to me, inspected and marveled at Englishman No. 4, then
took a picture of me in front of this Englishman. Just think: the king snapped
me! Some excellencies and generals then did the same.[344]

The pilots' military accomplishments were the key to their popularity and charismatic effect. But if they became models, idols or stars, it was also because of
the rewards and honor, the symbolic capital, that their feats earned them.[345] This
included their dealings with the aristocratic and military elite of Imperial Germany:

On November 14 I was summoned to the court table of Crown Prince Ruprecht
of Bavaria. The King of Saxony and Prince Ernst Heinrich were also there, as
were some familiar faces: His Excellency von Laffert, von Willsdorf, O'Byrn [sic]
(aide-de-camp), and a few other top people in the Saxon military. A few days
earlier, a telegram had come in which the war minister, von Falkenhayn, congratulated me and informed me that the Emperor had awarded me the Knight's
Cross of the Royal House Order. It was, of course, something extraordinary, for
the Order of the House of Hohenzollern is a very rare thing.[346]

Aviator-heroes made it into the upper echelons of Imperial Germany. While the
glitter of high society rubbed off on the figure of the hero, the Imperial elite
bathed in the warriors' popularity. The glamor emanating from them and their
successes in the air distracted attention from the harsh reality at the front; it
boosted confidence in victory and thus served the purpose of "wartime opinion
formation."[347]

The war books and reports suggested that military heroism and self-sacrifice for
the nation were worthwhile in themselves, while at the same time bringing further comforts, luxury, fame and popularity. The hero encouraged others to imitate
him, since the regard in which he was held had an objective character. In fact,
aviator-heroes were singled out many times over:

To K.S. [Royal Saxony] Reserve Lieutenant Max Immelmann, Pilot in the Field
Aviation Detachment ..., Commander of the Order of St. Heinrich, Knight
of the Iron Cross 1st Class and 2nd Class, Knight of the Order of the House
of Hohenzollern, Knight of the Order of St. Heinrich, Knight of the Order of

---

[344] Immelmann, *Meine Kampfflüge*, pp. 85–6.
[345] On the applicability of Pierre Bourdieu's concept of symbolic and social capital to historical research, see Sven Reichardt, "Bourdieu für Historiker? Ein kultursoziologisches Angebot
an die Sozialgeschichte," in Thomas Mergel/Thomas Welskopp (eds.), *Geschichte zwischen
Kultur und Gesellschaft. Beiträge zur Theoriedebatte*, Munich 1997, pp. 71–93. Reichardt's article also provides further bibliographic references.
[346] Immelmann, *Meine Kampfflüge*, pp. 84–5.
[347] See Lipp, *Meinungslenkung im Krieg*.

Albrecht with Swords, Knight of the Bavarian Order of Military Merit with Swords, Knight of the Order Pour le Mérite, Holder of the Friedrich August Silver Medal, Holder of the Hanseatic Cross, Hamburg. Forces Postal Service, Station 406.[348]

The highly decorated heroes received the coveted *Dank des Vaterlands* [Gratitude of the Fatherland], which further augmented their social privileges and symbolic capital.[349] As heroes of the nation, their military equipment was put on display at war exhibits: "My old 80 P.S. Fokker (factory number E. 13), with which I brought down my first five Englishmen, is being exhibited at the Zeughaus (former armory). Isn't that wonderful? One honor is coming on top of another."[350]

Between the beginning of 1917 and September 1918, the Exhibition of German Air Booty – which displayed not only German and enemy materiel but medals, clothing and trophies belonging to various heroes – made its way from Berlin to Dresden, Munich and Dortmund.[351] Immelmann records among his trophies: "Two English pistols, one burned; an engine cylinder and a piston, both hit by bullets. The pistol comes from my first Englishman, the burned revolver from my second, the cylinder and piston from my aeroplane that was shot down in combat on September 23."[352]

The director of the Delka (*Deutsche Luftkriegsbeute-Ausstellung*), as the travelling exhibition was known, was in fact Ernst Friedrich Eichler, the publisher of Udet's book *Kreuz wider Kokarde*. The organizer, however, at least in Munich, was the Bavarian airmen's association, the Bayerischer Luftfahrerdank e.V. Eichler reported to the Royal Bavarian Military Aviation Inspectorate on the great success of the exhibition in Munich; when it closed on June 16, 1918, before moving on to Dortmund, it had received 320,000 visitors.[353]

Like Field Marshal Hindenburg, aviator-heroes became military and national symbols, indeed "branded articles." Their portraits and names came to stand for the nation itself and another (victorious) war – a significance promoted not only by the media but also by jingoistic kitsch ranging from Willy Sanke's postcards through aircraft decoration for Christmas trees to the Skat card game.[354] However, the cult of the aviator-hero was not purely militarist in inspiration, nor was it

---

[348] Immelmann, *Meine Kampfflüge*, p. 118.

[349] See Ralph Winkle, *Der Dank des Vaterlandes. Eine Symbolgeschichte des Eisernen Kreuzes 1914 bis 1936*, Essen 2007, p. 91.

[350] Immelmann, *Meine Kampfflüge*, pp. 110–11.

[351] On war exhibitions, see Christine Beil, *Der ausgestellte Krieg. Präsentationen des Ersten Weltkriegs 1914–1939*, Tübingen 2004, pp. 178–93.

[352] Immelmann, *Meine Kampfflüge*, p. 80.

[353] BayHStA-KA, Iluft 212, Brief des Bayerischen Luftfahrerdank e.V. an die K.B. Inspektion des Militär-Luftfahrt-Wesens, 6.18.1918.

[354] See, for example, exhibits VII/41 and VII/43, in Rainer Rother (ed., for the German Historical Museum), *Der Weltkrieg 1914–1918. Ereignis und Erinnerung*, Berlin 2004, pp. 213–14.

mainly imposed from the top down. Rather, popular media forming part of the "culture industry" satisfied a demand for it on the part of "the masses."

The hero-narrative inscribed Horace's famous saying into popular discourse: to die for one's country was desirable, since it preserved the life of the nation. But there were other reasons why the hero's life was sweet and honorable: he enjoyed the reward of personal popularity; his trophies and everyday objects were honored as relics; he received fan mail and good luck charms from "sensitive young ladies"; various decorations conferred on him the grateful recognition of the fatherland; and he was able to interact on cordial terms with the military and aristocratic elite.[355] These too were part of the compensatory function of the popular hero-narratives. The war books offered their readers a heroic matrix as well as role models that embodied "possible forms of personal identity."[356] Against the background of trench warfare, media presentations of the aviator-hero "projected collective ideas of a 'proper' war."[357] But a "proper war" had above all to end in victory. The hero-narratives were able to compensate even for defeat.

---

[355] It should be borne in mind that the flier's image of luxurious and glamorous living developed in a climate of wartime austerity. The kind of life depicted in the war books looked attractive not only from a mud-filled trench on the Western Front, but also from a hungry, shivering home front in the "turnip diet" winter of 1916–1917.

[356] Lowry, "Star," p. 442.

[357] Reimann, *Der große Krieg der Sprachen*, p. 70.

3

Transitional Heroes and the Order of the *Gemeinschaft*

The fall of ruling dynasties and the ensuing political convulsions were only the most visible result of the fracturing of the old order induced by the First World War. Despite his emphatic, emotional prose and his political attitude, Ernst Jünger will be taken as a witness to this seismic shift; he did after all clearly identify the shocks surrounding it.[358] In the original German edition of *The Adventurous Heart*, published in 1929, he wrote:

> What then struck me as so strange about those battles of materiel – the glowing horizon that seemingly welded the enemy frontlines together – now appears to make more and more sense. It was a destructive war, a concentric raging against a secret center, an event on the western surface. We worked hard nihilistically for a few years with dynamite, and, dispensing with the most inconspicuous fig leaf, blew the nineteenth century – and ourselves – to smithereens; only right at the end did means and men of the twentieth appear in dim outlines.[359]

The great "battles of materiel" were an epicenter of the long earthquake that shook the old order to its foundations, but they were also the harbinger of a new order whose antecedents had first shown themselves in the nineteenth century. The aviation-hero constructs may help us to trace these transformative processes, which obviously involved mingling and interpenetration of prior and the latter order rather than a simple succession in time.

In his *Aufzeichnungen bei Tag und Nacht*, the original version of *The Adventurous Heart*, Jünger created the type of the "Prussian anarchist" based on his own self-image. It was a deliberate contradiction in terms, which made sense none the less. As the sociologist Hans-Georg Soeffner has shown, "paradoxes illustrate

---

[358] Although Jünger was by no means only the "seismograph" he made himself out to be, and although he vigorously urged on the collapse of the liberal order, his "seismographic" faculty should not be underestimated. On this image, see Morat, *Von der Tat zur Gelassenheit*, p. 386.

[359] Ernst Jünger, *Das Abenteuerliche Herz*, p. 186.

one fundamental form [of] dealing with contradictoriness and oppositionality: They represent individual contradiction and the process of its harmonization *simultaneously.*"[360] They therefore "mark a considerable step within a process of overcoming problems that begin in the experiencing of insecurity." Paradoxes like the "Prussian anarchist," for example, point to "practically achieved processes of harmonization" and to "the necessity of living with and in contradiction, and at the same time point to one of the means of dealing with these contradictions."[361]

In the idea of the "Prussian anarchist," Jünger harmonized the Prussian (which stood for order and stringency) and "the anarchistic" (which signified chaos and hence the dissolution of order). He transferred the image not only to himself but also to his readers. In the section from which we have already quoted, he addressed his "Prussian reader" as

the man [...] to whom (with Fichte) the world is only material of the heroic will, and for whom (with Nietzsche) the most resolute escape from the boundaries of the only-human is a delight. More: at a time when all order suffered shipwreck, he became perhaps the strangest manifestation of the Prussian anarchist, who, armed only with the categorical imperative of the heart and responsible to it alone, roams the chaos of the powers in search of the rules of new orders.[362]

The "Prussian anarchist" was the paradoxical embodiment of a widespread longing to overcome the "shipwrecked" order and to establish a more stable one in its place. The present chapter will seek to clarify what roaming through the "chaos of the powers in search of the rules of new orders" actually implied. It will show how interpretive models developed during the war were later transferred to the postwar societies: that is, how these social orders were derived from wartime "standards." According to Jünger, hopes in this new order rested

upon young people, who suffer from a raised temperature because the green pus of revulsion eats away inside them. [...] It [hope] rests on the rebellion that opposes the rule of comfort, and that needs the weapons of destruction aimed against the world of forms – explosive material so that living space is cleared for a new hierarchy.[363]

In tracing how this "new hierarchy" was created, we shall clarify how the "explosive material" "cleared living space" – not to destroy order as such, though, but only the bourgeois-liberal order. What was supposed to appear in its place was not "the anarchistic" but "the Prussian."

---

[360] Hans-Georg Soeffner, "Flying Moles (Pigeon-Breeding Miners in the Ruhr District): The Totemistic Enchantment of Reality and the Technological Disenchantment of Longing," in Soeffner (ed.), *The Order of Rituals: The Interpretation of Everyday* Life, New Brunswick, NJ 1997, p. 96.
[361] Ibid.
[362] Jünger, *Das Abenteuerliche Herz*, p. 257.
[363] Ibid., pp. 222–23.

The popular aviation narratives served to establish an "imagined order" for the community, with the nation at its center.[364] We shall see how the shift in nationalism during the First World War also crystallized in the hero-narratives, and how the New Right took the image of a military aviator-elite as the prototype for the new nobility and new man that it set out to create. In the last section of the chapter, we shall look at the sacralization and superelevation of the hero in popular culture, and thus at the spread of the palingenetic myth in society. The hero-narratives justified the longing for, and actual offering of, self-sacrifice and situated it in a "higher" context that brought consolation. When the postwar period turned into another prewar period, the aviator-hero who had transcended the stalemate of the trenches became the herald of the "rebirth of heroic man," as well as of the nation or *Volk*.[365] The preconditions of this reinterpretation will be dealt with in this chapter.

The aviator-hero and the unity of the wartime nation

All strata of society were involved in the First World War. This was a fundamental difference in relation to earlier wars, entailing a need for legitimation that was valid in the whole of society. The prevailing hero-images reflected this shift, with the spirit of self-sacrifice as the clearest example. But the new technological heroes – submarine captain, assault leader and especially fighter pilot – also testified to the change.[366] The hero-narratives generated an order within which it seemed understandable that the war affected one and all. They legitimized war by means of a conception of the nation that made each individual part of the fighting collective, and made the latter into a community distinct from the enemy collective. We shall see how the hero-narratives legitimized the war by postulating the identity of the individual with the nation. The experience of national defeat as personal failure rested not least upon this foundation.

The late summer of 1914 marked a boundary for many protagonists of the war books in question here. In the months of July and August, rites of passage not only turned boys into "men," but also made "heroes."[367] And like the passage

---

[364] On the concept of an "imagined order [*gedachte Ordnung*]," see Mario Rainer Lepsius, "Nation und Nationalismus in Deutschland," in Mario Rainer Lepsius (ed.), *Interessen, Ideen und Institutionen*, Opladen 1990, pp. 232–46. Lepsius's definition of "nation" (p. 233) underlies our own investigation: "The nation is first of all an *imagined order*, a culturally defined idea that sees a particular human collective as a unit. The type of unit this will be derives from the criteria for defining the collective within an idea of national order. If these are ethnic criteria, the nation is defined as a unit of ethnic descent; if they are cultural criteria, the nation appears as a community of language; if they are criteria of citizenship under the law, the nation is a unit of citizens. The collective of people who form national ties of solidarity with one another will differ according to the specific mix of these criteria."

[365] See Schwarz, *Die Wiedergeburt des heroischen Menschen*.

[366] See Schilling, *"Kriegshelden,"* pp. 252–88.

[367] On "rites of passage," see Arnold van Gennep, *Rites of Passage* [orig. 1909], Chicago, IL 1961.

from boy to hero, war was supposed to be a dividing line for the whole nation. It brought civil peace, and all party strife, all particular class interests, became as nothing in the face of great national tasks.

After the defeat, the New Right tried to revitalize this "Spirit of 1914" and the true nation that was its bearer.[368] The war and its sacrifices mark the threshold – to remain with the imagery of passage – beyond which Germany became a different nation; it was sacralized, "regenerated," and reborn through the war.[369] At the symbolic-semantic level, a cathartic effect was ascribed to the war. Notwithstanding defeat and revolution, the nation came back purified and reformed. The impurity had been "burned to slag" in the fire of technological warfare, and a new man had emerged from the experience.[370] In fact, it was mainly nationalism that came back changed from the war.[371] First, it was more aggressive and prone to violence. Second, the underlying concept of the nation had become increasingly ethnicized and loaded with *völkisch* overtones.[372] And third, it was more and more separated from monarchism and conservatism. The novelty in this "new nationalism" or "populist ultranationalism" was the fact that it was a nationalism of mass society.[373] The war had "democratized" it, turning it into a mass experience with a real impact on large numbers of people. Nationalism was now a cause of the *Volk*.[374]

The popular hero-narratives were both an indicator of this change and a factor in it. Their protagonists yearned for war, so that it would make real men – that is, soldiers and officers – out of them. In Max Immelmann's *Meine Kampfflüge*, an alleged letter written on November 16, 1914, features under the heading "Training Period." "In the heady August days," he writes, "I found it dreadful to have to go

---

[368] On the legendary "Spirit of 1914" and the development of a national community, see Verhey, *The Spirit of 1914. Militarism, Myth, and Mobilization in Germany*, New York, NY 2000. Also Bruendel, *Volksgemeinschaft oder Volksstaat*; Peter Fritzsche, *Germans into Nazis*; Wildt, *Hitler's Volksgemeinschaft and the Dynamics of Racial Exclusion*.

[369] See esp. the chapter on initiation rites in Van Gennep, *Rites of Passage*, pp. 65–115.

[370] See Wolfgang Schivelbusch, *The Culture of Defeat*, pp. 233 and 311.

[371] For an introduction to the new research on nationalism and a survey of the extensive literature, see Echternkamp/Müller (eds.), *Die Politik der Nation*; Rolf-Ulrich Kunze, *Nation und Nationalismus*, Darmstadt 2005; Dieter Langewiesche, *Nation, Nationalismus und Nationalstaat in Deutschland und Europa*, Munich 2000; Dieter Langewiesche, "Nation, Nationalismus, Nationalstaat. Forschungsstand und Forschungsperspektiven," *Neue Politische Literatur* 40/1995, pp. 190–236; Walkenhorst, *Nation – Volk – Rasse*, pp. 11–33. See also the "modern classics": Benedict Anderson, *Imagined Communities. Reflections on the Origin and Spread of Nationalism*, London 2006 [repr.]; Ernest Gellner, *Nations and Nationalism*, Oxford 1990 [repr.]; Eric J. Hobsbawm, *Nations and Nationalism Since 1780. Programme, Myth, Reality*, Cambridge 1990.

[372] See Walkenhorst, *Nation – Volk – Rasse*.

[373] On the new nationalism of the "conservative revolution," which was mostly an elite phenomenon, see *inter alia* Stefan Breuer, *Ordnungen der Ungleichheit – die deutsche Rechte im Widerstreit ihrer Ideen 1871–1945*, Darmstadt 2001, pp. 92ff. On the "populist ultranationalism" of the fascists, see Griffin, *The Nature of Fascism*.

[374] See Fritzsche, *Germans into Nazis*, pp. 63ff.

around as a civilian."[375] The account in *Kreuz wider Kokarde* of the experiences of Ernst Udet, who by the end of the war was Germany's second most successful fighter pilot and later became head of ordnance in the Nazi air force, makes a similar point: "When the Great War broke out in August 1914 and hundreds of thousands hurried enthusiastically to the flag, an urge to serve the fatherland awakened in me. [...] Finally came the longed-for day when my wishes would be fulfilled." And a few pages further on: "In August 1915 – a full year after my first attempts to take the field – I finally achieved my aim. I was posted to a newly formed artillery direction unit that would soon see action."[376]

A chance to serve the fatherland was the hero's most burning desire, and it was held up as a virtue for readers to imitate. In Gunther Plüschow's *Der Flieger von Tsingtau*, August 1914 is seen through the eyes of a navy pilot condemned to inactivity in the Far East:

> Again and again we said to ourselves: "While we sit here in faraway Tsingtao our brothers, our comrades, are there back home; they are lucky enough to live through the wonderful days of the mobilization, to set out against a world of enemies, to defend our sacred, beloved fatherland and our women and children, and we poor souls sit here and are unable to help!" [...] The English, Russians, and French, who greatly outnumbered us, would not find the courage to attack us out here. Yet we repeatedly had a spark of hope that they would come after all! Ah, what a reception we would have given them![377]

Many of those hungry for action speak similarly of the "days of hope" in early August – hope that the war will finally come and give them an opportunity to prove themselves. Emil Schäfer's book *Vom Jäger zum Flieger*, first published in 1918, has a section heading: "Hurrah, Now We're Off!"[378]

It has recently been shown a number of times that in August 1914 the coming of war did not arouse enthusiasm among all sections of the population.[379] More than 750,000 people, including 100,000 in Greater Berlin alone, turned out for the anti-war demonstrations of July 28 organized by the SPD;[380] and official figures long exaggerated the numbers of young men volunteering for duty. But although the picture of masses of enthusiastic volunteers should not be confused with the reality,[381] it needs to be taken seriously as recruitment propaganda and as a vision of a German Reich united in its readiness for struggle and sacrifice. The

---

[375] Immelmann, *Meine Kampfflüge*, p. 7.

[376] Eichler (ed.), *Kreuz wider Kokarde*, pp. 12ff. and 27.

[377] Gunther Plüschow, *Die Abenteuer des Fliegers von Tsingtau*, Berlin 1916, pp. 28–9.

[378] Schäfer, *Vom Jäger zum Flieger*, p. 9.

[379] See Müller, *Die Nation als Waffe und Vorstellung*, pp. 56–70; Verhey, *The Spirit of 1914*; Ziemann, *War Experiences in Rural Germany*, pp. 15–28.

[380] See Verhey, *The Spirit of 1914*, pp. 47–57.

[381] Verhey points out that, whereas the talk in August 1914 was of 1,300,000 volunteers, the real figure was "only" around 185,000. See ibid., pp. 97ff.

"August experience" trumpeted by the conservative press mainly gripped young educated people of middle-class origin.[382] The philosopher Karl Löwith, who was born in 1897, later described his mood at the time:

> The yearning to escape from the bourgeois narrowness of school and home, the feeling of being at odds with myself after the break-up of my first friendship, the passion for "living dangerously," which Nietzsche had instilled in us, the desire to throw ourselves into adventure and to test ourselves, and not least the easing of the burden of one's own existence by participating in the wider universe, of which Schopenhauer had made us conscious – these and similar motives caused me to welcome the war as an opportunity to live and die more intensely.[383]

Although the reference to Nietzsche and Schopenhauer was not necessarily typical, Löwith's motives were shared by many in his generation. The break-out from a constrictive bourgeois existence, the attractions of "living dangerously," the chance to prove oneself "a man," the lust for adventure, the need to serve a common cause higher than self-interest: these all drew bourgeois youth into the war, "as cheerfully as on a festive day," to quote Ernst Jünger.[384] On the other hand, fear and above all agitation were the main feelings among large parts of the population.[385]

The extent to which young bourgeois like Immelmann and Udet were really enthusiastic at the outbreak of war, seeing it as the opportunity for adventure and "a great time," cannot be determined. As the above quotations show, however, they thought it fitting in the books that recounted their "experiences" to use and pass on the tropes of war enthusiasm, self-testing and yearning for adventure. To question any of these, or to express feelings of fear or dejection, was not discursively acceptable. The hegemonic narrative that structured the war, and many of the models used in its interpretation, were later packaged and given declamatory expression in Hitler's *Mein Kampf*.

> Those hours seemed to me like a deliverance from the irksome feelings of youth. Even today I am not ashamed to say that, carried away by wild enthusiasm, I fell on my knees and thanked heaven from an overflowing heart that I had been granted the good fortune to live in such times.[386]

The myth of "the spirit of 1914" was of far greater significance than the question of who really experienced what. War enthusiasm, self-testing and the emergence

---

[382] On young people in 1914, see Wohl, *The Generation of 1914*, pp. 42–84.

[383] Karl Löwith, *My Life in Germany Before and After 1933 – A Report*, Chicago, IL 1994, pp. 1–2.

[384] Ernst Jünger, "Kriegsausbruch 1914," in Jünger *Sämtliche Werke. Erste Abteilung*, vol. 1, *Tagebücher 1, Der Erste Weltkrieg*, Stuttgart 1978, pp. 539–45; here p. 544.

[385] See Ziemann, *War Experiences in Rural Germany*, p. 15.

[386] Adolf Hitler, *Mein Kampf*, vol. 1, *Eine Abrechnung*, 19th edition, Munich 1933, p. 177.

of the national community were key pillars of the narrative that centered on the absolutized nation. In *Fliegerleutnant Heinrich Gontermann* we can read:

> But the waves of patriotic enthusiasm immediately rose in every part of the country. [...] Germany's sons recognized that it was a "to be or not to be" situation for the fatherland. [...] The German *Volk* had only one answer: to fight to the death until victory was assured. The enemies' numerical superiority was a challenge to summon up every bit of strength and self-sacrifice, together with a fiery enthusiasm of the will. All discord and narrow party spirit disappeared. They were the finest times Germany had ever seen.[387]

As Peter Fritzsche reminds us:

> 1914 is the crucial date because it set in motion tremendous political aspirations. The triumph of National Socialism has to be sought as much in the realm of ideas and loyalties as in the convergence of economic and military crisis. Because the war so thoroughly revised the national imagination and recombined 60 million people in novel and often dangerous ways, 1914 is an appropriate point of departure for an account of how and why the Nazis came to power.[388]

Decisive for the following years of war, as well as for the Nazi dictatorship, was the fact that "in August 1914 Germans could read that they had all experienced the outbreak of war in the same way, that through the August experiences a national identity had come to replace the various local and class identities as the most important social identity."[389]

This idea of a united national community was the core of the palingenetic myth of fascism. Its aim was to revitalize the entity whose "good basic elements" – to quote from the section of *Mein Kampf* on "the first signs of German rebirth" –

> had not been completely lost to our people; they were slumbering in the depths, and many a time they could be seen flashing like sheet lightning on the black-draped firmament of virtues, to be remembered by the future Germany as early signs of recovery. More than once, thousands upon thousands of young Germans were found with a determination as in 1914 to sacrifice their young lives freely and joyfully on the altar of the beloved fatherland.[390]

This hallowed nation, for which laying down one's life was justifiable, became the Archimedean point of an "imagined order." Bearing in mind the definition of myth that informs this study, we should stress that the "life context," but

---

[387] Müller (ed.), *Fliegerleutnant Heinrich Gontermann*, p. 21.
[388] Fritzsche, *Germans into Nazis*, p. 7.
[389] Verhey, *The Spirit of 1914*, p 112.
[390] Adolf Hitler, *Mein Kampf*, vol. 2, *Die nationalsozialistische Bewegung*, 19th edition, Munich 1933, p. 713. See also Fritzsche, *Germans into Nazis*, pp. 5–9.

especially death and suffering and conduct in war, were now grounded upon their relationship to the sacred sphere of the united nation. But if the nation lost this position as absolute reference, the justification for death in war ceased to be valid. The meaning given to suffering and action was then in danger of collapse.

The popular hero-narratives were a factor in this "spiritual mobilization" and helped to absolutize the conception of a national order. They were a constitutive part of the national myth developing around "the spirit of 1914": both because they created a narrative order into which actual experience could be integrated, and because they helped to define national identity by putting across the character qualities and virtues of the hero. They expressed what it meant to be "German" and what it meant to be "a German hero." The war books told of a spontaneous readiness to volunteer for action, in a spirit of duty, patriotism, and self-sacrifice:

> I don't want to be a regular officer in order to live, but so that I can become something virtuous! Meanwhile I shall do what I owe to the fatherland and hold my ground to the last drop of my blood. If I did not believe in this war with all my conscious being, I would never be able to atone for it in peacetime, and I would lose the power of my inner fortitude and assurance.[391]

To be an officer, to lay down one's life for the nation: this, we are told, perfects and ennobles human beings. In the popular war books, the hero's identity as an individual was inseparably bound up with the nation; personal and national goals merged into one. A passage from Plüschow's *Die Abenteuer des Fliegers von Tsingtau* is further evidence of this. After an adventurous journey back to the homeland from the German "protected territory" of Kiautschou, the author reports,

> I received the thanks of my Emperor. And I went proudly home to my loved ones decorated with the Iron Cross, First Class. Then, after a few weeks of rest, came my greatest reward. I became a pilot again and could join the great work of Germany's struggle and victory. When my most gracious Lord and Emperor visited the hydroplane station on the Eastern front that was under my command, and when he clasped my hand and personally expressed his imperial appreciation, I gazed into his eyes, and the words blazed in my soul: "With God for Kaiser and Reich!"[392]

To serve the nation is the greatest reward, true self-fulfilment. And the protagonists of the war books get their satisfaction from personal commitment, from struggle and victory over the enemy. For Immelmann, for instance: "Since I have been a fighter pilot, I have always thought that I'd like sometime to finish off two enemy planes in one day. That wish was fulfilled on March 13."[393]

---

[391] Müller (ed.), *Fliegerleutnant Heinrich Gontermann*, pp. 40–1.
[392] Plüschow, *Die Abenteuer des Fliegers von Tsingtau*, pp. 246–47.
[393] Immelmann, *Meine Kampfflüge*, p. 107.

This amalgam of individual ambitions and consciousness with collective national ones, this unity of personal success and war aims, forms the basis for any potentially radical and aggressive nationalism. The "nation" as an ordering concept played a decisive role in the make-up of these protagonists; the narratives of their books served readers in turn as a matrix for the interpretation of their own lives, and subsequently as guides to action.

The world war was a catalyst for radical nationalism, which became more and more virulent and influential as it spread into parts of society yearning for order. Before the war, the idea of the nation as the focus of order – an idea with particularly strong roots in the bourgeoisie – had been hedged around with other framing concepts. But the totalization of the war made it necessary to legitimize the latter in the eyes of the population as a whole, while also denying legitimacy to the enemy. Neither liberal conceptions of a universal, rational, emancipated humanity nor the Marxist class concept provided a solid basis for the erection of an order in which the mechanized slaughter of human beings appeared meaningful. Indeed, such universalist ideas actually made a legitimation of the war more problematic.[394] Only the idea of the nation was able to form a coherent unit with a common purpose out of the belligerent collective, and to exclude the enemy so categorically that he became a legitimate object of violence.

The war gave absolute prominence to the idea of order centered on the nation, but it was the traumatic defeat and (in the Italian case) the "mutilated victory" that prevented the rehabilitation of other concepts that might have helped to organize society. Any attempt to limit the absolutist claims of the nation was interpreted by the growing swarm of radical nationalists as an "enemy" attack from "without." The war thus became the starting point of a vicious circle: the search for a basis on which to legitimize violence and suffering led to a negation of ideas of order that tended to delegitimize violence and suffering; but, on the other hand, the exercise of violence and the experience of suffering increasingly authenticated the idea of order centered on the nation, whose claims to absoluteness grew as a result. The vicious circle consisted in the growing interdependence of violence and nation. Radical nationalism became more aggressive and paved the way for the fascist worldview: that is, for the "palingenetic form of populist ultranationalism" characterized by a cult, and the actual practice, of violence.[395]

---

[394] The Entente faced a dilemma that German propaganda gladly seized upon. How could the war be convincingly portrayed as a struggle for freedom and against autocracy, when the Russian Empire was one of the main Allied powers? The legitimation pressure was even greater for the European Left, however, since it was scarcely possible to overcome the conflict between nation and class. The universalism of the Catholic church also created serious difficulties, to which German Catholics responded by presenting the war against Catholic (but also secular) France as a war against a godless nation that had turned away from religion; on this whole problem, see Claudia Schlager, "Zwischen Feindesliebe und Erbfeindschaft. Deutsche und französische Katholiken im Ersten Weltkrieg," in Reinhard Johler/Freddy Raphaël/Claudia Schlager/Patrick Schmoll (eds.), *Zwischen Krieg und Frieden. Die Konstruktion des Feindes*, Tübingen 2009, pp. 177–206.

[395] On the palingenetic form of ultranationalism, see Griffin, *The Nature of Fascism*, p. 26. On the cult, but above all the practice, of violence, see Reichardt, *Faschistische Kampfbünde*.

As Peter Walkenhorst has shown, Imperial Germany also displayed a stronger ethnicization and biologization of the idea of the nation.[396] Although there is often talk of "Teutonic blood" (*Germanenblut*) in the popular hero-narratives, an analysis of the war books allows of no clear-cut conclusion regarding the spread of the biologically charged *völkisch* concept of the nation.[397]

The semantic ethnicization of the concept can be established, however:

> The German national spirit is being put to the test as never before. So far it has passed it brilliantly, and we shall persevere until the wicked enemy realizes that a national spirit of godliness and efficiency will never tire so long as noble Teutonic blood courses through a single vein. In this we have a model stock that has preserved and deepened the Teutonic national spirit for hundreds of years: the Hohenzollerns.[398]

The popular hero-narratives testify to a radicalization of the friend–enemy dichotomy. By handing down national stereotypes, the war books helped to define what was distinctively German, and therefore to be preserved, and what was alien, inimical, and to be excluded. They shaped self-perception and self-understanding by delimiting the Other:

> Much depends on whether we have for opponents those French tricksters or those daring rascals, the English. I prefer the English. The Frenchman backs out; the Englishman rarely. [...] In my opinion all that counts is being bold, and that spirit is very strong in us Germans. [...] The French have a different character. They like to set traps and to take their opponents unawares. [...] Sometimes, however, the Gaelic blood asserts itself. The Frenchman will then attack. But the French attacking spirit is like bottled lemonade: incredible pluck for a while, but it soon disappears completely. With the English, on the other hand, one notices now and then that they still have something of their Germanic blood. [...] They take a perfect delight in looping the loop, flying on their back, and indulging in other stunts for the benefit of our soldiers in the trenches. All these tricks may impress people who attend a Sports Meeting, but the public at the battle-front is not as appreciative of these things. It demands higher qualifications than trick flying. It wants the blood of English pilots to keep raining down.[399]

The perception of Germans as a "spirited" people who like to get "stuck in" is contrasted to that of the French as "tricksters" who often "back out" and like "to

---

[396] Walkenhorst, *Nation – Volk – Rasse*, pp. 102–19; and Müller, *Nation als Waffe und Vorstellung*, p. 359.

[397] On the *völkisch* movement in Imperial Germany, and for a further reading list, see esp. Uwe Puschner, *Die völkische Bewegung im wilhelminischen Kaiserreich. Sprache – Rasse – Religion*, Darmstadt 2001; and Uwe Puschner/Walter Schmitz/Justus H. Ulbricht (eds.), *Handbuch zur "Völkischen Bewegung" 1871–1918*, Munich 1996.

[398] Müller (ed.), *Fliegerleutnant Heinrich Gontermann*, pp. 55–6.

[399] Richthofen, *The Red Fighter Pilot*, pp. 115–16; translation modified.

set traps." The description of "the" Englishman is often more positive, however, because of the "Germanic blood" he has in him, and especially because of his "sportsmanship" and "chivalry."[400] In *Kreuz wider Kokarde*, for instance, we read:

> Again we had Frenchmen as opponents, and we would so much have liked to fight the English at last. The Frenchman flies in a more cautious and refined manner than the Englishman and is therefore not such an agreeable opponent. I have always found that the average Englishman gives battle even when he is outnumbered, which does not happen often, and that he does not pull away like the Frenchman.[401]

For all this misanthropic talk of "English pilots' blood," the war against the Western Entente powers was portrayed as a struggle between more or less worthy adversaries. The Italian armed forces, however – an enemy seen as having "betrayed" the Triple Alliance – were arrogantly dismissed as not worth taking seriously. Hans Joachim Buddecke, for instance, writes in *El Schahin*: "The further north we went, the more teeming it was with soldiers. There were so many of them that in Verona we seriously asked ourselves: Are they mobilizing here, or are the buggers planning some stage show?"[402]

The sense of superiority grew according to the rank given the enemy in the hierarchy of nations and peoples. Gunther Plüschow's *Abenteuer des Fliegers von Tsingtau* clearly voices the Germans' dominance over "primitive" peoples. The hero recalls the time when, fleeing from Japanese-conquered Tsingtao, he was forced to land in China.

> Truly, my appearing in darkest Africa could not have produced such horror. [...] The amazement of the Chinese! The touching and fingering! The gabbling and laughing! Only someone who knows the Chinese and is aware how child-like they can be will be able to imagine the deadly situation in which I found myself. There I was in my pilot's seat, bright-eyed and bushy-tailed, with children of nature swarming all around me.[403]

Such constructs of phylogenetic primitivism tend to feature in the rare accounts of enemies and theaters of war in Eastern and Southeastern climes, with the result that "the East" and the "uncivilized, barbarous" Russians are placed semantically at the same stage as "primitive" colonial peoples.[404] The amazement of Plüschow's

---

[400] On chivalry and sportsmanship in Germany and Britain during the First World War, see Goebel, *The Great War and Medieval Memory*, pp. 187–230.

[401] Eichler (ed.), *Kreuz wider Kokarde*, p. 65.

[402] Buddecke, *El Schahin*, p. 41.

[403] Plüschow, *Die Abenteuer des Fliegers von Tsingtau*, pp. 100–1.

[404] On the experiences of German soldiers and experts on the Eastern front during the First World War, as well as the effects in radical Nazi conceptions of order, see *inter alia* Jörg Baberowski/Anselm Doering-Manteuffel, *Ordnung durch Terror. Gewaltexzesse und Vernichtung im nationalsozialistischen und im stalinistischen Imperium*, Bonn 2006, pp. 19–24 and 32–8.

"children of nature" at German advanced technology is echoed in Killinger's *Die Abenteuer des Ostseefliegers*, when the protagonist reports his capture by Russian soldiers: "My flare gun made a great impression on them: they passed it from hand to hand and probably thought it was a fearsome instrument for killing people. Then they divided up with brotherly honesty the various belongings they had taken from us."[405]

The continuum linking these tropes of Russian "uncivilized backwardness" and exploitable labor power with the racist practice of the Nazis is not difficult to discern: "It's OK by me if they [the Russians] come. I could do with a couple of Ivans to repair the road to the airfield a little."[406] It is true that "scientific," biologistic racism was already developing in expert subcultures, and that the hero-narratives testify to the spread of Social Darwinist theories of inferior savages, but it seems unreasonable to project later, biologically charged, racist prejudices back into the discourse of aviation.[407] In fact, the latter tended to be more inclusive than exclusive in character.

Analysis of the hero-narratives suggests that the idea of order centered on the nation acquired absolute dominance, either subsuming or delegitimizing other ideas of order. Whether the nation had a *völkisch*-racist charge, as in National Socialism, or was conceived more along statist lines, as in Italian Fascism, is a secondary issue in the present context. The central point seems to be the transformation undergone by, for example, the liberal category of progress. This did still play a role in interpretations of "primitivism," only it was no longer conceived in a universalist manner. Progress now simply denoted the superiority of "us" over "them." As Mario Rainer Lepsius has pointed out, "the idea of order centered on the nation," rather than on other "ties of solidarity," involved a claim that the nation had a "higher, more general significance than class, religion, ethnicity, kinship, or social-cultural group."[408]

The aviation-hero narratives further illustrate how the "imagined order" of the nation was discursively constructed through the generation of an idea of community in the reader, as well as in the speaker or author.[409] The war books passed on, reproduced, and updated a catalog of common aims, qualities and virtues that differentiated the national collective from inferior enemy ones. This catalog, represented by the hero and conveyed by the media, became a guide to action when the reader in search of an orientation shaped his own personal identity through an *imitatio heroica*. The updating of mediated patterns of behavior and interpreta-

---

[405] Erich Killinger, *Die Abenteuer des Ostseefliegers*, Berlin 1917, p. 18.

[406] Heydemarck, *Doppeldecker "C 666,"* p. 52.

[407] On the "biologization of the nation," and for further bibliographic references, see Walkenhorst, *Nation – Volk – Rasse*, pp. 102–19; and Fatima El-Tayeb, *Schwarze Deutsche. Der Diskurs um "Rasse" und nationale Identität 1890–1933*, Frankfurt/Main 2001.

[408] Lepsius, "Nation und Nationalismus in Deutschland," p. 233.

[409] On the importance of the novel and newspaper, or the commercial printing industry as a whole, in the "imagined community," see Anderson, *Imagined Communities*, pp. 22–46. "For these forms [novel and newspaper] provided the technical means for 're-presenting' the *kind* of imagined community that is the nation" (pp. 24–5.).

tion meant that community – and therefore common characteristics – was not only imagined but actually produced.

For the war effort to succeed at the level of the whole society, it was thus necessary to imagine an inner homogeneity that overcame differences within the nation. "Total mobilization" therefore went together with the delegitimation of other "imagined orders" productive of loyalty and solidarity.[410] If the war was to be won, the individual person as well as disparate collectives fitted together by alternative "imagined orders" had to merge into the national community. Individualism and particular interests had to be sacrificed to the common good, and personal freedom traded in for supposed security. It was this requirement that led to the absolutization of the nation, making it the exclusive idea of order that brooked none other beside it. This meant suppressing not only any parallel national or ethnic "imagined orders," but also any orders that rested upon class, social origin or religious affiliation. Uniformity could be assured only if a single category of order was valid.[411] The attempted homogenization of the nation, which would eventually culminate in mass murder, had its roots in the drive to mobilize all the resources of society for war and in the resulting demand for total solidarity within the nation. By means of this absolutization, the nation became the "intersubjectively unchallenged" sacred entity at the center of the mythical order.

"Total mobilization" and total solidarity

In his *The Revolt of the Masses*, first published in 1930, the Spanish philosopher José Ortega y Gasset stated: "There is one fact which, whether for good or ill, is of utmost importance in the public life of Europe at the present moment. This fact is the accession of the masses to complete social power."[412] This rise of mass society – and, closely related to it, of class society – was a result of the "melting of everything solid into air" and the "dual revolution" that got under way at the beginning of the "long nineteenth century."[413] In the decades between 1880 and 1930, this class-divided mass society was "in *Aufbruch*," that is setting sail for new seas.[414] And it was the First World War that once more brought home the revolutionary significance of this development and the "need" to find a way to order mass society.

---

[410] Ernst Jünger, "Total Mobilization," [orig. 1930], in Richard Wolin (ed.), *The Heidegger Controversy*, Cambridge, MA 1993.

[411] See Zygmunt Bauman, *Modernity and Ambivalence*, Cambridge, UK 1991.

[412] José Ortega y Gasset, *The Revolt of the Masses* [orig. 1930], New York, NY 1957, p. 11.

[413] Karl Marx/Friedrich Engels, "Manifesto of the Communist Party," in Marx, *Political Writings*, vol. 1, *The Revolutions of 1848*, Harmondsworth 1973, p. 70. On the "long nineteenth century" and "dual revolution," see Eric J. Hobsbawm, *The Age of Revolution: Europe 1789–1848*, London 1962.

[414] Rüdiger vom Bruch, "'Der Zug der Millionen.' Massengesellschaft im Aufbruch," in August Nitschke/Gerhard A. Ritter/Detlev J. K. Peukert/Rüdiger vom Bruch (eds.), *Jahrhundertwende. Der Aufbruch in die Moderne 1880–1930*, vol. 1, Reinbek bei Hamburg 1990, pp. 92–120. On the importance and the plural meanings of the German term "Aufbruch" see Roger Griffin, *Modernism and Fascism. The Sense of a Beginning Under Mussolini and Hitler*, Basingstoke 2007, pp. 9ff.

The First World War ushered in the "nationalization of destruction" that would reach a provisional climax in the 1939–1945 war: that is, "organization of all the productive forces of the nation for the exercise of violence, and direct or indirect reorganization of all areas of life under the dictate of the application of violence."[415] In his "Total Mobilization" essay from 1930, Ernst Jünger noted that "the times are long gone when it sufficed to send a 100,000 enlisted subjects under reliable leadership into battle." Now "the people's representatives" have to be involved in the war, and the "armed defense of the state is no longer exclusively the duty and prerogative of the professional soldier, but the responsibility of everyone who can bear arms."[416]

It was not only the *levée en masse*, however, that had decisively changed the character of war; the spread of military space to the whole of society, its "totalization," also differentiated this war from previous ones.[417] War had become a "gigantic labor process," and, "in addition to the armies that meet on the battlefields," there were now the "modern armies of commerce and transport, foodstuffs, and the manufacture of armaments":

> In the final phase, which was already hinted at toward the end of the last war, there is no longer any movement whatsoever – be it that of the homeworker at her sewing machine – without at least indirect use for the battlefield. [...] In order to deploy energies of such proportion, fitting one's sword-arm no longer suffices; for this is an arming [*Rüstung*] that requires extension to the deepest marrow, life's finest nerve. Its realization is the task of total mobilization.[418]

As early as August 1916, the newly formed Third Army Command began to work toward "total mobilization" by means of the "Hindenburg Program" and the "Fatherland Auxiliary Service."[419] In any case, mobilization "to the deepest marrow" required the formation of a "community" out of a class-divided society marked by divergent and competing interests. War, or the pressure for its legitimation, thus impelled the establishment of an order that tended to homogenize society. But it was the defeat and the revolution of 1918–1919 that further radicalized these efforts at homogenization.

In the view of many people at the time, the defeat had been due to a lack of fully fledged "total mobilization"; German society had remained heterogeneous, divided and permeated with alien elements. While the army had been "undefeated on the field of battle," failings on the home front and a "stab in the

---

[415] Michael Geyer, *Deutsche Rüstungspolitik 1860–1980*, Frankfurt/Main 1984, p. 98.
[416] Jünger, "Total Mobilization," pp. 125–26.
[417] On the First World War as a "totalizing" war and its nineteenth-century predecessors, see Chickering/Förster (eds.), *Great War, Total War*; and Stig Förster/Jörg Nagler (eds.), *On the Road to Total War. The American Civil War and the German Wars of Unification, 1861–1871*, Cambridge 1997.
[418] Jünger, "Total Mobilization," p. 126.
[419] See Meschnig, *Der Wille zur Bewegung*, pp. 209–48; and Wehler, *Deutsche Gesellschaftsgeschichte*, vol. 4, pp. 114–22.

back" by the internal enemy had led to the final collapse.[420] The country had lost its nerve, self-interest had again taken precedence over the common good, and the will to victory had vanished because the "Spirit of 1914" had not been maintained. Thus, for those who relayed the "stab in the back" legend, internal division had undermined the community spirit of August 1914 and the comradeship of the front, and the ensuing revolution and civil war simply confirmed the necessity of a homogenization of mass society.

For victory, the "eternal Germany" that manifested itself in that August had to be revived, and the heterogeneous society remolded into a solidaristic national community ready for war. In fact, the Nazis placed themselves in a continuum stretching back to August 1914. When Hindenburg appointed Hitler Chancellor in January 1933, they characterized their "taking of power" as a "national rising" that revived "the Spirit of 1914" and the wartime years. Thus, on January 31, Herbert Seehofer wrote in the Nazis' *Völkischer Beobachter*: "Then as now incandescent signs of a national rising. Then as now the spell broken, the floodgates opened, the people risen up."[421] And on August 2, 19 years after the outbreak of the war, he asserted: "The German soldier's march into the Third Reich began on August 2, 1914."[422]

Inspired by "the spirit of 1914," the wartime society and the "socialist" community at the front became the guiding ideas for the Nazis.[423] The *Volk* of 1914, united in civil peace and soldierly comradeship, became the paradigm of social order:

> This *Volk* shaped itself first. In the trenches, skies, and waters, wherever the greatest commitment was required on a daily and hourly basis, amid machine-gun bursts, grenade fragments, and gas vapors, a new front grew up that was

---

[420] See Boris Barth, *Dolchstoßlegenden und politische Desintegration. Das Trauma der deutschen Niederlage im Ersten Weltkrieg 1914–1933*, Düsseldorf 2003; Schivelbusch, *The Culture of Defeat*, pp. 203–12; Verhey, *The Spirit of 1914*, pp. 219–23; and Wehler, *Deutsche Gesellschaftsgeschichte*, vol. 4, pp. 155–60.

[421] Herbert Seehofer, "Das erwachte Berlin marschiert," *Völkischer Beobachter*, January 31, 1933, No. 31, Year 46, unpaginated.

[422] Herbert Seehofer, "Das Gesetz, nach dem sie angetreten," *Völkischer Beobachter*, August 2, 1933, No. 214, Year 46. The article continues: "Never has soldierly sacrifice turned out to be such a shaper of events. In the thousand days and nights of the endless war at the front, the ideas were born for a new world – and, moreover, the will to impose them. The front experience was the creator of a new attitude in the militant souls that the war plowed up with its plowshares, more brutally and ruthlessly than ever before. Knowledge and wisdom going back centuries fell apart in the space of an hour. [...] Naïve supporters of feeble bureaucracies thought it would be possible to take frontline soldiers, who had known the apocalypse daily in close-up detail, and to put them again on the safe tracks of everyday life. But instead, the politicized soldier of the war years began his tough, dogged struggle for the souls of his compatriots and for state-led reorganization. [...] National Socialism, the new life form for the whole nation, grew out of the soldiers' practical experience of socialism at the front."

[423] See Bruendel, *Volksgemeinschaft oder Volksstaat*; Fritzsche, *Germans into Nazis*; and Verhey, *The Spirit of 1914*.

separated by more than defense lines from what was going on at home. Perhaps this is the meaning of November 1918: that men betrayed by everyone and spat upon by the dregs of society had to make their way back undefeated from defeat, bearing in themselves the image of a new people that had appeared to them at the front.[424]

It was at the front that the equality and solidarity of "comrades" had proved to be thoroughly compatible with hierarchical order.[425] Now that order was to be transferred to society as a whole; all social ambivalence would be overcome in favor of clarity and homogeneity.[426]

A study of the aviation hero-narratives is only partly capable of tracing this process of reinterpretation and remodeling, in which a divided, mechanistic society, based upon particular interests and held together only by the "cash nexus,"[427] was supposed to be transformed into a "solidaristic," organic and homogeneous national community. The actual quest to reshape the social order – the conversion of a threatening mass of erstwhile subjects into a militarily disciplined following, the devaluation of the individual in relation to community or nation – largely took place in other discourses. In addition to the general political conflicts of the Weimar years, the key areas were the economy and the market,[428] the armed forces and military theory,[429] as well as the specialist discourses of philosophy[430] and sociology,[431] "economics and administration, development planning, statistics and agronomy, labor input theory, and demographics."[432]

Nevertheless, the transformation of society into community, and hence the break with liberalism and the quest for a new order, can be identified in certain aspects of the hero-narratives. As we saw in the last section, these too helped to remold a *Gesellschaft* marked by plural "imagined orders" into a homogeneous *Gemeinschaft*. But above all they mirrored the ongoing absolutization of the nation. In them it became clear how the interests of individuals, as well as of

---

[424] Karl Pfeifer, "1914–1933," *Völkischer Beobachter*, August 2, 1933, No. 214, Year 46.

[425] See Thomas Kühne, *Kameradschaft. Die Soldaten des nationalsozialistischen Krieges und das 20. Jahrhundert*, Göttingen 2006, p. 51.

[426] See Bauman, *Modernity and Ambivalence*.

[427] On the concept of the "cash nexus," see Martin H. Geyer, *Verkehrte Welt. Revolution, Inflation und Moderne, München 1914–1924*, Göttingen 1998, pp. 385ff.

[428] Ibid.

[429] See Reichherzer, *"Alles ist Front."*

[430] See Peter Hoeres, *Der Krieg der Philosophen. Die deutsche und die britische Philosophie im Ersten Weltkrieg*, Paderborn 2004.

[431] See Paul Nolte, *Die Ordnung der deutschen Gesellschaft. Selbstentwurf und Selbstbeschreibung im 20. Jahrhundert*, Munich 2000. Ferdinand Tönnies was the first, in his *Community and Society* (originally published in German in 1887), to provide a sociological underpinning for the distinction between *Gemeinschaft* and *Gesellschaft*.

[432] Götz Aly/Susanne Heim, *Vordenker der Vernichtung. Auschwitz und die deutschen Pläne für eine neue europäische Ordnung*, Hamburg 1991, p. 13. On the development planners, see Ariane Leendertz, *Ordnung schaffen. Deutsche Raumplanung im 20. Jahrhundert*, Göttingen 2008.

religious groups, occupational strata or classes, were becoming subordinate to the imagined "common weal" of the nation. Thomas Nipperdey notes: "Radical nationalism raised the nation into the supreme value, superior to all other moral norms and absolutely superordinate to the individual. Right is what serves the *Volk*; you are nothing, your nation is everything, it would be said later."[433]

Individual interest had no meaning in comparison with the "general interest." This was also the view of Hermann Göring, the Nazi air minister and head of the newly founded Luftwaffe, in a speech he gave on May 20, 1936 to an oath-taking assembly of 1,000 lieutenants:

> The aim is the same for all. There are no oppositions: this interest over here, another over there. All have the same interest in serving and fighting for the cause, so the fact that everyone is marching in the same direction already means that there is no opposition of interests.

"We must all try hard," Göring lectured his audience,

> to hold back on our own selves and to put our duty first and foremost. That duty is: to serve for Führer, Volk, and Reich. If you all see that as your true aim and purpose, if you all strive for that and hold back on selfish urges, then the soil will be created for strong comradeship. [...] It makes quite a difference if someone works for their own profit, [...] or if he does his duty out of a sense that he is fighting for and serving a great common cause higher than his own self. That is why the highest performance of duty is expected of soldiers too, because it is a great and splendid duty to serve the fatherland; you are not doing it for your own account but for the future of your nation.[434]

This soldierly comradeship, this "holding back on selfish urges" and subordination to a "higher," "common" goal, was meant to differentiate the *Gemeinschaft* to which fascists aspired from *Gesellschaft*. Individual consciousness and national consciousness were superimposed on each other; the individual interests of the hero merged with the common good of the nation. This unity of individual and nation, class and *Volk*, was supposedly achieved in the national community, the *Volksgemeinschaft*. If the collective goal replaced "all individual quests for meaning," the masses lost their threatening character. And once the masses changed into a united, solidaristic *Volk*, they knew "why and for what they are here on earth."[435]

---

[433] Thomas Nipperdey, *Deutsche Geschichte 1866–1918*, vol. 2, *Machtstaat vor der Demokratie*, Munich 1992, p. 607.

[434] Hermann Göring, "Kameradschaft, Pflichterfüllung und Opferbereitschaft. Ansprache an 1000 Fliegerleutnants am Tage ihrer Vereidigung in Berlin am 20. Mai 1936," in Erich Gritzbach (ed.), *Hermann Göring. Reden und Aufsätze*, Munich 1938, pp. 226–44; here pp. 228–30.

[435] Arthur Moeller van den Bruck, *Das ewige Reich*, vol. 1, *Die politischen Kräfte*, ed. Hans Schwarz, Breslau 1933 [orig. 1923], p. 334.

The amalgamation of self and nation in the figure of the hero foreshadowed the devaluation of individualism and self-interest in favor of the common good that culminated in the Nazi maxim, "You are nothing, your *Volk* is everything." But this pedigree was apparent mainly to later interpreters. During the war itself, the popular hero-narratives owed their success to a perception of the aviator-hero as an autonomous, sovereign individual. His popularity rested precisely on his individualism, which set him apart from the anonymous mass. Yet this individual found his fulfillment in serving the fatherland. He was already a community-oriented individual.

As a role model created for the bourgeoisie, principally by bourgeois spokesmen and opinion leaders, the aviator-hero offset the ideal of self-sacrifice that reached its peak in the Langemarck narrative. For the heroic Langemarck myth drew on "expectations among the nationally minded bourgeoisie, which saw hope and rebirth for the nation in the optimistic German [student] youth marching off to war."[436] But with its emphasis on the anonymous masses, the myth ran counter both to bourgeois individualism and to the elitist self-conception of the bourgeoisie. Whereas a stress on individualism was the element linking the hero-narratives with the liberal order of society, the merging of the individual with the nation as its ideal-typical representative or embodiment created the bridge to fascist ideas of an order centered on *Gemeinschaft*. The aviator-hero could still be read as a bourgeois individual, but also – especially after the war – as the harbinger of an antibourgeois, collectivist warrior-type, whose ego merged with the nation or *Volk*. The latter figure had shed his socially divisive individualism and served the common good rather than his own self-interest. Admittedly he took the stage as an individual raised above the mass, but he embodied the collective and was as one with it.

In the reinterpretation and remolding of a threatening mass society, the new technological hero thus functioned as a pivot toward an ordered national community. He was a transitional figure between past and future, between bourgeois individualism and antibourgeois antiindividualism. During the war, the aviator-hero gave meaning to the *de facto* collapse of the value of the individual, while at the same time smoothing the way to the "heroic," "soldierly" attitude that would welcome the postwar decline of individualism and assume war as a "destiny":

> What we are experiencing is the complete collapse of individualism, the absolute bankruptcy of humanitarian thought. There are some quite remarkable examples of this process of dissolution: one of the most noteworthy is the way in which the elemental outbreak of war and its pervasion of the moral sphere provoke a sense of stunned horror. [...] The born warrior cannot get mixed up with humanitarian perspectives. He cannot do that because he is completely permeated with the fatefulness of the war. He knows he is engaged in a

---

[436] Schilling, *"Kriegshelden,"* p. 252. On the popularity of the Langemarck myth, which increased in the Weimar Republic, and on the counterfigure of the technological hero, see ibid., pp. 252–88; and Hüppauf, "Schlachtenmythen."

necessary task, and he performs it untroubled by all the opinions and formulas that people may bring to it.[437]

This was written by Friedrich Georg Jünger in 1930, in a book edited by Ernst Jünger, his elder brother by three years. "Total mobilization," which by then was his model of social order, required from individuals "total solidarity" with the nation – which, in the last analysis, meant the dissolution of self and the end of individualism.[438] Subsequently, National Socialism "derationalized and mystified" the idea of solidarity with national comrades, "turning both it and the idea of community on which it was based against civilization. [...] Whereas liberalism, after centuries of struggle, considered it settled that state and society were built upon individuals, nation and community were now the 'supreme value'."[439]

This demand for "total solidarity" and the sublation of private existence was not the least of the differences between the *Volksgemeinschaft* and the liberal order of society.[440] In Ernst Jünger's *Der Arbeiter*, we read "What dies or falls away is the individual as representative of weakened, doomed orders."[441] The brutal implication was that it mattered little whether the individual died or not: "The relationship to death has changed; its close proximity spares any mood that can still be

---

[437] Friedrich Georg Jünger, "Krieg und Krieger," in Ernst Jünger (ed.), *Krieg und Krieger*, Berlin 1930, pp. 51–67; here pp. 62–3.

[438] See Benito Mussolini, *The Doctrine of Fascism*, New York, NY 2006, pp. 31ff.: "If the nineteenth century was the century of the individual (liberalism means individualism), one may think that this will be the century of 'collectivism,' the century of the State. [...] The capital point of the Fascist doctrine is the conception of the State, its essence, the work to be accomplished, its final aims. In the conception of Fascism, the State is an absolute before which individuals or groups are relative. Individuals and groups are 'conceivable' inasmuch as they are in the State. [...] If liberalism signifies the individual, then Fascism signifies the State." See also Mussolini, "The Political and Social Doctrine of Fascism," in Mussolini, *My Autobiography*, New York, NY 2006, pp. 227–40. The constitutional theorist Reinhard Höhn, who headed the "Areas of German Life" department at the headquarters of the Nazi SD (Security Service), once wrote: "Today another principle has appeared in the place of the individual: the principle of community. The individual member of the state [*Staatsperson*] is no longer the cornerstone of constitutional law; the national community [*Volksgemeinschaft*] is the new starting point." Reinhard Höhn, *Otto von Gierckes Staatslehre und unsere Zeit*, Hamburg 1936, quoted from Michael Stolleis, *Geschichte des öffentlichen Rechts in Deutschland*, vol. 3, *Staats- und Verwaltungsrechtswissenschaft in Republik und Diktatur, 1914–1945*, Munich 1999, p. 327.

[439] Uwe Volkmann, *Solidarität – Programm und Prinzip der Verfassung*, Tübingen 1998, pp. 197–98.

[440] See E. Wolter, "Die Organisation des Sieges," *Deutsche Wehr* 19/1935, pp. 218–20; here pp. 218–19, quoted from Meschnig, *Der Wille zur Bewegung*, p. 239: "The total nature of mobilization does not mean that, in addition to frontline soldiers, this or that category in the civilian population is drawn behind the war aims; it means that, as soon as war breaks out, all merely private existence ceases, and that until the end of the war the only content of life for the whole nation rests on the idea that there is no longer a civilian population."

[441] Ernst Jünger, *Der Arbeiter*, pp. 105–06.

interpreted as festive in character. His [the individual's] fighting strength is not an individual but a functional value; he no longer falls but malfunctions."[442]

In the "organic community" that is supposed to have replaced "mechanistic society," the individual becomes a cog with purely functional value. But this functional value has a meaning by virtue of its connection with the nation. In the First World War, the state's demand for total solidarity already found expression in hero-images, but also in talk of "national socialism" or "war socialism."[443] The economist and sociologist Johann Plenge observed in 1915:

> The war has made us more of a socialist society than before. But socialism as a form of organization is the only fully conscious molding of society into the peak of strength and health; socialism as an ethos is only the freeing of the individual to become a conscious part of the state and society understood as a living totality.[444]

This special kind of "freeing of the individual" was what the hero-narratives conveyed into German society. In his portrait of Manfred von Richthofen, Friedrich Georg Jünger writes as follows:

> This best blood of the nation, these hearts of metal, knew of a higher meaning of life than self-preservation; the capacity for self-sacrifice was ingrained in them as a silent conviction that a higher life was at hand. To be sure, they believed in the immortality of their actions; how else could they in the prime of life have cheerfully thrown themselves into the arms of death.[445]

The heroes were increasingly honored for this dissolution of self: they had subordinated the self-interest of their own survival to a higher common good, to victory; and this "altruism" assured eternal life to them and the nation. The shift from the individualist to the collectivist hero points ahead to the task of building the *Volksgemeinschaft*. In the following sections, it will be seen how this change also involved a reinterpretation of categories of social order specific to particular classes or strata. The aviator-hero of the First World War was a transitional figure on the road to a "new nobility," but also to a new "worker" embodying the unity of the community.

The aviator-hero and the *Volksgemeinschaft*

But what did this order of the *Gemeinschaft* look like? Its basic requirements were a "new hierarchy" and a homogenized community. With regard to the first, the aviator-hero served as the model for a "new aristocracy" at the top of the hierarchical

---

[442] Ibid., p. 106.
[443] See Bruendel, *Volksgemeinschaft oder Volksstaat*, pp. 118–24 and 262–66.
[444] Johann Plenge, *Der Krieg und die Volkswirtschaft*, 2nd edn, Münster 1915, p. 172, quoted from Meschnig, *Der Wille zur Bewegung*, p. 213.
[445] Friedrich Georg Jünger, "Manfred von Richthofen," in Ernst Jünger (ed.), *Die Unvergessenen*, Berlin 1928, pp. 279–86; here pp. 281–82.

community; his nobility derived not from his status or lineage but from his achievements and willingness to sacrifice himself for the nation. He was the epitome of a new martial, meritocratic functional elite, which partly supplanted the traditional warrior caste of noblemen and occupied their position in society. With regard to the second requirement, the technically adept aviator-hero theoretically cut across social boundaries and functioned as a figure capable of integrating the industrial proletariat. On the one hand, the homogeneous community was supposed to be created through the exclusion, and ultimate annihilation, of ambivalence and indeterminacy – hence of that which was "alien to the community." On the other hand, unity and disambiguation were ostensibly achieved through the inclusion of "national comrades (*Volksgenossen*)."[446] The example of Jünger shows how, in the postwar period, the "worker" and his relationship to the aviator-hero came to symbolize the unity of a community intended to be in a permanent state of "total mobilization" and "total solidarity." But before we examine the function of the aviator-hero as a bridge from *Gesellschaft* to *Volksgemeinschaft*, we should take a brief look at this dystopian utopia itself.

The Nazi vision of a "national community" had its roots in the radical nationalist and *völkisch* movements of the Kaiserreich, and it was the war and then the defeat that first assured mass support for this fascist utopia.[447] The original title of Peukert's *Inside Nazi Germany* – *Volksgenossen und Gemeinschaftsfremde* [National Comrades and Community Aliens] – already points to the Janus face of the Nazis' utopian objective. The *Volksgemeinschaft* has a "double thrust": the inclusion of "national comrades" goes hand in hand with the segregation and eventual eradication of "community aliens," the creation of a new man with the elimination of human beings.[448] This dualism of inclusion and exclusion is purely analytic, but our investigation cannot conjure it away. As we saw in the last section, the popular hero-narratives did draw boundaries to the alien and gave voice to enemy-images current at the time. But as the heroes personified the "Spirit of 1914," they especially stressed the inclusive aspect of the idea of community. Therefore the following will deal only with the promise of inclusion: its roots in the war and its attraction for considerable numbers of people.

The *Volksgemeinschaft* offered a means of overcoming social division, of "bringing bourgeoisie and proletariat together."[449] Moreover, it was supposed to lead to victory over both the liberal and the Marxist model of society. As Hans-Ulrich

---

[446] See Peukert, *Inside Nazi Germany*.

[447] See Bruendel, *Volksgemeinschaft oder Volksstaat*; Fritzsche, *Germans into Nazis*; Müller, *Nation als Waffe und Vorstellung*, pp. 81–96; Verhey, *The Spirit of 1914*; Walkenhorst, *Nation – Volk – Rasse*; Wildt, *Hitler's Volksgemeinschaft*, pp. 26–68. See again here Frank Bajohr/Michael Wildt (eds.), *Volksgemeinschaft. Neue Forschungen zur Gesellschaft des Nationalsozialismus*, Frankfurt/Main 2009, as well as Martina Steber/Bernhard Gotto (eds.), *Visions of Community in Nazi Germany. Social Engineering and Private Lives*, Oxford 2014.

[448] Peukert, *Inside Nazi Germany*, p. 209.

[449] Norbert Frei, "'Volksgemeinschaft.' Erfahrungsgeschichte und Lebenswirklichkeit der Hitler-Zeit," in Frei (ed.), *1945 und wir. Das Dritte Reich im Bewusstsein der Deutschen*, Munich 2005, pp. 107–28, p. 110. See also Dietmar Süß/Winfried Süß, "'Volksgemeinschaft' und

Wehler has shown, the image of a harmonious, hierarchical, homogeneous and consensual national community had two dimensions:

> On the one hand, it could take on board the widely held theory of Marxism and convert it into an impetus for the allegedly superior counterutopia. On the other hand, the building up of the idea of *Volksgemeinschaft* for propaganda purposes had a sharp antibourgeois edge, since it counterposed a collectivist objective resting upon subordination and heteronomy ("Public interest before self-interest") to the vision of "civil society" based on performance and "natural" hierarchy, individual development, political freedoms, and autonomous lifestyle choices.[450]

In view of the perceived threat from Marxism and the association of Britain, France, and the United States with liberalism, a growing proportion of the German (and Italian) "interpretation elite" was incapable of theorizing the idea of an order centered on the nation (necessary to legitimize the war) in a way that was consistent with bourgeois-liberal categories. Consequently, they set their sights on a "national socialist" third way to modernity – a way which, though sharing a degree of "kinship," was opposed to Marxist and liberal New Deal ways of ordering the masses as it sought to transcend the concept of class.[451]

The promise of fascism was that, in the national and "socialist" *Gemeinschaft*, the antagonism between capital and labor, property-owning and working classes, would be overcome in the common struggle for the survival of the nation. The internal conflict between classes was turned around into competition between "rich and proletarian nations" or "superior and inferior races." The end goal, "national socialism," forced a redefinition of the criteria of differentiation and equality, but also of the prevailing hierarchy.[452] This entailed a radical reordering of "bourgeois society," on the basis of criteria which had nothing in common with the prevailing liberal or Marxist social categories of occupation, class, status group and lineage. The latter were not *supposed* to play any role in the fascist hierarchy of the community.[453] How does the popular figure of the aviator-hero relate to the ideology of *Volksgemeinschaft*, to its "sharply antibourgeois edges" and its ideologemes "radically opposed yet related to Marxism?"[454] The aviator-hero represented a transitional figure, who reflected the break with the old order then taking place as well as the quest for a new order. Therefore we have to take a closer look at these transmutations of old and new.

Vernichtungskrieg. Gesellschaft im nationalsozialistischen Deutschland," in Süß and Süß (eds.), *Das "Dritte Reich." Eine Einführung*, Munich 2008, pp. 79–100.

[450] Wehler, *Deutsche Gesellschaftsgeschichte*, vol. 4, p. 681.

[451] See Schivelbusch, *Three New Deals*, Chapter 1: "Kinship?" pp. 17–48.

[452] Nolte, *Die Ordnung der Gesellschaft*, p. 195.

[453] This is not to say that differences did not exist or were overcome, nor that they were insignificant for the corporatist economic order at which fascists aimed. It means that they were irrelevant for the community they sought to establish, since this was organized in accordance with nationalist and/or *völkisch* criteria. See Nolte, *Die Ordnung der Gesellschaft*, p. 192.

[454] See Ernst Nolte, *Three Faces of Fascism: Action Française, Italian Fascism, National Socialism*, New York 1969, p. 40.

In 1939 Hanns Haller, the head of the youth literature department of the National Socialist Teachers' Association of the "Bavarian Eastern March," published a book entitled *Der Flieger von Rottenburg* [The Flier from Rottenburg].[455] In it he describes the advancement of Max Müller from locksmith's apprentice to war pilot and eventually to Ritter Max von Müller, ending with a scene at the hero's graveside that explains the promise of the *Volksgemeinschaft* and testifies both to propagandistic integration of "the proletariat" and to the birth of a "new nobility" independent of ancestry. His story also illustrates the palingenetic, mythical core of fascism.

> He is the prodigious son of all the farmers and artisans who are dotted around the cemetery hillside: their brother, their blood, their fellow-worker, risen up from their land like a comet. [...] An officer speaks above the grave. [...] He said that, if a benevolent god were to come this way to bring Lieutenant Max Müller back to life, he would ask him not to do that. For, just as Germany now needs living heroes, eternal Germany needs these incomparable dead heroes. [...] When the gravestone was put in place a few weeks later, it had written on it: *Max Ritter von Müller*. It was the highest order and the final honor that a soldier could be given: the Military Order of Max Joseph. Iron became steel in the forge of action – a plain man raised his plain name into the arduous heights of the nobility and carved a noble coat of arms on his locksmith's hammer. But in the next few years it will be quiet around the escutcheoned grave of this exceptional man. As it will be quiet in the field when the seed is laid in the earth.[456]

The book closes with another gathering at the hero's grave, where a "young leader" speaks:

> With the couple of hundred, he said, many more are standing in the circle: all those who march behind the flags and pennants inscribed with the great soldier's name; those who live in the work camps before which this worker's gilded name stands; those who dwell in the streets that bear the glorious name; and the millions who look proudly up again when there is a roar through the air. [...] And now words begin to come from a hundred mouths, and they speak like a chorus from on high and speak of the eternity in front of the hero. [...] Four hundred boys stand singing in the garden of remembrance: they stand like young plants risen from the seed that Germany sowed in the last year of the war.[457]

For the National Socialists, the seed sown in the war sprouted in the "Third Reich." The national unity invoked in August 1914, when there were no parties but only

---

[455] On the NS Teachers' Association and its youth literature department, see Petra Josting, *Der Jugendschriftums-Kampf des nationalsozialistischen Lehrerbundes*, Hildesheim 1995.

[456] Hanns Haller, *Der Flieger von Rottenburg*, Bayreuth 1939, pp. 104–05. Emphases in the original.

[457] Ibid., pp. 107–08.

Germans, was now revived and "eternal Germany" was reborn.[458] A new man had come into being in the fire of war, in the "forge of action," and a new nobility had developed out of heroes akin to that simple locksmith's apprentice.

With the life story of the hero Max Müller, Haller drew attention to the fact that, in the National Socialist *Volksgemeinschaft*, ennoblement reflected what the individual had achieved and sacrificed for the community, not his ancestry. The new nobility distinguished itself by its "blood," by its superior character, will and leadership. The key question here is not whether "Hitler's social revolution" (to quote David Schoenbaum, writing in 1966[459]) actually happened, but rather the role of the fiction as a guide to action. Even if the Nazi regime did not "give the masses their right," it did give them "a chance to express themselves," not least in the hero constructs.[460] The moral of Haller's story is that, so long as a "young worker" had "German blood" in his veins, the National Socialist community judged him independently of his social origin. His achievements for the nation testified to his superiority; they showed that he was among the best and the most competent, and was for that reason capable of leadership. Under National Socialism,

> the position of leaders, both the one and the many, [was based] on their usefulness to the nation. They stood out from other national comrades because they had absorbed more "community content" and therefore, by virtue of their charisma, were more effective in binding the community together. This dynamic element of the National Socialist orientation to performance promised that anyone who respected the absolute priority of the community would have the opportunity to become a leader.[461]

This prospect inspired many at the time and motivated their contribution to the community. Leadership, the new man, the new type of worker and the new nobility that Haller saw sprouting from precious seed that Germany sowed during the war: all these had one of their sources in the "knights of the air."

We shall now attempt to trace the currents that flowed into National Socialism. In so far as they propagated "selection," the hero-narratives served the elitist yearnings of society and anticipated the "nobility of action" and the types of leaders that large sections of the population would pine for in the postwar period.[462] By revaluing the sphere of technology, the "heroes of the air" did much to promote acceptance of these "men of steel" and "worker-soldiers." As we shall see,

---

[458] In his speech of August 1, 1914, Kaiser Wilhelm II said: "In the struggle that lies ahead, I shall no longer recognize any parties in my nation. There are only Germans among us." *Vorwärts*, August 2, 1914, No. 208.

[459] Schoenbaum, *Hitler's Social Revolution*. See also Frei, "Volksgemeinschaft"; and Jill Stephenson, "Inclusion: Building the National Community in Propaganda and Practice," in Jill Caplan (ed.), *Nazi Germany*, New York, NY 2008, pp. 99–121.

[460] Benjamin, "The Work of Art in the Age of Mechanical Reproduction," p. 250.

[461] Kühne, *Kameradschaft*, p. 104.

[462] On the "new noble as man of action," see Alexandra Gerstner, *Neuer Adel. Aristokratische Elitekonzeptionen zwischen Jahrhundertwende und Nationalsozialismus*, Darmstadt 2008,

the aviation heroes represented not only the "particular contradiction" between egalitarian community and hierarchical order that defined the *Volksgemeinschaft*, but simultaneously also "the process of its harmonization."[463]

The figure of the aviatior-hero and the martial nobility of action

The quest for tested leaders and a fighting new nobility was grounded on the social transformations that had been taking place since around 1890, which in large sections of the bourgeoisie had led to repudiation, and subsequent redefinition, of their bourgeois way of life.[464] Here it is not possible – nor is it intended – to trace the "erosion of specifically bourgeois social forms" that occurred during those decades.[465] We may say, however, that the figure of the aviator-hero constructed during the First World War was an expression of that process. Although it "had scope to develop only on the soil of distinctively bourgeois lifestyles,"[466] its coding as a "noble" figure already pointed clearly to antibourgeois features.

As we have seen, the success of the figure was based on the achievements and individualism of the aviator-heroes – hence on motifs that enabled the "bourgeois" to be differentiated from the mass. But the war catalyzed a process that transformed these principles into elements of an antibourgeois way of thinking, so that the bourgeois virtue of achievement became detached from individualism and associated with the collective. It is true that the antibourgeois bourgeois continued to stand out from the mass by virtue of his achievement. But the achievement that underpinned the new nobility and the Führer principle was no longer an end in itself; it was achievement for the sake of the collective.[467]

In order to clarify the nexus linking the figure of the aviator-hero with the "new nobility of action" and the *Volksgemeinschaft*, we first need to take a more general look at the period after the First World War.[468] Since national homogenization was

---

pp. 197–294. The author sees the noble men of action as belonging more to a spiritual than a military aristocracy.

[463] Hans-Georg Soeffner, "Flying Moles (Pigeon-Breeding Miners in the Ruhr District): The Totemistic Enchantment of Reality and the Technological Disenchantment of Longing," in Soeffner (ed.), *The Order of Rituals: The Interpretation of Everyday Life*, New Brunswick, NJ 1997, p. 96.

[464] On the transformation processes affecting the bourgeoisie as a valuable subject for research, see Manfred Hettling, "Eine anstrengende Affäre. Die Sozialgeschichte des Bürgertums," in Sven Oliver Müller/Cornelius Torp (eds.), *Das Deutsche Kaiserreich in der Kontroverse*, Göttingen 2009, pp. 219–32; here p. 231.

[465] Hans Mommsen, "Die Auflösung des Bürgertums seit dem späten 19. Jahrhundert," in Jürgen Kocka (ed.), *Bürger und Bürgerlichkeit im 19. Jahrhundert*, Göttingen 1987, pp. 288–315; here p. 291.

[466] Ibid.

[467] The conflict between "equality" and elite that defined the ideology of *Volksgemeinschaft* rested precisely on the individualism of the bourgeois society from which the antibourgeois protagonists of the New Right themselves originated. See Gerstner, *Neuer Adel*, p. 21.

[468] On the "new nobility," see Eckart Conze, "Die Idee eines Neuadels in den Gesellschaftsvorstellungen der SS," in Eckart Conze/Monika Wienfort (eds.), *Adel und Moderne. Deutschland im europäischen Vergleich im 19. und 20. Jahrhundert*, Cologne 2004, pp. 151–76; Gerstner, *Neuer Adel*; Stephan Malinowski, *Vom König zum Führer. Sozialer Niedergang und politische Radikalisierung im deutschen Adel zwischen Kaiserreich und NS-Staat*, Berlin 2003², esp. pp. 293–320, 531–52; Stephan Malinowski, "'Führertum' und 'Neuer Adel.' Die Deutsche Adelsgenossenschaft und der Deutsche Herrenklub in der Weimarer Republik," in

supposed to elicit a different hierarchy from the characteristic one of nineteenth-century liberal society, it was necessary to establish a *new* system of ranks and a *new* elite. "Rule by the inferior" had to be prevented at all costs, but it also had to be negotiated who "the inferior" actually were.[469]

For the conservative revolutionary Edgar Julius Jung,[470] who was shot by the Nazis in the Night of the Long Knives,[471] it was necessary to change "the mass" back into a "structured *Volk*."[472] "The great task of politics," he wrote, was to generate "order among unequals, whether people or nations."[473] The paradigm for this was the wartime community at the front; the "comradeship" that supposedly prevailed there between the classes and between officers and subordinates was to replace, in the *Volksgemeinschaft*, the "equality" tainted by mass society. Instead of "rule by the inferior," there would be "rule by the best" – that is, an aristocracy providing real leadership, such as ostensibly existed at the front. In an essay "Schöpferische Kritik des Krieges [Creative Criticism of the War]," which he published in 1930 in Ernst Jünger's collection *Krieg und Krieger* [War and Warriors], the officer and rightist Freikorps member Wilhelm Ritter von Schramm lauded the "self-respecting democracy of frontline troops":[474]

> Never openly recognized, but tacitly tolerated, an informal democracy that admittedly appeared as comradeship was very soon winning out in the field. [...] In the trenches and shellholed terrain, there was no longer any artificial distance; the strict, rigid hierarchy present in the army throughout the long peace was destroyed. A new, natural hierarchy took shape, based on deeds and genuine achievement, and certainly not on insignia of rank.[475]

Heinz Reif (ed.), *Adel und Bürgertum in Deutschland*, vol. 2, *Entwicklungslinien und Wendepunkte im 20. Jahrhundert*, Berlin 2001, pp. 173–211; and Heinz Reif, "Einleitung," in ibid., pp. 7–27.

[469] Edgar Julius Jung, *Die Herrschaft der Minderwertigen. Ihr Zerfall und ihre Ablösung durch ein neues Reich*, Berlin 1930 [orig. 1927].

[470] The term "conservative revolution" was first established by the Swiss theorist Armin Mohler. See his *Die konservative Revolution in Deutschland 1918–1932. Ein Handbuch*, Stuttgart 1999. After the German attack on the Soviet Union, Mohler tried unsuccessfully to enlist in the Waffen-SS; he was later interned in both Germany and Switzerland, and after writing his dissertation (published under the above title) he became Ernst Jünger's private secretary. Mohler's *parti pris* means that a degree of caution is required in applying the term "conservative revolution," but it has been widely adopted and is therefore useful in the present context. On Mohler, see Dirk van Laak, *Gespräche in der Sicherheit des Schweigens. Carl Schmitt in der politischen Geistesgeschichte der frühen Bundesrepublik*, Berlin 1993, pp. 256–62.

[471] For more on Edgar Julius Jung and further literature, see Gerstner, *Neuer Adel*, pp. 88–99. On the following points, cf. ibid., pp. 240–41.

[472] Jung, *Die Herrschaft der Minderwertigen*, p. 348.

[473] Edgar Julius Jung, *Sinndeutung der deutschen Revolution*, Oldenburg 1933, p. 103; quoted from Breuer, *Ordnungen der Ungleichheit – die deutsche Rechte im Widerstreit ihrer Ideen 1871–1945*, Darmstadt 2001, p. 11.

[474] Wilhelm von Schramm, "Schöpferische Kritik des Krieges. Ein Versuch," in Ernst Jünger (ed.), *Krieg und Krieger*, Berlin 1930, pp. 31–49; p. 45.

[475] Ibid., p. 44.

This "new, natural hierarchy based on deeds and genuine achievement" was supposed to structure the *Volksgemeinschaft*.[476] Its distinction between genuine leaders tested in action and those who followed them was to be transferred to society in a kind of *trincerocrazia*.[477] National Socialists and neoaristocratic conservative revolutionaries were in agreement about that.

Of course, although we cannot go into them in any detail here, there were many nuances and differences within the New Right that originated in Imperial Germany. In the NSDAP and Hitler's "prerogative state (*Maßnahmenstaat*)," as well as in the Partito Nazionale Fascista and Mussolini's apparatus of power, the most varied currents were present under the banner of fascism. Italian Fascism as an ideology was (within limits) "pluralistic" and diffuse, and Fascists were therefore intrinsically heterogeneous. National Socialism too, especially after the takeover and through the threat and use of terror, was capable of integrating members of ideologically diverse groupings who had once kept their distance from it. Cohesion was ensured by several factors: (1) a vague utopianism, or the common goal of establishing an alternative modern order, with a palingenetic, community-generating myth based on the sacralized nation or *Volk*; (2) shared hostility to Marxism, liberalism, traditional conservatism and so on; (3) links with the war and a resulting cult and practice of violence; and (4) a charismatic leader at the top.

As far as the *Volksgemeinschaft* is concerned, it may be claimed with a degree of simplification that differences within the New Right related not so much to the core of this idea of order – national renewal in a clearly defined, homogeneous community – as to the question of suitable leaders, the importance of racism, and the role of the mass-become-*Volk*. Should the rulers be a spiritual nobility, a technocratic elite, or an aristocracy of blood? Should one make "common cause with the masses" – as elitist revolutionaries, rightly stigmatized as "conservative," often accused the "plebeian" National Socialists of doing? Some of the differences therefore centered on the elitism of particular groupings and the "selection criteria" for the new hierarchy.[478] In Jung's view, anyway, the task was the following: "The highest law for state-building is that the state should be held and led by the elect of the people. But selection presupposes living stratification, traditional virtues, solid concepts, and social demarcation that the most competent repeatedly breach from below."[479]

---

[476] See Bruendel, *Volksgemeinschaft oder Volksstaat*, pp. 117–18.

[477] The Italian term *trincerocrazia* (literally "trench rule," but modeled on *aristocrazia*) comes from an article by Mussolini in *Il Popolo d'Italia*, December 15, 1917. See Gentile, *The Origins of Fascist Ideology*, p. 34; and Sternhell/Sznajder/Asheri, *Die Entstehung der faschistischen Ideologie*, p. 277.

[478] See the "Definition of Central Analytic Categories" in the "Introduction" above, and the works especially of Stefan Breuer: *Die Völkischen in Deutschland. Kaiserreich und Weimarer Republik*, Darmstadt 2008; *Nationalismus und Faschismus. Frankreich, Italien und Deutschland im Vergleich*, Darmstadt 2005; *Ordnungen der Ungleichheit – Grundpositionen der deutschen Rechten 1871–1945*, Tübingen 1999; and *Anatomie der konservativen Revolution*, Darmstadt 1993.

[479] Jung, *Die Herrschaft der Minderwertigen*, p. 331.

Even though there might have been differences over "selection criteria," there was consensus in regard to the need for a new elite and new leaders at the summit of the *Volksgemeinschaft*.[480] These had to have demonstrated their achievement and capability, and to have shown evidence of charisma.[481] Such tried and tested leaders were to replace the old nobility and to fill the vacuum of power and symbolism it left behind.

The aviator-hero narratives of the First World War fostered belief in such charismatic soldierly leader-figures and a meritocratic elite that, though largely bourgeois in origin, had the attributes of a nobility. As the contrast between land and air war has clearly shown, the aviator-hero embodied an elite and an antagonism to the masses; he stood both for high quality in the military and for upward striving in society, precisely because he had the aura of those "perennially aristocratic elements of the soldier's craft" that Wilhelm Ritter von Schramm missed in the unattractive army of the "age of democracy" and the "mass state":[482]

> The ethos of the age, which was anything but high-minded, trusting, and heroic, pressed ever more decisively into the mass of the military. In the materialist spirit of equalization, with no instinct for blood differences and qualities, there was a wish to make the military conflict between nations everyone's affair, but all that did was level down the lofty spirit of soldierly strife. People became incapable of leaving the elite conflict to the top national leadership and its entourage, who had an inherited disposition for it by virtue of their blood and ethos.[483]

Depicted as battling knights, fliers in particular embodied this "lofty spirit of soldierly strife." They seemed to herald what Edgar Julius Jung predicted: "If the age of mass democracy is waning, so too is the age of mass armies. As the aristocratic component in the state becomes more pronounced, so does it also in the armed forces. A modern knighthood is dawning on the horizon."[484] Jung's "modern knighthood" appears to Ernst Jünger, the former assault officer decorated Pour le Mérite, in the form of "steel characters":[485]

> They are the steel characters whose eagle gaze searches the clouds ahead over whirring propellers, who sit wedged in among the confusion of tank engines, who risk the Descent into Hell across roaring fields of craters, who squat all day long facing certain death, half-fainting behind blazing machine-guns in nests

---

[480] See Gerstner, *Neuer Adel*, p. 530.
[481] See Stefan Breuer, "Das Charisma der Nation," pp. 110–43; and "Das Charisma des Führers," pp. 144–75.
[482] Schramm, "Schöpferische Kritik des Krieges," p. 40. Cf. Jung, *Herrschaft der Minderwertigen*, p. 348.
[483] Schramm, "Schöpferische Kritik des Krieges," p. 40.
[484] Jung, *Herrschaft der Minderwertigen*, p. 349.
[485] Schramm, "Schöpferische Kritik des Krieges," p. 46.

surrounded by enemies and mounds of their corpses. This is the new man. The storm pioneers, the elect of Central Europe. A whole new race – clever, strong, and willing.[486]

For Jünger, this vanguard produced by war should "tomorrow be the axis around which life whirrs faster and faster."[487] In his *Der Arbeiter* [The Worker], first published in 1932, he considered the conventional criteria of selection and distinction to be relics of the liberal nineteenth century. The war had made them obsolete.

> Just as mere traces remain of the [medieval] estate-structure and the host of people representing it, we may say that the differentiation of individuals by class, caste, or even occupation has become difficult, to say the least. Whenever people try to arrange and classify one another ethically, socially, or politically in terms of class, they do not focus on the key frontline positions; instead, they move in a nineteenth-century landscape that liberalism, over decades of activity, has more or less flattened by means of universal suffrage, conscription, general education, the alienability of real estate, and other principles.[488]

And a few pages later:

> For example, towards the end of the war it became more and more difficult to make out who was an officer, because the entire labor process tends to blur class and status group differences. On the one hand, battlefield activity produces a single breed of tried-and-tested team leader; on the other hand, the number of important functions has been increasing so much that a new type of selection is essential. Flying, for example, especially in the case of fighter aircraft, is something that depends on race rather than social group. The number of individuals in any nation who are capable of such high performance is so limited that sheer aptitude must suffice.[489]

In the *Volksgemeinschaft*, as at the front, the key factor is not class or group differences but individual aptitude and performance. But though not tied to any social class, the new nobility of action still displays many attributes of the old nobility and is linked to it in a number of ways. Thus, in his *Wäldchen 125*, published in 1925, Jünger described his visit to a fighter squadron that was throwing a party nearby, in a "little castle" where it had its quarters. The men were an "elite"

> who had been brought together by the pressure for bolder forms of battle involving greater nervous tension. There were cavalrymen present too: gaunt jockey types, with smug faces frozen by gleaming monocles. They had grown tired

---

[486] Jünger, *Der Kampf als inneres Erlebnis*, p. 76.
[487] Ibid.
[488] Jünger, *Der Arbeiter*, p. 98.
[489] Ibid., p. 108.

of being in villages and castles to the rear and waiting for another advance to begin. You could see they belonged to a breed that had had the spirit of cavalry warfare in its blood for centuries, and that looked down on this combat from behind engines and automatic weapons as inappropriate to their station.[490]

But there was also another type,

who, having grown up in the centers of modern industry, are thoroughly representative of the new century: twenty-year-olds with grim faces wrought by brute facts, for whom the bounce of city railroads, the tempo of the factory, and the poetry of steel and reinforced concrete were childhood experiences they take for granted. [...] They find technology fun: they have command of their aircraft as an Aborigine does of his boomerang. They are used to machines making life more intense.[491]

In Jünger's vision, these two sharply drawn manifestations of the aviator-hero, one noble, the other bourgeois, merge into a single human type destined for leadership. The soldierly ideals of heroism and manhood associated with particular estates, classes or social strata were unified and transcended in a new paradigmatic type representing a martial nobility of action or a meritocratic "composite elite."[492]

If we follow the change in the male soldierly model back to Wilhelmine Germany, the mediating function of the First World War aviator-hero becomes apparent: he combined the technological combatant from the war period (who was "used to machines that make life more intense") with the Wilhelmine ideal of heroism and manhood from the figure of the noble cavalryman. The "knightly" ideal of yesteryear was updated to fit the conditions of mechanized mass warfare.

At issue here is not the social make-up of the "composite elite" but its semantics. The fusion of the habitus of the noble knight with the developing habitus of the mechanized warrior created the semantic "composite elite," thereby continuing a process under way since the "desirable sections" of the bourgeoisie (the so-called "nobility of spirit or conviction") were incorporated into the Prussian officer corps. As officers became more professional and socially heterogeneous, they acquired a noble *esprit de corps*, which, like the "feudal" tradition, differentiated and homogenized the officer corps.[493]

---

[490] Jünger, *Das Wäldchen 125*, p. 78.

[491] Ibid., pp. 78–9.

[492] On the concept of a "composite elite," see Reif, "Einleitung."

[493] Here and on the following, see Marcus Funck, *Feudales Kriegertum und militärische Professionalität. Adel und Bürgertum in den preußisch-deutschen Offizierskorps 1860–1935*, Berlin 2005; and Mark R. Stoneman, "Bürgerliche und adlige Krieger. Zum Verhältnis zwischen sozialer Herkunft und Berufskultur im wilhelminischen Armee-Offizierskorps," in Heinz Reif (ed.), *Adel und Bürgertum in Deutschland*, vol. 2, *Entwicklungslinien und Wendepunkte im 20. Jahrhundert*, Berlin 2001, pp. 25–63.

Between 1860 and 1913, the bourgeois proportion of officers in the Prussian army doubled from 35 to 70 per cent.[494] Yet although one cannot speak of a "feudalization" of these officers,[495] nor was the result a "bourgeoisification" of the officer corps.[496] Despite the greater heterogeneity and professionalism, the "noble" habitus of the knightly warrior was preserved as its defining culture. Still, this culture was not rigidly fixed, but remained subject to historical change. The "'feudal' aspect of the officer corps" was "a creative response to its growing social and functional heterogeneity. The officer corps required a cultural binding agent to bridge its internal differences with a strong *esprit de corps*. Recourse was therefore had to familiar cultural models, which gained new meaning in the new context."[497]

Also "creative" was Wilhelm II's solution for the opening up of the officer corps to the bourgeoisie, which had become necessary because of the prior's enlargement. His Cabinet Order of March 29, 1890 insisted: "Today, unlike in the past, it is not just nobility of birth that may claim the privilege of providing officers for the army. But the nobility of spirit that has always filled the officer corps should and must remain intact and unchanged."[498] By participating in "aristocratic culture" – in the cult of honor and dueling practices, for example – this new "nobility of spirit or conviction" (*Adel der Gesinnung*) helped to bridge milieu-specific distinctions and to homogenize the officer corps as a whole with a new *esprit de corps*.[499] At the same time, in the period between 1890 and 1918, the term nobility and its connotations became more diffuse and open-ended.[500] This led to the very construct of a meritocratic, non-milieu-specific new nobility that we have seen *in statu nascendi* in the writings of the postwar period.

Since the aviator-hero was open in both directions, representing both the old type of the noble knight and new steel-clad nobility, he was an ideal transitional figure to bridge the gap between the two orders. Already during the war he conveyed the idea of a new elite:

Actually, *everyone out there* wants to become a flier. For everyone, the wall of clouds hiding the sky from us is a door that opens at will for them to

---

[494] Stoneman, "Bürgerliche und adlige Krieger," p. 29.

[495] On the debate over a supposed "feudalization" of the (big) bourgeoisie, see Hartmut Kaelble, "Wie feudal waren die deutschen Unternehmer im Kaiserreich? Ein Zwischenbericht," in Richard Tilly (ed.), *Beiträge zur quantitativen vergleichenden Unternehmensgeschichte*, Stuttgart 1985, pp. 148–71. For more recent research, see Stephan Malinowski, "Ihr liebster Feind. Die deutsche Sozialgeschichte und der preußische Adel," in Sven Oliver Müller/Cornelius Torp (ed.), *Das Deutsche Kaiserreich in der Kontroverse*, Göttingen 2009, pp. 203–18.

[496] See Nipperdey, *Deutsche Geschichte 1866–1918*, vol. 2, pp. 219–26. On the representation of the bourgeoisie and nobility in the different branches and positions, see ibid., pp. 220 and 222.

[497] Stoneman, "Bürgerliche und adlige Krieger," pp. 51–2.

[498] "Kaiser Wilhelm II, Cabinet Order on the Officer Corps (March 29, 1890)," http://germanhistorydocs.ghi-dc.org/sub_document.cfm?document_id=663, p. 1.

[499] On the significance of the duel, see Ute Frevert, *Men of Honor. A Social and Cultural History of the Duel*, Cambridge 1995.

[500] See Gerstner, *Neuer Adel*, esp. pp. 24–33.

fly to the sun and the stars. But *only the best* will be chosen, and those who are chosen must have the best heart, the keenest hearing, the sharpest eyes and strongest nerves, the mature gravity of age and the fresh audacity of youth.[501]

This elite excelled by its physical and spiritual qualities, such as strong nerves and youthful audacity. According to the psychiatrist Otto Binswanger writing in 1896, war brought out "Nietzsche's violent, brutal strongman" that Wilhelmine society longed for.[502] In the "steel bath of war," "weakness" and "nervousness" were conquered, and the tough soldierly ideal of manhood that had already been longed for in the prewar "nervous age" finally imposed itself. The many-sided reinterpretation and relinking that underlay this process cannot be considered here in detail, since it involved a basic restructuring of the symbolic order. The tropes of nobility and knighthood went through a metamorphosis, while the connotations of technology and the machine also changed, as did the ascriptions of "weak" and "hard."

Mechanized warfare created a nexus between the tough ideal of fighting masculinity and the fast pace of "modern" technological existence that had previously been associated mainly with the decadence and nervousness of "the big city." In parallel to this tendency, which enhanced the status of the mechanized combatant, the "multipolar noble ideal of manhood" centered on the "courtier" and gentleman fell increasingly into disrepute after the turn of the century.[503] Among other things, the scandal that broke in 1906–1908 around Philip, Prince of Eulenburg-Hertefeld and his round tables in Liebenberg suggested that the aristocratic "courtier" was effeminate and homosexual. The strong-nerved, steel-hard, technologically competent soldier and the cool, professional, technocratic type were now placed side by side with the "decadent," aristocratic type of man and officer – placed side by side, because those other models of military manhood appeared precisely as "mixed and mutated."[504] In this way, the "non-feminine" attributes of the gentleman or knight could be extrapolated and blended with the attributes of the new types. Precisely this mingling and mutation are apparent in the figure of the aviator-hero.

The accounts of the aviator's luxurious, glamorous lifestyle that we examined in the last chapter did not only have the function of offsetting the arduousness

---

[501] Alfred Marquard, "Wesen und Bedeutung der Luftwaffen," in Deutscher Luftflottenverein (ed.), *Das Fliegende Schwert. Wesen, Bedeutung und Taten der deutschen Luftflotte in Wort und Bild*, Oldenburg i. Gr. 1917, p. 15. Emphases in the original.

[502] Otto Binswanger, *Die Pathologie und Therapie der Neurasthenie*, Jena 1896, p. 49, quoted from Radkau, *Das Zeitalter der Nervosität*, p. 447. According to Radkau, Binswanger, who would treat the ailing Nietzsche as well as Warburg, saw the "pathological backdrop to the doctrine of the Superman." On the following, as well as the discourse of neurasthenia and the changing image of masculinity in general, see ibid., esp. pp. 389–455.

[503] See Funck, "Vom Höfling zum soldatischen Mann," p. 207. On the persistence of the aristocratic ideal of manhood, see also Mosse, *The Image of Man*, pp. 17–24.

[504] Funck, "Bereit zum Krieg?," p. 86.

of the land war in the public imagination. Rather, they served to extrapolate the "courtly" qualities of the officer in a new and updated form. The aviator-hero's links with the "upper crust" of the Kaiserreich, his *savoir-vivre* in the midst of death, his numerous decorations, titles and honors integrated the mechanized warrior into the symbolic order of the aristocratic ideal of fighting masculinity and thereby considerably enhanced his status. To some extent, the pilot displaced the knightly figure of the cavalryman, who had previously embodied that ideal, but was obsolete in the positional warfare on the Western Front. The figure of the aristocratic mechanized warrior now took over, demonstrating that "combat from behind engines and automatic weapons" could not be simply looked down upon as "inappropriate to [the officer's] station." Indeed, people should look up to the tech-savvy troops for whom the Wilhelmine military had previously had scant regard. This new elite with nerves of steel displayed the very virtues needed in the new age and for industrialized warfare.

The "aerial duel" trope, the hunting metaphors and the armorial bearings, insignia and decorations on the side of aircraft also testify to use of the symbolic capital of "feudal warriordom." This knightly-aristocratic coding of the air war heroes manufactured a tradition for a new branch of the armed forces. At the same time, it reconciled the old nobility with mechanized warfare and "ennobled" a new meritocratic elite which, though largely bourgeois in its origins and material existence, increasingly defined itself along antibourgeois lines.

Contrary to a still widespread image of the fighter pilot, most of the men in question were by no means noblemen or former cavalry officers. At least this is what Harald Potempa's study of the Royal Bavarian Air Force suggests, based on the official register of pilots (*Namentliche Verzeichnis der Flugzeugführer*) for the period from August 1, 1915 to September 1, 1917. Of the 1,338 pilots whose names appear in it, a total of 334 – just 25 per cent – were officers of any kind;[505] 539 (40 per cent) were NCOs and 461 (35 per cent) came from lower ranks. This picture is reinforced by another list, for the period from the start of the war until spring 1918, which, according to Potempa, points to an even higher proportion (42 per cent) of pilots from lower ranks. The list records 3,021 Bavarian military pilots by name, of whom 557 (18 per cent) were officers; a mere 21 (3.8 per cent) of these were of aristocratic origin. Thus, only 0.7 per cent of the listed military pilots were officer-noblemen.

Potempa further ascertains that only 43 (13 per cent) of the 334 officers in the *Namentliche Verzeichnis der Flugzeugführer* had been in the cavalry before they were transferred to the air force, and that the largest percentage (43 per cent) of the 557 officers in the second list had come from the infantry, with 137 (25 per cent) from the artillery. Only 56 officers (10 per cent) had transferred from the cavalry to aviation. It is true that most of the 21 noblemen had previously served in cavalry regiments, but all this shows is that the cavalry continued to be attractive to the

---

[505] For this and the following data, see Potempa, *Die Königlich-Bayerische Fliegertruppe*, pp. 90ff.

nobility.[506] In purely numerical terms, aristocratic ex-cavalry officers were of no significance in the Bavarian air force. Even if all the 21 noble officers are counted as cavalry officers, they make up no more than 3.8 per cent of the officer total and 0.7 per cent of the military pilots.

Although the Prussian air force and the Imperial fleet air arm generally had a considerably higher proportion of noble officers from the cavalry or elsewhere,[507] the success of the trope of the "knightly" noble cavalryman clearly did not depend so much on the actual social composition of the air force. Rather, it was the symbolic function of the trope and the chivalrous image it conveyed that rooted it in the discourse of the time and the cultural memory. The figure of Manfred von Richthofen – who still dominates the popular iconography of aerial combat and even continues to serve as a role model – featured prominently in the process of converting the class-specific, aristocratic knight into the non-class-specific mechanized warrior belonging to the "new nobility."[508] In this construct, the paradox of knight and mechanized warrior, nobleman and bourgeois, old and new was resolved in a collective symbol.[509]

Georg Wegener's report *Bei der Jagdstaffel Richthofen* throws a clear light on this amalgam and the way in which it reconciled contradictions. The author especially emphasizes the venerable noble family into which Manfred von Richthofen was born; it had previously served the Kaiser and Reich mostly in diplomatic or "scholarly" positions, but then it distinguished itself in the military too.[510] Wegener goes on to describe the physical and mental qualities of Richthofen, who saw himself as Boelcke's successor:

> Like Boelcke, he was of medium height and powerfully built; his head with its arched brow and bright-blue Teutonic eyes – whose expression strongly reminded one of Boelcke – astonished people because of the almost rosy freshness of its color. None of the huge nervous tension that one associates with his daily life-and-death combat had left its mark on him. His whole being was of a surprisingly calm, reserved, almost gentle nature; […] exceedingly genial,

---

[506] Marcus Funck, "Schock und Chance. Der preußische Militäradel in der Weimarer Republik zwischen Stand und Profession," in Reif (ed.), *Adel und Bürgertum in Deutschland*, vol. 2, *Entwicklungslinien und Wendepunkte im 20. Jahrhundert*, Berlin 2001, pp. 127–72; here p. 164.

[507] In 1913, when the proportion of noblemen in the Bavarian officer corps stood at "only" 15 per cent, the corresponding figure in Prussia was 30 per cent. On this and for further bibliographic references, see Gerstner, *Neuer Adel*, p. 228.

[508] In 1961, on the 43rd anniversary of the hero's death, Heinrich Lübke – President of the Federal Republic at the time – bestowed the name "Richthofen" on the 71st Fighter Wing of the Bundeswehr. Two more of the seven aerial combat units of the Bundeswehr bear the name of a First World War "hero": the 51st Reconnaissance Wing "Immelmann," and the 31st Fighter-Bomber Wing "Boelcke." On the "Richthofen" Fighter Wing, see http://www.luftwaffe.de/portal/a/luftwaffe/org/luftm/jg71, accessed on August 6, 2014.

[509] On this symbolic resolution, see Soeffner, "Flying Moles."

[510] Georg Wegener, *Der Wall von Eisen und Feuer*, vol. 3, *Die beiden letzten Jahre*, Leipzig 1920, pp. 39–40.

refined, and simple, without a trace of boastfulness, although one could detect in him a joyful pride at his youthful radiance.[511]

The physical inconspicuousness mirrored the modesty and reserve that were commonly attributed to the hero-figures, but also an ability to endure "huge nervous tension" with calm composure. One prominent feature, though, immediately pointed to his sense of comradeship and his capacity for leadership: "Only the strongly formed chin perhaps betrayed the man's uncommon energy. So too did the effect he had on those around him, who visibly clung to their champion with an extraordinary mixture of cheery comradeship, enthusiastic admiration, and absolute attachment."[512]

Von Richthofen led "his men" from the front, so that a "cheery comradeship" and "absolute attachment" prevailed between them. Wegener then describes him as a "sportsman," who nevertheless performed noble deeds:

> Like Boelcke, Manfred Freiherr von Richthofen has been a passionate lover of sport since the days of his youth. He is first of all a horseman and hunter. He started his military service as an active officer in the Ulanen regiment. But then cavalry activity no longer satisfied him in the war of position, and he became an aviator. [...] When I asked him whether he put his successes in aerial combat down to a special technique, he emphatically denied it. [...] Of course you have to be in control of your machine [he said, but] otherwise you just have to be bold. [...] He did not feel the slightest physical disturbance in even the sharpest and fastest descents. – Was he exceptionally agitated after a difficult battle in the air? "No," he replied, "I can't say that at all." [...] Afterwards, his comrades told me something that they saw as the key to his superiority. Above all, he had a fabulous eye – a real phenomenon. [...] That hunter's eye helped him also in flying and shooting.[513] A second thing was his decisiveness and tenacity. He always flies straight at an enemy he has spotted and does not slacken off until he has dealt with him; the idea that he too might be hit never seems to cross his mind.[514]

On the one hand, the Richthofen construct epitomizes the venerable nobleman who has loyally served the monarchical state, with "the spirit of cavalry warfare in his blood for centuries" – the traditional masculine ideal of the noble warrior, with the associations of "superior quality, honor, composure (affect and body control), and politeness or good manners (toward women, for example)."[515] On the other hand, these "chivalrous soldierly attributes" were mixed in with a strong-nerved

---

[511] Ibid., p. 40.

[512] Ibid., pp. 40–1.

[513] On the importance of sharp eyesight for a "cool persona," see Helmut Lethen, *Verhaltenslehren der Kälte. Lebensversuche zwischen den Kriegen*, Frankfurt/Main 1994, pp. 187–98.

[514] Wegener, *Der Wall von Eisen und Feuer*, vol. 3, pp. 41–2.

[515] Funck, "Vom Höfling zum soldatischen Mann," p. 212.

"cool persona," ruthless, dynamic and death-defying. Richthofen stood out by his "decisiveness" and "tenacity"; he hunted the enemy until he had "dealt with him." Aerial combat did not excite him: he remained calm and objective, with "nerves of steel." He also had "uncommon energy," and charisma that led people to trust and follow him. The yearning for such leaders would grow after the war, but they already had mass appeal during it. Richthofen's own popularity was certainly based on his actions in the air, and on hopes of victory fueled by his "exceptional talent," but it also rested upon the mediation he offered between different symbolic orders and the bridge this created between *Gesellschaft* and *Gemeinschaft*.

It is impossible not to see Richthofen's wartime hero-image in the light of experiences later in the twentieth century. Here, for example, is an account of a bombing mission and "shooting match" that makes one think of the Second World War and conjures up the famous/notorious scene in Francis Ford Coppola's *Apocalypse Now* with Wagner's "Ride of the Valkyries" on the sound track:[516]

> At last we get into a quiet atmosphere. Now comes the enjoyment of bombing. [...] After having thrown one's bombs, one has the feeling that one has achieved something, while frequently, after searching for an enemy to give battle to, one comes home with a sense of failure at not having brought a hostile machine to the ground. Then a man is apt to say to himself, "You have acted stupidly." It gave me a good deal of pleasure to throw bombs. [...] We reserved a bomb, hoping to make particularly good use of it on our way home. [...] We went home by roundabout ways and looked for camps. It was particularly amusing to pepper the gentlemen down below with machine guns. Half savage tribes from Asia are even more startled when fired at from above than are cultured Englishmen. It is particularly interesting to shoot at hostile cavalry. [...] My observer fired energetically into the crowd down below with his machine gun and we enjoyed it tremendously.[517]

War is great fun: such is the tenor of all the aviator-hero literature. For the Richthofen construct, to be sure, there was also the need to climb to the top of the rankings, but aerial combat is deadlier work than a sports match. The German version of *The Red Fighter Pilot* published in 1933 contains a passage that was not yet there in the 1917 edition:

> The commander of a fighter wing must live among his men. It is not acceptable for him to hang around somewhere in the rear. [...] The commander of a fighter

---

[516] On the links between *Apocalypse Now*, the German official newsreel of June 4, 1941 (also accompanied with "The Ride of the Valkyries") in which Stuka bombing prepares the way for a paratroop landing on Crete, and a passage from Marcel Proust's *À la recherche du temps perdu*, see Ulrich Fröschle/Helmut Mottel, "Medientheoretische und mentalitätsgeschichtliche Probleme filmhistorischer Untersuchungen. Fallbeispiel *Apocalypse Now*," in Bernhard Chiari/Matthias Rogg/Wolfgang Schmidt (eds.), *Krieg und Militär im Film des 20. Jahrhunderts*, Munich 2003, pp. 107–40.

[517] Richthofen, *The Red Fighter Pilot*, pp. 99–101.

wing must know how to sort the wheat from the chaff. He can do that only if he is constantly together with the people he commands. [...] The commander of fighter squadrons must be a fighter pilot himself, and a good one at that: in other words, he must be successful. He must go up with his men. [...] The squadrons that achieve something at the front consist of comrades who know one another inside out, who work as a team in combat, and who know for sure that no one will leave the others in the lurch if things get a bit uncomfortable.[518]

The principles of comradeship and charismatic leadership govern these lines, which appeared one and a half decades after the first publication of Richthofen's book. He appears here as the yearned-for leader, whose support rests neither on birth and ancestry nor on "abstract 'rules' and positions," but is "sustained by great deeds, resounding successes, and notable achievements."[519] With a calling to lead, this combative, charismatic and meritocratic noble of action would stand at the peak of the "order of inequality" that the fascists sought to structure the community. The price for this community would be paid by those excluded from it. "The price is paid in the currency of freedom. [...] Missing community means missing security; gaining community, if it happens, would soon mean missing freedom."[520]

The "worker-soldier" and the age of the fourth estate

The overlaps between the aviator-hero narrative and the "new nobility" discourse have demonstrated the process of the hierarchization of community. Now we need to clarify, through the associations between aviator-hero and "worker," the semantic inclusion of the proletariat and the homogenization of society. The aviator-hero was a symbol that, though coded mainly along elitist lines, was sufficiently polyvalent to be read also in an egalitarian-homogenizing sense. This reading was based on his properties as mechanized man. To be sure, the propagandistic integration of the industrial proletariat into the *Volksgemeinschaft* – a process illustrated by the social rise of the worker Max Müller – was not achieved by the aviator-hero narratives of the First World War. But they did help to enhance the status of the indispensable technological-industrial domain, which represented a semantic bridge to the industrial proletariat that enabled it to be discursively included in the national community. Although the "worker" had embodied the fracturing of society, he changed in the Weimar Republic into a symbol of the inner unity and cohesion of the community.[521]

This transformation of "the worker" may be cursorily sketched by reference to the figure of the aviator-hero. It had its roots in the war and eventually led, via the Third Army Command and Jünger's *Der Arbeiter*, to fascism.

---

[518] Manfred von Richthofen, *Der rote Kampfflieger*, Berlin 1933, pp. 196–97.
[519] Kershaw, *The "Hitler Myth,"* p. 9.
[520] Zygmunt Bauman, *Community. Seeking Safety in an Insecure World*, Cambridge 2001, p. 4.
[521] See Nolte, *Die Ordnung der Gesellschaft*, pp. 92–5.

The vision of "national socialism," writes Sven Oliver Müller, functioned as a "promise" of integration and "participation."[522] Contrary to the legend of resistance on the part of Marxist industrial workers, the backdrop of mass unemployment following the world economic crisis allowed the Nazi regime to bind workers to itself and the nation.[523] The terror against Communists, Socialists and labor unions, but also foreign policy successes, full employment, social policies, activities of the *Schönheit der Arbeit* [Beauty of Labor] and *Kraft durch Freude* [Strength through Joy] organizations, and the symbolic value given to manual workers, succeeded in eroding the class-specific "imagined order" and tying the industrial proletariat to the *Volksgemeinschaft* and the "Führer".[524]

At the center of this propaganda was the firm cohesion of the nation, as proclaimed in Alexander M. Cay's poster of March 1918: *Durch Arbeit zum Sieg! Durch Sieg zum Frieden!* [Through Work to Victory! Through Victory to Peace!] (Figure 3.1). The handshake it shows between a soldier and a worker expresses not only the unity of the home front and the battle front but also the integration of "workers without a fatherland" into the struggling nation in its "hour of need."[525] Recognizing the importance of "labor" for victory, the poster thus points to the future equation of frontline soldiers and workers. In fact, the transclass aspects of the wartime community were seldom explicitly addressed in the aviation war books. Occasionally reference was made to cooperation between the "observer," who was always an officer, and the "chauffeur" or "driver" (often an NCO or private) responsible for actually flying the aircraft.[526] For example: "Good understanding between pilot and observer is the key to success. Indeed, not only understanding but a telepathic friendship, easily formed in common experiences of danger, makes the hardest tasks hugely easier for the crew."[527]

In contrast to the class arrogance of officers in other branches of the armed forces, airmen preserved this "informal democracy," symbiotic cooperation and solidaristic comradeship – at least that is what the following passage from *The Red*

---

[522] Müller, *Die Nation als Waffe und Vorstellung*, p. 349.

[523] See Wehler, *Deutsche Gesellschaftsgeschichte*, vol. 4, pp. 731–41, and pp. 1116–117. For further literature on the question, see also Lutz Niethammer (ed.), *"Die Jahre weiß man nicht, wo man die heute hinsetzen soll." Faschismuserfahrungen im Ruhrgebiet*, Berlin 1983, esp. the contribution by Michael Zimmermann, "Ausbruchshoffnung. Junge Bergleute in den Dreißiger Jahren," pp. 97–132.

[524] See Götz Aly, *Hitler's Beneficiaries. Plunder, Racial War, and the Nazi Welfare State*, New York, NY 2006; David Welch, "Nazi Propaganda and the *Volksgemeinschaft*. Constructing a People's Community," *Journal of Contemporary History* 39/2004, pp. 213–38. For an overview of older literature and debates, see Ulrich Herbert, "Arbeiterschaft im 'Dritten Reich.' Zwischenbilanz und offene Fragen," *Geschichte und Gesellschaft* 15/1989, pp. 320–60; and Timothy W. Mason, *Nazism, Fascism and the Working Class*, Cambridge 1995.

[525] See Verhey, *The Spirit of 1914*.

[526] On the cooperation between pilot and observer, see Kehrt, "'Schneid, Takt und gute Nerven'," pp. 177–201.

[527] Eddelbüttel, *Artillerieflieger*, p. 31.

Figure 3.1 Alexander M. Cay's propaganda poster of March 1918 declared the unity of the nation

Fighter Pilot suggests. An officer in the engineers hurries to the place where the author has been shot down and offers him a lift:

> I assured him that I felt quite well, jumped down from the side of my machine, and introduced myself to him. Of course he did not understand a particle of my name. [...] My host was still extraordinarily excited. Suddenly he jumped up and asked: "Good Lord, but where is your chauffeur?" [...] Probably I looked puzzled. Then it dawned upon me that he thought I was the observer of a two-seater and that he asked after the fate of my pilot. I pulled myself together and said in the driest tones: "I always drive myself." [...] In the eyes of the kind gentleman I had obviously lost caste when he discovered that I "drove" my own aeroplane.[528]

While the mechanized warrior was still looked down upon, the handling of engines was thought inappropriate for a gentleman and tended to be left to more tech-savvy "drivers" from the common people. Friedrich Wilhelm Eddelbüttel's book *Artillerieflieger* is written from the point of view of one such "driver." Exceptionally it also tackles the relationship with mechanics, more than eight of whom were needed to keep a man and his machine in the air:[529]

> I also liked getting to know the mechanics. How much depends on them! Not just the carrying out of orders: no, our own life is always in their hands. A little oversight, a forgotten screw, or a badly fitted cable can lead to a fatal crash, especially here at the front, where aircraft are so often put under extreme stress. You cannot pay enough attention to the work of mechanics, and if possible you must try to lighten their difficult work by describing your flights and the performance of their machines, by praising them if it is at all appropriate, and by showing enthusiasm for the common task.[530]

It is rare for the huge importance of ground personnel to be acknowledged so clearly in popular accounts of the air war. Recognition was given to technicians and mechanics, "the men in the background," only in fascist accounts that played

---

[528] Richthofen, *The Red Fighter Pilot*, p. 120.

[529] See Potempa, *Die Königlich-Bayerische Fliegertruppe*, pp. 88–9. and 170–82. Potempa points out that it is essential for mechanics to have a civilian qualification. "Of the 73 privates [in a Royal Bavarian tactical squadron at the Army Command], only five did not need to have had specific prior training in the civilian world. In addition to 18 machine fitters, there was a requirement for eight aircraft mechanics, eight assistant mechanics, eight electricians, 16 joiners, eight upholsterers, one tailor, and one shoemaker (p. 89)."

[530] Eddelbüttel, *Artillerieflieger*, p. 30.

up their "silent heroism" and emphasized "common tasks." For example, in the air force magazine *Der Adler*, published by Scherl Verlag in association with the Reich Aviation Ministry, an article from 1940 prescribes:

> This report must sound a note of thanks. It concerns thousands and millions of German soldiers who face the enemy every day and every hour – yet are unable to grapple with him. It concerns the Luftwaffe ground personnel, who, using few words, tirelessly and inconspicuously fulfill their duty; a duty that is very hard for a soldier who wants to fight the enemy with a gun in his hand. It concerns the young and the old who keep the sword ever sharpened for their comrades – the men in the background, unmentioned in any army report, who haul the bombs without dropping them, fill drums, and are unable to shoot at the enemy. [...] They are the fliers who do not fly. But they belong to our Luftwaffe; they are the casing around the iron core of the crew, without whom no mission would be possible.[531]

Although the development of technology had by then further increased their importance, the "men in the background" also played no central role in the fascist aviator-hero narratives. The ground personnel were essential, of course, but their activity was too unspectacular to be singled out in the media, and so their "heroism" usually remained as silent as before. The "fliers who do not fly" were presented, however, as a solid, indispensable and integral part of the wartime community of comradeship. The previously mentioned article continues:

> But if it fails to happen [the crew's return from a mission], there is a break in the small, stable community of men who belong together. This sense of belonging together finds its strongest expression at the moment when a member of the crew goes missing and a man from the ground personnel immediately takes his place, usually as the rear gunner.

Naturally, this promise of integration applied not only to the ground personnel but also to workers in the aircraft factories:

> Yet they too [civilian workers] are soldiers: they wear the yellow armband of the German Wehrmacht and cooperate in its work. Numerous mechanics in the great German aircraft factories – such as Junkers, Messerschmitt, Dornier, or Heinkel – are always ready to help in words and action on a servicing job. They too are among the men who go unmentioned when great successes are spoken of. They are all soldiers and workers, like the millions in the workshops of the homeland.[532]

---

[531] Siegfried Kappe, "Männer im Hintergrund. Tag und Nacht unermüdlich – so arbeitet das Bodenpersonal unserer Luftwaffe," *Der Adler*, October 1, 1940 in: DMM, LR05437.
[532] Ibid.

All were soldiers and all were workers, whether in cockpit or workshop, in the factory or behind a machine-gun. In 1931 Lieutenant-General Max Schwarte, the editor of the ten-volume *Der große Krieg 1914–1918*, noted that "weapons production *by workers*" was on a par with "weapons handling by *fighting men*."[533] This equation of the two, which would later reach a peak in the phrase "soldiers of labor,"[534] served to integrate the industrial proletariat; it stemmed from the realization that "there is no longer any movement whatsoever – be it that of the homeworker at her sewing machine – without at least indirect use for the battlefield."[535]

The militarization of labor, and the simultaneous transfer of the concept of labor to the battlefield, became the starting point for Ernst Jünger's metaphysically grounded theory of "total mobilization," which he developed in 1932 in his essay *Der Arbeiter. Herrschaft und Gestalt*.[536] In this transformation of labor, Jünger saw a revolutionization of existence and the world that had to be accepted unreservedly if man was not to perish. The growing dominance of technology in all spheres of life indicated "total mobilization" on a "planetary" scale, the task of which was "the conversion of life into energy, as it reveals itself as fire and movement in industry and commerce, technology, and transport, in the spinning of wheels or on the field of battle."[537]

"Total mobilization," understood as "will to power," converted everything into work by means of technology. In *Der Arbeiter*, this nexus developed not simply into a fascist utopia but into a full-blown philosophical sketch resting upon the contemporary understanding of Nietzsche.[538] We cannot examine that sketch

---

[533] Max Schwarte, *Der Krieg der Zukunft*, Leipzig 1931, p. 35. Emphases in the original.

[534] The phrase "soldiers of labor" referred to the Reich Labor Service, but the militarization of labor expressed in it was characteristic of National Socialism in general. See Kiran Klaus Patel, *"Soldaten der Arbeit." Arbeitsdienste in Deutschland und den USA 1933–1945*, Göttingen 2003.

[535] Jünger, "Total Mobilization," p. 562.

[536] On Jünger's "worker" and for further bibliographic references on the subject, see *inter alia* Uwe K. Ketelsen, "Ernst Jüngers *Der Arbeiter*. Ein faschistisches Modernitätskonzept," in Ketelsen (ed.), *Literatur und Drittes Reich*, Schernfeld 1992, pp. 258–85; Kiesel, *Ernst Jünger*, pp. 384–99; Lars Koch, *Der Erste Weltkrieg als Medium der Gegenmoderne. Zu den Werken von Walter Flex und Ernst Jünger*, Würzburg 2006, pp. 287–330; Peter Koslowski, *Der Mythos der Moderne. Die dichterische Philosophie Ernst Jüngers*, Munich 1991, pp. 35–76; Rohrkrämer, *Eine andere Moderne?*, pp. 301–41; Harro Segeberg, "Regressive Modernisierung. Kriegserlebnis und Moderne-Kritik in Ernst Jüngers Frühwerk," in Harro Segeberg (ed.), *Vom Wert der Arbeit. Zur literarischen Konstitution des Wertkomplexes "Arbeit" in der deutschen Literatur (1770–1930)*, Tübingen 1991, pp. 338–78; Friedrich Strack (ed.), *Titan Technik. Ernst und Friedrich Georg Jünger über das technische Zeitalter*, Würzburg 2000; and Marianne Wünsch, "Ernst Jüngers *Der Arbeiter*. Grundpositionen und Probleme," in Lutz Hagestedt (ed.), *Politik – Mythos – Kunst*, Berlin 2004, pp. 459–75.

[537] Jünger, *Der Arbeiter*, p. 210.

[538] See Martin Heidegger, *Zu Ernst Jünger. GA 90*, ed. by Peter Trawny, Frankfurt/Main 2004. In Heidegger's manuscript for a "talk on Jünger" that he gave to a group of colleagues in 1939–1940, he writes: "All being, nature, and history is 'work' (will to power) in what it is and how it is. The human 'representative' of the will to power is called 'the worker.' [...] The name 'worker' refers metaphysically-anthropologically to the form of mankind,

here. Rather, we shall draw on Jünger's major essay to clarify the reinterpretation of the "worker" within the rightist political spectrum – a process that made it easier for the industrial proletariat to be integrated into the *Volksgemeinschaft*, permitted the acceptance of technology, and indeed promoted its radical affirmation. Our aim will be to demonstrate the link that technology, "work," and war created between the aviator and the "worker."

The third Supreme Army Command, which began its "work" on August 28, 1916, completed a radical "transition to mechanized warfare in Germany" that placed it on an industrial basis.[539] Industrialized warfare, particularly in Ludendorff's understanding of it, meant "not simply that industrially produced weapons were employed, but also that the conduct of soldiers (their war) was determined by what they did with weapons rather than by the dictates of officers. The cross-class frontline community was synthesized by war machines. War was 'work'."[540]

In the same way, the national community as a whole was synthesized through work – that is, shaped through participation in a technologically driven "total mobilization." The individual thereby lost his individuality and became a representative of the "worker-type." By using machines, the soldier or industrial worker himself turned into part of a higher-order machinery of the "workers' state" striving for power.[541] Within this "organic construction," man appeared "in a higher unity with his instruments"; only then was the "tension between nature and civilization, organic and mechanical world" dissolved.[542]

In the early 1930s Jünger's writings switched from "sober description to outlines of a totalitarian political order,"[543] and it was then that he saw "total mobilization" heralding the end of bourgeois civilization:

> In the unlimited marshaling of potential energies, which transforms the warring industrial countries into volcanic forges, we perhaps find the most striking

---

which finds completion in the mastery of beings as a whole, and whose being generally is 'will to power.' Therefore, 'worker' is not the verbal term for an idea that generalizes an already known phenomenon, the so-called 'estate' and 'class' of 'workers.'" And he explains in a footnote: "Nevertheless, difficulties [arise] here: the fourth estate [is] not 'bourgeois' but [conceived] in its relation to the elementary [!]. What Jünger means by the name 'worker' is detached from any evaluation merely in terms of estate or *Volk* and especially from any 'social' concern. 'Work' and 'worker' are metaphysical concepts. The soldier is a 'worker', and also 'thinker'; not because both 'work,' whether with 'fist' or 'brain' – i.e., perform things of benefit to the common good – but because, in standing up to being in general as will to power, each in its way is that being. Jünger calls this standing-up attitude 'heroic realism.'"

[539] Geyer, *Deutsche Rüstungspolitik*, p. 102.

[540] Ibid., p. 100.

[541] Jünger, *Der Arbeiter*, p. 235.

[542] Ibid., pp. 259 and 216–17.

[543] Daniel Morat, *Von der Tat zur Gelassenheit. Konservatives Denken bei Martin Heidegger, Ernst Jünger und Friedrich Georg Jünger 1920–1960*, Göttingen 2007, p. 80.

sign of the dawning age of the fourth estate. It makes the world war a historical phenomenon at least equal in significance to the French Revolution.[544]

This signified a revolutionary transformation of bourgeois society and its political-social categories – first of all, the category of the individual. In any event, "total mobilization" was the expression of "life in the age of the masses and machines," and man had to subordinate himself to their "mysterious and pressing claim." For Jünger, "total mobilization" was a process and a revolution, which executed itself and issued into the "age of the fourth estate." The *"first fact of the revolution"* had been the war.

> The function of the warrior – hence the frontline soldier – corresponds at the human level to this function of war as revaluer of values, destroyer of what has come into being, progenitor of things yet to come. [...] He [the soldier] is the epitome of the modern worker and fighter, the bearer of a new opening out into the world.[545]

Technology was the engine of this revolution that manifested itself in the war. Here it showed its "Janus face,"[546] at once destructive and constructive; it had to be affirmed, in accordance with Jünger's "heroic realism," or else man would perish. "But if one accepts – and this is very important – one becomes not only the subject of technological processes but also their object. The use of tools involves a distinctive mode of living, which extends to the big as well as the small things in life."[547]

Only the "worker" who said yes to technology could make himself the subject as well as the object of technological processes and therefore the "bearer of [his] destiny."[548] The bourgeois, on the other hand, had rejected this active and passive enlistment of technology, since it was a threat to his individuality. This rejection had entailed both an "incomplete technologization of the war"[549] and a failed "total mobilization." Faith in progress also prevented the bourgeois from wholeheartedly accepting technology:

> In the bourgeois sphere, technology appears as an organ of progress moving toward rational-virtuous perfection. It is therefore closely bound up with the value placed on knowledge, morality, humanitarianism, commerce, and comfort. The martial side of the Janus face does not fit well into this schema. But it

---

[544] Jünger, "Die totale Mobilmachung," p. 562; cf. Jünger, "Total Mobilization," p. 126.

[545] Ernst Jünger, "Der Kampf um das Reich. Vorwort [Dezember 1929]," in Jünger, *Politische Publizistik. 1919–1933*, ed. by Sven Olaf Berggötz, Stuttgart 2001, pp. 527–36; here p. 529. Emphases in the original.

[546] Jünger, *Der Arbeiter*, p. 155.

[547] Ibid., pp. 158–59.

[548] Ibid., p. 63.

[549] See Radkau, *Technik in Deutschland*, pp. 239ff.

is incontestable that a locomotive can move a company of soldiers instead of a dining car, or an engine a tank instead of a luxury vehicle. [...] Since the use of progressive instruments of "civilization" is inescapable in battle, bourgeois thought displays a tendency to excuse them. This happens because it lays the ideology of progress over the procedures of war, so that armed force appears a regrettable exception, a means of taming unprogressively minded barbarians. [...] The aim of using them [technological instruments] is not victory but the liberation of peoples, their integration into a community that is more civilized in its behavior.[550]

The inability of the bourgeois to accept the "power dimension inherent in technology" stemmed from his faith in progress – one thinks of Warburg's coding of the airplane – and spelled his downfall.[551] The "worker," however, spoke the "elemental language" of technology; he placed it "truly and unquestioningly in [his own] service," while at the same time placing himself in its service.[552] In Jünger's essay, the "worker" and technology are not only tightly interlinked; each becomes the *definiens* of the other. "Technology is how the figure of the worker mobilizes the world. The degree to which man stands decisively in relation to it, being assisted rather than destroyed by it, depends on the degree to which he represents the form of the worker."[553]

Who or what a "worker" was, could be ascertained from his relationship to technology, to "the material." On the basis of his experience of war, Jünger achieved a reinterpretation of "materialism." Martin Heidegger, who developed his understanding of "work" and technology in an encounter with Jünger's writings, argued in his *Letter on Humanism* (1946) that the essence of materialism "does not consist in the assertion that everything is simply matter but rather in a metaphysical determination according to which every being appears as the material of labor."[554] The idea that every being has become "material of labor" was Jünger's main discovery in the war. He saw that war was an immense production process, which reached its peak in "battles of materiel." In *Feuer und Blut* (1925), Jünger made it quite clear what he meant by "materiel" or "the material":

This is the material. Vast industrial districts with coal mine headframes and the nocturnal glow of blast furnaces appear before the eyes – machine halls with transmission belts and sparking flywheels, massive goods stations with gleaming track systems, the flurry of multicolored signal lamps, and white,

---

[550] Jünger, *Der Arbeiter*, pp. 156–57. Cf. above: Chapter I.1, *"Idea non vincit."*

[551] Jünger, *Der Arbeiter*, pp. 155, 158, 161.

[552] Ibid., p. 162. See Wünsch, "Ernst Jünger's *Der Arbeiter*," p. 469.

[553] Jünger, *Der Arbeiter*, p. 150.

[554] Martin Heidegger, "Letter on Humanism," in Heidegger *Basic Writings*, rev. edn, New York, NY 1993, pp. 213–266, here p. 243. He continues: "The essence of materialism is concealed in the essence of technology." See Günter Seubold, "Martin Heideggers Stellungnahme zu Jüngers 'Arbeiter' im Spiegels einer Technikkritik," in Friedrich Strack (ed.), *Titan Technik. Ernst und Friedrich Georg Jünger über das technische Zeitalter*, Würzburg 2000, pp. 119–32.

geometrically shaped arc lights. Yes, it is produced back there in the meticulously ordered operations of a gigantic labor process, whose aggregated output will then roll as stored energy along the major transport routes to the front, to be discharged with kinetic destructiveness against human beings. That battle will be a fearful measurement of production on both sides, and victory will mark success in the competition to work out how to produce in ways that are cheaper, faster, and fitter for purpose.[555]

Already in the preface to *In Stahlgewittern*, Jünger had pointed out that "not only machines, iron, and explosives" but also "man himself" were "evaluated as materiel."[556] But whereas the "bourgeois" despaired that he had become an object and function within a huge production process, the "worker" welcomed the fact that this was destroying "the old era" and ushering in a new age and a new man:[557] "the master of material and master of himself, who has turned from sorcerer's apprentice to craftsman in his own right."[558]

The new man resulting from the "battle of materiel" was both master and slave. He shaped material and was material himself. Thus, in assuming his function within the total "labor process," this man-become-material again became a subject and therefore "worker." The equation of frontline soldiers with "blue-collar work" [*Arbeitertum*], fliers with industrial workers, assault leaders with seamstresses, had its basis in the universalization of labor. While the metalworker carried out his tasks at a workbench – one thinks of Alexander M. Cay's "Through Labor to Victory" poster – the soldier performed his with a special instrument: "We are soldiers, and our weapon is the tool with which we shape things. Our work is called killing, and it is our duty to do this work well and properly."[559]

In the present context, the advance echoes of Himmler's infamous speech of 1943 in Poznan are less important than Jünger's identification of the soldier with the "worker." This rests upon three points: (1) war does not offer heroism but consists of "work";[560] (2) both soldier and "worker" serve technology as well as being served by it; and (3) both "unquestioningly fuse with the instruments at their disposal."[561] This amalgamation of man and machine is most clearly embodied in the "steel character-type" of the flier. For example, in the preface to *Luftfahrt ist not!* [Aviation Is Necessary], a collective volume he edited in 1928, Jünger writes:

Yes, the airman is perhaps the sharpest manifestation of a new manhood. He represents the type that was already showing signs of itself in the war. [...]

---

[555] Jünger, *Feuer und Blut*, pp. 30–1.
[556] Jünger, *In Stahlgewittern*, p. vii.
[557] Jünger, *Der Arbeiter*, p. 106.
[558] Jünger, *Feuer und Blut*, pp. 38–9.
[559] Ibid.
[560] Earlier, in *Der Kampf als inneres Erlebnis* (p. 41), Jünger had written: "Ah, how long ago we exchanged that skin gleaming with heroism for the laborer's dirty overall!"
[561] Jünger, *Der Arbeiter*, p. 178.

Here, under the aegis of war, was combined every element of energy, distinction, and technical intelligence that characterizes modern civilization, as well as the secret categorical imperative that lends the final hardness to the alloyed metal of machines [...]. [...] Perhaps he illustrates most clearly the profound link between the soldier's and the worker's condition. For although they have remained the same, the forms of the soldier and the worker are here mingled with each other. [...] The path that led across the heroic landscapes of war continues through the more sober fields of labor, and in both cases it is the flier's heart that gives the activity its real value.[562]

On the one hand, the flier had come to epitomize the "melting of the distinction between organic and mechanical world," and hence the "organic construction" that the whole labor state was understood to be.[563] On the other hand, however, he gave the clearest expression to the link between the worker and the soldier.

For Jünger, the "worker" was a "form" that appeared clearly in the flier, but that was to be found in many different manifestations. In any event,

the worker is not to be understood either as an estate in the old sense of the term, or as a class in the sense of the revolutionary dialectic of the nineteenth century [...]. [...] In particular, one will never get accurate results if one simply identifies the worker with the class of industrial workers: that is, if one rests content with one manifestation instead of seeing the form itself. [...] What is true is that in the industrial worker one can see an especially hardened type, whose existence makes clear the impossibility of continuing life in its old forms.[564]

The "bourgeois" categories of estate and class were to be rejected – after all, they contradicted the new order of society. What Jünger understood by "form [*Gestalt*] of the worker" was thus more in tune lexically than semantically with current usage at the time.[565] But, as we can now clearly see, the "type of the worker" covered many semantic possibilities for the integration and incorporation of the industrial proletariat. The semantics of "the worker" therefore served the new order of bourgeois society, but in accordance with antibourgeois principles. As early as January 1927, Jünger wrote in "Der neue Nationalismus," an article for the *Völkischer Beobachter*:

It no longer has the real ring of conviction when people call themselves bourgeois. A new estate is appearing on the scene and readying itself to take the helm of state. It is the fourth estate. The estate of the workers! [...] *Today's worker [Arbeitertum] is something different from "the workforce" [Arbeiterschaft], that old*

---

[562] Ernst Jünger, "Vorwort," in Jünger (ed.), *Luftfahrt ist not!*, Leipzig 1928, pp. 9–13; here pp. 11–12.
[563] Jünger, *Der Arbeiter*, p. 169.
[564] Ibid., p. 74.
[565] See Wünsch, "Ernst Jüngers *Der Arbeiter*," p. 465.

term of historical materialism that the bourgeoisie too came up with. [...] But just as the class state abolished the dynastic state, the nationalist state will abolish the class state.[566]

While the bourgeois "class state" had been abolished by the "nationalist state" and the *Volksgemeinschaft*, the working class like the bourgeoisie had become "workers." Jünger linked not only the future of the new nationalism but also his praise of Fascist Italy to this reinterpretation of the "worker":

> The workers [*Arbeitertum*] in the new sense are the blood-related community of everyone working within the nation for the nation. Only this community is capable of triumphing over the deformities of capitalism. The most urgent task for the new nationalism is therefore to grow into the form of a workers' movement, just as it is the task of workers to recognize that their justified struggle for existence can lead to victory only in the framework of the nation. Up to now we in Europe have only one state shaped by the nationalist workers: the Italian state.[567]

Jünger registered the collapse of the bourgeois order then under way, but he also forced its pace. In fact, fascism attempted to replace the individual with the type and to transform the mass into an "organic construction." This process of reinterpretation and symbolic-semantic inclusion of the industrial proletariat into the community becomes evident in Jünger's sketch of "the form of the worker." Yet what his *Der Arbeiter* makes especially visible is the bourgeois vision of an anti-bourgeois society and an antiliberal modernity.

Jünger's sketch of the "workers' state" anticipates much that the Nazis tried to translate into reality, and it expresses the technocratic tendencies underlying the most diverse "distantly related" regimes of the 1930s.[568] In 1932, according to Helmuth Kiesel, Jünger moved away from his radical nationalism, "downplaying it in favor of a 'planetary' perspective."[569] But since the absolutized nation or *Volk* was lacking in Jünger's "fascist conception of modernity" (Uwe Ketelsen), the latter was not, despite all the anticipations and parallels, a fascist conception in the sense of the mythical modernity under investigation here.[570] This did not, however, prevent the Nazis from drawing on Jünger's idea. As Wulf Dieter Müller emphasized in his 1934 biography of the writer,

> German youth is grateful mainly to Ernst Jünger for the fact that it no longer finds technology a problem. It has taken on board his fine tributes to it in *Feuer*

---

[566] Ernst Jünger, "Der neue Nationalismus," *Völkischer Beobachter*, January 23–24, 1927, in Jünger, *Politische Publizistik*, pp. 285–91; here p. 286. Emphases in the original.

[567] Ibid., pp. 286ff.

[568] Schivelbusch, *Entfernte Verwandtschaft*.

[569] Kiesel, *Ernst Jünger*, p. 397.

[570] On Jünger's worker as a "fascist conception of modernity," see Ketelsen, "Ernst Jüngers *Der Arbeiter*." On the various Nazi reactions to *Der Arbeiter*, see Kiesel, *Ernst Jünger*, pp. 394–99, and Morat, *Von der Tat zur Gelassenheit*, pp. 102ff.

und Blut; it lives in harmony with it. It no longer needs any ideologies to overcome technology, which it now understands as a branch of the Idea. This was
new for us: the incorporation of materiality into the sense we have of events.
Jünger has freed us from a nightmare.[571]

Müller also notes the necessity of taking "total mobilization" seriously:

> We must now move as fast as possible out of the transitional period between
> epochs. Unquestionably, the nation that is "ready" first, that adjusts all its life
> expressions first to the essence of the form [of the worker] and thereby becomes
> a total unity, will have the greatest chance to display its power and, in the unity
> of being, action, and will, to fight until it achieves victory.[572]

Many years later, writing to Walter Patt in February 1980, Jünger observed that,
although his "worker" was the "manifestation of the mythical form," the totalitarian order he outlined was not a blueprint for mythical modernity. For it lacked
an absolute center lying beyond work or the will to power itself.[573] In *Der Arbeiter*
we read the following

> Now the question of legitimation arises: a special and necessary, yet by no means
> volitional, relationship to power, which may also be described as a task. This
> legitimation brings forth a being that no longer appears as pure elemental power
> but as historical power. The degree of legitimacy differs from the degree of domi-
> nation that can be attained through the will to power. We call domination a state
> of affairs in which the limitless sphere of power is applied at a point from which
> it appears as a sphere of law. Pure will to power, on the other hand, possesses
> legitimacy as little as it does the will to belief. It is not fullness but a sense of lack
> that is expressed in these two attitudes into which Romanticism broke up.[574]

Since there was no Absolute, a lack of legitimacy prevailed. Therefore Jünger saw
"heroic realism" as the only way out – a solution that, for Heidegger, represented
"active nihilism." In Heidegger's notes of 1939–1940 on Ernst Jünger, the "heroic
posture [says] 'yes' [...] to what is – to 'the real,' without regard for itself."[575]
As Peter Koslowski has pointed out, Jünger already refers in *Der Arbeiter* to a
Nietzschean statement of Léon Bloy's that he would later often invoke:[576] "*Tout ce
qui arrive est adorable* [Everything that happens is adorable.]"

---

[571] Wulf Dieter Müller, *Ernst Jünger. Ein Leben im Umbruch der Zeit*, Berlin 1934, p. 42. Cf.
Morat, *Von der Tat zur Gelassenheit*, p. 104.
[572] Müller, *Ernst Jünger*, p. 59.
[573] Letter from Ernst Jünger to Walter Patt, February 6, 1980, quoted from Koslowski, *Der
Mythos der Moderne*, p. 64.
[574] Jünger, *Der Arbeiter*, pp. 67–8.
[575] Heidegger, *Zu Ernst Jünger*, p. 240.
[576] See Koslowski, *Der Mythos der Moderne*, pp. 74ff.

This nihilist stance – widespread among intellectuals from the postwar generation, but also in the "uncompromising" war youth generation[577] – based itself upon "knowledge" of the "death of God," the lack of an Absolute, and the resulting impossibility of a stable myth. And although this "knowledge" differentiated Jünger and his like from the fascists, it by no means kept them from participating in attempts to create a new order or from involvement in the crimes of the Third Reich. Rather, it seems as if these nihilists who knew of the death of God yet longed for a god to appear among them sought a way out from the paradox of modern existence in the destruction of order and the millionfold annihilation of human life. For all their "heroic realism," they were not willing to put up with contingency and ambivalence.

Some saw in fascism an opportunity to remove the decadent old order they despised and to celebrate the dawning of a new age. For Jünger, it was "one of the great and terrible pleasures of our time to be involved in these blasting operations."[578] What was supposed to follow them was still unclear. Meanwhile, one had to wait and hope that – in a mantra that Heidegger borrowed from Hölderlin – "*Wo aber Gefahr ist, wächst Das Rettende auch*" ["But where danger threatens / That which saves from it also grows.]"[579] Others thought that, given the lack of legitimation, fascism allowed the possibility to practice power for its own sake and struggle for its own sake. The problem of "transcendental homelessness," of the empty heavens above, was solved simply by positing an Absolute or by doing without one altogether. In the collective volume *Krieg und Krieger* that Jünger published in 1930, there is an essay by the jurist Werner Best (b. 1903), who, according to Ulrich Herbert, was the later "architect of the extermination policy."[580] In this essay, entitled *Der Krieg und das Recht* [War and Law], he writes:

> "Salvation" from the unbearable present, here on earth or beyond reality [...], relief from the unrest of this reality through the peace of a state of fulfilment or completion: this is the (conscious or unconscious) leaning of human beings who find the equation for the puzzle of their existence in the formulas of the utopian-rationalist and moral-idealist conception. [...] The inner stance here described as nationalism affirms the restless reality of the world around us, which is filled with struggle and tension. It does not strive for "salvation" or a condition that will end the present unrest. For it knows with immediate certainty that all life, the dynamic of the cosmos, consists in tension, struggle, and unrest.[581]

---

[577] See Michael Wildt, *An Uncompromising Generation. The Nazi Leadership of the Reich Security Main Office*, Madison, WI 2009; also Herbert, *Best*.

[578] Jünger, *Der Arbeiter*, p. 40.

[579] Friedrich Hölderlin, "Patmos," in Hölderlin, *Poems and Fragments*, translated by Michael Hamburger, Cambridge 1980, p. 499. See Martin Heidegger, "Einblick in das, was ist," in Heidegger, *Bremer und Freiburger Vorträge*. GA 79 ed. by Petra Jaeger, Frankfurt/Main 1994, pp. 3–77; here p. 72.

[580] Herbert, *Best*, p. 528.

[581] Werner Best, "Der Krieg und das Recht," in Ernst Jünger (ed.), *Krieg und Krieger*, Berlin 1930, pp. 135–61; here pp. 149–50.

To the striving for salvation, Best counterposed a different stance and "morality":

> The morality of the new stance cannot prescribe a "what," since it knows of none. It is not geared to an objective and serves no fulfilment or completion. Each moment calls the preceding one into question. No value that is struggled for at any given time has a claim or prospect of long-term certainty. So, what remains as measure of morality is not a content, not a "what," but the "how," the form. "The key thing is not what we fight for but how we fight" (Ernst Jünger [...]). Struggle is what is necessary and eternal; the goals of struggle are time-specific and change. [...] Others for whom only belief in a final goal makes struggle here and now endurable are also capable of struggle, in the expectation that it will actually be victorious or that the "good cause" will one day triumph. But the affirmation of struggle at lost positions for a lost cause is the defining criterion of the new posture: what matters is the good fight, not the "good cause" and success.[582]

Since no "good cause" existed, since "God was dead," all that was left for men like Werner Best was "struggle at lost positions for a lost cause." It would need to be discussed whether a suspicion or "knowledge" of the impossibility of mythical modernity, of an ultimate ordering of chaos, did not further intensify the vehemence and inhumanity with which that struggle would be waged.

"I want to become a Boelcke!" The inscription of the hero on the community and the national life

At the beginning of his previously quoted *Bei der Jagdstaffel Richthofen*, Georg Wegener tells of a heroes' "changing of the guard" at the end of 1916:

> When Boelcke fell, a profound grief came over the German *Volk*, and it was felt: "We shall never see his like again." – But new fighters crowned with success arose out of the multitude of his disciples, and with them one whom the nation in recent months has seen shoot up to the same solar heights of fame as Boelcke, one who, like Boelcke, has trained up a glittering circle of close comrades imbued with the same spirit and spurred on by him to the most successful emulation. I need not mention him by name; today everyone in our nation acclaims the fighter pilot Baron von Richthofen.[583]

In the course of 1917, Richthofen took the place in the German pantheon left vacant by Boelcke's death, because his numerous victories in the air had made him the most successful fighter pilot. A far-reaching change or break with the old order came together with this hero replacement in 1916–1917, so that the "long nineteenth century" was metaphorically laid to rest.

---

[582] Ibid., pp. 151–52.
[583] Wegener, *Der Wall von Eisen und Feuer*, vol. 3, p. 39.

The "dead hero cult" around Boelcke and Richthofen, involving participation in the national mythology, shows once again that the idea of a homogeneous, hierarchically structured community stretched far into German society. In an age when people's customary lives and values were in chaos, there was a widespread search for order and secure bearings. A society in mourning for the dead tried as best it could to cope with its losses and suffering. The hero cult afforded comfort, gave a meaning to death and promised life everlasting to the departed.

The worship of heroes may become clearer if we look at Boelcke's funeral ceremony and obituary.[584] Rudolf Oskar Gottschalk's book on Boelcke opens with a poem by Albert Markmann:

Hail, Boelcke, hardy seasoned aviator. / Hail, awesome crusher of the foe! / [...] If you with will and eagle-eyes do foil / The enemy's wild battle lunges / We fold our hands for you / That God may protect you from a sudden end. / You, o hero grown to human size, / Has God chosen for eternal life. / Hail, Boelcke, hail! To you as just reward / Does the *Volk* raise the crown of life / That will ever deck your hero's deeds / And honor you with *immortality*![585]

Another poem in the same volume, which one Prof D. Gerhard Heine-Dessau wrote on the occasion of Boelcke's visit to his home town, goes as follows: "Today was a great day in Dessau-Ziebigk: / Flight-Lieutenant Boelcke was there – / And did become a captain; / He had descended from the skies / And sat like an ordinary mortal / In the little house on the corner."[586]

The large number of such poems is one reason to take them seriously. For even if the language is "only metaphorical," it testifies to imaginative worlds of reverential worship and sacralization. Presented as a creature under God's protection, blessed by Him and chosen for heroic immortality, Boelcke enjoys the gift of "divine grace" that is the essence of Weberian charisma. At the same time, he is a godlike being himself: no "ordinary mortal," but "honored with immortality." The semantic affinity between the flier "descended from the skies" and the heavenly "gods" helps to strengthen this impression. But the eschatological dimension – the German "*Heil*," here translated as "Hail," can also mean "salvation" – also has the effect of sacralizing the hero.[587]

Boelcke's death on October 28, 1916, following a collision with another German airplane, was not detrimental to this sacralization: "No enemy can boast of triumph over our flier-hero," a fellow-officer in Boelcke's fighter squadron wrote the next day in a letter published in the *Anhaltischer Staats-Anzeiger*.[588] The

---

[584] The cult began with the burial rites: see Behrenbeck, *Der Kult um die toten Helden*, p. 67.

[585] Rudolf Oskar Gottschalk, *Boelcke. Deutschlands Fliegerheld. Schilderung seines Lebensweges und seiner Heldentaten im Luftkampf*, Leipzig n.d. [circa 1916–1917].

[586] Ibid., p. 81.

[587] On the idea of salvation, see Sabine Behrenbeck, "Heil," in Etienne François/Hagen Schulze (eds.), *Deutsche Erinnerungsorte*, vol. 3, Munich 2001, pp. 310–27.

[588] Adolf Victor von Koerber, *Deutsche Heldenflieger*, Bielefeld n.d. [circa 1916–1917], p. 29.

war reporter Georg Wegener, who accompanied Boelcke's corpse on its way back from the Western Front to Germany, spoke of a conversation with his parents, who obviously treasured the "unconquered by the enemy" trope:

> "Of course," he [the father] said, "but his cheerful sense of security communicated itself to us too in the end; we hardly had any worries, but thought that God would see him through to the end of the war. It was not to be. But what gives me and my wife heart is that he remained unconquered."[589]

It is irrelevant whether the father of the 25-year-old actually said those words; the point is that the sentiment could be voiced at all.

Before the repatriation in the company of his parents and a guard of honor, a magnificent ceremony of remembrance took place on October 31 in Cambrai cathedral. Boelcke's corpse was then

> laid on a large gun carriage drawn by six black horses and bedecked with flags and laurel wreaths. A flier from his fighter squadron, Lieutenant Manfred von Richthofen, carried a velvet cushion with the decorations that used to adorn the young hero's chest, and the endless procession, accompanied by lancers and foot-soldiers, made its way to the sound of a funeral march in the direction of Cambrai station, through a city bathed in bright sunshine after several days of stormy weather.

The soldiers who formed the guard "were trucked over early today from the trenches and later taken back to the front, from which the gunfire of the dreadful ongoing battle thundered across, dull yet powerful, during the funeral procession."[590]

Wegener here reminds us that, while a lavish funeral was being staged in Cambrai for a single dead man, the Battle of the Somme was raging 40 kilometers away – a conflagration that cost 465,000 German casualties between the British attack on July 1, 1916 and the provisional cessation of hostilities on November 25.[591] This (mis)match between the pompous farewell to one airman (who a few days earlier had shot down his 40th enemy plane) and the incomparably larger slaughter of anonymous troops should certainly be noted. For the importance of the hero-figure lay not least in this *pars pro toto* function. The honor shown to the exemplary hero Boelcke extended in principle to all the fallen. But above all Boelcke's own narrative could be transferred to thousands upon thousands of "unknown soldiers."

A train carried Boelcke's mortal remains and "a mountain of stacked laurel wreaths" back to his native Dessau, where the actual burial ceremony took place.[592] Already in Magdeburg, "a serious crowd numbering thousands" gathered

---

[589] Wegener, *Der Wall von Eisen und Feuer*, vol. 2, p. 355.

[590] Ibid., p. 357.

[591] On the losses at the Battle of the Somme, see Hirschfeld, "The Battle of the Somme 1916," pp. 63–81.

[592] Gottschalk, *Boelcke*, p. 93.

to greet the dead hero, and when the coffin was taken the next day from the church to the Dessau Cemetery of Honor a squadron flew along overhead, as at Cambrai, and dropped more wreaths. Thousands of local people lined the route "in rows three to four deep," and "the entire Dessau garrison formed a guard of honor, along with various associations, schools and youth brigades."[593] (Figure 3.2)

The soldiers' and veterans' associations gathered outside the cemetery lowered their banners as a mark of honor, and then a number of addresses were made in the presence of the family and local dignitaries, representatives of the Kaiser and the German Crown Prince, the Prince of Anhalt and his entourage, and delegations from the Austro-Hungarian, Ottoman, Bulgarian and German armies.

The speeches by Mayor Ebeling and Wing Commander Thomsen explained the charismatic relationship between the hero and his admirers, as well as the hero's function as a role model. The Mayor noted:

> Even he, the incomparable flier-hero accustomed to victory, had his life cut short by a sad stroke of fortune. [...] But soaring above the pain and suffering is enthusiasm for the greatness of the fatherland for which he who has passed away sacrificed everything. [...] Your spirit – the spirit of intrepid self-sacrifice, irresistible attacking power, and indomitable tenacity – still has its effect in all the young blood that has freely placed itself in the service of the fatherland. [...] From such seed will sprout a fine harvest. May the name "Bölcke [sic]" be a glowing beacon that shows us the way to the future.[594]

Figure 3.2 The funeral procession to the cemetery in Dessau

---

[593] Ibid., p. 96. This is also the source of the accompanying photograph.
[594] Otto Weddigen, *Deutschlands Luftkrieg und Helden-Flieger, 1914–1917*, Regensburg 1918, pp. 69–71.

Central here is the idea of sacrifice as the seed of a future that the young will inherit. This palingenetic trope of rebirth following sacrifice was the core of the fascist myth; it would be worked hard in the years ahead, albeit in a number of different variants. Parallels with the Christian narrative are evident not only in the resurrection after death but also, as Wing Commander Thomsen illustrated in his funerary address, in the theme of emulation:

> Today there is no fresh-faced youth anywhere in the homeland for whom the wish does not silently burn in his heart: "I would like to become a Boelcke!" That is a mighty consolation that [...] we all take home with us today. [...] And so I set down these words at the grave of the departed, as a farewell greeting and solemn vow from each and every one of our German airmen: "I want to become a Boelcke!"[595]

We can only speculate about the effect of this vow, especially on the young people assembled there. The grief and reverence really might have made his death for the fatherland appear "sweet and fitting" in the eyes of those schoolchildren and of others present. Abruptly though his life came to an end, Boelcke was assured a media "afterlife." As the *Deutscher Luftflottenverein* announced:

> And now he is dead and has long been resting beneath the mound strewn with flowers. But inwardly he shall live [within] us and our whole nation for ever and ever, and the thankful German *Volk* shall honor him as great nations have always honored their greatest heroes. Thousands of brave young airmen out there vow: we shall become and be like Boelcke! Often though weary swordsmen / Fall in bloodiest battle, / New generations shall come / And bravely fight it out.[596]

The hero-narratives served to give meaning to death in combat, but also to fortify and discipline young people in the spirit of self-sacrifice. The vow "I want to become a Boelcke!" was the key spur to the *imitatio heroica* that marked all the popular war literature. The same impulse to emulate heroes may be found in a somewhat different form in *Fritz der Flieger. Ein neues Kriegslesebuch für die Unter- und Mittelstufe* [Fritz the Flier. A New War Reader for Lower and Middle Grades], a book published in 1917 by Karl König, the head of a teacher training college. One chapter is entitled "What Fritz Dreams":

> "I want to become a pilot," Fritz said to his mother at home. [...] But at night he dreamed he was sitting in an aeroplane and flying from Freiburg to the Rhine. [...] There he turned around and headed back to the Black Forest, as far as Freiburg. But there he knocked into the cathedral, fell from the plane, and

---

[595] Koerber, *Deutsche Heldenflieger*, p. 29.
[596] Deutscher Luftflottenverein (ed.), *Das fliegende Schwert*, p. 20.

tumbled to the ground. And that hurt. He wanted to cry, but then he awoke from his dream and ... really was lying on the ground. He had fallen dreaming out of bed. He quietly slipped back into bed and murmured half-asleep: "I want to become a pilot!"[597]

The storybook ends with Fritz volunteering for war service and becoming a pilot. When he comes down behind enemy lines, he is rescued by his father and cured of his severe wounds; a letter has already arrived from the Kaiser promoting him to lieutenant. "Don't you think that was good medicine?" the narrator asks. "The poor locksmith's apprentice became an officer. And I don't think it will be long before he also receives the 'Pour le mérite' Order. I won't begrudge him it. How about you?"[598] The promise of integration into the community is extended through the spirit of self-sacrifice that also turned the humble Max Müller into a knight. The extracts from *Fritz der Flieger* further make it clear that pupils starting secondary school already had it suggested to them, "in a way suitable for children," that they should act like Boelcke and other heroes.

In stepping into Boelcke's shoes, Manfred von Richthofen came above all to personify self-sacrifice for the nation. It may possibly be anachronistic to draw a dividing line between the two hero-constructs; the boundary becomes visible only in the light of what then was still to come. For the Richthofen of the First World War can be separated to only a limited extent from the "Spirit of Richthofen" and the "invention of tradition" that the Nazis instrumentalized for their new Luftwaffe and cult of heroes.[599] Nevertheless, by analogy with the transition from spiked leather helmet to steel helmet, it seems meaningful to see Boelcke as personifying a demise (not least of individualism) and Richthofen the beginning of a new development. The figure of Richthofen contributed many features of the deindividualized "steel character" and new man that the discourse of the 1920s and 1930s would bring to the fore. In point of fact, the "battles of materiel" in 1916 (at Verdun and the Somme) did mark a dividing line for people at the time. But what at first appeared as apocalypse, "the last days of mankind," would be reinterpreted after the war as a new beginning and the basis for the palingenetic myth.

Despite the large print runs of Richthofen's *The Red Fighter Pilot*, it would seem that he or the hero-image he personified had not been as popular during the war as his "mentor" Oswald Boelcke.[600] The lasting fame of the Richthofen hero-construct owes much to Anglo-American popular culture, as well as to the adoption, remold-

---

[597] Karl König, *Fritz der Flieger. Ein neues Kriegslesebuch für die Unter- und Mittelstufe*, Strasbourg 1917, p. 7.

[598] Ibid., p. 32.

[599] On the "invention of tradition," see Eric J. Hobsbawm/Terence Ranger (eds.), *The Invention of Tradition*, Cambridge 1983.

[600] One indication of this is the number of books on Boelcke the hero: Aviaticus, *Boelcke. Der Held der Lüfte. Ein deutsches Heldenleben*, Chemnitz n.d.; Boelcke, *Hauptmann Bölckes Feldberichte*; Gottschalk, *Boelcke*; Koerber, *Deutsche Heldenflieger*; Anton Luebke, *Hauptmann Boelcke. Ein Gedenkblatt für den ruhmbedeckten Heldenflieger*, Leipzig n.d. [circa 1916]; Friedrich Albert Meyer, *Immelmann und Boelcke. Deutsche Helden der Luft*, Warendorf i. Westfalen n.d.

ing and instrumentalization of "Richthofen remembrance" on the right of the political spectrum in the Weimar Republic, and especially in the Nazi period.[601] The post-1933 harping on the "Spirit of Richthofen" successfully projected him as the sought-after heroic leader, but also, in the words of René Schilling, as the "charismatic-martial 'national or popular hero' (*Volksheld*)." After all, the first fighter squadron of the new Luftwaffe did bear his name rather than Boelcke's, and in 1935 it was the anniversary of Richthofen's death, April 21, that was declared Luftwaffe Day.

Shortly after Richthofen's death in 1918, a local historian by the name of Adolf Wasner put together a commemorative book that was published later that year. This contains not only obituaries to the hero's memory but also, as in Boelcke's case, numerous poems:

> No one outfought him, though. / Molded perhaps by an angel / To end the praises for the youth / Without grief or hatred, / A stray bullet it was / That lightly touched his heart / and carried him the godlike off to God. / No more is he here among us. / His deeds tower as a mountain / Eternal is his selfless work. / The boldest did come unstuck, / But his people, his bold German *Volk*, / Remains as ever alert! / Fly, o nation, fly / Full of youthful heroes, / To glory, to victory! / Do you not at sundown see / The red pilot wave to you aglow? / Yes, every evening on the sky's edge / He soars gently and looks down on the fatherland! / He did but fly into the setting sun... / He is not dead.[602]

The amateurish verse testifies to the popular dimension of sacralized nationhood, but it also highlights the attempts to endow death in war (which affected countless families) with meaning as life of the nation. The reinterpretation of the hero's death in combat was paradigmatic in so far as it supplied narrative models that could be applied to a personal context. The hero served to reproduce heroic virtues, especially the willingness to sacrifice one's life, but he also enabled people to adopt interpretations that gave meaning and comfort in relation to death. The consolation was that a "more meaningful life" followed a profane death: both as a hero in people's memory and as part of the nation's continued existence, which seemed to be guaranteed precisely through exemplary self-sacrifice of the part of the hero and his successors. These narratives transcended profane time and embedded death in sacred time. Another poem in Wasner's volume, which places the German aviator-heroes in a mythological continuum, also testifies to this temporal order. Its author is one Alfred Wlotzka and its title is *Ikaros – Richthofen*:

In the preparatory work for this study, I came across no books on Richthofen published during the war years apart from his *Der rote Kampfflieger*.

[601] On the Richthofen cult in the Weimar Republic and the Third Reich, see Schilling, "Kriegshelden," pp. 295–315, 326–33, 343–48, 365ff. On the persistence of the cult, see Joachim Castan, *Der Rote Baron. Die ganze Geschichte des Manfred von Richthofen*, Stuttgart 2007.

[602] Alfred Hein, "Zum Gedächtnis des Rittmeisters Freiherr von Richthofen," in Adolf Wasner (ed.), *Rittmeister Manfred Freiherr von Richthofen. Ein Lebensbild nach Zeitungsberichten*, Diesdorf bei Gäbersdorf 1918, pp. 5–6.

Mighty time! From tales and myths / Your reality sprouted blossom upon blossom, / A sunny race of heroes! – / Young Icarus fell but his name has remained, / Written with the sun's gold into history; / Even Daedalus found his way consoled. / Hero Immelmann fell. To read his name intact / Was a radiant symbol for hero Boelcke, / Who then shone in gold yet brighter. / Ye conquerors of storms, ye eagle comrades, / Your names gleam molded in steel, / Recorded too in the enemy's rolls of fame! – / Hero Richthofen dead! – Who blazed so wondrous, / Whose star rose to the starry sky whence it came! / His death a loss? Come on him too soon? / Nay, such glorious deeds glow forth / And announce the coming of giants!! / Hail, Richthofen! Hail to your fatherland!!

The dilettantism should not blind us to the fact that the war diffused perceptual, conceptual and interpretive patterns in which the national order was sacralized and absolutized. The mythical temporal order accompanying it is well expressed in the poem *Deutschland und die Welt* [Germany and the World] that closes the book:

O land filled with blood and wounds, / Wronged and derided, / Of all your friends remained but one: your God! / One only, though the strongest, / Who does not abandon you. – / Germany, you land of faith, / Hold fast yet to your faith! / You bore it well indeed / What no other nation endured, / When the scourge of war struck you / For thirty long years. / Such tears as you wept / Has no other people wept; / In such mortal sorrows / Has none other been fixed. / Yet amid the sorrows, / amid death's frightful pangs, / The light of the stars / Never faded in your heart. / Your sweet countenance arose / Out of all the horrors, / Stirring the play of children's laughter / Exciting sacred trust, / And of what they took from you / One thing will never be stolen, / The future remains yours, O Germany, / Because in it you believed. So you have risen again / Alive from the grip of death. / So will you now resist / This time of trouble too.[603]

It was precisely such resurrection after a time of defeat and troubles that the fascists would preach, using aviation to conjure up national rebirth and the dawning of mythical modernity. It may be no more than a coincidence, but the timing of the first Day of the Luftwaffe on April 21, 1935 – the 17th anniversary of Richthofen's death – was certainly not inopportune: it was also Easter Sunday, the day of the Resurrection.

---

[603] Ernst von Wildenbruch, "Deutschland und die Welt," in Adolf Wasner (ed.), *Rittmeister Manfred Freiherr von Richthofen. Ein Lebensbild nach Zeitungsberichten*, Diesdorf bei Gäbersdorf 1918, pp. 87–8; here p. 88. The striking parallels with Paul Gerhardt's hymn *O Haupt voll Blut und Wunden* [O Head Full of Blood and Wounds] again point to a link with the Christian trope of death and resurrection.

4
Fractured Order – Summary

This chapter has focused on central tropes of the fascist mythical order as they developed in the First World War – in particular, the contexts within which they arose and acquired their meaning; the media channels that fostered their diffusion in society; and the metamorphoses they underwent in the postwar period. The Brescia air show, where entertainers, acrobats and sportsmen of the machine age were given an enthusiastic reception, made clear the fascination that aviation held for large numbers of people and its tailor-made capacity to convey messages such as D'Annunzio's vision of the flying Superman. The aviation discourse was able to spread within the network or "dispositif" that had evolved by the time of Brescia.[604] Initially the chief components of this network were a curious and captivated public, popular icons and prominent figures, media in the course of becoming a consumerist "culture industry," major events and their organizers, representatives of economic and national interests, the aviation industry, the military and the government.

The mythical tropes then spread during the world war, by means of a "culture industry" that had grown out of the commercialization of the book market and the "literary mobilization." The war books in question were products of the "grassroots nationalism" of bourgeois publishers and authors, which reproduced the needs and ideas of a largely bourgeois public and disseminated images of "the war as it really was" for general consumption. Paratexts located the war books within an ostensibly factual context, which, corresponding neither to the process of their production nor to their external appearance, satisfied the demands of consumers searching for models, orientation, meaning and consolation in times of war. For the price of just one mark, readers, probably drawn mostly from the war youth generation, were promised a true-to-life picture of the air war, as well as excitement, adventure and images of "real men" worthy of emulation. The war books, like Cervantes's *Don Quixote*, structured the chaotic reality of war by means

---

[604] On the "dispositif," see "The Confession of the Flesh" (1977), interview, in Michel Foucault (ed.), *Power/Knowledge. Selected Interviews and Other Writings*, ed. by Colin Gordon, New York, NY 1980, pp. 194–228; here p. 194.

of literary tropes and interpretive models, thereby contributing in turn to what Paul Fussell called the "curious literariness of real life."[605]

In the midst of a war that marked the triumph of "modern industrialism, materialism, and mechanism" epitomized by the airplane,[606] the books spread a mythical way of thinking that culminated in the hero cult. The heroes were a projection surface that displayed the values and norms of society and the virtues it regarded as desirable. The popular accounts of heroes offered a narrative as a guide to action, which gave meaning to suffering and death in war. They implied that death in combat was by no means vain but helped the nation to go on living. The self-sacrificing hero transcended profane time and entered the sacred time of the nation or *Volk*. This overcoming of death was assured to anyone who acted in the same way.

While the success of the hero-narratives was generally based on this provision of a meaning beyond death, the high media profile of the newly created air force rested upon the compensatory function of the aviation narratives. However, the focus of these was not the "workhorse" planes that carried out the main task of the air force in the First World War – that is, aerial observation – but the fighter pilots, the "knights of the air," who preserved the traditional male matrix of the warrior that was being simultaneously destroyed by the "battles of materiel" on the ground. In this respect, the aviator-hero was a product of positional warfare and represented a countermodel to it; he offered a possibility of imaginative escape from the trenches into a form of mobile warfare where the positions were not frozen – and where individual talents and achievements, far from being obsolete, could help to achieve honor and glory.

The aviator-hero also rose above terrestrial impotence in the face of technology: he represented control over technology as a means to power in the name of the nation. The conventional tokens of bourgeois-heroic masculinity – individualism, efficiency, mobility, the overcoming of spatial limits, control of nature through technical know-how – continued in the shape of a mechanical-organic hybrid, but were adapted to the antibourgeois, especially antiindividualist, context of the new age. The war in the air, then, was the "proper war" people had been longing for when they summoned up the purifying thunderstorm.

In the air, it seemed as though the war still offered scope for men to prove themselves, but also the potential for social distinction and the amassing of symbolic capital. The hero's elitist and exclusive status, for which he was much envied, and his comparatively glamorous lifestyle made him a popular cultural icon – evidence that "going off to war" could still be worthwhile, glorious, and meaningful. The aviator-hero therefore became a model especially for young people still growing up, who for the time being were not "granted" the possibility to "prove themselves."

The aviator-hero narratives also assisted the popularization and radicalization of nationalism. They helped to define a national identity by conveying desirable

---

[605] Paul Fussell, *The Great War and Modern Memory*, New York, 2000 [repr.].
[606] Ibid., p. 115.

virtues and qualities of character, by suggesting that self-sacrifice and the surrender of individual goals for the common good ennobled human beings. In this way, the aviation war books spread the demand for total solidarity and total identification with the nation. They served to absolutize the nation, which increasingly became the sole authority making action legitimate, the source of all norms and values, and the objective to which they were all directed. The nation thus became the "sacred realissimum" of a mythical order; no other "imagined order" would be tolerated beside it.

By disseminating the ideas of a united, homogeneous nation, a new elite and a new hierarchy, the war books in question here created a disposition for the palingenetic fascist visions based on the "Spirit of 1914." Aviator-heroes served as transitional figures between the liberal order of society and the national order of the *Volksgemeinschaft*. They could be read both as symbols of the bourgeois, autonomous individual raised above the masses and as embodiments of the newly forming antibourgeois, collectivist warrior-type. The individual hero protruding from the crowd became the leader striding ahead of his followers. The meritocratic, military-technological elite of 1914–1918, endowed with the symbolic capital of the old warrior caste, became a new nobility that did not owe its status to its origins. Its achievements and self-sacrifice for the nation were supposed to legitimize its place at the head of the homogeneous, hierarchically structured, *Volksgemeinschaft* or national community. The figure of Manfred von Richthofen reconciled the old with the new soldierly ideal of manhood, the aristocratic knight and feudal cavalryman with the strong-nerved, ruthless, dynamic, death-defying "cool persona."[607] He thus formed a semantic-symbolic bridge between the orders.

The new-noble "steel character-type," for which the aviator-hero paved the way by enhancing the status of the technological-industrial sphere, became a symbol of the new age and community, acquiring a watershed function in their his own right. The mechanized warrior created a bridge to "the workers," permitting their adoption and inclusion in the national community. A transformation of interpretive models and ideas of order, expressing a will to remold bourgeois society from the roots, underlay this inclusion of "the workers." Ernst Jünger's sketch of the "workers' state" clearly illustrates the transformation of the bourgeois order that gathered pace during the war, as well as its antibourgeois and antiliberal thrust. In the writer's vision of a totalitarian modernity, there was no place for the bourgeois categories of class and individuality. In the face of defeat and "mutilated victory" at the end of the war, the flier became the herald of the New Age and national rebirth, paradigmatically embodying the spirit of a new beginning. The flier as symbol of movement and ascent thus came to epitomize the renewal and rebirth of the nation, while the aviator-hero narratives functioned as templates for future palingenetic myths. In the context of the radical quest for order, the one who had overcome stasis in the war developed into the one who would overcome the liberal order.

---

[607] On the "cold persona," see Lethen, *Verhaltenslehren der Kälte*.

Part III
Eternal Order

1

Volare! The Fascist Take-off Towards Eternal Order

On August 20, 1919, almost exactly five months after the founding of the *fasci di combattimento* on the Piazza San Sepolcro in Milan, a full-page article on aviation appeared in *Il Popolo d'Italia*, the newspaper founded by Mussolini in 1914. A propeller driven, six-cylinder engined biplane, framed the heading of the *pagina dell'aviazione*. And beneath an italic *Volare!* [Flying!] came the following:

> *Volare!* Ever higher, in an extraordinary effort of the nerves, will, and intellect that only man's small mortal frame can support. *Volare!* To fly over the ghastly, continual trench skirmishes that are the stuff of daily life. *Volare!* To fly for the beauty of flight, almost *l'art pour l'art*; to fly so that the collective has a new weapon to make intellectual, moral, and commercial relations easier among distant nations. Once the sky is ploughed by airships that shorten the distance from one nation to another, we shall be able to claim that we have fused all souls into one. The closer we come to the infinite, the more we shall feel capable of uniting our own law with that of others. [...] *Volare!* To fly, because although Icarus died, it was man's first bold act when he stole a little of heaven's glory, and because Prometheus taught that the human heart can be stronger than any adverse fortune.[1]

Flying, the future Duce implied, holds out the promise of new shores and higher goals. Aviation goes beyond the everyday, embodying man's link to "the eternal and divine," as well as his Icarian spirit, his Promethean or Faustian powers. In Germany as in Italy, fascism presented itself as a new beginning – as the rebirth of an eternal nation buried by the decadent bourgeois-liberal order. Mussolini's "March on Rome" and his "seizure of power" were choreographed as a national revolution opening a new age, which required and would bring forth a new man.

Aviation stood for the dawn of an eternal order. The aviator, not least because of his boldness, love of danger, indomitable will and Faustian power, was the new man corresponding to this order. Flying thus became a metaphor for fascism itself.

---

[1] Benito Mussolini, "Volare!," *Il Popolo d'Italia*, August 20, 1919, quoted from Ministero dell'Aeronautica (ed.), *L'aviazione negli scritti e nella parola del Duce*, Rome 1937, p. 31.

In this chapter, we shall see how aviation – which during the First World War had symbolized movement in space – became a symbol of movement in time and hence a medium of the fascist palingenetic myth.

The fascist new departure went hand in hand with a break from the past, and with the establishment or invention of a "history" befitting the new fascist man. This was conceived as a discontinuous series of the times when the "eternal nation" had manifested itself. In the following we shall resume our reading *inter alia* of Ernst Jünger and then show how the First World War acquired the significance of a pivotal point for the future. The "Eternal" had disclosed itself during the war; the task now was to renew it. In keeping with Nietzschean "monumentalist history," Jünger placed the past – the past war – in the service of future life.[2] For fascists, it had precipitated the downfall of an execrable, decadent world and the despised liberal order; it had sown the seeds of a different modernity. Only this orientation to the future had made a palingenetic myth out of the narratives of sacrifice so common during the war years. Heroes had died so that the nation would live. It was this forward-looking perspective – this futurist will, so to speak, with the young and the new at its core – that radically differentiated fascists from the conservative Right.

The fascist ordering of time was not, however, just a question of war as the "beginning" of the future. For the future to begin, the past first had to be destroyed. We shall see from the Italian Futurists how war was declared on the past and its cult of history (what they called "passéism").[3] This assault on historicism – or rather, on processual thinking and the principle of historicity – was at the same time a campaign against the "bourgeois," against the liberal order and its underlying belief in progress.[4] Antihistoricist condemnation of the past changed into celebration of the future, finding a potent symbol in the airplane and aviator. It is true that this celebration of the future defined, and gave a name to, only a small group of Italian *avant-garde* intellectuals. But for all the distinctions in detail, the "futurist" will directed to the future was paradigmatic for fascism and its chronopolitical orientation in general. Although it also won the adherence of

---

[2] On "monumentalist" history, see Friedrich Nietzsche, "On the Uses and Disadvantages of History for Life (1874)," in *Untimely Meditations*, Cambridge 1983.

[3] On the Russian Futurists and their appropriation of Nietzsche, see Bernice Glatzer Rosenthal, *New Myth, New World. From Nietzsche to Stalinism*, University Park, PA 2002, pp. 94–111.

[4] On antihistoricism, see Anselm Doering-Manteuffel, "Mensch, Maschine, Zeit. Fortschrittsbewusstsein und Kulturkritik im ersten Drittel des 20. Jahrhunderts," in *Jahrbuch des Historischen Kollegs 2003*, Munich 2004; Friedrich Jaeger, "Theorietypen der Krise des Historismus," in Wolfgang Bialas/Gérard Raulet (eds.), *Die Historismusdebatte in der Weimarer Republik*, Frankfurt/Main 1996, pp. 52–70; Kurt Nowak, "Die 'antihistoristische Revolution.' Symptome und Folgen der Krise historischer Weltorientierung nach dem Ersten Weltkrieg in Deutschland," in Horst Renz/Friedrich Wilhelm Graf (eds.), *Umstrittene Moderne. Die Zukunft der Neuzeit im Urteil der Epoche Ernst Troeltschs*, Gütersloh 1987, pp. 133–71; Otto Gerhard Oexle, "Von Nietzsche zu Max Weber: Wertproblem und Objektivitätsforderung der Wissenschaft im Zeichen des Historismus," in Otto Gerhard Oexle (ed.), *Geschichtswissenschaft im Zeichen des Historismus. Studien zu Problemgeschichten der Moderne*, Göttingen 1996, pp. 73–94; and Otto Gerhard Oexle, "Krise des Historismus – Krise der Wirklichkeit. Eine Problemgeschichte der Moderne," in Oexle (ed.), Krise des Historismus – Krise der Wirklichkeit: Wissenschaft, Kunst und Literatur 1880–1932 [Veröffentlichungendes Max-Planck-Instituts für Geschichte, vol. 228], Göttingen 2007.

various reactionary forces, fascism was by no means backward-looking: it set its sights on a different future, understood as an epiphany of eternity.[5]

For the Futurists, as for Jünger, it was technology, especially the airplane, that characterized the future and the new times; it defined the rhythm and pulse of life. The acceleration that technology brought with it was to be welcomed, since it both permitted and epitomized the break with the past. Abolition of the past was the guarantee of a new future. The Futurists and Jünger further serve to illustrate the mythical dimension of the fascist order. In so far as technology was "enchanted" – that is, formed part of a narrative centered on the sacral nation and productive of community – they could accept it not despite, but *because* of its peculiar destructiveness. The destruction of profane time, accelerated and radicalized by technology, granted entry to the sacred time of eternity.

Although fascism practiced extreme forms of violence from the beginning, few suspected that it would exhaust itself in its outpouring of destructiveness. Rather, after the First World War, many Italians and Germans were attracted to it because of its vision of a new order. Support for fascism was sustained by a promise to replace the age of decadence with a new, vital epoch. As the *Esposizione dell'aeronautica italiana* (EAI) eventually showed in 1934, the aim of the fascists was not only to redeem through masterly propaganda this promise of ascent and renewal, but also to inscribe it within society. Through such exhibitions – which evoked a collective, public experience at once sacred and artistic – the fascist state demonstrated its powers of social organization, and a new ordering of time and mythical modernity developed within society.

Jünger's monumentalist remembrance of war and the beginnings of the future

In 1919, the German-Jewish philosopher and cultural critic Theodor Lessing (b. 1872 in Hanover) published his polemical work *Geschichte als Sinngebung des Sinnlosen* [History as Giving Meaning to the Meaningless].[6] The author, regarded as an intellectual outsider himself, was strongly influenced by Nietzsche's *On the Uses and Disadvantages of History for Life*, and also by his own wartime experiences as a medical orderly. Lessing was an antihistoricist, but his antihistoricism differed from that of the fascists. Although both shared a rejection of progress as

---

[5] See Griffin, "'I am no longer human. I am a Titan. A God!' The Fascist Quest to Regenerate Time," in Griffin, *A Fascist Century. Essays*, New York, NY 2004, pp. 3–23; "Modernity Under the New Order. The Fascist Project for Managing the Future," in ibid., pp. 24–45; and *Modernism and Fascism*. See also Peter Osborne, *The Politics of Time: Modernity and the Avant-garde*, London 1995.

[6] On the biography of Lessing, see Rainer Marwedel, *Theodor Lessing 1872–1933. Eine Biographie*, Darmstadt 1987; and Julius H. Schoeps, "Der ungeliebte Außenseiter. Zum Leben und Werk des Philosophen und Schriftstellers Theodor Lessing," in Walter Grab/Julius H. Schoeps (eds.), *Juden in der Weimarer Republik*, Stuttgart 1986, pp. 200–17. Unless otherwise indicated, the quotations from Lessing's book in this section are taken from the reprint of the first edition: Theodor Lessing, *Geschichte als Sinngebung des Sinnlosen oder die Geburt der Geschichte aus dem Mythos*, Munich 1983 [orig. 1919].

the principle of temporal organization, the fascists believed it was possible to step outside the flow of profane time and to enter a sacred time. Lessing did not in any way deny the historicity of human existence. Lessing, as Ulrich Kittstein put it, "combated the naïve epistemological optimism [...] according to which past events could be objectively reproduced, and also rejected the belief [...] that such events constituted a meaningful structured process."[7]

In Lessing's view, the war had debunked the faith in progress and ruled out any approach to history as a "meaningful structured process." Where the fascists hoped to replace the "false" faith in progress with another meaning of time, Lessing argued that mythical as well as historical consciousness always involved an attempt to redeem reality.[8]

Using concepts comparable to those of Warburg or Blumenberg that we examined in Chapter I.1, Lessing saw historiography as a defensive action against reality. Despite, or perhaps because of its claim to *scientific* truth, it had the same function as myth: to bestow meaning on the meaningless. History freed man from real life; chaos was retrospectively shaped into history, which read like an exciting novel. Historiography, for Lessing, was a "salutary act of self-deception."[9]

Lessing drew attention to the fact that histories were *constructs*. He also pointed to their narrative character and their dependence on the present, in whose light the past always appeared. Like Nietzsche he therefore denied that historiography was scientific, although he did not dispute its functional necessity. His aim was not to do away with history as bestower of meaning, but to expose the naïve faith in objectivity, in the realistic mapping of the past.[10] "Giving meaning to the meaningless" might be a fictional construct, but it was still necessary.[11] Also like Nietzsche, Lessing called for a history in the service of life, and therefore life within an ahistorical or mythical frame of meaning.

This conflict, which is characteristic of Lessing's approach, also marks the spirit or ethos of modernity. The paradoxical state of consciousness is determined by the dialectic of freedom and order. A critical, order-dissolving dimension is complemented by a dimension that yearns for and generates order. The "bestowing of meaning on the meaningless," and hence the success of the attempt to overcome the contingency of human existence, was possible only through suppression of the thinking that delegitimized order – the thinking that made "giving meaning to the meaningless" necessary in the first place.

---

[7] Ulrich Kittstein, *"Mit Geschichte will man etwas." Historisches Erzählen in der Weimarer Republik und im Exil (1918–1945)*, Würzburg 2006, pp. 108–09. See also Jürgen Grosse, *Kritik der Geschichte. Probleme und Formen seit 1800*, Tübingen 2006, esp. pp. 208–09 and 256ff., and Schoeps, "Der ungeliebte Außenseiter," p. 206.

[8] Lessing, *Geschichte als Sinngebung des Sinnlosen*, pp. 155ff.

[9] Ibid., p. 156.

[10] Ibid., p. 85: "The old faith in natural development, progress, an immanent world order, God, and divine providence in history therefore breaks down. So much the better! We expect all future salvation to come from the breakdown of historical madness."

[11] See Kittstein, *"Mit Geschichte will man etwas,"* pp. 110 and 112.

Historical thought is capable of both: it dissolves orders and builds others in their place. But as soon as it loses the capacity to historicize and question the foundations of the posited order, it becomes supra-historical or mythical thought. The spirit of modernity, and above all the First World War, exposed the temporal order constructed by the narrative of progress as an illusory and arbitrary "giving of meaning." Another temporal narrative, another fiction, another myth appeared instead of the narrative of progress. On the night of August 30, 1933, the Nazis cold-bloodedly murdered Lessing in Mariánské Lázní, his Czechoslovak city of refuge. But their "giving of meaning to the meaningless," their "fictional mythmaking," had begun earlier.[12] As we shall see, war for them was the occasion for "consolatory self-healing and the inevitable overlaying of human need with wishes or ideals," along the axis of their temporal order.

On June 8, 1928 a review of five war books appeared in the sports magazine *Sport im Bild*. Among these were two by Ernst Jünger that have been discussed in Chapter II.2: *In Stahlgewittern. Aus dem Tagebuch eines Stoßtruppführers* [Storm of Steel], first published in 1920 under his own imprint, which was then into its eighth edition at the Berlin military publishing house Mittler & Sohn; and *Das Wäldchen 125* [Copse 125]. The reviewer writes:

Both of Jünger's books are of a salutary objectivity. Precise, serious, robust, and powerful, they keep rising in intensity, until the harsh face of war, the grimness of the battle of materiel, the huge, all-conquering force of vitality and the heart, truly find expression. The "storms of steel" depict the course of events with all the power of the experience at the front; they reproduce without any emotionalism the soldier's dogged heroism, drawn by a man who, like a seismograph, captures all the vibrations of battle. [...] Jünger, one of the few young infantry officers to have received the Pour le Mérite order, has better credentials than almost anyone else to speak of battles and the war. He does it plainly and simply, and therefore with great force.[13]

The commendation came from the pen of Erich Maria Remarque. As is well known, Jünger did not conceal any of the horrors of war, while Remarque's *All Quiet on the Western Front* not only exuded a spirit of "comradeship" but testified to a kind of heroism that consisted of enduring a lack of meaning.

Thus, were it not for a certain consensus that they reveal in the Weimar Republic about the exceptional importance of the war, no further attention would be due to Remarque's words in praise of a man who in 1928 was still in his radical-nationalist or even fascist phase. The appearance of a review of war literature in a sports magazine further underlines this consensus, even if we consider that Scherl Verlag (which published *Sport im Bild*) belonged to the nationalist Hugenberg

---

[12] Lessing, *Geschichte als Sinngebung des Sinnlosen, oder die Geburt der Geschichte aus dem Mythos*, 4th edn, Munich 1927, p. 3.
[13] *Sport im Bild*, vol. 12, June 8, 1928, unpaginated.

corporation. We should bear in mind, however, that the agreement concerned only the importance of the war; its actual meaning was an object of contention in the Weimar Republic. But how did the war become the basis for the palingenetic myth of fascism then taking shape?

In the fascist myth of community, the war had meaning as a pointer to the future. In Remarque's bestseller, on the other hand – and this lay beneath the raging controversies over the book and the ensuing film – the war had lacked any purpose or meaning, it had cost untold human lives and ruined those who were spared and eventually returned home. According to Gertrude Stein, it turned them into "a lost generation." And Remarque himself wrote in a foreword: "This book is to be neither an accusation nor a confession. [...] It will simply try to tell of a generation of men who, even though they may have escaped shells, were destroyed by the war."[14]

Nevertheless, *All Quiet on the Western Front* came to be seen as an accusation. Here was an author who used "critical history, a history that judges and condemns," who was "oppressed by a present need and [wanted] to throw off this burden at any cost."[15] To remain with Nietzsche's terminology, this "critical" way of recalling the war stood opposed both to the "antiquarian" mode (as in the memoirs of generals or regimental histories) and to the "monumentalist" mode that Ernst Jünger cultivated in his war books and *a fortiori* in the many articles he published between 1920 and 1933.[16]

Although Jünger's political journalism commanded attention in conservative-revolutionary circles, his books – unlike Remarque's – were scarcely noticed by the Weimar "public." Of course, the review by the then little known Remarque did nothing to change this. When the ninth edition of *In Stahlgewittern* appeared in 1929, it had sold only some 26,000 copies;[17] and even after the "boom in war literature" in 1928–1929 and the Nazi seizure of power in 1933, the total had risen to "only" 51,000.[18] As to Remarque's *Im Westen nichts Neues*, which first appeared in book form in 1929, it sold 160,000 copies in March of that year, 150,000 in April and 100,000 each in May and June. Until the book was burned and banned,

---

[14] Erich Maria Remarque, *All Quiet on the Western Front*, New York 1982 [orig. 1929], unpaginated.

[15] Friedrich Nietzsche, "On the Uses and Disadvantages of History for Life (1874)," in *Untimely Meditations*, Cambridge 1983, p. 72.

[16] According to Bruno Reimann, Jünger published 138 articles in newspapers and magazines between 1920 and 1933. Sven Olaf Berggötz, however, puts the total at 144. The most prolific period, between 1925 and 1930, overlaps with Jünger's national-revolutionary phase or his commitment to the "new nationalism." See Sven Olaf Berggötz, "Nachwort," in Ernst Jünger, *Politische Publizistik 1919 bis 1933*, ed. Sven Olaf Berggötz, Stuttgart 2001, pp. 834–69; and Bruno W. Reimann, *"...die Feder durch das Schwert ersetzen..." Ernst Jüngers politische Publizistik 1923–1933*, Marburg 2001, esp. pp. 71–80.

[17] See Helmuth Kiesel, *Ernst Jünger. Die Biographie*, Munich 2007, pp. 206–07.

[18] The word "boom" (*Hausse*) was already used to refer to it at the time, as in the title of one contribution in the *Literarische Welt*. See Thomas F. Schneider, "'Die Meute hinter Remarque.' Zur Rezeption von *Im Westen nichts Neues* 1928–1930," *Jahrbuch zur Literatur der Weimarer Republik* 1/1995, pp. 143–70; here p. 144.

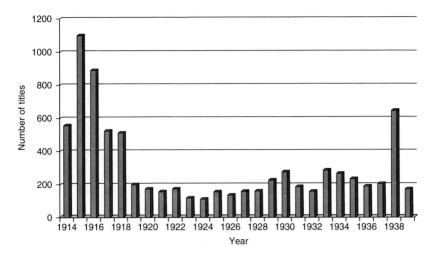

Figure 1.1 War literature published in Germany: numbers of titles per year
Source: Based on data in Schneider et al., *Die Autoren und Bücher der deutschsprachigen Literatur zum 1. Weltkrieg 1914–1939*, p. 9

it appears to have sold a total of 1,200,000 copies.[19] Jünger's omnipresence in academic research therefore owes more to the aesthetic quality of his work than to the size of his readership. But it is still worth focusing on him once again, since he is an excellent illustration of the change in attitudes to the First World War. It was a change from a past-centered to a future-centered perspective, and the latter increasingly asserted itself against any "critical" approach. In Nietzsche's terms, "monumentalist" remembrance of the war would triumph over the "antiquarian" and the "critical".[20]

A cursory glance at the output of war literature in the Weimar Republic will help to contextualize this shift (Figure 1.1). Thomas F. Schneider has shown that, even in the early Weimar period, the public was not exactly weary of the genre: between 1919 and 1923, despite the turmoil of revolution and galloping inflation, an average of 164 such titles were published per annum.[21] In the years from 1925

[19] Thomas F. Schneider, "'Krieg ist Krieg schließlich.' Erich Maria Remarque: Im Westen nichts Neues (1928)," in Thomas F. Schneider/Hans Wagener (eds.), *Von Richthofen bis Remarque. Deutschsprachige Prosa zum 1. Weltkrieg*, Amsterdam 2003, pp. 217–32; here p. 231; and Thomas F. Schneider et al., *Die Autoren und Bücher der deutschsprachigen Literatur zum 1. Weltkrieg 1914–1939*, p. 10.

[20] Remembrance is here understood as social construction of the past. The terminology used here borrows from Jan Assmann's study of "cultural memory," which in turn draws upon Maurice Halbwachs among others. See Jan Assmann, *Cultural Memory and Early Civilization. Writing, Remembrance, and Political Imagination*, New York, NY 2011. There was a similar tendency toward monumentalist history or myth in the Soviet Union under Stalin: see Rosenthal, *New World, New Myth*, esp. pp. 383–87.

[21] Schneider "Einleitung," in Schneider et al., *Die Autoren und Bücher der deutsch sprachigen Literatur zum 1. Weltkrieg 1914–1939*, p. 8) defines the genre as follows: "Texts dealing

to 1928, an annual total of 111 to 160 books appeared on the war. The following year, 1929, the number reached 226, and 1930 marked the peak of the "war literature boom" with a total of 276.[22]

In comparison with the glut of the war years themselves – 1,099 titles in 1915, for example – the war book boom of 1928–1930 does not appear as extraordinary as has previously been thought. Nevertheless, the "boom" of 1929/1930 requires an explanation that goes beyond the often noted "commemoration" of the end of the war.[23] The tenth anniversary of the end of the war may have offered an occasion for increased interest in the war, but it was not the cause of it.

The huge impact of Remarque's book should not be underestimated in this context. As we have seen in relation to the aviation war books of 1914–1918, the Weimar publishing houses were also the driving forces behind a consumerist "entertainment industry." In part, Ullstein Verlag owed the million-plus sales of *All Quiet on the Western Front* to a massive marketing campaign.[24] Smelling a profit, other publishers followed suit and brought out their own war novels and plays, personal accounts, battlefield stories and the like, but the market impulse and increased public demand simply reflected the greater presence in the book world of the controversial question of the meaning and purpose of the war – a controversy that had been simmering all the time just beneath the surface.[25]

The mounting crisis of the liberal order, amid world economic shockwaves that seemed to mock the belief in progress, blighted the widespread hopes of the mid-1920s. The internal stability of the Republic between 1924 and 1929 now appeared to be no more than an "illusion."[26] "Transcendental homelessness," the meaninglessness and contingency of "modern" existence, broke through more powerfully

---

primarily with the war experience of an individual or group of individuals at the front, behind the lines, or 'back home'; it includes novels, plays, and poetry, as well as 'firsthand reports,' memoirs, anthologies, collections of letters, volumes of textual/visual material, 'authentic' diaries, regimental histories, campaign chronicles, etc."

[22] Hans-Harald Müller, *Der Krieg und die Schriftsteller. Der Kriegsroman der Weimarer Republik*, Stuttgart 1986, pp. 9–10.

[23] On the "anniversary-related renaissance" of war literature, see Michael Gollbach, *Die Wiederkehr des Weltkrieges in der Literatur. Zu den Frontromanen der späten Zwanziger Jahre*, Kronberg/Ts. 1978; Ann P. Linder, *Princes of the Trenches. Narrating the German Experience of the First World War*, Columbia, SC 1996; Müller, *Der Krieg und die Schriftsteller*; Karl Prümm, *Die Literatur des Soldatischen Nationalismus der 20er Jahre. Gruppenideologie und Epochenproblematik*, 2 vols., Kronberg/Ts. 1974; Schneider, "Einleitung," pp. 7–14.

[24] See Modris Eksteins, *Rites of Spring.: The Great War and the Birth of the Modern Age*, New York 2000 [repr.], pp. 275–99, esp. 276–77. Cf. Eksteins, "*All Quiet on the Western Front* and the Fate of a War," *Journal of Contemporary History* 15/1980, pp. 345–66; here p. 353.

[25] Schneider, "Die Meute hinter Remarque," p. 143. For Schneider, the disputes over Remarque's book were one of the "central and most consequential literary, cultural, and political controversies in the late Weimar Republic."

[26] See the chapter "The Illusion of Domestic Stability," in Peukert (ed.), *The Weimar Republic*, pp. 207–21.

than ever.[27] *All Quiet on the Western Front* expressed this sense of abandonment, disorientation, and hopelessness:

> Had we returned home in 1916, out of the suffering and strength of our experience we might have unleashed a storm. Now if we go back we will be weary, broken, burnt out, rootless, and without hope. We will not be able to find our way anymore. [...] But perhaps all this that I think is mere melancholy and dismay, which will fly away as the dust, when I stand once more beneath the poplars and listen to the rustling of their leaves. It cannot be that it is gone, the yearning that made our blood unquiet, the unknown, the perplexing, the oncoming things, the thousand faces of the future, the melodies from dreams and from books, the whispers and divinations of women; it cannot be that this has vanished in bombardment, in despair, in brothels.[28]

In 1928–1929 these "melodies from dreams and from books," the cautious hopes for the future, had once more scattered to the winds. Left behind were countless people who felt "weary, broken, burnt out, rootless, and without hope." Remarque was setting down a literary testament to the prevailing mood in society – this was the secret of his success and the reason why spirits became so inflamed.[29] None of his critics on the right could accept that the war had been without meaning or purpose. Conservatives – they who, to quote Nietzsche, "preserve and revere [...] the familiar and the revered of old,"[30] – were incensed that Remarque sat in judgment over the Empire and its representatives. The New Right, on the other hand, turned on him because it was trying to "unleash a storm." It saw the exhaustion and disillusionment he highlighted as symptoms of the "rotten, antiquated" liberal order, which it sought to replace with a "brave new world" that had roots in the past but also a promising future.

Thus, the reason for the "war literature boom" lay in the struggle over the meaning and purpose of the war. Fought out at the level of memory, this became more explosive during the stabilization period of the mid-1920s, finally bursting into violent arguments in 1930 over *All Quiet on the Western Front* and its American film version,[31] in a political situation already radicalized by the world economic crisis. The struggle also opposed the old and the young, the past and the future.

Although the conservative and reactionary forces of the Empire typified by the aged President Hindenburg still held the main levers of power, only a dwindling

---

[27] See Michael Makropoulos, "Crisis and Contingency: Two Categories of the Discourse of Classical Modernity," *Thesis Eleven* 111/2012.

[28] Erich Maria Remarque, *All Quiet on the Western Front*, New York 1987, pp. 294–95.

[29] Cf. Eksteins, "*All Quiet on the Western Front* and the Fate of a War," p. 362.

[30] Nietzsche, "On the Uses and Disadvantages of History," p. 72. On the reception of Remarque's book, see *inter alia* Schneider, "Die Meute hinter Remarque."

[31] On the reception of Lewis Milestone's film of *All Quiet on the Western Front*, see Modris Eksteins, "The Fate of the Film *All Quiet on the Western Front*," *Central European History* 13/1980, pp. 345–66. And, more generally, on Weimar cinema and the First World War, see Bernadette Kester, *Film Front Weimar. Representations of the First World War in German Films of the Weimar Period (1919–1933)*, Amsterdam 2003.

circle of the older generation longed for a return to the *status quo ante*.[32] Especially among anti-Weimar forces in the younger, politically active generations – the "front generation" and the "redundant" or "war youth generation" – the war became a defining factor for visions of the future and revolutionary moods on both the left and the right.[33]

The New Right placed the war casualties in the context of a movement of renewal: that is, a palingenesis of the nation or *Volk*. They thus took on the significance of a *sacrificium*, which could only be followed by rebirth.[34] The discourse and cult of youth that had developed in life reform movements around the turn of the century was linked with a "cult of the fallen soldier,"[35] and it may be assumed that palingenetic ideas received a significant impetus from it. By virtue of its emphasis on youth and innovation, such ideas found great resonance in the war youth generation, which now let loose against everything "old" and "rotten," against political gerontocracy and "the past" in general.[36]

In a labor market that offered little or no prospect of personal advancement, those born between 1900 and 1910 now demanded "the rights of the younger generation"[37] – including, not least, an opportunity to "prove themselves," as their fathers and elder brothers had done in the war. Future-oriented palingenetic ideas, which received an anchorage in the present through the association between the cults of youth and the fallen soldier, gave meaning, direction and purpose to this increasingly violent activism on the part of young people. "The youth," whose actual age was not of central importance, functioned here as a "code for new beginnings and future-oriented activity" and a related "belief in salvation."[38] Rüdiger Graf confirms this: "The idea of youth [...] did not exist alongside other

---

[32] On the continuity of the conservative elites in the Weimar Republic, see esp. Fritz Fischer, *From Kaiserreich to the Third Reich. Elements of Continuity in German History, 1871–1945*, Boston, MA 1986.

[33] On the "four political generations" of the Weimar Republic, see Detlev J. K. Peukert, *The Weimar Republic: The Crisis of Classical Modernity*, London 1991, pp. 14–18.

[34] See inter alia George L. Mosse, *Fallen Soldiers: Reshaping the Memory of the World Wars*, New York, 1990, esp. pp. 70–80. Thus (p. 73): "Youth and death were closely linked in that myth [of the war experience]: youth as symbolic of manhood, virility, and energy, and death as not death at all but sacrifice and resurrection. The differences between generations became part of the mythology of war: the fallen symbolized the triumph of youth."

[35] On the cult of youth, see *inter alia* Barbara Stambolis, *Mythos Jugend. Leitbild und Krisensymptom. Ein Aspekt der politischen Kultur im 20. Jahrhundert*, Schwalbach 2003; Detlev J. K. Peukert, "'Mit uns zieht die neue Zeit...' Jugend zwischen Disziplinierung und Revolte," in August Nitschke/Gerhard A. Ritter/Detlev J. K. Peukert/Rüdiger vom Bruch (eds.), *Jahrhundertwende*, pp. 176–202.

[36] On the war youth generation, see Ulrich Herbert, *Best. Biographische Studienüber Radikalismus, Weltanschauung und Vernunft, 1903–1989*, 3rd edn, Bonn 1996, pp. 42ff.

[37] See Detlev J. K. Peukert, *The Weimar Republic: The Crisis of Classical Modernity*, London 1991, p. 18. Cf. Michael H. Kater, "Generationskonflikt als Entwicklungsfaktor in der NS-Bewegung vor 1933," *Geschichte und Gesellschaft* 11/1985, pp. 217–43.

[38] See Jürgen Reulecke, "Neuer Mensch und neue Männlichkeit. Die 'junge Generation' im ersten Drittel des 20. Jahrhunderts," in *Jahrbuch des Historischen Kollegs 2001*, Munich 2002, pp. 109–38; here p. 111.

political ideologemes but cut across them: 'youth' and 'youthfulness' concretized and materialized the formal conception that the future was already here in the present, that it would grow and realize itself with a natural necessity."[39]

The hopes placed in "youth" were based *inter alia* on the spirit of self-sacrifice paradigmatically associated with Langemarck, in Belgian Flanders, the site of a major battle in August 1917. The young people who died there were the seeds of the new, but also eternal, nation. Or, as Hans Schwarz put it in his speech:

> Youth too has its point of maturity, and, if it must die before reaching fulfilment, all it has left is rebirth in the spirit of those yet to come! [...] For the dead of Langemarck are but the advance guard of an army. Not only have they resisted one world; they have ushered in a new one with their weapons.[40]

In any event, the army following behind the vanguard grew constantly toward the end of the Weimar period and fought for a "new world" and a new order. In this struggle, it also turned its fire against "antiquarian" and "critical" remembrance of the war; the latter, as in Remarque, put the past on trial, while the former sought to honor and preserve it.

In epigonal, reactionary and nostalgic visions, "everything new and evolving is rejected and persecuted."[41] Thus, "antiquarian" reverence for the past went together with laments over lost grandeur, withering denunciation of everything new, and a picture of the future drawn in the darkest colors. Right-wing "youth" had nothing but contempt for such attitudes, as it did also for any condemnation of the past in principle. Opposing both "critical" and "antiquarian" approaches, the New Right placed the war and other "great" past events in the service of the future life of the nation. In an article entitled "Die Reaktion," which appeared on November 1, 1925 in a special issue of the *Stahlhelm, Die Standarte*, Ernst Jünger wrote:

> The soldier at the front cannot be reactionary, because he is a man of reality and wants to learn from history. He cannot be reactionary because, although he recognizes the greatness and splendour of Bismarck's creation, he has a new and greater Germany before his mind's eye, and because in the breaking of the old ties he sees the possibility of greater ties.[42]

The war, which broke the "old ties," appears here as a godsend: Jünger does not regret the fall of the German Empire founded in 1871, but sees it as the prerequisite for a greater Germany to come. A week later he published another article, "Die Tradition," also in *Die Standarte*:

> To be an heir is not the same as being an epigone. And to live within a tradition does not mean to restrict yourself to that tradition. To inherit a house obliges

---

[39] Rüdiger Graf, *Die Zukunft der Weimarer Republik. Krisen und Zukunftsaneignungen in Deutschland 1918–1933*, Munich 2008, p. 238.

[40] Schwarz, *Die Wiedergeburt des heroischen Menschen*, pp. 13–14.

[41] Friedrich Nietzsche, "On the Uses and Disadvantages of History for Life (1874)," in *Untimely Meditations*, Cambridge 1983, p. 74.

[42] Ernst Jünger, "Die Reaktion," in *Politische Publizistik. 1919 bis 1933*, p. 124.

you to maintain that house, but not to make of it a museum in which the household goods of your forefathers are left untouched.[43]

The very term "household goods of your forefathers" is an indication that Jünger – who considered Nietzsche one of his greatest influences – was targeting the "antiquarian" relationship to the past; the same term appears in the parallel passage in Nietzsche's essay on history.[44] In any event, by the "life of tradition" Jünger did not mean the cozy upholding of things past. The same article continues:

> We do not live in museums, but in an active hostile world. It is not living tradition when Old Fritz [Frederick the Great] is painted on every cigarette pack, or when every ashtray or pair of suspenders bears his black, white, and red seal. That is the worst kind of advertising, as most of our parades, commemorations, and anniversaries are just tasteless advertising – cast iron kitsch from which you gain nothing if you go along with it. [...] Do not write "Fridericus" novels but the national novel of our times.[45]

Jünger advanced his radical new nationalism in place of the old black-white-red nationalism. One of the main differences between the two was that, far from regretting the collapse of the old order, the new nationalism regarded its demise as the precondition for the possibility of a new order. The article continues:

> The hero goes under, but it is like the blood red sundown that promises a new and more beautiful morning. We should recall the Great War as a glowing red sunset in which a splendid morning already begins to take shape. We should think of our fallen friends and see their going under as the sign of a completion, the most intense affirmation, of life itself. As if from a disgusting heap of garbage, we should keep away from the shopkeeper's judgment that "all was in vain" – if, that is, we wish to find happiness living in the hall of destiny and flowing in the secret currents of the blood, if we wish to work in a useful and meaningful landscape, rather than vegetate in a time and space where the luck of birth has placed us.[46]

---

[43] Ibid., p. 129.

[44] This applies to the German text and not to the English translation. Nietzsche, "On the Uses and Disadvantages of History for Life," pp. 72–3: "By tending with care that which has existed from of old, he [the 'antiquarian'] wants to preserve for those who shall come into existence after him the conditions under which he himself came into existence – and thus he serves life. The possession of ancestral goods changes its meaning in such a soul: *they* rather possess *it*. The trivial, circumscribed, decaying and obsolete acquire their own dignity and inviolability through the fact that the preserving and revering soul of the antiquarian man has emigrated into them and there made its home."

[45] Jünger, "Die Tradition," p. 130.

[46] Ibid., pp. 126–27.

Jünger's "monumentalist" relationship to the recent past consisted in his seeing the war as "inspiration to imitate or to do better"[47]; only that made it possible to escape the sense of forlornness and lack of foundations that had arisen from the war. For the present and the future to take shape, a purpose had to be wrested from it. Only if the war was understood to herald a "new and more beautiful morning" would it be useful as a "specific against resignation."[48]

Fascists, then, sought to make a present reality out of the "splendid morning" foretold in the "glowing sunset." When the Nazis took power, the Germanist Julius Petersen exclaimed:

> Now tomorrow has become today; an apocalyptic mood has changed into a sense of dawn; the end goal appears in the present field of vision, and all belief in miracles is applied to the active shaping of reality. [...] The entrance of the infinite into the finite creates a foundation into which the anchor of hope can sink.[49]

What Petersen calls "the entrance of the infinite into the finite" may be understood precisely as the replacement of profane time with the sacred time of the "Third Reich." His book – which discusses "the six universal concepts of salvation": "golden age," "kingdom of God," "world empire," "kingdom of the spirit," "springtime of the peoples or pentecost of mankind" and "empire of the people" – ends on an unctuous note:

> The new Reich is planted. The Leader, longed for and foretold, has appeared. His words say that the Third Reich is coming about: it is no longer a dream for which men yearn, but nor is it accomplished fact; it is a task facing Germans in the course of their renewal.[50]

The struggle for tomorrow in today, and for the new German man, was waged by Jünger and other fascists during the Weimar Republic. They sought to place war "in the service of life," seeing themselves as the "men of deeds and power." They believed – to quote Nietzsche again – "that the greatness that once existed was in any event once possible and may thus be possible again."[51] This belief in the renewal of "greatness," the basis of the "monumentalist" relationship to the past, was transferred to the new man, who was seen as a reincarnation or epiphany of an eternal German essence. This too becomes clearer from Nietzsche's key work

---

[47] Nietzsche, "On the Uses and Disadvantages of History for Life," p. 68.

[48] Ibid.

[49] Julius Petersen, *Die Sehnsucht nach dem Dritten Reich in deutscher Sage und Dichtung*, Stuttgart 1934, p. 1. See Jost Hermand, *Der alte Traum vom neuen Reich. Völkische Utopien und Nationalsozialismus*, Weinheim 1955, pp. 205ff. I am grateful to Roger Griffin for the reference to Julius Petersen.

[50] Petersen, *Die Sehnsucht nach dem Dritten Reich*, p. 61.

[51] Nietzsche, "On the Uses and Disadvantages of History for Life," pp. 68, 69.

on history: belief in the new yet eternal man[52] rested upon the idea that "that which in the past was able to expand the concept 'man' and make it more beautiful must exist everlastingly, so as to be able to accomplish this everlastingly."[53]

For Jünger, writing in the *Stahlheim-Jahrbuch* in 1926, the "new type of German man" had already appeared during the war. Since aviation was associated with new departures and the future, this new man had been embodied above all in the pilot, but also in the tank driver or assault troop leader. And, in whatever form or expression, the new man of the First World War was an updated version of the Superman who had the capacity to "expand" human existence and "to make it more beautiful" – a type that had repeatedly appeared in the past and present, and would continue to do so in the future. This was a radically unhistorical way of thinking.

The mythical idea of an eternal return (or return of the eternal) substituted itself for the processual or developmental course of history.[54] It implied "that the great moments in the struggle of the human individual constitute a chain, that this chain unites mankind across the millennia like a range of human mountain peaks, that the summit of such a long-ago moment shall be for me still living, bright, and great."[55] Thus the war acquired a meaning as part of a chain of "great moments" and through its function for present and future "life." Monumentalist remembrance of the war served to "give meaning to the meaningless," which Theodor Lessing described as the function of both history and myth.

Jünger inserted this "monumentalist" reading of the war into his struggle against the present and against "apathetic habit, all that is base and petty," which, "billowing up around all that is great like a heavy breath of the earth, casts itself across the path that greatness has to tread on its way to immortality and retards,

---

[52] The new man of the interwar period is ubiquitous in historical research, but no detailed study has yet been made of this trope and of the change it underwent in fascism. Some investigations, though, within various perspectives, look at the theme of the fascist new man. See Matthew Biro, "The new man as Cyborg. Figures of Technology in Weimar Visual Culture," *New German Critique* 62/1994, pp. 71–110; Peter Fritzsche/Jochen Hellbeck, "The new man in Stalinist Russia and Nazi Germany," in Michael Geyer/Sheila Fitzpatrick (eds.), Beyond Totalitarianism. Stalinism and Nazism Compared, New York, 2009, pp. 302–41; Gentile, "L' 'uomo nuovo' del fascismo"; Alexandra Gerstner (ed.), *Der Neue Mensch. Utopien, Leitbilder und Reformkonzepte zwischen den Weltkriegen*, Frankfurt/Main 2006; Friedrich Wilhelm Graf, "Alter Geist und neuer Mensch. Religiöse Zukunftserwartungen um 1900," in Ute Frevert (ed.), *Das Neue Jahrhundert. Europäische Zeitdiagnosen und Zukunftsentwürfe um 1900*, Göttingen 2000, pp. 185–28; Hüppauf, "Schlachtenmythen"; Mosse, The Image of Man; Reulecke, Neuer Mensch und neue Männlichkeit; Bernd Wedemeyer-Kolwe, "*Der neue Mensch.*" *Körperkultur im Kaiserreich und in der Weimarer Republik*, Würzburg 2004. On the socialist-Soviet new man, see the bibliographic references in Fritzsche and Hellbeck. Also: Günther, Der sozialistische Übermensch; Manfred Hildermeier, "Revolution und Kultur. Der 'neue Mensch' in der frühen Sowjetunion," in *Jahrbuch des Historischen Kollegs 1996*, Munich 1997, pp. 51–67; Hellbeck, Revolution on My Mind; and Rosenthal, New Myth, New World.

[53] Nietzsche, "On the Uses and Disadvantages of History for Life," p. 68.

[54] See Mircea Eliade, The Myth of the Eternal Return. Cosmos and History, Princeton, NJ 2005 [orig. Paris 1949].

[55] Nietzsche, "On the Uses and Disadvantages of History for Life," p. 68.

deceives, suffocates, and stifles it."[56] This struggle concerned not only liberalism and Marxism but also the black-red-white nationalism that Jünger saw as an equally restrictive force of "apathetic habit." The greatest danger was

> that we become all too fixed and static, not understanding how to insert ourselves enough into the dynamic forces of the age. For example, although we can understand it, we should not go along with the stubborn insistence that everything since 1918 is mere nonsense – which is similar to Louis XVIII's dismissal of everything since the Revolution as nonexistent.[57]

The new "human type" that "returned from the war" did not remain stuck in the old ways. It had understood,

> notwithstanding all its fighting capacity, how to plug its insights into the great currents of the age, so that they were capable of actually having an effect. Our present life is a continuation of the war with other means, and we everywhere come across natures that are able to transfer the impetus and energy of the war to other forms. Those at the head of the Freikorps, for example, were exceptionally modern men who grasped the problems of their times. [...] They [the "old frontline soldiers" whom Jünger was here addressing – FE] see in our fighter pilots how the old spirit expresses itself in new forms.[58]

This theme of the "old spirit" repeatedly expressing itself in "new forms" and marking a return of former greatness involved a cyclical notion of time and mythical mode of thought; it incorporated the war and the fallen soldiers into a meaningful narrative that grounded a temporal order. As early as the preface to *Das Wäldchen 125*, Jünger wrote:

> The spirit of those days [in August 1914] will always be a highpoint and a model for us. [...] It should not be attacked with the question: "What was the point of all that?," since it displays a greatness that outgrew boundaries and had certain purposes within it. An immortal deed is absolute and does not depend on its outcome; it is an eternal source of strength for a people. We survivors will always be proud to have been part of such a youth.[59]

The Remarquean interpretation of the war as something without meaning or purpose was felt as a threat by the former *Frontsoldaten* and the war youth generation. They vehemently combated "critical" recollection of the war, on the grounds that it endangered the coherence of their own life histories and conjured up the specter of contingency. If the war had been pointless, the sense of their self-sacrifice would be lost and the temporal order of their lives would collapse. The fascist

---

[56] Ibid.

[57] Ernst Jünger, "Der neue Typ des deutschen Menschen," in *Politische Publizistik. 1919 bis 1933*, pp. 167–72; here p. 170.

[58] Ibid., pp. 171–72.

[59] Ernst Jünger, *Das Wäldchen 125. Eine Chronik aus den Grabenkämpfen 1918*, Berlin 1925, pp. vii–viii.

palingenetic narrative therefore also established a personal-existential time that helped people to get over the many breaks in their lives and created a meaningful unity out of them. Experiential space was thereby linked up with their horizon of expectation, and their seemingly useless and meaningless past was endowed with a meaning for the future. "Monumentalist" recollection of the war served precisely this purpose. Indeed, for Jünger, "giving meaning to the war" became a "sacred duty" to the fallen, and more generally to "youth" and the future:

> We who have a responsibility, not to erase memory, but to preserve and uti-lize it must strive to make it [the war] our property in this respect too [as a mental or spiritual problem]. To give meaning to that which lower-level per-ception may regard as an absurdity or an expression of human imperfection is a sacred duty to the fallen and to the coming generation; they it is who will have to build on work already done, the inner unity of which they must recognize in order to set about their task with real conviction. For one day it will be up to complete what we were unable to complete. They will be able proudly to enter into their inheritance if the wonderful eternal core of these times – that which is absolutely German – has survived the fog of everyday vulgarity.[60]

This passage displays several themes of the palingenetic thinking that was then rife in Germany. "Youth" has a responsibility to "complete" the "work" and to take up its "inheritance." The future, whose face was already visible in the war, stands under the sign of an "absolutely German eternal core." This utopian renewal of the eternal is concentrated in the vision of a "Third Reich."[61] According to Julius Petersen, who is here following Oswald Spengler and ultimately Nietzsche, the "Third Reich" is an "eternal dawn" that directs "the shafts of yearning to the other shore."[62] This "other shore" was a hybrid that comprised novelty, youth and the future as well as the

---

[60] Ibid.

[61] On the trope of a "Third Reich," see Claus-Ekkehard Bärsch, *Die politische Religion des Nationalsozialismus. Die religiöse Dimension der NS-Ideologie in den Schriften von Dietrich Eckart, Joseph Goebbels, Alfred Rosenberg und Adolf Hitler*, Munich 1998, pp. 45–135. Bärsch notes (on p. 133) that the concept derives from an interpretation of the Book of Revelation. He con-tinues: "The 'Third Reich' is an empire of the future, whose main feature is the 'redemption' or 'salvation' of the Germans. Present and future are divided by a qualitative leap, which is preceded by a time of crisis and even catastrophe. Hardship and misery are a necessary transitional stage on the way to future salvation. A struggle must take place in order to achieve salvation – not any old conflict but one interpreted, within a substantive dualism, as a struggle against evil."

[62] Petersen, *Die Sehnsucht nach dem Dritten Reich*, p. 1. In *The Decline of the West*, the spirit of antiquity wants to know the future, whereas the Western spirit wants to build it. In this connection, Spengler describes the Third Reich as "the Germanic ideal," "an eternal morn-ing to which every great man from Dante to Nietzsche and Ibsen – arrows of longing for the other shore, as Zarathustra says – has linked his life" (Oswald Spengler, *The Decline of the West*, vol. 1, New York 2000 [orig. 1926], p. 363); translation modified. The "arrows of longing for the other shore" refer to "Zarathustra's Prolog" that presents the transition

"eternal core" already revealed in the past. In so far as fascism produced this nexus between the future and an eternity including the past, it was capable both of satisfying the need for awakening and "revolution" and of stilling the yearning for roots.

It is useful to conceive of fascism as a syncretic, variegated ideology, which in fact united contradictory "theories" and awakened contradictory hopes. Quite different people found their respective desires and imaginings reflected in the mythical fascist models for the ordering of time; the common element was a pressure to overcome the present by helping to bring about a different future.[63] It was necessary to escape the very past on which the decadent present rested, a past conceived as a history of decline, and to posit a new beginning in which the "eternal core" that had shown its face in the war would be renewed. In this respect, the "conservatism" of Jünger the "conservative revolutionary" would be understandable only in Moeller van den Bruck's definition of the term: "It is reactionary," he wrote, "to seek a political way out only where there was a historical ending. It is conservative to posit a beginning over and over again."[64]

The New Right – that is, German fascists who, though not necessarily seeking salvation in the NSDAP, had a worldview compatible with National Socialism – certainly did not seek a way out "where there was a historical ending." They sought a new beginning. And that beginning, as we have seen in the case of Ernst Jünger, was the war. It was there that the longed-for break with the past had taken place. Instead of the present, which was in a dual sense profane time, there would appear a *sacred* time both new and eternal. Mircea Eliade sees in such a regenerative order of time an "archaic conception" that offered a means of tolerating the "increasingly powerful pressure of history."[65] People sought to defend themselves against history by abolishing it through a "regeneration of time." This made it possible to endow historical events with an extra-historical – in Nietzsche's terms, "suprahistorical" – or even mythical meaning. In so far as he located in war the birth of the new (or regenerated) man and the new Germany, Jünger bestowed such a suprahistorical meaning on the war and its attendant sacrifice, suffering and death. War became the prelude to a new age and a different future. For the Futurists too, war was the source of renewal. But above all it provided their longed-for opportunity to abolish history.

The fascist awakening: abolishing history from the air

The Italian Futurists were the most radical proponents of a destruction of the past in order to open up a mythical future.[66] Like Jünger, though earlier than him, they

---

from the human to the superhuman: Friedrich Nietzsche, *Thus Spake Zarathustra*, trans. by Adrian del Caro, New York, NY 2006, p. 7.

[63] Here and on the following, see Griffin, "'I Am No Longer Human. I Am a Titan. A God!' The Fascist Quest to Regenerate Time."

[64] Arthur Moeller van den Bruck, *Das Dritte Reich*, Berlin 1926[2], p. 217.

[65] Eliade, *The Myth of the Eternal Return*, p. 141.

[66] Along with its head, Filippo Tommaso Marinetti (1876–1944), the first generation of the Futurist movement included the graphic artists Giacomo Balla (1871–1958), Umberto Boccioni (1882–1916), Carlo Carrà (1881–1966), and Gino Severini (1883–1966); the poets

glorified war as the seed and herald of the future, as "the world's only hygiene" (i.e. cathartic force).[67] Even before the lights went out in Europe, the Futurists made known their contempt for the "museumization" of the present and declared war on historicism and "passéism," the "antiquarian" cult of the past. They saw the destruction of history – that is, of profane time and historicity *per se* – as a precondition if "tomorrow" was to "become today." And because the aviator embodied tomorrow in today, they proclaimed him to be the paradigm worthy of veneration, the new man of the eagerly awaited tomorrow.

Before we look more closely at the metaphorical function of aviation in Futurism, we need to justify the attention given to the movement in this study. In the period following the Second World War, it was discredited as the art of the Fascist regime in Italy.[68] Only in the 1960s did it become the object of fresh attention, albeit at first mainly in the context of art history, where some argued that it was impossible to understand modernism unless one appreciated the paradigmatic role of Futurism as an artistic *avant-garde*. This led to attempts to save Futurist art by highlighting the anarchist and leftwing elements in the movement.

This ambiguity of Futurism, and above all of the study of it, was already reflected in the title of Günther Berghaus's book *Between Anarchist Rebellion and Fascist Reaction* (1996), one focus of which was Futurist resistance to attempts by the regime to incorporate the movement.[69] Meanwhile, the publication of Marinetti's diaries in 1987 had shown that his public declarations in support of Mussolini had not always been in harmony with his "private" assertions. Further attempts to "rescue" Futurism downplayed its involvement in politics and highlighted its "real" aesthetic objectives.[70] But this sharp distinction between political and aesthetic Futurism was doomed to failure, since the *avant-gardes* of the early twentieth century had been concerned precisely to overcome such a separation of art from life.[71] Through the aestheticization of life – including the political sphere – and

---

Libero Altomare (1883–1962) and Paolo Buzzi (1874–1956); the composer Luigi Russolo (1885–1947); and the architect Antonio Sant'Elia (1888–1916).

[67] See Point Nine of the *Manifesto of Futurism*, which first appeared in *Le Figaro* on February 20, 1909: Marinetti, *Fondazione e Manifesto del Futurismo*, p. 11 ["The Founding and Manifesto of Futurism," in R. W. Flint/Marinetti (eds.), *Selected Writings*, New York, NY 1972, pp. 185–90; here p. 187].

[68] See Emilio Gentile, "Political Futurism and the Myth of the Italian Revolution," in Günter Berghaus (ed.), *International Futurism in Arts and Literature*, Berlin 2000, pp. 1–14.

[69] Günter Berghaus, *Futurism and Politics. Between Anarchist Rebellion and Fascist Reaction, 1909–1944*, Providence, RI 1996.

[70] Claudia Salaris, *Artecrazia. L'avanguardia futurista negli anni del fascismo*, Florence 1992. See also Gentile, *Political Futurism and the Myth of the Italian Revolution*.

[71] See Cornelia Klinger, "Die Utopie der Versöhnung von Kunst und Leben. Die Transformation einer Idee im 20. Jahrhundert. Vom Staat als Kunstwerk zum life-style des Individuums," in Cornelia Klinger/Wolfgang Müller-Funk (eds.), *Das Jahrhundert der Avantgarden*, Munich 2004, pp. 211–45. Klinger writes (p. 212): "From the beginning of the modernization process, a plea is made for art 'to make poetry vibrant and social, and life and society poetic' – hence to bind together what is separate in reality. Complaints about the split between art and life, or ambitions to reconcile them, should be read as metaphors

through a vitalization of art, the *avant-gardes* sought to close the "tearing of the world into two" that Heine attributed to the spirit of modernity and lamented in the period before the 1848 Revolution.[72]

Not long ago Emilio Gentile remarked, with a touch of sarcasm, that it was perhaps a mistake to accord too much importance to Futurist politics, but that the first to make this mistake were the Futurists themselves.[73] Ever since the 1970s Gentile has been pointing to the affinity between Fascism and Futurism. The importance of political Futurism for Italian Fascism, he argues, was not just a question of its participation in the "cult of the lictor"; many Futurists were also among the founders of the Fascist movement, and some of them – not least the mastermind of the Futurists, Filippo Tommaso Marinetti – remained loyal to it even in the days of the "Republic of Salò," the Nazi satrapy that lasted from September 13, 1943 to April 25, 1945.[74] In a private conversation with his biographer Yvon de Begnac, Mussolini himself is even supposed to have said: "I formally declare that without Futurism there would never have been a Fascist revolution."[75]

Common to Futurism and Fascism, as Gentile has shown, were attempts not only to aestheticize politics but also to sacralize it. This involved a new form of political modernism, which counterposed a different modernity to that of the liberal order.[76] Futurism like Fascism aimed to overcome "the tearing of the world into two," by creating a myth of community. Gentile is therefore right when he argues:

> The Futurists were restless fascists and disagreed with some of the regime's political and cultural decisions. None of them, however, ever questioned the fundamental motifs of the totalitarian state: the primacy of mythical thought, the vitalist realism, the mystical exaltation of national community, the heroic and warlike pedagogy, the imperial ambitions, or the myth of the Italian nation as the vanguard of a new society. The Futurists were neither deceived nor misled by fascism; they were fascinated by its appeal for the total mobilization of culture to regenerate Italians in a religious cult of the nation and to construct a new society that would leave its mark upon the future in the style

---

for the process of modernity as a process of differentiation, for the resulting deficits, or for the endeavor to offset or overcome them."

[72] Heinrich Heine, "The Baths of Lucca," in Heine, *Travel Pictures*, New York, NY 2008, p. 107.

[73] Emilio Gentile: *"La nostra sfida alle stelle." Futuristi in Politica*, Rome 2009, p. 8.

[74] Emilio Gentile, *The Struggle for Modernity. Nationalism, Futurism, and Fascism*, Westport, CT 2003; *Politics as Religion*, Princeton, NJ 2006; and *The Origins of Fascist Ideology, 1918–1925*, New York, 2005 [repr.].

[75] Yvon de Begnac, *Taccuini Mussolini*, quoted in Gentile, *The Struggle for Modernity*, pp. 41 and 66.

[76] See Gentile, *The Struggle for Modernity*. Following Marshall Berman, he defines political modernism as follows (p. 58): "By political modernism [...] I mean those political ideologies that arose in connection with modernization, ideologies that seek to render human beings capable of mastering the processes of modernization in order not be overwhelmed by the 'vortex of modernity,' giving them 'the power to change the world that is changing them, to make their own way within the vortex and to make it their own.'"

of "Italian modernity." No Futurist believed that Fascism harbored the intention of realizing a world of reason, freedom, equality, or peace.[77]

The political oscillations only hinted at here have left their mark on the plethora of research literature on the Futurist *avant-garde*.[78] On the one hand, they high-light the intellectual barriers that long hindered an interpretation of Fascism as "modern" and "modernist," for it seemed that Futurism, being self-evidently "modern" or "modernist," could only lie outside the sphere of "reactionary" Fascism.[79] On the other hand, they point to the difficulties in principle of dealing with generic concepts. In any event, the ambiguity concerning the relationship between Futurism and fascism, which has also affected the political categorization of Ernst Jünger, seems more of a gain than a loss and certainly should not rule out generalization. Generic concepts serve to direct our attention to common features; this naturally pushes differences into the background, but without negating them.

Futurism, like fascism, embraced a complex, partly contradictory, jumble of thought patterns, interpretive models and concepts, motives for action, desires, longings and fears – and individual "Futurists" appropriated, interpreted and objectified these in a variety of ways.[80] Thus, although there may be reasons not to subsume particular Futurists, or Ernst Jünger, under the term fascism,[81] there is still a "kinship" in the longing for a different, alternative, mythical modernity. National palingenesis, albeit conceived of in various ways, is at the center of this mythical order, and rebirth or renewal of the nation went together with the destruction of profane time or history. This was true of Ugo Tommei, for instance, one of the lesser-known "angry young men" who gathered under the banner of Futurism. In May 1917 he published an article "Aboliamo la storia! [Let Us Abolish History!]" in the Florentine journal *L'Italia futurista*, the "organ of the Futurists at the front" – a programmatic goal for which, in the eyes of the Futurists, the war seemed to provide the best chance of success.[82] Destruction of the past and history thus went hand in hand with national renewal, the birth of modern Italy and the New Italian.[83] We now need to look more closely at Futurist antihistoricism and

---

[77] Ibid., p. 79.

[78] There is a detailed bibliography in Günter Berghaus, "Bibliography. A Futurist Reference Shelf," in Berghaus (ed.), *International Futurism in Arts and Literature*, pp. 487–596.

[79] Emilio Gentile was one of those who exploded this dichotomy of "reactionary" and "mod-ern": see his *The Origins of Fascist Ideology*, first published in Italian in 1975.

[80] See the section on the central categories of analysis in the Introduction above.

[81] Archival research is therefore necessary to verify, and to periodize, the applicability of gen-eral statements both to individuals who participated in the first, "heroic" phase of Futurism (from 1909 to 1916) and to those who were active in the second phase up to Marinetti's death in 1944.

[82] Ugo Tommei, "Aboliamo la storia," *L'Italia futurista*, May 7, 1917, quoted from Gentile, *"La nostra sfida alle stelle,"* pp. 25–6. See also Emilio Gentile, "Il futurismo e la politica. Dal nazionalismo modernista al fascismo (1909–1920)," in Renzo De Felice (ed.), *Futurismo, cultura e politica*, Turin 1988, pp. 105–59, esp. pp. 112ff.

[83] See Gentile, "Il futurismo e la politica"; and Griffin, *The Nature of Fascism*.

to explain its vision of the abolition of history through war. We shall then turn to the utopia of a modern Italy peopled with mechanically enhanced aviators.

Never before has the present seemed so separate from the genetic chain of the past.

At the head of the Futurist movement stood Filippo Tommaso Marinetti. He was born on December 22, 1876 in Alexandria, Egypt, where his father worked as a lawyer for foreign companies.[84] Before he went to Paris in 1893 on account of his baccalaureate, Marinetti was educated at the Saint François Xavier College, whose French Jesuit teachers disapproved of his literary inclinations. He then studied law in Pavia and Genoa, but poetry remained his passion in life. In 1907 he was still composing verse in the spirit of Symbolism, and after his father's death that year he disposed of a large inheritance that allowed him to devote himself entirely to artistic activity. This included the publication of a journal, *Poesia*, which he had founded in 1905 with a likeminded group of poets.

By 1909 Marinetti had made a name for himself as a poet, but it was his publication of the "Futurist Manifesto" in *Le Figaro*, on February 20, 1909, that really caused a stir. He fueled this interest in masterly fashion, by submitting the Manifesto simultaneously to a number of international papers.[85] This already showed that perhaps his greatest talent was in causing scandals and finding a place in the limelight for himself and his disparate group.

Marinetti had operated in anarcho-syndicalist and socialist circles, but his political visions centered on the dream of a Greater Italy and an Italian revolution.[86] This *italianismo*, which was the common denominator of "modernist nationalism" in general as well as the basis of Futurism, fed on a sense of inferiority *vis-à-vis* more "successful" nations such as France, and on dissatisfaction with the outcome of the Risorgimento and the ruling caste of liberal politicians.[87] The dream of a greater "true" Italy was a spur to irredentism, but also to the First World War interventionism of the Futurist movement taking shape around Marinetti. The struggle against the Austro-Hungarian monarchy was supposed to enlarge Italy with newly "redeemed" territories. Participation in the war would bring about a national renewal and inaugurate Italy's rise to become a major European power.[88]

---

[84] On the following, see Berghaus, *Futurism and Politics*, pp. 15–46; Peter Demetz, *Worte in Freiheit. Der italienische Futurismus und die deutsche literarische Avantgarde 1912–1934*, Munich 1990, pp. 42ff.; Marja Härmänmaa, *Un patriota che sfidò la decadenza. F. T. Marinetti e l'idea dell'uomo nuovo fascista 1929–1944*, Helsinki 2000, pp. 31–2; Christine Poggi, *Inventing Futurism. The Art and Politics of Artificial Optimism*, Princeton, NJ 2009, pp. 11ff.; and Hansgeorg Schmidt-Bergmann, *Futurismus. Geschichte, Ästhetik, Dokumente*, Reinbek bei Hamburg 1993, pp. 52–72. On Marinetti's biography in general, see Gino Agnese, *Marinetti. Una vita esplosiva*, Milan 1990.

[85] See Poggi, *Inventing Futurism*, pp. 4–5.

[86] Here and on the following, see Gentile, "Il futurismo e la politica," pp. 10–24; and Emilio Gentile, *La Grande Italia. Il mito della nazione nel XX secolo*, Rome 2006, pp. 95–108.

[87] See Berghaus, *Futurism and Politics*, pp. 6–7.

[88] See Gentile, *The Struggle for Modernity*, p. 58.

The Futurists conducted irredentist propaganda right from the first *serate* that launched the movement in 1910. Like D'Annunzio, they agitated in 1915 for Italy to enter the war on the side of the Entente, and many of their number volunteered to fight when Rome declared war on the Habsburg monarchy in May of that year. Marinetti, their 38-year-old leader, was second to none in this respect, having already celebrated Italy's involvement in the Libyan conflict of 1911.[89] But the new world war, which had already been raging for three-quarters of a year, was of a different order from that colonial expedition against the Ottoman Empire. It fulfilled the longing for a violent, cathartic force that had marked the development of Futurism in the first place.

As is well known, in his Futurist Manifesto, Marinetti sang "the love of danger" and praised war as "the world's only hygiene."[90] But the hopes and fancies of Marinetti and his fellow-combatants went beyond the "purifying storm" for which many of his contemporaries yearned. The Futurists wanted to see a veritable shaking of the foundations, a Nietzschean revaluation of all values. Their verbal cannonades were aimed especially at "passéism" or "antiquarian" history; they even wanted "to destroy museums, libraries, academies of every kind." These were cemeteries that should be visited, if at all, only on All Souls Day. This rage against places of remembrance in the first Futurist Manifesto expressed Marinetti's contempt for the "historical disease" and the cult of the past that produced only epigones:

> We want to free this land from its smelly gangrene of professors, archeologists, ciceroni, and antiquarians. [...] We mean to free her from the numberless museums that cover her like so many graveyards. [...] In truth, I tell you that daily visits to museums, libraries, and academies (cemeteries of empty exertions, calvaries of crucified dreams, registries of aborted beginnings!) is, for artists, as damaging as the prolonged supervision by parents of certain young people. [...] But we want no part of it, the past, we the young and strong *Futurists*![91]

To want no part of the past, to abolish it, to long for tomorrow, to revere and experience dynamism and speed: such were the demands of the Futurists, Italy's most radical antihistoricist revolutionaries. The paradox of the longed-for temporality of mythical modernity becomes particularly clear from a consideration of Futurism.[92] It is true that speed, revolution and dynamism were central

---

[89] See the poetic war reportage that he published in 1912: Filippo T. Marinetti, *La battaglia di Tripoli*, Milan 1912. Cf. Isnenghi, *Il mito della Grande Guerra*, pp. 25–30.

[90] Marinetti, "The Founding and Manifesto of Futurism," pp. 41–2. In the "Technical Manifesto of Futurist Literature" (1912), the Futurists called for the abolition of punctuation. So as not to falsify the character of Futurist writings, translations – where the sense permitted it – followed the lack of punctuation in the original. Often, Futurist texts used a set of three period marks at the end of a sentence ("..."): not to mark an omission, however, but as a stylistic device.

[91] Ibid., pp. 42–3.

[92] See Manfred Hinz, *Die Zukunft der Katastrophe. Mythische und rationalistische Geschichtstheorie im italienischen Futurismus*, Berlin 1985, esp. pp. 43–56. On p. 43 he writes: "In the pair of

to the Futurist vision, but constant change was also transfigured as an eternal absolute into which people longed to flee, and through which they planned to suspend time. Thus, the first Manifesto of Futurism proclaims: "We stand on the last promontory of the centuries! ... Why should we look back, when what we want is to break down the mysterious doors of the Impossible! Time and space died yesterday. We already live in the absolute, because we have created eternal, omnipresent speed."[93] In paying homage to a speed shorn of the dimensions of time and space, the Futurists sought to kick open the famous Huxleyan "doors of perception" leading to the Impossible.[94] The timeless speed that they chose as the absolute, eternal, omnipresent principle served to establish a sacred time of religious festival.

These demands of the Futurist trailblazers should by no means be understood as mere fantasies of an artistic *avant-garde*. To renew art by dumping the ballast of past centuries was not their only concern; Futurist "modernism," like the modernism of all the *avant-garde* movements of the early twentieth century, was more fundamental than that.[95] In rejecting the past styles, ages and traditions out of which it had issued, Futurism was declaring war on its own

---

opposites comprising Futurism and 'passéism,' Futurism is the progressive element, but it is also the one that first erected this opposition in order to legitimize itself as a forward-striving force. [...] The Futurist *avant-garde* necessarily places itself at the peak of the time axis and claims to be able to delineate the course of history, but it also declares itself to be beyond history altogether by virtue of a new primitiveness; the former is supposed to be its historically progressive consciousness, the latter its suprahistorical *epochē*-consciousness. In its polemic against a frozen 'passéist' culture Futurism seems to want to regain a temporal perspective, but in its own programmatic declarations it presents itself as establishing a new condition outside history. [...] The Futurist *epochē*-consciousness tends to neutralize its *avant-garde* consciousness. The *avant-garde* knows it is at the top of the time pyramid and says that it will not grow anymore; even the future is past vis-à-vis the new state of the Absolute. Futurism thereby loses any relationship to it: to stand on the peak of history means to stand outside and above it. Futurism therefore cannot be conceived as a unique novelty – Merinetti [sic] was right to differentiate it sharply from fashion – but makes itself the law of motion of constant replacement. The new age of Futurism no longer has a history; its unchanging static equilibrium is at once preserved and ideologically disguised by the mastery of space in speed. The view of history as a continent that cannot be left behind [...] presupposes that the past too appears as not subject to change. Only when no vector points any more from the past into the future can this spatialization of time be successful."

[93] Ibid., p. 187.

[94] Aldous Huxley's essay *The Doors of Perception* (London 1954) alluded to a line in Blake's *The Marriage of Heaven and Hell*: "If the doors of perception were cleansed, everything would appear to man as it is, infinite." See William Blake, *The Marriage of Heaven and Hell*, with an introduction and commentary by Geoffrey Keynes, London 1975, pp. xxii.

[95] On the theory and practice of the *avant-garde*, see esp. Peter Bürger, *Theory of the Avant-Garde*, Minneapolis, MN 1984; and Wolfgang Asholt/Walter Fähnders (eds.), *Der Blick vom Wolkenkratzer. Avantgarde – Avantgardekritik – Avantgardeforschung*, Amsterdam 2000. On the *avant-garde* and the question of temporality, see esp. Bernd Hüppauf, "Das Unzeitgemäße der Avantgarden. Die Zeit, Avantgarden und die Gegenwart," in Asholt/Fähnders (eds.), *Der Blick vom Wolkenkratzer*, pp. 547–82.

epoch. In place of "passéist" art were supposed to appear a new poetry, *parole in libertà* (liberated words), and later *aeropoesia* (aerial poetry), as well as a new painting, sculpture and architecture, even a new clothing and new foods.[96] The Futurists sought to bring about a radical renewal, by merging life with art, and by plugging life into the temporality peculiar to art and religion. The New that they wanted to put in place of the Old had to be definitively new. Hence the palingenetic project that the Futurists pursued was all-embracing. It envisaged the creation of a New Italy and a new man through a truly anthropological revolution.[97]

With such a comprehensive vision of renewal, the Futurists left no stone unturned in bringing their radical ideas to the masses. Their manifestos were printed in many newspapers, and their posters put up in city streets or thrown out of car windows to passers-by. And then there were all the exhibitions and the notorious *serate futuriste*, such as those held in Austro-Hungarian Trieste on January 12, 1910 and at Milan's Teatro Lirico on February 15 of the same year.[98] These evenings often ended in uproar; after all the Futurists wound up their audience not only with undisguised irredentism but also with strange-sounding verse and unsettling appeals. The real effect of these was incidental, since their significance lay in their anticipation and overdrawn articulation of palingenetic longing and the sense of a new departure.

The Futurists and their works were the shrillest, most extreme expression of the wish for a different modernity – including a new temporality – that was then widespread among the bourgeoisie. A sacralized nation also stood at the core of this utopian new order. Earlier, in the first decade of the century, the Turin sociologist Mario Morasso preached a new nationalism in the periodicals *Il Marzocco* and *Il Regno*, as well as in two books published in 1905: *L'imperialismo nel secolo XX* [Imperialism in the Twentieth Century] and *La nuova arma* [The New Weapon]. Similarly, the Associazione Nationalista Italiana (ANI),[99] founded in 1910 by Enrico Corradini (the publisher of the above two journals), and the Florentine intellectuals around the reviews *Leonardo*, *La Voce* and *Lacerba*, yearned for a renewal of the nation.[100] As Emilio Gentile noted, these "modernist nationalists" varied in their conceptions of the "Italian revolution" and the ways in which national renewal should be achieved. Yet they shared the idea that Italy had an important role to play in the twentieth century, in its path to national renewal.

---

[96] For an overview of the movement's variegated art see: Vivien Greene (ed.), *Italian Futurism 1909–1944. Reconstructing the Universe*, New York 2014.

[97] See Gentile, *"La nostra sfida alle stelle,"* p. 4.

[98] For an account of the *serata futurista* at the Teatro Lirico in Milan, see Berghaus, *Futurism and Politics*, pp. 48–52; and Schmidt-Bergmann, *Futurismus*, pp. 65ff.

[99] On the ANI see Alexander J. De Grand, *The Italian Nationalist Association and the Rise of Fascism in Italy*, Lincoln, NE 1978.

[100] On the Florentine *avant-garde* around Giovanni Papini, Giuseppe Prezzolini, and Ardengo Soffici and *Leonardo*, *La Voce*, and *Lacerba*, see Walter L. Adamson, *Avant-Garde Florence. From Modernism to Fascism*, Cambridge, MA 1993.

This core of "modernist nationalism" would also form the basis of Futurism and Fascism.[101]

Some of the most important differences among the variants of "modernist nationalism" concerned their relationship to history. In contrast to the ANI around Enrico Corradini, and the later Fascist cult of the renewal of ancient Rome, the Futurists sought to legitimize their nationalism without resorting to past greatness or glorious traditions.[102] The mythical age for which the Futurists yearned was quite empty of historical "ballast." In the words of Umberto Boccioni, author of the *Manifesto of Futurist Painters* (February 1910) and the *Technical Manifesto of Futurist Painting* (April 1910): "We modern Italians have no past."[103] And Marinetti wrote in an open letter to the Belgian painter Aimé Felix Mac Del Marle, which was published on August 15,1913 in the new review *Lacerba* founded by the heads of *La Voce*, Ardengo Soffici and Giovanni Papini:

> Since a famous past was stifling Italy and an infinitely glorious future was brewing in its breast, it is precisely here, beneath our overly voluptuous skies, that Futurist energy was born four years ago, and is now being organized and channeled, and finding in us its driving forces, its apparatuses of enlightenment and propaganda. Italy needed Futurism more urgently than any other country, because it was dying of passéism. We are its fortuitous doctors.[104]

So, Marinetti understood Futurism as a cure for the oppressive burden of the past; the Futurists were the fortuitous medium through which the treatment would be administered. Despite its ambition to have no history, however, the Futurist cure was always associated with a vehement nationalism. The open letter continues:

> We therefore admit to an ultraviolent anticlerical and antisocialist nationalism, which has its roots in the inexhaustible vitality of Italian blood. Our Futurist nationalism therefore fights with fierce determination against ancestor worship, which does not strengthen the race but makes it bloodless and pitiful in its decay.[105]

---

[101] Gentile, "Political Futurism and the Myth of the Italian Revolution," p. 4. See also Emilio Gentile, "The Conquest of Modernity. From Modernist Nationalism to Fascism," *Modernism/Modernity* 1/1994, pp. 59ff. On p. 59 he writes: "Prior to World War I these movements gave birth to a generational revolt, conducted under the banner of the creative role assigned to youth, that involved a radical contestation of parliamentary government, one in which divergent visions of modernity, even though posed as alternatives to one another, were aligned in a common front against rationalist, liberal, and bourgeois modernity."

[102] See Emilio Gentile, "Il futurismo e la politica," p. 112.

[103] Umberto Boccioni, *Opere complete*, quoted from Gentile, "Il futurismo e la politica," p. 112.

[104] Filippo T. Marinetti, "Lettera aperta al futurista Mac Delmarle" [orig. 1913], in Marinetti, *Teoria e invenzione futurista*, ed. by Luciano De Maria, Milan 1988, pp. 91–4; here p. 92. Further quotations from this edition will be referenced as TIF.

[105] Ibid.

The Italy of the Futurists was evidently rooted only in a sacred, absolute time, with no history or progression. The paradox seems insuperable: what does a nation without history consist of? How can a nation rooted in "the inexhaustible vitality of Italian blood" refuse ancestor worship? All the more vehement, however, was the Futurists' call for a new beginning or awakening, their direction of the will toward the future. They sought to express this not only in their art, but also – from the beginning – in a political program that appeared as a declaration of war on the bourgeoisie, on "yesterday's world" with its security and resulting *ennui*.

The Futurists did also preach progress, of course, but they understood this as progress beyond the bourgeois world and the past in which they themselves had originated. This call for the bond linking past, present and future to be broken, and for the course of history to be violently interrupted, makes them paradigmatic representatives of the temporality peculiar to modernism, in which the new is preferred to the old, and the horizon of expectation to the space of experience.[106] For the Futurists, the aim was to complete this devaluation of the space of experience and to widen the fissure with the past into a full-scale rift. A *tabula rasa* was to be created once and for all. Or, as Marinetti wrote in his "Futurist preface" to Gian Pier Lucini's verse collection *Revolverate* (1909):

> The hour is auspicious. Men will become mythical again! The bowels of the earth will disgorge the monster of speed. Scrap iron will look for sudden fire. People will honor athletes – those who race in stadiums and the skies above. [...] Never before has the present seemed so separate from the genetic chain of the past, son of itself and magnificent father of future powers.[107]

The aspiration of Marinetti and his earliest comrades for self-generation, for *creatio ex nihilo*, was the expression of an antihistoricist longing for a world without history. It was to the "son of itself" that Marinetti dedicated his "African novel" *Mafarka le futuriste* [Mafarka the Futurist] (1909), which interwove attacks on "passéism" and historicity with mythical images of a flying Superman.

Marinetti completed the original French manuscript of *Mafarka le futuriste* in 1908, before the publication of the founding manifesto of Futurism.[108] Situated in a continent shrouded in legend, the book may be understood as a trip into Conrad's

---

[106] See the remarks on modernism in the Introduction above, and in Peter Osborne, *The Politics of Time: Modernity and the Avant-garde*, London 1995.

[107] Filippo T. Marinetti, "Prefazione futurista a 'Revolverate' di Gian Pietro Lucini" [orig. 1909], in TIF, pp. 27–33; here p. 28.

[108] On the context and interpretation of *Mafarka the Futurist*, see Luci Ballerini, "Introduzione," in Filippo T. Marinetti, *Mafarka il futurista*, Milan 2003, pp. vii–xlviii; Hansgeorg Schmidt-Bergmann, "*Mafarka le Futuriste* – F. T. Marinetti's literarische Konstruktion des futuristischen Heroismus," in Filippo T. Marinetti, *Mafarka der Futurist. Afrikanischer Roman*, Munich 2004, pp. 261–84; and Barbara Spackman, "Mafarka and Son. Marinetti's Homophobic Economics," *Modernism/Modernity* 1/1994, pp. 89–107. For an English translation of the novel, see Filippo Tommaso Marinetti, *Mafarka the Futurist. An African Novel*, London, 1998.

"heart of darkness" or Jean Paul's "inner Africa." Its core concern is parthogenesis: that is, the all-male generation of the Futurist superman, Gazurmah, by Mafarka, the king of the North African fortress Tel el-Kebir. Like D'Annunzio's *Forse che sì, forse che no*, it involves the appropriation and adaptation of Nietzsche's vision of the *Übermensch* in *Thus Spake Zarathustra*. Unlike Zarathustra, however, both Tarsis, D'Annunzio's superman and alter ego, and Marinetti's Gazurmah are fliers or winged creatures.

As Marinetti himself implied in his preface to the 1910 edition, *Mafarka the Futurist* is based on two sets of ideas corresponding to Point Nine of the first Futurist Manifesto of 1909: that is, glorification of war, and "scorn for woman."[109] The first chapter of the novel, "Lo stupro delle negre" (The Rape of the Negresses), ends with an unprecedented dramatic account of a mass rape. This "offense against morals" led to a trial and, in the court of second instance, to a suspended prison sentence of three months.

But our main focus here is not so much the novel itself and the misogyny it portended as the role of its imagery. For already in *Mafarka* it is clear that flight is being used as a metaphor for a new awakening, and the flying creature as the polar opposite of the "passéist" woman or bourgeois. The flier is not only the Superman but also the new man. This encoding of the aviation metaphor is absolutely central to both Futurism and Fascism. The violated, abused women in *Mafarka* are to be understood as ciphers for "sentimental" passéism. As Marinetti explains in his preface, what he contests is not the "animal value" of woman but the "sentimental importance" ascribed to her. "I want to fight the gluttony of the heart," he continues, and "the tyranny of love."[110]

The main aim of the novel is, as Barbara Spackman puts it, "to bypass the vulva and impregnate the ovary that is the male spirit."[111] In the preface dedicated to his "Futurist brothers," Marinetti writes:

> So are you resigned to remaining, like them, the miserable sons of the vulva? Do you want to strangle the roaring Future and incalculable Destiny of man? In the name of the human Pride that we adore, I tell you that the hour is near when men with broad foreheads and chins of steel will give birth prodigiously, by one effort of flaring will, to giants infallible in action ... I tell you that the mind of man is an unpracticed ovary ... It is we who are the first to impregnate it.[112]

There is no doubt that the cult of virility and extreme misogyny play an important role in *Mafarka* and many other Futurist novels. The brutal, pornographic rape scene in the first chapter makes this all too plain. But for this study the

---

[109] Marinetti, "The Founding and Manifesto of Futurism," p. 42.
[110] Marinetti, *Mafarka the Futurist*, p. 2.
[111] Spackman, "Mafarka and Son," p. 90.
[112] Marinetti, *Mafarka the Futurist*, p. 3.

forward shadows cast by such violent "fascist virility"[113] are less important than the vision of "giants" or flying supermen begotten by an effort of the will and impregnation of the "male spirit."

"Sexless procreation" (a term Barbara Spackman borrows from Jean-Joseph Goux) or "womanless reproduction" anticipates the second Futurist Manifesto, "Let's Murder the Moonshine."[114] The birth of the Superman, it is suggested in *Mafarka*, presupposes annihilation of "the feminine," or rather of love, sentimentality and the past. The "son of themselves and magnificent fathers of future powers" must have no mother – and that, as we shall see, means no history. The vision of *Mafarka*, then, is of a new man created out of ahistorical nothingness, devoid of any "ballast" from the past (identified with female sentimentality and the figure of the bourgeois). And that vision is embodied in the winged creature Gazurmah.

The creation of Gazurmah the flying superman is successful because Mafarka resists romantic, "passéist" temptations. Mafarka's joy at having given birth to Gazurmah, "free of all the blemishes that come from the inefficient vulva and bias us to old age and death," is based upon the overcoming of the "passéism" that Marinetti associated precisely with woman or the vulva.[115] Shortly before Gazurmah's birth, Mafarka hurls the following words at Colubbi, the female figure who tempts him and was once his beloved:

> Back, you foul hyena keeper! ... Take yourself far away, with your pack that thrives on rotting sexual organs! ... I won't allow you to see my son! He is mine alone! It is I who made his body. It is I who engender him through sheer exertion of my will! ... And I didn't call on you to help me! ... I did not lay you on your back and pump the divine seed into your ovaries, with heaves of pleasure! ... The seed is still there, in my heart, in my brain! I have to be alone to bring my son to life![116]

The born-to-rule new man is begotten by the "male" heroic spirit alone. Nothing ties him any more to the "sentimental" past that the Futurists rape and kill. This is the reason why Gazurmah, having been brought to life by Mafarka, also kills his creator. Only after this patricide does he begin his flight and his ascent. Then, however, he encounters Colubbi floating on the ocean and awaiting death at his hands.

> A gush of blood splashed a red spray over Gazurmah's breast, as with a surge of his wings he rose into the open sky. So quickly that he scarcely heard Colubbi's dying voice beneath his feet: "You've crushed my heart under your ribs of

---

[113] On "fascist virility," see Barbara Spackman, *Fascist Virilities. Rhetoric, Ideology, and Social Fantasy in Italy*, Minneapolis, MN 1996; and Klaus Theweleit, *Männerphantasien*, vol. 1, *Frauen, Fluten, Körper, Geschichte*, Frankfurt/Main 1977.

[114] Cf. Spackman, "Mafarka and Son," pp. 91–2.

[115] Marinetti, *Mafarka the Futurist*, p. 188.

[116] Ibid., p. 183.

bronze! It is the Earth you've killed in killing me! ... Soon you will hear her first death-throes." [...] [Then Gazurmah:] "Pah! This odor of mummies, this stench of dead centuries, makes me sick! Let's soar higher!"[117]

Flight is a metaphor for setting off into a mythical realm outside history. Gazurmah's wings enable him to soar upward to a new humanity, leaving behind the "stench of dead centuries."

We cannot retrace here Marinetti's complex anthropogony in *Mafarka*. For our purposes, the main point is the mythical quality of the airborne new man: the metaphorical-pictorial narrative of *Mafarka* is mythological rather than rational-explanatory in form, and the novel also has the functional quality of a myth. After all Marinetti seeks to establish the Futurist community and its values through this narrative. The abolition of the past is justified by the promise of a great future and the birth of a Superman. Profane time has to be conquered and sacred time brought into existence. Also underlying *Mafarka* is the sequence of old age and youth, the basic palingenetic structure of sacrifice and rebirth. Mafarka himself must die, in order to be reborn in the flying giant Gazurmah:

Soon I am going to die, to be reborn in the body of my son! ... I shall start my life again in his powerful limbs, whose youthful splendour will strike dead with amazement and pleasure all those who look at him! ... I shall live again in him, without the remorse, the heavy mistakes, or the wounds of the first defeats; in his veins, I shall regain the hope I felt when I was twenty. ... My son will have melodic wings to fly above the curve of the earth![118]

Only through an effort of will can the airborne new man create "youthful splendour" and usher in the New Age. Mafarka proclaims in his "Futurist speech": "I've had enough of your despicable life, you bundles of weakness, flaws, and slow leprosy, men doomed to decrepitude and death! I want to surpass myself by creating, through the sheer effort my heart, [...] an eternal youth!"[119] The new beginning must be wanted, however; then it will be possible "to produce from one's flesh an immortal giant with unfailing wings, [...] without the support and stinking collusion of the woman's womb."[120]

Mafarka preaches a Nietzschean will to power that will make it possible to seize reality and to shape it as one sees fit. To initiate the new beginning and to create the new man, all that is necessary is to "hear the mysterious word of my religion! ... I teach you to despise death, to feed on danger, and to gamble your life, as you are doing now, for an idea, a glance, a performance!"[121] The new beginning will come about through belief in a new myth, through participation in the truth

---

[117] Ibid., p. 202.
[118] Ibid., p. 138.
[119] Ibid., p. 141.
[120] Ibid., p. 145.
[121] Ibid., pp. 144–45.

of art-religion. This goes together with the "heroic" values of dangerous living, self-sacrifice and contempt for death. Glorification of "violent death" is even counterposed to "love," which is "poisonous to life."[122]

In the second Futurist manifesto, "Let's Murder the Moonshine" (April 1909), Marinetti found other imagery for his declaration of war on passéism and the bourgeois world. No longer did "woman" have to be raped and killed; it was the inhabitants of Paralysias and Podagras, the cities of "paralysis" and "gout" (symbolizing immobility and gluttony), who had to be wiped out. A war had to come, but it would offer the opportunity to eradicate the "paralytic," "gout-stricken" bourgeoisie.[123] In any event, Marinetti ends the second Futurist manifesto with a grotesque chase, in which Futurists riding "airplanes" institute a kind of Last Judgment.[124] Aviators fly ahead of the Last Judgment and serve as standards for the destruction of the old world. Marinetti's phantasmagoria ends in a bloodbath. The apocalyptic fliers and their surreal army have pushed the inhabitants of Paralysias and Podgaras "against the high walls of Gorisankar":

> Finally! Finally! So there you are ahead of us, great swarming populace of Paralysis and Gout, disgusting leprosy devouring the mountainsides ... Swiftly we fly against you [...]. But you are numberless! ... And we might use up our ammunition and grow old in the slaughter! ... Let me direct the fire! [...] Oh! The joy of playing billiards with Death! [...] It's ours, the victory ... of that I am sure, because the madmen are already hurling their hearts toward heaven, like bombs! [...] Up 800 meters! Ready! ... Fire! ... Our blood? – Yes! All our blood, in waves, to recolor the sick dawns of the Earth![125]

Won over to the Eros of technology and death, elated by the thrill of speed and violence, the Futurists saw the world war (which they helped to extend to Italy) as the chance to destroy the whole of "yesterday's world." It would, they hoped, begin to eliminate the contemptible "passéist antiquarians" who made up the

---

[122] Ibid., p. 148.

[123] "Let's Murder the Moonshine," in Marinetti, *Selected Writings*, ed. and with an introduction by R. W. Flint, New York 1972, pp. 45–54.

[124] The manifesto continues (ibid., pp. 52–3.): "We need wings! – Then let's make airplanes. [...] We cut our Futurist planes from the ochre-colored cloth of sailing ships. Some had balancing wings and, carrying their motors, rose like the bloody vultures that lift thrashing heifers into the sky. Here it is: my own multicellular biplane steered by the tail; 100 HP, 8 cylinders, 80 kilograms. ... Between my feet I have a tiny machine gun that I can fire by pushing a steel button. ... And we take off, in the intoxication of a keen manoeuvre, a lively, snapping flight, rhythmic and graceful like a song of invitation to drink and dance. Hurrah! Finally we're worthy to command the great army of the mad and the unchained beasts! Hurrah! We master our rearguard, the Ocean with its tangle of foaming cavalry! ... Forward, madmen, madwomen, lions, tigers, and panthers! Forward, you squadrons of waves! For you our biplanes will be as war banners and passionate lovers."

[125] Ibid., pp. 53–4.

population of Paralysias and Podagras; the bloodthirsty dream-imagery of 1909 would become reality in 1915, and even without the assistance of the Futurists the "bourgeois" and the liberal order would now be under fire. As in a previously quoted passage from Jünger's *The Adventurous Heart*, they would "work hard nihilistically for a few years with dynamite, and, dispensing with the most inconspicuous fig leaf, bl[ow] the nineteenth century [...] to smithereens."[126]

The old and bygone had to give way to the young and new: such was the main thrust of Futurism and the reason why it welcomed the war. Take the following passage, for example, which again recalls the second of Nietzsche's "Untimely Meditations":

> The word Futurism contains the most extensive formula of renewal, which, because it is at once purifying and exciting, reduces doubts, destroys skepticisms, and rallies forces in a splendid celebration. All innovators will gather under the banner of Futurism, because Futurism proclaims the necessity of constant advance, and because it urges the destruction of all bridges to cowardice. Futurism is a form of artistic optimism opposed to all chronic pessimism; it is ongoing dynamism, unstoppable development, indefatigable will.[127]

In his *Appunti per un diario* [Notes for a Diary, c. 1909], Umberto Boccioni expressed even more clearly the discontent with time and history that underlay this urge for renewal:

> This eclecticism, this dilettantism disturbed me and made me suffer. They gave me the malaise of him who feels attached to nothing; they gave me the doubt of him who does not believe he will triumph; it [sic] gave me apathy, skepticism, the intolerance of the scientific temperament. I have gathered together all that I have observed of the character of the time, and I have found that what renders us uncertain is the lack of a faith, that is, of an indisputable principle. We who are always at the same point before the infinite lack a new finitude that would be a symbol of our new conception of the infinite.[128]

The Futurists would see in the nation the indisputable principle capable of putting an end to apathy, skepticism and intolerance, and of grounding a new finitude, a new closed order. Another passage from Boccioni's *Appunti* clearly shows that the discontent felt by himself and his Futurist fellow-campaigners can be traced back to Nietzsche's philosophy and is partly explicable in terms of it: "Philosophically speaking, we have abolished the concept of God as creator and judge, with the

---

[126] Ernst Jünger, *Das Abenteuerliche Herz*, p. 186.

[127] Filippo T. Marinetti, "Guerra sola igiene del mondo" [Or. 1915], in TIF, pp. 233–341; here p. 329.

[128] Umberto Boccioni, "Appunti per un diario," in Boccioni, *Altri inediti e apparati critici*, ed. by Zeno Birolli Milan 1972, p. 62. See Poggi, *Inventing Futurism*, pp. 266ff., from which the English translation is taken.

social consequence that our respect for his representatives on earth has gone by the board. Art naturally senses these demolitions and proceeds blindly."[129]

The Futurists and their contemporaries suffered from the "death of God" and the resulting loss of values that the "historical disease" had brought forth. What was missing was a "horizon," as Nietzsche called it in "The Uses and Disadvantages of History."[130] In the *italianismo* preached by the Futurists, the nation constituted such a value-producing horizon once more. The Futurists hoped to gain from the war the "youth" and the "innovators" necessary for "lifegiving" nationalism and for their revolt against historicity and the "gravediggers of the present."[131]

Like D'Annunzio, with his vision of a cathartic holocaust, the Futurists saw the war as a purifying fire from whose ashes a greater Italian phoenix would immediately arise;[132] it was sweeping away everything rotten and decadent. For this Italian *avant-garde*, the "great world war" was thus the fulfillment of "dynamic and aggressive Futurism."[133] Only the Futurists – Marinetti claimed in 1915 – had foreseen the war and extolled it in advance, and now that it was a reality they considered it "the finest Futurist poem yet to appear." The "passéist" poets, among whom Marinetti included the "war prophet of May" D'Annunzio, were pacifists at heart. They were fighting against Germany and Austria in order to kill war. For Marinetti, however, war could not die:

> since it is a law of life. Life = aggression. Universal peace = senile decay and death agony of races. War = bloody and necessary testing of a people's strength. [...] We [...] have always regarded war as the sole inspiration of art, the sole purifying morality, the sole leaven in the human dough. Only war is capable of rejuvenating and accelerating, of stimulating human intelligence, aerating the nerves, freeing us from everyday chores, giving life a thousand flavours and bestowing spirit on the feeble-minded. War is the only horizontal rudder of the new aerolife we are preparing.[134]

The next section will examine more closely this *aerovita* that the Futurists hoped to bring about through war. First, though, we should note that the Futurist new awakening had a primarily negative charge. The Futurists were the most prominent representatives and (formally and stylistically) the boldest vehicles of a widespread antihistoricist longing to destroy the past, for the sake of the future, but also purely for its own sake. The destructive rage of both Futurists and fascists was directed against historicity and temporality *per se*.[135] In essence, the Futurist

---

[129] Boccioni, "Appunti per un diario," p. 62.

[130] Friedrich Nietzsche, "On the Uses and Disadvantages of History for Life (1874)," in *Untimely Meditations*, Cambridge 1983, pp. 62–3.

[131] Ibid.

[132] Filippo T. Marinetti, *Taccuini 1915–1921*, ed. Alberto Bertoni, Bologna 1987, p. 77; diary entry for April 24, 1917.

[133] Marinetti, "Guerra sola igiene del mondo," p. 333.

[134] Ibid., pp. 334–35.

[135] Hanno Ehrlicher, *Die Kunst der Zerstörung. Gewaltphantasien und Manifestationspraktiken europäischer Avantgarden*, Berlin 2001, pp. 156–57.

utopia of renewal boiled down to destruction of the old, demolition of the past, wreaking havoc and leaving behind a heap of rubble.

The Futurist vision of "progress" inevitably calls to mind the much-discussed Benjaminian image of the "angel of history," albeit in the reverse direction. The Futurist "angel," Marinetti's Gazurmah, has his wings outstretched – like Paul Klee's *Angelus Novus* – but he is actually trying to unleash the storm of war. In contrast to Benjamin's angel, he would not "like to stay, awaken the dead, and make whole what has been smashed"; nor would he like to close its wings.[136] For he expects the storm to impel him into the future at which he gazes, while behind him the pile of ruins grows heavenward.

Aerovita: aerolife in a new future

In 1933 Marinetti's "proconsul," the Futurist journalist and publisher Mino Somenzi, proclaimed in the journal *Futurismo*:[137] "*Aerovita* must be the Futurist program for the new Italian generation."[138] But what concepts did the Futurists subsume under *aerovita*, or the "aerolife" of the new Italian generation?

Mino Somenzi was one of the signatories of the *Manifesto of Aeropainting* (1929), along with the Futurists Giacomo Balla, Marinetti's wife Benedetta Cappa Marinetti, Fortunato Depero, Gerardo Dottori, Luigi Colombo (aka Fillia), Enrico Prampolini and Guglielmo Sansoni (aka Tato). The manifesto begins with a short genealogy of the Futurist preoccupation with aviation:

> In 1908 F. T. Marinetti published *The Pope's Airplane*, the first lyrical free-verse exaltation of flight and the aerial prospects of our peninsula from Etna to Rome Milan Trieste. Aeropoetry was further developed in Paolo Buzzi's book *Aeroplani* [Airplanes] and Luciano Folgore's *Ponti sull'Oceano* [Bridges over the Ocean] and Mario Carli's *Caproni*. In 1926 the Futurist painter and aviator Azari created the first work of aeropainting, *Prospettive di volo* [Perspectives of Flight], exhibited in the Futurist Great Hall at the Venice Biennale.[139]

This minihistory is neither accurate – Marinetti first published his *The Pope's Airplane* in 1912 – nor complete. But it is significant in that it closely associates Futurism with aviation from the very beginning of the movement. Although Marinetti, in predating *The Pope's Airplane*, tried to imply his prophetic gifts and

---

[136] Walter Benjamin, "Theses on the Philosophy of History," in Walter Benjamin, *Illuminations*, rev. edn, New York, NY 1969, pp. 253–64; here p. 257.

[137] Mino Somenzi's role as Marinetti's "proconsul" was due first of all to his publishing activity; he was responsible for the journals *Futurismo*, *Sant'Elia*, and *Artecrazia*. See Claudia Salaris, *Storia del futurismo. Libri, giornali, manifesti*, Rome 1985, p. 191.

[138] Mino Somensi, in *Futurismo*, November 12, 1933, quoted (and translated) from Härmänmaa, *Un patriota che sfidò la decadenza*, p. 135.

[139] "Manifesto of Aeropainting," in Lawrence Rainey/Christine Poggi/Laura Wittman (eds.), *Futurism. An Anthology*, New Haven, CT 2009, pp. 283–86; here p. 283.

future-oriented vision, there was no disputing the fact that Futurism had arisen in the slipstream of emergent aviation.

Admittedly, the first Futurist manifesto had hailed "a roaring car that seems to ride on grapeshot" as "more beautiful than the *Victory of Samothrace.*"[140] But it was less the automobile than the airplane and the language of flight that inspired the spirit of Futurism.[141] As we saw in the last section, the nexus of flying, awakening, ascent and the future established the airplane and the aviator as the central symbol, and aviation as a key metaphor of both fascism and Futurism. We must now consider how the Futurist aestheticization and sacralization of technology further underpinned and supplemented these connotations.

The aestheticization and sacralization of technology, and of its use in war, made it possible to work these connotations into a narrative centered on community and values – hence into the palingenetic myth of the nation. What people at the time often perceived as a threat could thus be understood as an integral part of a renascent national order. Whereas the mechanized slaughter at the front blew sky-high the liberal narrative of progress and the temporal order rooted in the "long nineteenth century." The Futurists succeeded in inserting technology and its destructive military power into a different narrative order.[142] Technological devastation was integrated into the palingenetic narrative as a necessary condition for the emergence of the new order; only removal of the fragile bourgeois order, through purification and sacrifice, could clear the way for rebirth. In the fascist myth that ordered time and society, the First World War battles of materiel and mechanized warfare appeared as a necessary purgatory. The cleansed nation and the new man arose out of the cathartic storm of steel.

As we saw in Chapter I.1, liberals such as Aby Warburg hoped that scholarly work would wrest some "space for thought" from the "serpent" of technology. Both the Futurists and Jünger showed, however, that "the modern Prometheus and the modern Icarus" – that is, the "fateful destroyers of our sense of distance who threaten to lead the world back into chaos" – resemble the Indians observed by Warburg, in that they employ mythical ways of thinking to answer the question "how do elemental destruction, death, and suffering come into the world?"[143] They too met the "absolutism of reality" with myth and thereby became capable of mastering technological chaos.[144] And as if wanting to confirm Warburg's motto – "Es ist ein altes Buch zu blättern; Athen-Oraibi alles Vettern" [It is an old, old story; Athens to Oraibi always cousins][145] – the

---

[140] Marinetti, "The Founding and Manifesto of Futurism," p. 41.

[141] For a more detailed discussion, see Felix Philipp Ingold, *Literatur und Aviatik. Europäische Flugdichtung 1909–1927*, Frankfurt/Main 1980, esp. pp. 59–82 and 279–99.

[142] On the "long nineteenth century," see Eric J. Hobsbawm, *The Age of Revolution: Europe 1789–1848*, London 1962.

[143] Warburg, *Schlangenritual*, p. 55; cf. Warburg, "A Lecture on Serpent Ritual," p. 292.

[144] On the "absolutism of reality," see Hans Blumenberg, *Work on Myth*, Cambridge, MA 1985.

[145] Warburg, *Schlangenritual*, p. 9; "A Lecture on Serpent Ritual," p. 277. [Warburg's motto, adapted from Goethe's *Faust*, Part Two, Act 2, refers to Oraibi, the location in Arizona where he studied the Hopi serpent ritual. *Trans. note*].

Futurists too aimed to enchant technology, the "steeliest serpent of knowledge," so that it would not choke them to death.[146] Technology was aestheticized, sacralized and made the central element in their utopia of the new man of regenerated Italy.

Although it was mainly the third Futurist generation that used the airplane as a medium, the aviator always constituted, from the earliest days of the movement, the paradigm of the longed-for awakening and the Futurist new man.[147] Thus, in an article "Aeroteatro," which appeared in February 1935 in the journal *L'Ala d'Italia* published by the Regia Aeronautica [Royal Airforce], the opportunistic Futurist Anton Giulio Bragaglia wrote: "The first bards of modern aviation were, as everyone knows, the Futurists. [...] The mechanical aesthetic, the glorification of the machine, the passion for speed: all this belonged to the deification of the phenomenon of aviation."[148]

As Marinetti's *Mafarka the Futurist* has already suggested, "deification of the phenomenon of aviation" defined the early works of the Futurists. This fascination was based not least on the symbolic or metaphorical possibilities that aviation seemed to offer. The airplane was not only the expression of ascent, of man's advance into the heavens; it also combined destruction with reawakening and death with rebirth, and as such it was an excellent vehicle for the idea of *guerra festa*. A diary entry for September 23, 1917, describing an incident of aerial combat, makes it clear that for Marinetti flight made war a festival and an aesthetic experience.[149]

---

[146] Jünger, *Das Abenteuerliche Herz*, p. 224. Cf. Rohkrämer, *Die Verzauberung der Schlange*, pp. 848–74, here p. 865; and Helmuth Kiesel, *Ernst Jünger. Die Biographie*, Munich 2007, p. 359.

[147] The literature postulates three Futurist generations, within two "phases" of Futurism. The distinction between a first and a second Futurism goes back to the art historian Enrico Crispolti, for whom the former, "heroic" phase stretched from 1909 until the First World War or the death of Umberto Boccioni in August 1916, while the latter came to an end with Marinetti's death in 1944. Within this periodization, the "third generation" of Futurists – which devoted itself to the creation of a new man fit for the *aerovita* – was prominent from the late 20s until the 40s. Cf. Maurizio Scudiero, "Die Metamorphosen des Futurismus, von der futuristischen Rekonstruktion des Universums zur mechanischen Kunst. Die Kunst tritt ins Leben ein," in Ingo Bartsch/Maurizio Scudiero (eds.), *...auch wir Maschinen, auch wir mechanisiert!... Die zweite Phase des italienischen Futurismus 1915–1945*, Bielefeld 2002, pp. 15–29. Scudiero's "first generation" is more "earthbound," more tied to the speed-producing means of transportation of the nineteenth century (train, streetcar, bicycle), whereas his second and third generations are born "under the sign of 'liberation from the earth'." See ibid., p. 24.

[148] Anton Giulio Bragaglia (1890–1960) was linked with Futurism because of his photodynamism. He never joined the movement, however, and associated himself with it only when it seemed opportune to do so. See Mario Verdone/Günter Berghaus, "*Vita futurista* and Early Futurist Cinema," in Günter Berghaus (ed.), *International Futurism in Arts and Literature*, pp. 398–421; here p. 398.

[149] On Marinetti's view of war as festival, and on the importance of "celebrating" the war after it has ended, see Isnenghi, *Il mito della Grande Guerra*, pp. 179–83; Mosse, "Futurismo e cultura politicha in Europa," pp. 13–31; and George L. Mosse, "The Political Culture of Italian Futurism. A General Perspective," *Journal of Contemporary History*, XXV/April–July 1990, 2–3, pp. 253–68.

A "bolide," a "rocket," fall from the sky like a "flaming mass." "Limitless joy" and an "explosive" welling up of vengeful rancor define what he feels:

> Joy that one belongs to the same race as those magnificent fliers, who now have the know-how to keep absolute control of the skies. While the fuel tank and the tangle of frame and undercarriage came rushing down, our two fighters spiraled around it in pursuit, as if they wanted to be sure of the kill, but also to enjoy the sight of the falling, burning wreck. In the circling and swaying of those two fighters, I thought I could make out the excess of divine joy shooting forth from those two men.[150]

The blazing descent of an enemy airplane sends Marinetti into ecstasy, filling him with pride that he too is Italian. Nor is the combat only a festival for himself; the two victorious pilots also enjoy the spectacle as they circle excitedly around the condemned man. Marinetti feels no pity: "One villain less. Long live Italy! I have tears of joy in my eyes." *Schadenfreude* also grips the men on the ground, who, as at a festival, race to the hilltop in search of a trophy or souvenir from the shot-down enemy. It is the lasciviousness of Marinetti's description that makes it stand out from his (otherwise rather typical) enthusiasm over a successful duel in the air, affording fresh insight into the eroticism of war that he cultivated:

> There it is amid the nontoxic, suffocating, nauseating reek of burned flesh and fat – an iron ball entangled with the barbed wire and wire fencing. [...] The fumes choke me. Wood flesh bones fat and aluminum are on fire. A leg without a foot, but still with its puttees, is carbonized and half-burned to ashes. The arm, wrapped around some iron object, displays a roasted elbow the color of lacquered mahogany. It reminds me of the bone of a barbecued leg of mutton. The gutted tank is burning among the jumble of iron rods, knotted tubes, and rusty wire fencing, and on top sits the skull with its exposed brain that boils and fries. I cannot help thinking of a fragile little machine, made up entirely of silvery nickel-plated tubes, over greased and full of steam. I take an aluminum rod and we go back down.[151]

One would therefore have to agree with Walter Benjamin: war here serves

> the artistic gratification of a sense perception that has been changed by technology. This is evidently the consummation of *l'art pour l'art*. Mankind, which in Homer's time was an object of contemplation for the Olympian gods, now is one for itself. Its self-alienation has reached such a degree that it can experience its own destruction as an aesthetic pleasure of the first order.[152]

---

[150] Marinetti, *Taccuini 1915–1921*, p. 125, entry dated September 23, 1917.
[151] Ibid., pp. 125–26.
[152] Benjamin, "The Work of Art in the Age of Mechanical Reproduction," p. 242.

The Futurists derived "aesthetic pleasure" from their "own destruction," hence from the decline and fall of the bourgeoisie. Like Ernst Jünger, they extolled war because it substituted materiel and machinery for the bourgeois individual. Already in the "Technical Manifesto of Futurist Literature" (1912), Marinetti longed for the modern bourgeois individual to be replaced with matter:

> Destroy the *I* in literature: that is, all psychology. [...] We must abolish him [man] in literature and replace him once and for all with matter, whose essence must be seized by strokes of intuition, something which physicists and chemists can never achieve. Capture the breath, the sensibility, and the instincts of metals, stones, woods, and so on, through the medium of free objects and capricious motors. Substitute for human psychology, now exhausted, the lyrical obsession with matter.[153]

In Marinetti's staging, it is an airplane propeller that announces this call for elimination of the ego and its replacement with matter.[154] The manifesto opens with the following paragraph:

> Sitting astride the fuel tank of an airplane, my stomach warmed by the aviator's head, I felt the ridiculous inanity of the old syntax inherited from Homer. A raging need to liberate words, dragging them out from the prison of the Latin period! Like all imbeciles, naturally, this period has a prudent head, a stomach, two legs, and two flat feet: but it will never have two wings. Just enough to walk, take a short run, and come up short, panting! This is what the swirling propeller told me as I sped along at two hundred meters above the powerful smokestacks of Milan.[155]

Marinetti himself became the mouthpiece of the propeller and the machine kingdom, which was dissolving the animal kingdom:

> After the reign of the animal, behold the beginning of the reign of the machine. Through growing familiarity and friendship with matter, which scientists can know only in its physical and chemical reactions, we are preparing the creation of the *mechanical man with interchangeable parts*. We will liberate man from the idea of death, and hence from death itself, the supreme definition of the logical mind.[156]

"Multiplied man" consisting of interchangeable parts, Marinetti's dream figure whom he identified with the aviator, overcame earthbound existence and death.

---

[153] "Technical Manifesto of Futurist Literature," in Lawrence Rainey/Christine Poggi/Laura Wittman (eds.) *Futurism. An Anthology*, pp. 119–25; here p. 122.
[154] See Jeffrey T. Schnapp, "Propeller Talk," *Modernism/Modernity* 1/1994, pp. 153–78.
[155] "Technical Manifesto of Futurist Literature," p. 119.
[156] Ibid., pp. 124–25.

This was his superhuman essence. In *Guerra sola igiene del mondo*, Marinetti declared that the Futurists – in keeping with the transmutation thesis of the French naturalist Jean-Baptiste Lamarck, Darwin's predecessor – aimed to achieve a transformation of the human species; or rather, the creation of a "nonhuman type," who knew neither "moral pain, goodness, and affection" nor "love."[157] They wanted to transform man, because they believed

> that wings rest in the flesh of man. [...] The nonhuman, mechanical type, built for ubiquitous speed, will naturally be cruel, omniscient, and aggressive. He will have unexpected organs, adapted to the demands of an environment involving constant clashes. We can foresee the development of a nose, a protuberance of the breast bone, all the more important as future man will be an ever better aviator.[158]

This "multiplied" man of the Futurist imagination, one better suited to flying, will not know "the tragedy of old age." It is a secondary matter whether this heralds an anthropological revolution and the actual construction of a "cyborg," or whether Marinetti's vision is of a purely literary nature. The key point is again the ideas, imagery and associations connected with the figure of the aviator.

The hero of *l'aerovita*, who was supposed to be the model for Italy's younger generation, had not only reconciled himself to technology but actually fused with it. Futurist man used technology as a metallic extension of his own capacities, a mechanical prosthesis and instrument of power that enabled him to control the world. For the Futurists, the mechanically enhanced aviator was the epitome of a completely transformed new man. Not only because of its destructive potential to sweep away the old order, but also because of its capacity to increase power, technology became a key element in the palingenetic narrative. It would enable Italy to rise higher; it was the guarantee of heroic deeds and triumphant victories.

The Futurists stripped the "serpent" technology of its horror only in so far as horror became the precondition of the new awakening. Horror and destruction were to be loved, not softened. The Futurists sought to counter the alienation from technology that had increased as a result of the war, by exalting it as a means to power and the coming of the heroic *Übermensch*. They also enchanted the serpent of machine culture in their artworks, by literally integrating technology into human beings, or even fusing the two together. The parallels with the serpent rituals observed by Warburg should not be overstated, of course, but this incorporation of the "steeliest serpent of knowledge" does resemble the magical "entry of human beings into the deity in order to share in its superhuman strength."[159]

The aviator became the symbol of a transformation that Marinetti, referring to his own maiden flight of October 1910, described in the manifesto "The New

---

[157] Marinetti, "Guerra sola igiene del mondo," p. 299.
[158] Ibid., pp. 299, 301.
[159] Warburg, *Schlangenritual*, p. 40; cf. Warburg, "A Lecture on Serpent Ritual," p. 286.

Religion-Morality of Speed" (May 1916).[160] Here Marinetti glorified places with a divine presence: trains, city squares, automobile race tracks and telegraph stations, as well as battlefields, and proclaimed the divinity of machine-guns, rifles, cannons, bullets and gasoline. He spoke of the "religious ecstasy" produced by 100 horsepower, called patriotism the "direct velocity of the nation," and hailed war as the "natural testing ground of an army, the central motor of a nation."[161] But this new religion of speed could be experienced most intensely in flight:

> To hurry to hurry to hurry to fly to fly. Danger danger danger danger to left to right below above inside outside to scent to breathe to drink death. Militarized revolution of gears. [...] To enjoy more coolness and more life than you can find in rivers or the sea you have to fly into the coolest headwind at full speed. When I flew for the first time with the aviator Bielovucic, I felt my chest opening like a great hole through which all of the sky – smooth, fresh, and torrential – was deliciously plunging. Instead of the slow watered-down sensuality of walks under the sun and amidst flowers, you should prefer the ferocious and blood-tingling massage of the raging wind. Increasing lightness. An infinite sense of pleasure. You get out of the plane with an elastic and springy bounce. You've gotten something heavy off your back. You've triumphed over the trap of the road. You've triumphed over the law which forces man to crawl.[162]

The airborne new man left behind his link to the earth, shook off the burden of the terrestrial. Man had grown beyond himself. The new man had risen from insect to divine being, had become superhuman. He owed this ascent to the new vantage on the world provided by aviation, to the technological expansion of his body and the technological generation of speed. Aerial velocity, Marinetti writes a few lines later, includes: "hatred of the earth (perpendicular mysticism) spiraling ascension of the 'I' toward Nothing-God = Aviation, the cleansing agility of castor oil."[163] Marinetti conceives of flying as a purifying, "cleansing" act, which he, anticipating Fascist punitive expeditions to come, associates with castor oil. The decisive point here is the deification of technologically induced speed, which the Futurists equated with "disdain for obstacles" and "desire for the new and unexplored" – therefore with modernity.[164] This love of speed and acceleration, of the new and unexplored, is what the Futurists meant by "being modern" and in tune with the times; it is a major part of the nexus between aesthetic modernism and Futurism. But it would be wrong to suppose that "modernism" can be simply

---

[160] See Ingold, *Literatur und Aviatik*, p. 61.
[161] "The New Religion-Morality of Speed," in Lawrence Rainey/Christine Poggi/Laura Wittman (eds.) *Futurism. An Anthology*, pp. 224–28; here p. 226. Translation modified.
[162] Ibid., pp. 227–28.
[163] Ibid., p. 228.
[164] Ibid., p. 225.

reduced to this longing for the new and unexplored, with its basis in the experience of acceleration.[165]

Acceleration, which caused this sense of temporal discontinuity, brought about both: The fracturing of past, present and future occasioned not only the love of speed and the unknown or unexplored but also the longing for a lost past – a past that, in fact, was constituted for the first time as an object of that longing.[166] Acceleration triggered a yearning for distant times as well as "homesickness." The turn to novelty and the faith in progress were certainly "modern" – in the sense of the ethos of modernity – but so too was the longing for the old and for an unattainable timeless stability.

Not only the *Sattelzeit* between 1770 and 1830, which brought the onset of modernity,[167] but the whole of the nineteenth century had unfolded under the aegis of acceleration. To long for deceleration and stasis correlated with devaluation of the "space of experience," whereas to hail dynamism and speed corresponded to valorization of the "horizon of expectation" that had taken shape since then.[168] And whereas the stasis of positional warfare that prevailed in the Alps and Karst, as well as on the Western Front, catalyzed the longing for movement, the revolutionary unrest shaking Europe had the effect of sharpening the desire for deceleration and standstill.[169] After the war, fascism produced a synthesis that reconciled these contradictory longings: it satisfied the need for both acceleration and deceleration by promising a new awakening as well as an eternal order.

The airplane and aviator functioned in fascism as a symbol in which this paradox was harmonized.[170] By virtue of its various connotations – that is, its polyvalence taking in the concepts of awakening and transcendence – aviation

---

[165] For an extensive definition of modernism, see Roger Griffin, *Modernism and Fascism. The Sense of a New Beginning under Mussolini and Hitler*, Basingstoke, UK 2007, pp. 55–6. and 116–17. See also Peter Fritzsche, *Nazi Modern*, p. 12: "Modernism, which has usually been conceived in literary or artistic terms, has remarkable social and political implications. It is the apprehension of the malleable: the dark acknowledgement of the fragility and impermanence of the material world allied with the conviction that relentless reform could steady collapsing structures. In this perspective, modernism breaks with the past, manufactures its own historical traditions, and imagines alternative futures."

[166] Osborne, *The Politics of Time*, p. 164. Cf. Griffin, *Modernism and Fascism*, pp. 221, 257.

[167] On the concept of the *Sattelzeit*, or "saddle period," and the historical discontinuity that it involved, see Reinhart Koselleck, *Futures Past: On the Semantics of Historical Time*, Cambridge, MA 1985 [orig. 1979], p. 247.

[168] See Koselleck, "'Space of Experience' and 'Horizon of Expectation' – Two Historical Categories." On the divergent reactions to technologically induced acceleration, see Stephen Kern, *The Culture of Time and Space, 1880–1918*, London 1983; Rosa, *Beschleunigung*, pp. 79ff.; Wolfgang Schivelbusch, *The Railway Journey: tThe Industrialization of Time and Space in the 19th Century*, Berkeley, CA 1986; and Paul Virilio, *Revolutionen der Geschwindigkeit*, Berlin 1993.

[169] On the catalytic effect of positional warfare on the longing for movement, as well as the Nazis' translation of this into the political sphere, see Alexander Meschnig, *Der Wille zur Bewegung. Militärischer Traum und totalitäres Programm. Eine Mentalitätsgeschichte vom Ersten Weltkrieg zum Nationalsozialismus*, Bielefeld 2008, pp. 287–300.

[170] On the harmonization of paradoxes through symbols, see Hans-Georg Soeffner, "Flying Moles (Pigeon-Breeding Miners in the Ruhr District): The Totemistic Enchantment of Reality and the Technological Disenchantment of Longing," in Soeffner (ed.), *The Order of Rituals: The Interpretation of Everyday Life*, New Brunswick, NJ 1997, pp. 131–56.

symbolism was able to satisfy the yearnings for both dynamism and permanence. There seems to have been an imbalance between the two, not only in respect of their dissemination in society. In fact, they overlapped with each other in eternity: the longing for dynamism and the longing for deceleration were reconciled in a longing for an eternal. For it was not only the champions of deceleration and permanence who hoped to escape the model of historical progression and time itself.[171] As we have seen in the Futurists, the proponents of a faster tempo also sought an exit from temporality.

"Accelerators" and "decelerators" alike saw in fascism an echo of their longing to escape meaningless, desacralized time, and, as far as they were not machine breakers, they found in aviation a compressed symbol of their wishes and dreams. The "racing standstill" of the Futurists was merged into the eternal order promised by fascism, as was Moeller van den Bruck's vision of a "forward restoration"[172] [*Wiederanknüpfung nach Vorwärts*] or Himmler's Teutonic medieval romanticism expressed in the chain of generations.[173] To use again the thread imagery with which Wittgenstein explained "family resemblances," we might say that Futurism is at the far end of a long "fascism thread" consisting of the most diverse, partly contradictory, "temporal fibers." The conceptions of time underlying the Futurist cult of machinery, speed and movement did conflict with the cult of *romanità* that the Italian Fascist regime pursued during the 1930s.[174] But various "fibers" – for example, a common *italianismo*, vitalism and cult of violence – intertwined with one another and helped to strengthen the "thread."

Furthermore, both the cult of *romanità* and the Futurist "religion-morality of speed" underlay the longing to escape from desacralized, profane time.[175] In the cult of *romanità*, the Fascists also sought to revive or repeat a mythical time, a sacred time of "mythical origins," thereby participating in immortality and eternity.[176] The Futurists, however, wanted to bring about eternity by means of

---

[171] See Anselm Doering-Manteuffel, "Mensch, Maschine, Zeit. Fortschrittsbewusstsein und Kulturkritik im ersten Drittel des 20. Jahrhunderts," in *Jahrbuch des Historischen Kollegs 2003*, Munich 2004

[172] Moeller van den Bruck, *Das Dritte Reich*, p. 236.

[173] See Frank Lothar Kroll, *Utopie als Ideologie. Geschichtsdenken und politisches Handeln im Dritten Reich*, Paderborn 1998, pp. 230–55.

[174] On the cultist "renaissance" of ancient Rome and its aspirations to empire, see Emilio Gentile, *The Sacralization of Politics in Fascist Italy*, Cambridge, MA 1996, pp. 73–9; Romke Visser, "Fascist Doctrine and the Cult of the *Romanità*," *Journal of Contemporary History* 27/1 (January 1992), pp. 5–22; and Vollmer, *Die politische Kultur des Faschismus*, pp. 551–626.

[175] On desacralized and profane time, see Eliade, *The Sacred and the Profane*, pp. 68–115.

[176] Gentile, *The Sacralization of Politics in Fascist Italy*, p. 78. See also Hans Poser, "Zeit und Ewigkeit. Zeitkonzepte als Orientierungswissen," in Hans Michael Baumgartner (ed.), *Das Rätsel der Zeit. Philosophische Analysen*, 2nd edn, Freiburg 1996, pp. 17–50. Poser (on pp. 25–6) elucidates the relationship between profane and sacred time with the help of Kurt Hübner's study of Greek myth: "*A sacred cyclical time* [is differentiated from] a *profane, irreversible, linear time* in which the cyclical time is embedded. What appears as repetition in profane time is identical in sacred time. Thus, in the perspective of profane time, sacred time is eternity in the sense of an eternal return of the same, but in its own perspective sacred time is the timeless Now. This gives a completely different relational and modal structure

acceleration. Speed would enlarge the hiatus separating past, present and future. Frenzy would link up with standstill. The totalization of speed – or its eternal omnipresence, as the first Futurist manifesto put it – would lead into the nonspatial, nontemporal Absolute.[177]

In any event, the mechanically enhanced new man loved and worshiped speed because it expressed the departure for new shores, new skies and seas, and the realm of eternity. The Futurists took this theme of entry into eternity or infinity from the cultural criticism of the nineteenth century, but we may suppose that they borrowed the metaphors from Nietzsche and translated them into the technological (and aeronautic) imagery of the twentieth century. Take, for example, Nietzsche's poem "Nach neuen Meeren" [Toward New Seas], in the *Songs of Prince Vogelfrei*:

> Dorthin – will ich; und ich traue
> Mir fortan und meinem Griff.
> Offen liegt das Meer, in's Blaue
> Treibt mein Genueser Schiff.
> Alles glänzt mir neu und neuer,
> Mittag schläft auf Raum und Zeit
> Nur dein Auge – ungeheuer
> Blickt mich's an, Unendlichkeit![178]

The Futurists saw the decline of the old (order of) values, the "death of God," as an opportunity. They thought of themselves as the "fearless ones," those "free spirits" whom Nietzsche addressed in Book Five of *The Gay Science*:

Indeed, we philosophers and "free spirits" feel, when we hear the news that "the old god is dead," as if a new dawn shone on us; our heart overflows with gratitude, amazement, premonitions, expectation. At long last the horizon appears free to us again, even if it should not be bright; at long last our ships may venture out again, venture out to face any danger; all the daring of the lover of knowledge is permitted again; the sea, *our* sea, lies open again; perhaps there has never yet been such an "open sea."[179]

---

from the one familiar to us since Plato and Aristotle, with its modal opposition of the past as immutable-necessary, the present as real and capable of being shaped through action, and the future as possible and open (since both temporal dimensions, the eternal and the profane, interpenetrate). The clear succession of earlier and later is thereby interrupted, since the eternal Now of sacred time also occurs in the past and future of profane time."

[177] Marinetti, "The Founding and Manifesto of Futurism," pp. 41–2.

[178] "That way is my *will*; I trust / In my mind and in my grip. / Without plan, into the vast / Open sea I head my ship. / All is shining, new and newer, / upon space and time sleeps noon; / Only *your* eye, monstrously, / Stares at me, infinity!" Friedrich Nietzsche, *The Gay Science*, trans. by Walter Kaufmann, New York, NY 1974, p. 371.

[179] Nietzsche, *The Gay Science*, 344, p. 280.

They welcomed the demise of the old order, worshiped speed and war for accelerating that demise and making it definitive. The task now was to set sail into "open sea" or – in the new technological paradigm – to "scale the heavens." The Futurist poet Remo Mannoni, who called himself Libero Altomare (Free High Seas), wrote in 1912 in "Scalata" [Scaling]:

> Vogliamo dare la scalata al cielo!
> strappare il velo azzurro
> Tuonare rulli di tamburi elettrici,
> che riveste l'androgino Mistero,
> saettare fluidici dardi
> su gli astri beffardi.
> Vengano dunque ì novi mostri alati:
> ali di tela,
> cuori di acciaio:
> lo spirito gaio
> dell'uomo l'inciela![180]

The Futurists' scaling of the heavens, their "defiance of the stars" (in the words of the first Manifesto), fell afoul of reality.[181] The Futurist "maximum program" failed because of the political realities in post-war Italy. Marinetti's "lyrical order," like D'Annunzio's, was doomed.[182] The Futurist manifesto of March 1923, "The Artistic Rights Defended by Italian Futurists," affirms:

> Vittorio Veneto [the Italian victory of October 1918] and the coming to power of Fascism mark the realization of the Futurist minimum program (the maximum program has not yet been achieved), which was brought into being some fourteen years ago by a group of bold young men [...]. This minimum program defends Italian pride, the endless trust that Italians have in the future, the destruction of the Austro-Hungarian empire, everyday heroism, love of danger, the rehabilitation of violence as the decisive argument, the glorification of war as the world's only hygiene, the religion of speed, of novelty, optimism, and originality, the future of youth in power against the parliamentary, bureaucratic, academic, and pessimistic spirit. [...] Futurism is a purely artistic and ideological movement. It gets involved in political struggles only in the nation's hours of need. [...] We Futurists, who announced and long prepared

---

[180] "We want to scale the heavens! / to tear the blue veil / that covers the androgynous Mystery / To beat electric drum rolls / and to hurl magic darts / at the mocking stars. / Let the new winged monsters come / with their wings of cloth / hearts of steel: / the joyful human spirit / lifts you to the skies!" Libero Altomare et al., *I Poeti futuristi*, with a declaration by F. T. Marinetti and a study of free verse by Paolo Buzzi, Milan 1912, p. 57.

[181] Marinetti, "The Founding and Manifesto of Futurism," p. 44.

[182] On lyrical versus political order, see Gentile, *The Origins of Fascist Ideology*, pp. 134–54, and *"La nostra sfida alle stelle,"* p. 95.

today's great Italy, are glad to acclaim a magnificent Futurist spirit in the not yet forty-year-old head of government. [...] With Fascism, Mussolini has rejuvenated Italy. It is His Task to help us renew the field of art, where sorry men and things still exist. The political revolution must support the artistic revolution – that is, Futurism and all avant-gardes.[183]

At the latest, after Mussolini and his Partito Nazionale Fascista had taken power, Futurism was compelled to limit itself to art and the Futurist-Fascist minimum program. The "Italian revolution" would take place without its fathers and prophets, D'Annunzio and Marinetti. The estrangement of Futurism from Fascism had gathered pace after the feeble result scored by their joint electoral list in November 1919 in Milan (where Marinetti too stood as a candidate).[184] A split then ensued at the Fascist congress of May 23–25, 1920 in Milan.[185] Political activity had worn down the Futurists, and their ideas had proven too *avant-garde* and remote from practical realities. Their totalitarian dream of an "Italian revolution" failed not least because of its radical antibourgeois dimension. In that respect, Mussolini had shown himself to be far more realistic and "open to compromise."

Political Futurism sought to make its mark on nationalist groups and movements that had their origins in the wartime spirit and *combattentismo*: the Arditi (elite assault troops), the ANC veterans' movement (Associazione Nazionale Combattenti), the forces involved in the Fiume breakaway state, and further "modernist nationalists."[186] For them the war was a central unifying point of reference, the opening act in a national revolution that was meant to sweep away the old bourgeois order and replace it with a new order based upon the purified, reborn, vital, combative nation. However, not only were the contours of this nation unclear, but there were many forks to negotiate along the way to its definitive renewal; the role of the Church and the conservative elites, the importance of the national past and the monarchy, the organization of the economy, the relationship to the wartime Allies. These and other issues had a divisive effect on those who had hitherto fought alongside one another.

---

[183] Filippo T. Marinetti, "I diritti artistici propugnati dai futuristi italiani. Manifesto al Governo fascista," in TIF, pp. 562–69; here pp. 562ff.

[184] Here and on the following, see Gentile "La nostra sfida alle stelle," esp. pp. 95–113.

[185] For Marinetti himself, the break with the Fasci di Combattimento was already clear before the congress. In his diary entry for May 20, 1919 he wrote: "I am thinking of the need to leave the Fasci di Combattimento together with Carli and Nannetti and to issue the following statement: Seeing that the divergences between ourselves and the Fasci di Combattimento are intensifying, with regard to (1) the active sympathy that genuine economic strikes deserve, (2) the antimonarchical position that for us is beyond dispute, and (3) the inadequate anticlericalism of the Fasci di Combattimento, we present our resignations from the Central Committee and as members of the Fasci di Combattimento." Marinetti, *Taccuini 1915–1921*, p. 486.

[186] See Gentile, "La nostra sfida alle stelle," pp. 126–27, and *The Origins of Fascist Ideology*, pp. 86–103.

Two factors that differentiated Mussolini from the artistic leaders D'Annunzio and Marinetti, ultimately led to the success of his movement-cum-party: one was his ability to widen the base of the movement and to integrate the (bourgeois) masses into the program of national renewal; the other was his willingness, for the sake of power, to reach an accommodation with political opponents, and to follow it through successfully. Whereas Marinetti, D'Annunzio and their "idealistic" supporters regarded political compromise and maneuvering as a betrayal of the dreamed-of national revolution,[187] Mussolini, the Blackshirts and the Fascist Party created facts on the ground that would revolutionize Italy in the medium term. The revolutionaries oriented to radical, utopian renewal, felt contempt for Mussolini's pragmatism or realism, which they wrongly characterized as opportunism. Emilio Settimelli, for example, the champion of "Ardito Futurism," wrote in his "Confession of a Former Sympathizer of Fascism" (published in the *Giornale di Milano* on June 18, 1920):

> The artwork is not enough for us modern artists. We want to renew the nation's whole way of thinking. With the venture in Fiume and the Fascist election campaign, the coalition of new Italians to fundamentally rejuvenate the life of the nation seemed to be taking shape. Admittedly we were few – but a select few. We could achieve something truly great. But reality told us that it was too early. The victory of the Socialists in the November elections isolated D'Annunzio and made Fascism lean in the direction of the conservatives and priests. Because of this strategic turn, which the powerful brain of our great friend MUSSOLINI SEEKS TO JUSTIFY, MARINETTI AND CARLI HAVE LEFT THE FASCI. The beautiful dream of renewal is suffering shipwreck.[188]

Thus, while the uncompromising artist-politicians spoke of national betrayal and withdrew from practical politics, Mussolini pursued a course that would result in the seizure of power and the establishment of a Fascist party dictatorship. As Mussolini put it in a speech in Udine, a month before the March on Rome: "Our program is simple: we wish to govern Italy."[189]

In the course of 1920, Mussolini's "political order" imposed itself against D'Annunzio's and Marinetti's "lyrical order." Many representatives of the groups that gathered in Milan's Piazza San Sepolcro in March 1919 – Arditi, syndicalists, anarchists, interventionists and indeed Futurists – had to give up their claim to be defining Fascism, because Fascism was gradually transforming itself as it shook off all "democratic" and "libertarian" elements that the Fiume venture had shown it

---

[187] In the course of the war, Marinetti's rejection of the "passéist" D'Annunzio gave way to recognition of his "Futurist life." Thus, a diary entry from July 1918 suggests that, although D'Annunzio is "boringly passéist and anachronistic when he writes or speaks, he is futurist in his life and an admirable Italian soldier." Marinetti, *Taccuini 1915–1921*, p. 280.

[188] Quoted [and translated] from Gentile, *"La nostra sfida alle stelle,"* p. 112.

[189] Benito Mussolini, *Opera Omnia*, vol. XVIII, p. 416; quoted from Ernst Nolte, *Three Faces of Fascism: Action Française, Italian Fascism, National Socialism*, New York 1969, p. 274.

still to have.[190] Fascism was turning more strongly "to the right."[191] The antibourgeois gestures and palingenetic nationalism that it shared with representatives of the "lyrical order" were now supplemented with a more concrete struggle against the Socialist victors of the elections of November 1919 – a struggle motivated by decidedly social and economic interests.[192] The Futurists had also been adversaries of the Socialist Party, one major reason being its opposition to Italy's involvement in the war. It is true that some Futurists subsequently flirted with the proletarian revolution, but they had no more confidence in the proletariat than in the bourgeoisie to bring about genuine national renewal.[193]

The Futurists' elitist vanguardism did much to ensure their political failure but it was above all their radical contempt for "passéism" that stood in the way of any alliance with the conservative elites. Mussolini, on the other hand, was not afraid of such an alliance so long as it helped him to gain power. The crunch came after the Rapallo Treaty between Italy and Yugoslavia was signed in November 1920, when Mussolini – despite expressions of solidarity with D'Annunzio and his adventures – failed to side with the revolutionaries when they were driven out of the port of Fiume/Rijeka in the *Natale di Sangue* (Bloody Christmas).[194]

The bloody anti-Socialist actions of the Fascist *squadre*, most notably in the Po Valley, changed the composition of the movement itself and enlarged the circle of those who supported it or at least thought it useful.[195] Not only for the Italian Left but even for the early "idealistic" Fascists, the Blackshirts had now mutated into agents of the large landowners. This evaluation of Fascism as an instrument of capital, which even after the collapse of the Eastern bloc still has adherents, obscures the changing relationship of forces between "master" and "servant" and fails to recognize the pragmatism of Mussolini's will to power. In retrospect, what seems to have happened is that Fascism pushed into the power vacuum which was left by the fall of the Liberal government, and which the Socialists were incapable of filling. Contrary to the views of early Fascist revolutionaries this did not signal a betrayal of the vision of national rebirth, but was a basic precondition for it to come about in practice.

Unlike Marinetti, D'Annunzio, and their supporters, but also unlike many of his own later interpreters, Mussolini does not seem to have been an idealist or essentialist. In his eyes, the road to power and a transformation of the Italian state and nation passed through a number of different tactical stages, contradictory positions and circuitous compromises. These tortuous routes would be successful in so far as they secured a variegated mass base for Fascism. This support base was inherently "pluralist": it pursued particular interests that were different and partly

---

[190] On the libertarian character of the Fiume venture, see Salaris, *Alla festa della rivoluzione*.
[191] On the "right turn" of Italian Fascism, see Gentile, *The Origins of Fascist Ideology*, pp. 155–211, and *Storia del partito fascista*, pp. 60–162.
[192] See Robert O. Paxton, *The Anatomy of Fascism*, New York 2004, p. 60.
[193] See Gentile, "*La nostra sfida alle stelle*," pp. 118–22.
[194] Ibid., p. 124.
[195] See Michael Mann, *Fascists*, Cambridge 2004, pp. 100–37.

contradictory. But in order to achieve the unshakable primary objective – that is, a new order that overcame "transcendental homelessness" within a single, greater national community – it was necessary to give multiple interests a sounding board. Fascism had to be as polyvalent and multifaceted as its supporters. Here too, in slightly modified form, Hans Frank's statement is applicable: there were basically as many Fascisms as there were Fascists.[196]

In this contradictory relationship of unity and multiplicity lies another nexus between fascism and mythical thought. Whereas scientific thinking rests upon a mutually exclusive opposition between true and false, so that in the end there is only one truth, mythical thinking is capable of allowing several "truths." It does not consist of logical propositions with a determinate truth content, but expresses the many-sided truth of the artwork.[197] Despite variations that arise with each telling, a myth remains the same and always permits of interpretations on the part of its audience. Precisely because myth is polymorphic and many-sided, it is capable of unifying people. Myth, like symbols – as Lévi-Strauss showed – serves to harmonize or mediate contradictions.[198] If we transfer this function to the underlying contradiction of modernity, between the longing for order and the urge to dissolve order, it becomes evident that myth is needed to come to terms with the contradiction or to overcome it at a symbolic level.

The ambiguity of Italian Fascism, made up as it was of overlapping fibers held together by family resemblances, was both the precondition for its success in mass society and the cause of its weakness.[199] Most dramatically on the Night of the Long Knives, Hitler managed to shake off the traveling companions he had needed for the conquest of power, or to silence them through his "successes." But Mussolini was not capable of that. His totalitarian pretensions, concentrated in the Fascist vision of an anthropological revolution, came up against hard reality and the partly political, but mainly apolitical, resistance of the population. In the PNF, the army and the state apparatus, as well as in the monarchy, a range of forces constantly opposed Mussolini's totalitarian ambitions. Both the monarchical and conservative elites, as well as currents pulling in a different direction within the Fascist Party, limited Mussolini's power until he finally dispensed with them in 1943.

When Mussolini and the PNF consolidated their rule in the years following the Matteotti crisis, between 1925 and 1928, a majority of the Futurists bowed to the *fait accompli*. Although D'Annunzio and Marinetti believed themselves to be

---

[196] Hans Frank, *Im Angesicht des Galgens. Deutung Hitlers und seiner Zeit auf Grundeigener Erlebnisse und Erkenntnisse. Geschrieben im Nürnberger Justizgefängnis*, 2nd edn, Neuhaus bei Schliersee 1955, p. 176. See the "Introduction" above.

[197] See Gadamer, Hans-Georg Gadamer, "Mythos und Vernunft," in *Gesammelte Werke*, vol. 8, *Ästhetik und Poetik*, Tübingen 1993, p. 168: "The fact that a distinctive truth becomes audible in myth naturally requires us to recognize the truth of modes of cognition that lie outside science. These cannot be merely brushed aside as vague fantasies. That artistic experience of the world has a binding character of its own, and that this binding character is similar to the artistic truth of mythical experience, is shown in their structural affinity."

[198] See Lévi-Strauss, "The Structural Study of Myth," p. 229.

[199] See Gentile, *The Origins of Fascist Ideology*, p. 192.

better, more radical leaders, they recognized that Mussolini had won the power struggle – and from then on, despite their private hostility, they courted his favor. After all, the Fascist regime, the orchestrated "Fascist spectacle,"[200] and the "cult of the Lictor"[201] based themselves on a style of politics created by D'Annunzio and the Futurists, which involved the conquest of public space, the aestheticization and sacralization of politics, and the outlines of a "lyrical order." The wartime tropes that structured the palingenetic myth – of which D'Annunzio, Marinetti and the Futurists were the clearest and most articulate preachers – lay at the basis of Fascism.

Nevertheless, no more than Mussolini should D'Annunzio and Marinetti be seen as autonomous historical actors, "great men," or geniuses who created realities. Rather, they were prominent mouthpieces or discursive hegemons, who gave linguistic expression (and thereby actual form) to the changes taking place around them. True, they did have considerable powers of definition; they too made "their own history, but not of their own free will; not under circumstances they themselves [had] chosen but under the given and inherited circumstances with which they [were] directly confronted."[202] They gave voice to the widespread longing for a stable order or "nomos," and for the war to have had a meaning. And they found an audience by virtue of the power they had in their respective spheres.

D'Annunzio, Marinetti and Mussolini moved more skillfully in the symbolic network created by the constant updating of language and the rules of everyday life, and by the various objectifications of thought. Their interpretive models, metaphors and tropes, the conceptual or terminological associations created by their narratives, were taken up by other actors and converted into an order of reality that made it possible to master the "tearing of the world." Or, to use another image: the prominent spokesmen struck out on paths in the symbolic landscape that others had taken before them and would continue to take after them, until a way appeared through the labyrinth of reality. The palingenetic myth was one such way through the labyrinth. It gave the war a deeper meaning and satisfied the hunger for an ordering of time and the community. It also united the lyrical and the political order.

D'Annunzio's and Marinetti's lyrical order proved too remote from the realities of power to grasp hold of the state, but Mussolini's "political order" was in many ways reliant upon it. It referred back to and supplemented it. Although economic and social interests as well as acts of violence helped Mussolini and the PNF to power, Fascism cannot be simply reduced to these. For the interests and the violence, the mundane motives of individuals and groups, needed to be embedded in a horizon of meaning that afforded them a solemn legitimacy. The bloody campaign of terror against the Socialists, striking workers and day laborers, as well as against ethnic minorities on the northeastern borders of Italy, acquired their

---

[200] See Simonetta Falasca-Zamponi, *The Fascist Spectacle. The Aesthetics of Power in Mussolini's Italy*, Berkeley, CA 2000 [repr.].

[201] See Gentile, *The Sacralization of Politics in Fascist Italy*.

[202] Marx, Karl Marx, "The Eighteenth Brumaire of Louis Bonaparte," in Marx (ed.), *Surveys from Exile*, London 1973, p. 146.

significance from that horizon.[203] The embedding of motives and interests in the "lyrical order" made the land owners' interest a national interest, fear of losing power a matter of national security, defense of international prestige a sacred duty to the war dead, and so on.

It would be wrong to think of this embedding of interests in a horizon of meaning as an instrumental device to manipulate the submissive masses. Rather, it involved a unity of meaning and interests that was inseparable in practice – although, whereas the former was always integrated into a symbolic network, the latter were motivated by needs, drives and compulsions. The success of the Fascist movement and the self-image of the regime were based upon the idea that it could satisfy the widespread hunger for community, clarity and transcendence, for a stable nomos and a different, eternal order and modernity. Fascism had to proclaim this myth in society and to ensure that it took root.

Although Marinetti himself, like D'Annunzio, ceased to appear on the political stage as such, artistic Futurism continued to aim at the creation of a new man and nation through art. To fulfill the Futurist maximum program, which was alluded to in the 1923 manifesto "The Artistic Rights Defended by Italian Futurists," Marinetti relied on Mussolini's goodwill and his support for Italy's renewal in the domain of art.[204] It is true that Mussolini never complied with Marinetti's wish to make Futurism the official art of the regime, which would have contradicted the aesthetic pluralism of its policy of divide and rule.[205] But the regime supported Futurist artists, and Futurism enjoyed a prominent position at all national and international exhibitions.[206] Whilst the proximity between Fascism and Futurist politics was never greater than in the two years after the world war, when the Futurists put forward a decidedly political program in combination with their artistic activity,[207] the proximity between Fascism and Futurist *art* was never greater than in the "aero" period of the second Futurism.

After the appearance of the *Manifesto dell'Aeropittura* (Manifesto of Aeropainting) in September 1929,[208] and at the latest after Italo Balbo's first transatlantic

---

[203] For a balance-sheet of the violence, see Sven Reichardt, *Faschistische Kampfbünde. Gewalt und Gemeinschaft im italienischen Squadrismus und in der deutschen SA*, Cologne 2002, pp. 53–81.

[204] Marinetti, "I diritti artistici propugnati dai futuristi italiani," pp. 563–64.

[205] Marla Susan Stone, *The Patron State. Culture and Politics in Fascist Italy*, Princeton, NJ 1998, pp. 61–94, esp. 69–70; and Emily Braun, "Sironi in Context," in Roger Griffin (ed.), *Fascism. Critical Concepts in Political Science*, vol. 3, *Fascism and Culture*, New York 2004, pp. 225–48; here p. 226. Now see also: Monica Cioli, *Il fascismo e la "sua" arte. Dottrina e istituzioni tra futurismo e Novecento*, Florence 2011.

[206] Härmänaa, *Un patriota che sfidò la decadenza*, pp. 15–16; and Stone, *The Patron State*, pp. 51–2.

[207] See Gentile, *"La nostra sfida alle stelle."*

[208] The *Manifesto dell'Aeropittura* was published in the Turin *Gazzetta del Popolo* on September 22, 1929 and (in French) in the Paris *Comoedia* on February 14, 1931. It then appeared again in *Artecrazia*, the supplement to the journal *Futurismo*, in July 1932 and (in French) in *Stile Futurista* in August 1934. The signatories were: Giacomo Balla, Benedetta (Cappa Marinetti), Fortunato Depero, Gerardo Dottori, Fillia (Luigi Colombo), Filippo Tommaso Marinetti, Enrico Pampolini, Mino Somenzi, and Tato (Guglielmo Sansoni). See TIF, p. cxxxiii. For an English translation, see "Manifesto of Aeropainting," in Lawrence Rainey/Christine Poggi/Laura Wittman (eds.) *Futurism. An Anthology.*

flight, Futurists used "aeropainting" as a means to shape the "new aviation consciousness"[209] of Fascist society.[210] But aeropainting was not the only means used to promote "aerolife." The Futurists also tried to shape Fascist man through *aeropoesia, aeroscultura, aeroarchitettura, aeromusica, aerodanza* and *aeropranzo* (i.e. "aeromeals"), as well as other "aeroplane" arts.[211] In 1929 Marinetti and Fedele Azari published the *Primo dizionario aereo italiano* [First Italian Dictionary of the Air], with the aim of establishing an "absolute *italianità* of all the vocabulary."[212] The book also served "to verbalize the already existing feeling for aviation"; it was "the first aviation dictionary to appear anywhere in the world, at a time when the age of flight is beginning and a characteristic discourse of aviation is arising for the new generations."[213]

After the publication of the *Manifesto dell'aeropittura*, the prefix "aero" began to appear more and more often instead of the epithet "Futurist." As Günter Berghaus pointed out, Futurist painting now became

> *aeropittura*, Futurist music *aeromusica*, etc., probably in an attempt to detract from the negative image Futurism possessed and to replace it with something that was more in tune with a popular fashion of the period. But *aeropittura/ scultura/poesia* etc. was also a way for many artists to ingratiate themselves with the regime. The aero genres offered plenty of scope for the promulgation of Fascist ideology, and since the subject matter, because of its modernist connotations, found little interest amongst traditionalist or Novecento painters, it offered a unique opportunity to prove that Futurism was after all more suited to become Fascist State Art than any of the other competitors in the field.[214]

Aeropainting was meant to give visible expression to the dissolution of space and time; the speed achieved in flight, as well as the flier's observation point "suspended everywhere in the infinite," would enable the creation of a new reality without space or time, together with an "extraterrestrial plastic spirituality."[215] Although very few observers of *aeropittura* understood the grandiloquent aesthetic-philosophical ideas that the Futurists derived from the aerial perspective, from

---

[209] The Fascist journalist Paolo Orano noted in his biography of Balbo that he had given Italy a *nuova coscienza aviatoria*: see Paolo Orani, *Balbo*, Rome 1940, p. 11.

[210] On "aeropainting," see Susanne von Falkenhausen, *Der Zweite Futurismus und die Kunstpolitik des Faschismus in Italien von 1922–1943*, Frankfurt/Main 1979, pp. 140–89.

[211] In his *Delizia! The Epic History of the Italians and Their Food*, New York, NY 2008, John Dickie described the courses at such an *aeropranzo* organized by the Futurist Fillia at the Taverna del Santopalato (Tavern of the Holy Palate) in Turin. On the Futurist attempt to revolutionize Italian cooking, see also Filippo T. Marinetti, *The Futurist Cookbook*, San Francisco, CA 1989 [orig. 1932]. Claudia Salaris has collected together many works of these aeroplane or aviation arts: see her anthology *aero... futurismo e mito del volo*, Rome 1985.

[212] Filippo T. Marinetti/Fedele Azari, *Primo dizionario aereo italiano*, Milan 1929, p. 9.

[213] Ibid., 12–13.

[214] See Berghaus, *Futurism and Politics*, p. 247.

[215] "Manifesto of Aeropainting," p. 285. See Asendorf, *Super Constellation*, Vienna 1997.

the so to speak "divine" view of the world, most were able to grasp the primal or natural layer of meaning involved in the new genre of painting. The flight theme condensed all the tropes that had fascinated the Futurists from the beginning and been reflected in their art: the cult of machines, virility and *Übermensch*, the love of speed and danger, the glorification of war. Aeropaintings became icons of the "new religion-morality of speed," proclaimed in 1916 in the manifesto bearing that title. They were cult icons in as much as the object of glorification and reverence appeared to be actually present in them – a status all the more fitting because the products of *aeropittura* preserved their original auratic role as state art in numerous public buildings, the new "temples" of Fascism.[216] Here the imagery had a representative function: it mediated between observer and object of reverence, between people and state or nation.[217] Marinetti and Fillia underlined this sacral significance of aeropainting in the "Manifesto of Futurist Sacred Art," first published in June 1931 in the *Gazzetta del Popolo*. Points 2 and 3 read as follows:

> 2. Only Futurist aeropainters, masters of aerial perspective and accustomed to painting while high in flight, can give plastic expression to the unfathomable fascination and blessed transparencies of the infinite. That is not something accessible to traditional painters, all of whom are more or less restricted by obsessive realism, all inescapably terrestrial in outlook and hence incapable of rising up to mystical abstraction. 3. Only Futurist aeropainters can make a canvas sing with the multiform and speedy aerial life of Angels and apparitions of Saints.[218]

But how was aeropainting able to establish itself? Fedele Azari's *Prospettive di volo* (Perspectives of Flight), the first work referred to as such in the *Manifesto of Aeropainting*, was exhibited at the Venice Biennale in 1926. Three years later, in 1929, Gerardo Dottori was awarded a public contract and produced an *aeropittura* mural for the arrivals hall at Ostia airport near Rome, which, according to the Manifesto, convinced the public and art critics "that traditional painted eagles, far from glorifying aviation, seem today like miserable chickens when set beside the torrid mechanical splendor of a flying motor, which would certainly disdain even roasting them."[219]

In any event, both the public contracts and the presence at national and international exhibitions continued to grow. As Susanne von Falkenhausen has shown, "none of the previous themes with which the Futurists tried to express

---

[216] On the Fascist "temples of the faith," see Gentile, *The Sacralization of Politics*. "As in the great periods of the church," he notes, "artists were called on to illustrate and glorify the myths of the Fascist religion" (p. 105). See also Emilio Gentile, *Fascismo di pietra*, Rome 2007.

[217] On the sacral function of icons, see Hans Belting, *Bild und Kult. Eine Geschichte des Bildes vor dem Zeitalter der Kunst*, Munich 1990.

[218] F. T. Marinetti and Fillia, "Manifesto of Futurist Sacred Art," in Lawrence Rainey/Christine Poggi/Laura Wittman (eds.) *Futurism. An Anthology*, pp. 286–87.

[219] "Manifesto of Aeropainting," p. 283.

their understanding of the *era fascista* had registered such prompt success."[220] Marinetti, who for more than 20 years had successfully drawn attention to himself and the Futurists, knew how to play the Fascist patronage system and to engage in consummate "public relations work."[221] But clearly the success had to do also with the themes of *aeropittura* and their link with Fascism. Marinetti never tired of emphasizing this association and extending it to Futurism. For example, in an interview in *L'Ala d'Italia* on "aeropainting and aviation," he said:

> Aeropainting is a purely Italian movement, which has its origins in the same spirit that gave life to the magnificent Fascist aviation: a revolutionary, aggressive, fervid spirit. [...] Aeropainting was prefigured by the early Futurists in the dawn of Fascism; their poetry, sculpture, and painting expressed their longing to break free of the earth and to produce a first aesthetic of flight and aerolife.[222]

The Futurists' best opportunity to express this close linkage of Fascism, Futurism and aviation came from Italo Balbo's organization of long-distance mass flights. Starting out in Ferrara, Balbo (b. 1896) had made a name for himself as a violent *squadrista* leader in the Po valley and beyond, and he went on to become undersecretary of state in the aviation ministry in 1926 and aviation minister in 1929.[223] Himself a symbol of the Fascist awakening and of the youthful, virile and combative new age, he set out to apply this spirit in the Regia Aeronautica, the Italian royal air force. On March 23, 1928 he announced to the Chamber of Deputies:

> Last year the Fascio Littorio emblem was attached to every machine. This year every flier who is entitled to it may decorate himself with the ribbon of the March on Rome, and the flier's salute without headgear – instead of the ungainly and absurd bow that was beginning to spread – is now the Roman salute. I am proud to announce that the martial spirit of our fliers is the same as their Fascist soul.[224]

The Fascist regime made a great effort to build the air force it had founded in 1923 into an institution in its own image, hoping that this would give it more leverage *vis-à-vis* the conservative elites in charge of the army and navy, but also

---

[220] Falkenhausen, *Der Zweite Futurismsus*, p. 149. She also notes: "Marinetti's introduction to the catalog of the Futurist exhibit at Milan in 1929, in which he presents *aeropittura* and its thematic possibilities for a truly Fascist (that is, new, virile, and aggressive) art, makes it clear that *aeropittura* would from them on be an offering of the Futurists to the regime."

[221] See Falkenhausen, *Der Zweite Futurismsus*, esp. pp. 311–12.; and Stone, *The Patron State*, pp. 51–2.

[222] "Aeropittura e aviazione," *L'Ala d'Italia*, November 1937, pp. 44–5; here p. 45.

[223] On the biography of Balbo, see Giorgio Rochat, *Italo Balbo. Lo squadrista, l'aviatore, il gerarca*, Turin 2003; and Claudio G. Segrè, *Italo Balbo. A Fascist Life*, Berkeley, CA 1987.

[224] Italo Balbo, *L'aeronautica italiana. Realizzazioni e proposti. Discorso pronunciato alla Camera dei deputati sul "Bilancio dell'aeronautica" il 23 Marzo 1928*, Rome 1928, p. 58.

that it would have a powerful propaganda effect at home and abroad.[225] The most modern and future-oriented branch of the armed forces was to appear the most Fascist – and conversely the qualities of the air force, the *arma fascistissima*, would carry over to the regime.

This campaign to equate the "spirit" of aviation with the "Fascist soul" also bred specialist journalists who came to be nicknamed *le penne di Icaro*, the feathers of Icarus.[226] They hammered out the same message in the expanding channels of mass culture, which the regime itself had partly created and placed at their disposal.[227] A book published in 1933 by one of "Icarus's feathers," Luigi Contini, contains this passage:

> Fascism and aviation are indicators of will, power, and total domination. Flying is not only an expression of material power. It has been said that the air force is the Fascist force *par excellence*, but this is so precisely at a spiritual-ethical level; just as the empire devised by fascism is not only a territorial expression. [...] Like Fascism, the developing air force is a sign of vitality. Whoever does not expand, whoever stands still, will submit and become enslaved. Flying is the weapon most identified with the race that is awakening from a long sleep, from sluggish dilapidation, just as Fascism is the doctrine that most fittingly represents our inclinations, our striving, our enthusiasm, our consciousness.[228]

Balbo strengthened the nexus between Fascism and the Regia Aeronautica and aviation in general by means of his four "mass flights" and their propaganda impact in the media.[229] The very fact that they were *mass* flights – that is,

---

[225] On the split between reality and propaganda, see the contribution by a military historian working at the Italian air force academy: Gregory Alegi, "'L'arma fascistissima:' il falso mito dell'Aeronautica come preferita del regime," in Massimo Ferrari (ed.), *Le ali del Ventennio. L'aviazione italiana dal 1923 al 1945. Bilanci storiografici e prospettive di giudizio*, Milan 2005, pp. 111–54. Cf. Giorgio Rochat, *Italo Balbo. Aviatore e ministro dell'aeronautica 1926–1933*, Ferrara 1979, pp. 69–75.

[226] Here and on the following, see Massimo Ferrari, "La stampa aeronautica italiana in epoca fascista," in Massimo Ferrari (ed.), *Le ali del Ventennio*, pp. 31–110.

[227] In addition to the significant "aviation page" in Mussolini's *Il Popolo d'Italia*, the following press organs helped to popularize the link between Fascism and aviation: *L'Ala d'Italia*, already quoted above a number of times; the less expensive weekly *Le Vie dell'Aria*; the illustrated *Ali di Guerra*; and the official review of the Regia Aeronautica, *Rivista Aeronautica*. Lastly, there was the magazine *L'Aquilone*, aimed at schoolchildren and young people.

[228] Luigi Contini, *La signoria degli stormi*, Milan 1933, p. 274; quoted from Alegi, "'L'arma fascistissima'," pp. 113–14.

[229] The propaganda battle was waged in all the existing media. Not only the air ministry press office but also Editoriale Aeronautica, its publishing spin-off founded in 1932, exploited Balbo's mass flights to the hilt in *L'Ala d'Italia* and other press organs. But the most important reports were in the daily papers, on the radio and in the weekly newsreels, while numerous posters also trumpeted the successes. Commemorative medals were struck and model airplanes put on sale. The Futurists made the flights a central theme in their "aeropainting," and, as we shall see, *Crociera del Decenale* was a key topic at the Italian Air

collective rather than individual achievements, though under the leadership of an outstanding "hero" – strengthened and underlined their fascist "communitarian" character.[230] Thus, in the previously mentioned speech to the Chamber of Deputies, Balbo declared:

> So I think that the competition in courage, will, and skill among our fliers now requires new objectives. Not so much individual ventures as collective flights of several squadrons, with no fewer than 80 airplanes flying together, will give our men valuable experience of long-distance flight to faraway climates and countries as well as the practice necessary for flying *en masse*. Closed fighter formations and closed reconnaissance and bomber groups, collective flying in general, are an excellent school of leadership and discipline, and the danger they represent also helps to form men's character.[231]

Sixty aircraft took part in Balbo's first mass flight, which was conducted in the spirit of Giulio Douhet, the visionary of great air fleets and aerial battles; it unfolded in six stages between May 25 and June 2, 1928, from Orbetello in Tuscany over the Balearics and along the Spanish and French coasts back to Italy.[232] The second mass flight, from June 5 to 19, 1929, took the 41 participating aircraft from Orbetello over Greece, Turkey, Bulgaria and Romania to Odessa in the Soviet Union. It served as an illustration of Fascist modernity to people at home and abroad, but also as an advertisement for Italian technology. The Bolshevik enemy ordered 30 Savoia S-55 hydroplanes.

Balbo achieved his greatest popularity and success, however, with the two transatlantic flights of 1930–1931 and 1933.[233] On December 17, 1930 a group of 14 airplanes headed by Balbo set off on the 10,400 kilometers to Rio de Janeiro, where 11 arrived on January 15, 1931. Despite the loss of three machines and five airmen, the flight was considered a great success. Mussolini congratulated Balbo in person: "Thanks to the preparation, courage, and technique of a band of brave men, sons of the New Italy, the wing of Italy and the regime are emerging greater from the beginning of Year IX."[234]

Exhibition. Balbo himself wrote reports on his flights for the prestigious Treves and Mondadori publishing houses: see Italo Balbo, *Da Roma a Odessa. Sui cieli dell'Egeo e del Mar Nero*, Milan 1929; *Stormi in volo sull'oceano*, Milan 1931; *Stormi d'Italia sul mondo*, Milan 1934; and *La centuria alata*, Milan 1934.

[230] Mario Isenghi, *L'Italia del Fascio*, Florence 1996, pp. 233–51.

[231] Balbo, *L'aeronautica italiana*, p. 55.

[232] On Balbo's mass flights, see Rochat, *Italo Balbo. Lo squadrista, l'aviatore, il gerarca*, pp. 126–35; Segrè, *Italo Balbo*, pp. 191–265; and Wohl, *The Spectacle of Flight*, pp. 69–78 and 88–102. The aircraft figures vary: Wohl and Segrè speak of 61 for the first Mediterranean flight, whereas Rochat puts the total at 60; there is also a discrepancy in estimates for the second Mediterranean flight. The figures used here are those of Rochat.

[233] The second transatlantic crossing, the *Crociera aerea del decennale*, will be discussed in the next section.

[234] Mussolini to General Balbo, January 15, 1930, quoted from Ministero dell'aeronautica (ed.), *L'Aviazione negli scritti e nella parola del Duce*, Rome 1937, p. 144.

Balbo's South Atlantic adventure also offered the Futurists an opportunity to praise the aviators and market themselves as Fascist artists. The First Exhibition of Aeropainting, held in Rome in February 1931, thus became a homage to the *transvolatori*, the Trans(-Atlantic) fliers.[235] Giacomo Balla, one of the signatories of the *Technical Manifesto of Futurist Painting* back in 1910, took Balbo's flight as the theme for a painting on display there: *Balbo e i transvolatori italiani*. Contrary to the *Manifesto of Aeropainting*, the work is not based on a cockpit view of the earth. However, the dynamic movement of the aircraft and of Fascism is the central focus (Figure 1.2).

The symbolism of Balla's picture was unambiguous: Italy's rise in the world and Fascist dynamism, were expressed by the upward movement of the planes on their transatlantic flight. Balbo and his select group of fliers were the New Men of Fascism, who went beyond familiar boundaries, took bold risks and, like Columbus, set off into a new world. The importance of the theme was confirmed by the article "L'aeropittura, nuova espressione italiana d'arte" [Aeropainting: New Italian Expression of Art], which appeared in the *Corriere Padano* on February 2, 1931. "Flying," it said, was "the most outstanding expression of our times."[236]

After 1936, when Fascist arts policy took a greater distance from modernism in the wake of the alliance with Nazi Germany, the "Fascist themes" of aeropainting ensured greater exhibition space for the Futurists.[237] As Marla Susan Stone has pointed out, with the wars in Ethiopia, Spain and the European theater, *Aeropittura* really did become the medium of Fascist struggle and Fascist "victories."[238] Typical of this role as state art were also the later works that portrayed Fascist fighters more realistically and had a more straightforward propaganda value.

Tullio Crali's *Incuneandosi nell'abitato (In tuffo sulla città)* [Penetrating the Built-up Area (Diving on the City)], in which the painter's and the pilot's viewpoint were fused into one, as the *Manifesto of Aeropainting* demanded, became an icon of this glorification of the new mechanized warrior (Figure 1.2). His depiction of the view from the cockpit inevitably evoked images of dive-bombing and its associations with death and destruction. In 1939, the year when it was first exhibited, the painting was seen as a celebration of war and the imperial pretensions of the new Greater Italy, symbolizing the virility and vitality of the regime, and the courage and fighting spirit of the new Fascist man. In April of that year, *L'Ala d'Italia* printed Carli's picture above a poem, *Volare* [Flying], by Luciano Folgore, who had joined the Futurists back in the days before the First World War, when he still bore the decidedly unfuturist name Omero Vecchi. The poem further associated the

---

[235] See Salaris, *aero...*, p. 19, which also has an illustration featuring the exhibition catalogue.
[236] Mino Lakatos, "L'aeropittura, nuova espressione italiana d'arte – I futuristi in onore dei transvolatori atlantici," *Corriere Padano*, February 8, 1931; quoted [and translated] from Falkenhausen, *Der Zweite Futurismus und die Kunstpolitik des Faschismus*, p. 154.
[237] Falkenhausen, *Der Zweite Futurismsus*, p. 156.
[238] Stone, *The Patron State*, pp. 51–2.

Figure 1.2 Giacomo Balla's *Balbo e I transvolatori* (1931) and Tulio Crali's *Incuneandosi nell'abitato (In tuffo sulla città)* (1939)

painting with the tropes of life, awakening, future and transcendence that have been the theme of this chapter:

> To release oneself from
> the earthly weight of habits.
> To soar and leave behind
> the prison of the world,
> and to free the whole of being
> from the pain and troubles of
> everyday enslavement.
> Life in the open sky above
> takes on the color of the sun,
> the freshness of the dew,
> the deep cadences of song.
> The blue presses into senses and mind,
> dissolves the murky dust
> of any lingering fear of death,
> makes everywhere spring,
> so that human beings are all buds of hope.
> It suffuses instinct and thoughts
> with an unquenchable longing
> for light, dignity, and space.
> Flying.
> The machine is alive, vibrates,
> becomes the wing of him who flies it.
> Its breath has the same rhythm as ours,
> and its warm music the same rhythm
> as our fast-coursing blood.
> Flying.
> Wood, metal, flesh,
> will and courage,
> are weightless as the wind.
> Who still remembers death?
> No one. In the intoxication of flight
> the only time is future time.[239]

During the Fascist years *volare* was synonymous with setting off into a different future and an eternal order. The flier represented man's outgrowing of himself, his striving for immortality, his longing for the transcendent and eternal embodied in the nation. The heroic aviator, death-defying, danger-loving, sacrificing himself for the renewal of the nation, using the machine to expand his own capacities, was the epitome of the new Fascist man. Nello Quilici, whom Balbo numbered among the "top five" journalists and who met his death together with Balbo

---

[239] Luciano Folgore, "Volare," *L'Ala d'Italia* 20/1939, April 1–15, 1939, p. 63.

in June 1940 above Tobruk, expressed this metaphorical relationship between Fascism and aviation in his book *Aviatoria*, published in 1934:[240]

> You cannot be a Fascist without feeling a little like a flier; you cannot be a flier without feeling yourself a Fascist. [...] The scorn for death, intoxication with the new, control over space, scope for achieving rapid fame, and most of all the awareness of serving a great cause through a discipline that does not kill off individual strengths but, on the contrary, asks you to give of them to the utmost: this is the sublime prospect that opens up to the willing young man who aspires to the realm of the eagles. What else has fascism taught? What else does the generation of Vittorio Veneto seek to achieve – against the dreadful banality of the nineteenth century, with its crazy ideas of social leveling and its envy expressed in democratic formulas?[241]

This synonymy or cross-reference between fascism and aviation is also a feature of *Aviatori*, a book published in 1935 by the Lucca-born Futurist Cristoforo Mercati, aka Krimer. Right at the beginning, under the heading: "The Duce has flown, is flying, and will fly in every sphere that arises," Krimer referred to the "wings of the tricolor" as the "quintessence of our Fascist age." And he went on:

> The fruitful seed coming from the wonderful sacrifice of countless heroes has found fertile ground in the minds of the new generation. Today's Italian is wonderfully aware of the motto: "To dare is to conquer." And conquer he shall. As he has always conquered before. [...] The instinct of Icarus, the human instinct to fly [...] has spread in no generation with such strong roots, with such strength and passion, as it has in ours. The will of the Roman eagle, which preceded the steady march of the legionaries across the earth, appears to be reborn in our youthful front lines. It is no longer the standard of foot soldiers; it precedes our eagle-like conquerors and gives them encouragement. [...] Today there is an Italy that cares for its belly, that bows its head, that wishes to be coarser, that seeks to graze in the fenced-off areas. But there is also an Italy that gazes upward, that aims far ahead, that listens to the winds in wide open spaces, that loves trackless routes and shelterless foreign parts. [...] There were times when we gave pilots and captains to the whole world. Today we can give it wings and teach it to fly.[242]

The aviation metaphors, which in Krimer's case also go back to D'Annunzio, illustrate the vision of another Italy. The flier was poles apart from the bourgeois

---

[240] On the "journalistic pentarchy," see Mario Isnenghi, *L'Italia del Fascio*, Florence 1996, p. 237. On the relationship between Italo Balbo und Nello Quilici, who was also a Fascist from the earliest days, see Ferrari, "La stampa aeronautica italiana in epoca fascista," p. 97; and Segrè, *Italo Balbo*, pp. 75 and 136.

[241] Nello Quilici, *Aviatoria*, Naples 1934, pp. 274f; quoted [and translated] from Isnenghi, *L'Italia del fascio*, p. 236.

[242] Krimer, *Aviatori*, Florence 1935, pp. 5ff.

Italian, whether Liberal or Socialist. Whereas "the Reds" (according to Guido Mattioli in *Mussolini aviatore*) "did not want to hear of work for aircraft production" and Liberals were well-fed, complacent and timorous, the new Fascist Italian was inspired by "the spirit of Icarus." The aviator-cum-Fascist was an adventurous soul ready for self-sacrifice: bold, valiant, combative and as such a virile, winged Columbus. This guaranteed him the admiration of the young, as an article "I bimbi, le ali e il domani" [Children, Wings, and Tomorrow] noted in *L'Ala d'Italia* in July 1931:

> Italy's young schoolchildren greet them [rulers of the skies] with the deep and warm appreciation of little Italians who behold the resurrection in the heavens, as we warriors once fearlessly stared death in the face up there. They greet in them that spirit of Italy, great, free, righteous, and upstanding, that provokes inevitable alienation from the conventional past, with its ordinary celebrity and ordinary art treasures – a past that they, scions of a race that prefers the future to the past, only value in so far as it affects the potential achievements and future destinies of the nation.[243]

Young people were expected to revere the past only when it was of use for the future, for the nation; the past was to be enlisted and renewed in the creation of a new world. That was the antihistoricist, monumentalist, and mythical spirit that underlay Fascism. To quote another article in *L'Ala d'Italia*, from its 20th anniversary edition:

> Fascism has created a new world. Mussolini has brought about a new era of history. [...] It is an ancient, rejuvenated race that sets itself against the old age of the world, a new faith that rises up against old habits, decrepit beliefs and ideologies: it is a new destiny. The caution of pen pushers, the weariness of "Europeans," the reverential sentimentality, the false, abhorrent, poisonous sentimentality of bourgeois all over the world, lashes out against the "world" that Mussolini has created in the last twenty years. Fascism aims straight and without respite, singling out the weak, the incapable, the losers; it is confident of attaining the goal it has set for itself. Flying is at the pinnacle of this new power. Italian fliers are Fascists by belief and by temperament: they are anti-bourgeois and anti-European; they are fighters by temperament and by tradition. [...] The air force developed by Mussolini has become the perfect creation of Fascism.[244]

In contrast to Liberals, the aviator-Fascist was also willing to subordinate and sacrifice his ego to the collective. Attilio Longoni, himself a First World War pilot, who in 1919 was appointed general secretary of the Milanese Fasci and editor of *Il Popolo d'Italia*, as well as founding *L'Ala d'Italia*, wrote as follows in his book

---

[243] "I bimbi, le ali e il domani," *L'Ala d'Italia* 10/1931, July 1931, pp. 519–20.
[244] "Ventennale," *L'Ala d'Italia* 20/1939, March 15–31, 1939, p. 3.

Fascismo ed aviazione. Gli aviatori nella rivoluzione fascista [Fascism and Aviation. Aviators in the Fascist Revolution]:[245]

> Sacrifice, faith, and passion produced by the righteous few for the many who, out of apathy or mistrust, have been blinded by pathological theories imported from the East or West. In subordination of the self to the collective interests and the collective salvation lies the glory of the race, confidence in certain development and the inevitable destiny of the fatherland.[246]

The airplane and aviator were a technoid totem of fascist order. As its central collective symbol, this totem contracted "the contradictory into a unity, the nonsimultaneous into the simultaneous, and the merely juxtaposed into a single form." The contradictory nature of fascism, the paradox of modernity, was resolved in the totemic figure of the aviator: "By exploiting this structure, central collective symbols represent a myth in which all details of lifeworld experience are bound together in an integrated higher significance, chance is eliminated, and each individual phenomenon is converted into a cipher of the power and effectiveness of this myth."[247]

The vehicles of this myth were manifold: they stretched from the high-cultural works of an artistic *avant-garde* such as the Futurists, to their popularization in a magazine like *L'Ala d'Italia*, from the writings and speeches of D'Annunzio to the slogans that caught people's eye in the streets and public buildings, at mass rituals and in the "liturgy of the 'harmonious collective'."[248] Thanks to media exploitation of Italo Balbo's successful mass flights,[249] as well as the growing number of movies about aviation, the palingenetic myth of the reborn or resurrected nation reached an ever wider audience.[250] The fascist coding of airplanes and aviators made them vehicles of myth and political practice:

> Myths were potent and indispensable engines of political action. A new state and civilization were founded and survived through myths that had become

---

[245] On Attilio Longoni and the Milan Fascio, see Gentile, *Storia del partito fascista*, pp. 26ff.

[246] Antonio Longoni, *Fascismo ed aviazione. Gli aviatori nella rivoluzione fascista*, 2nd edn, Milan 1931, p. 12.

[247] Soeffner, "Flying Moles," p. 97; translation slightly modified.

[248] On the "harmonious collective" and mass rituals, see Gentile, *The Sacralization of Politics*, pp. 109ff.

[249] On Balbo's flights and the use of them for propaganda purposes – one medium being the Editoriale Aeronautica publishing house, which was located in Balbo's ministry and produced literature for sale in special kiosks – see Ferrari, "La stampa aeronautica italiana in epoca fascista"; Isnenghi, *L'Italia del fascio*, pp. 233–51; Rochat, *Italo Balbo. Lo squadrista, l'aviatore, il gerarca*, pp. 126–39; Segrè, *Italo Balbo. A Fascist Life*, pp. 191–65; and Wohl, *The Spectacle of Flight*, pp. 86–102.

[250] On the Italian movies dealing with aviation, and the many documentaries produced by the Istituto Luce that showed the influence of aeropainting, see Raffaele De Berti, "Lo sguardo dall'alto. Percorsi incrociati tra cinema e aeropittura," in Annamaria Andreoli/Giovanni Caprara/Elena Fontanella (eds.), *Volare! Futurismo, aviomania, tecnica e cultura italiana del volo 1903–1940*, Rome 2003, pp. 175–80.

the collective faith of the masses. [...] Within Fascist culture, myths were not a form of thinking confined to an archaic world or to the primitive mind, but a structured form of human thought, expressed primarily in artistic creation and religious movements, but just as relevant in the world of politics. [...] Ways of visualizing and dramatizing myths, symbols, and rituals were necessary to make the mythology of the "Fascist religion" accessible to the masses and to convert them to the new faith.[251]

To gear the "Fascist religion" to the masses, the regime made use of official exhibitions in which the Futurists too played a part.[252] These made myth an accessible public experience. As at the Exhibition of the Fascist Revolution in 1932 and the Esposizione dell'aeronautica italiana (the Italian Air Exhibition) in 1934, myth was given concrete form and animated with objects. The Fascist memorabilia on display played no small part in generating the desired sense of community and transcendence – indeed, the pennants, medals, machine-guns, wrecks, airplanes and much else besides became holy objects, and as Barthes has pointed out – in the case of the Citroën DS [Déesse: i.e., "goddess"] – the "best messenger of a world above that of nature."[253]

The exhibition of myth: the Esposizione dell'aeronautica italiana

In June 1934 the public streamed into the Palazzo d'Arte in Milan. From then until January 1935, the newly opened halls designed by the architect Giovanni Muzio – the site of the Triennale since 1933 – housed the Italian air exhibition, the Esposizione dell'aeronautica italiana (EIA). The show was devoted to "the spirit, heroism, and labour of Italians who have helped to solve the problems of flight, to perform memorable deeds in war and peace, and to build and use a means of transportation destined to grow more and more extensive."[254]

According to an article by Colonel Francesco Cutry, one of the organizers, more than one million Italians visited the exhibition in a little over six months. On the two floors of the monumental building they were able "to relive the history of the Italian eagles and the magnificent poem of towering human conquest." The artists and curators succeeded in creating an atmosphere "in which the object was animated and brought back to life." It saw "little kids, ordinary women, grand ladies, young people and old, united before the pinion of war, in the same feeling of awe and joy to be Italians." After the Duce ordered an extension, "so

[251] Gentile, *The Sacralization of Politics*, pp. 83ff.

[252] See Stone, *The Patron State*, p. 18.

[253] Roland Barthes, *Mythologies*, New York 1972, p. 88. See also Gottfried Korff, "Vom Verlangen, Bedeutungen zu sehen," in Ulrich Borsdorf/Heinrich Theodor Grütter/Jörn Rüsen (eds.), *Die Aneignung der Vergangenheit. Musealisierung und Geschichte*, Bielefeld 2004, pp. 81–103; here p. 95.

[254] Fondazione Bernocchi/Palazzo dell'arte Milano (eds.), *Esposizione dell'aeronautica italiana. Giugno – Ottobre 1934–XII*, Catalogo Ufficiale, Milan 1934, p. 17: hereafter abbreviated as CEAI.

that the wish of countless Italians to undertake this pilgrimage of love could be fulfilled," the exhibition finally closed its doors. But then, again on the Duce's instructions, work began on a new building to house the exhibition on a permanent basis.[255]

The idea for the show originated with Marcello Visconti di Modrone, the Milanese *podestà* (Fascist governor), who entrusted the work of organization to Cutry (from the historical department of the air ministry), Carlo A. Felice and the architect Giuseppe Pagano. Some of the most renowned architects and designers of Italian modernism, including the one-time Futurist Mario Sironi, would also be involved.[256] In fact, Sironi was the artist most closely linked to Mussolini and the regime, not least because of his activity as a cartoonist for *Il Popolo d'Italia*:

> In both high and popular art forms, Sironi transformed the spirit of modernist nationalism into striking visual images, mythic narratives, and instrumental propaganda that animated the religion of the state. [...] In his own words, Sironi corresponded to the Fascist ideal of a militant artist who serves a moral idea and subordinates his own individuality to the collective cause."[257]

The same inspiration guided the painter Erberto Carboni in his work on the façade at the entrance and the rear side: a stylized *fascio littorio* towered over a map of the world, flanked by a dynamic swarm of fighter planes symbolizing "the primacy of the Italian air wing." (Figure 1.3)[258]

Once inside the entrance hall, the visitor saw a characteristically resolute Mussolini staring down at him from the far end. The portrait was mounted on a collage of a lictor sheaf and a wing, which merged into the Italian cockade. The caption, which underlined Mussolini's and therefore Italy's determination, read: "This wing that has resumed its flight shall not be broken again."[259] The quotation was taken from a speech that the Duce gave to a gathering of aviators on November 6, 1923 before the Grand Hotel in Rome:

> I affirm here, before this glorious, outstanding assembly of young men, that the hopes of Italian aviation will not be disappointed; so long as I hold the post of aviation commissar, there can be no doubt that I will devote all my strength to aviation. You shall have the necessary funds [...]. I, my government, and the whole Italian people shall give you the spirit. Not everyone can fly; it is not even desirable that everyone should. Flying must remain the privilege of an aristocracy; but everyone should have the desire to fly, everyone should long

---

[255] Francesco Cutry, "Bilancio morale dell'esposizione azzura," *L'Ala d'Italia*, January 1935, pp. 14–16.

[256] Antonella Russo, *Il fascismo in mostra. Storia fotografica della società italiana*, Rome 1999; and Jeffrey T. Schnapp, "Mostre," in Deutsches Historisches Museum (ed.), *Kunst und Propaganda im Streit der Nationen 1930–1945*, Berlin 2007, pp. 78–87.

[257] Emily Braun, *Mario Sironi and Italian Modernism*, New York, 2000, pp. 8–9.

[258] CEAI, p. 23.

[259] Ibid., pp. 24–5.

for it. This wing was banished for two or three years from our country's blessed skies. This wing shall not be broken again. As a pilot myself, and as head of the Italian government, I can solemnly and formally vouch for that.[260]

On the first anniversary of the "Fascist revolution," almost eight months after the dissolution of the air wing within the army and navy and the foundation of the independent Regia Aeronautica on March 28, 1923, Mussolini was awarded a symbolic *medaglia d'oro*, the highest Italian military honor.[261] It helped to establish the link between Fascism and aviation, which the exhibition a decade later was meant to underpin.

Mussolini himself probably saw flying as a metaphor for the Fascist order as well as the modernity of Fascism in the conventional sense of the term, and he tried to instrumentalize it for that purpose.[262] A desire and longing to fly was to be aroused in the whole population, and the EAI was supposed to help in creating a "nation of fliers" (as Göring put it in Germany[263]) – that is, a Fascist nation, imbued with the myth given concrete form at the exhibition, and animated by the Fascist awakening.

In an article in *L'Ala d'Italia* on "aviation and art" in the context of the EAI, Vincenzo Constantini emphasized that such commemorative exhibitions and the art on display at them allowed the visitor "to live in the spirit of the exhibits."[264] Myth became palpable there, as the author explained in relation to the Exhibition of the Fascist Revolution at the Palazzo delle esposizioni, opened by Mussolini on October 28, 1932 (the tenth anniversary of the "March on Rome"); four million Italians (one in every 11) visited it over the following two years, and it represented an "unforgettable example" by which the EAI had to be measured.[265]

---

[260] Speech of November 6, 1923; quoted from Guido Mattioli, *Mussolini aviatore e la sua opera per l'aviazione*, Rome 1935–36, p. 167. See also Ministerio dell'Aeronautica (ed.), *L'aviazione negli scritti e nella parola del Duce*, pp. 69–70.

[261] On the Italian air arms before the "March on Rome" and the founding of the Regia Aeronautica, see Roberto Gentilli, "L'aeronautica italiana nel primo dopoguerra," in Ferrari (ed.), *Le ali del Ventennio*, pp. 13–30; John Gooch, *Mussolini and his Generals. The Armed Forces and Fascist Foreign Policy 1922–1940*, Cambridge 2007, pp. 52–60 and 92–108; and Rochat, *Italo Balbo. Aviatore e ministro dell'aeronautica*, pp. 11–45.

[262] See Isnenghi, *L'Italia del fascio*, pp. 233–51; and Alegy, "'L'arma fascistissima'."

[263] See Hermann Göring, "Der alte Fliegergeist lebt. Rede zum Abschluss des Deutschlandfluges am 24. Juni 1934," in Erich Gritzbach (ed.), *Hermann Göring* pp. 121–24; here p. 122.

[264] Vincenzo Constantini, "Aviazione e arte," *L'Ala d'Italia*, June/July 1934, pp. 61–5; here p. 61.

[265] On the Exhibition of the Fascist Revolution, see *inter alia* Claudio Fogu, *The Historic Imaginary. Politics of History in Fascist Italy*, Toronto 2003, pp. 132–64; Gentile, *The Sacralization of Politics*; Jeffrey T. Schnapp, *Anno X. La Mostra della Rivoluzione fascista del 1932*, Pisa 2003, and "Epic Demonstrations. Fascist Modernity and the 1932 Exhibition of the Fascist Revolution," in Richard Joseph Golsan (ed.), *Fascism, Aesthetics, and Culture*, Hanover, NH 1992, pp. 1–37; and Marla Susan Stone," Staging Fascism. The Exhibition of the Fascist Revolution," *Journal of Contemporary History* 28/1993, pp. 215–43.

Figure 1.3 Façade of the Palazzo d'Arte and entrance to the exhibition
Note: CEAI, p. 23. See the room marked (1) in Figure 1.4 of this chapter. In the following paragraphs, all figures inside brackets refer to this floor plan.

The spiritual force of that magnificent historical moment [the Revolution], alive in the murals, architecture, and room arrangements where the dead become sacred and appear present, makes it possible not only to study the documents in the glass cases but also to live in the tragic spirit and the ideas of that revolution.[266]

The exhibitions served to renew and celebrate "great" historical moments; they sanctified the dead and brought them back to life, while permitting visitors to transcend profane time and to enter the realm of sacred time. Such was the power especially of the "art on display,"

> which in the end [is] more interesting than the usual "still lives" or "bathers." Why? Because "commemorative exhibitions" inscribe the power of artistic imagination within the life of society; [...] aesthetic contemplation [...] descends into the active life of the nation, gives itself to the people, who flock there *en masse* to look at works concerning ordinary life. That is why these exhibitions have such great success.[267]

Like public buildings and ceremonies, the great exhibitions were transmission belts: for example, they relayed Futurist tropes and D'Annunzio's narratives of sacrifice and salvation from high to popular culture, from the elites to society at large. The exhibition instilled not only "artistic" but Fascist ideas into the visitor. In so far as aviation was coded and understood as an expression of "Fascist" properties and goals, its display in museums served as a means of spreading Fascist ideas. The EAI conveyed the heady ascent of Fascism and its underlying ideals, and in so doing it structured time and history in accordance with Fascist conceptions. Although less prominently than the Exhibition of the Fascist Revolution, the EAI thus served the Fascist "invention of tradition"[268] and illustrated the dawning of sacred time. Overcoming profane time, the war dead or their spirit lived on in the sacred time of a New Italy and were heading for a glorious destiny. The exhibition thus served a Fascist "monumentalist" politics of history and a mythical politics of time. The once "great" Italy of the imagination was thus renewed and brought back to life in and through the exhibition.

Having passed Mussolini's orienting quotation in the entrance hall, the visitor entered an atrium (2) and caught sight of the showpiece on the first level: the SVA-airplane of the San Marco squadron (15), from which D'Annunzio dropped his cynical leaflets on the Austro-Hungarian capital, Vienna, on August 9, 1918. (Figure 1.4)[269] The spatial arrangement was making a clear statement: D'Annunzio's relic lay at the bottom of a flight of stairs, which ascended not only to the second level but to the reborn, rejuvenated Italy that Fascism had brought into being.

The teleological layout was unmistakable, and the telos was the New Italy. On the first level, the visitor moved through the eventful first decades of aviation,

---

[266] Constantini, "Aviazione e arte," p. 61.

[267] Ibid.

[268] On the "invention of tradition," see Hobsbawm/Ranger (eds.), *The Invention of Tradition,* Cambridge 1983.

[269] See the account in Chapter I.2 above.

338 *Fascism, Aviation and Mythical Modernity*

also taking in significant Italian "precursors" (3) such as Leonardo da Vinci and the inventor Enrico Forlanini (4). The chronological sequence continued through the "first flights" (5) to a hall on the Italo-Turkish war of 1911 (6), when the Italians became the first to use airplanes for military purposes. Then followed the "sanctum," dedicated to the 14 dead and 12 living pilots awarded the *medaglia d'oro* (8), and the D'Annunzio Hall (9). In the latter case, the word hall (*sala*) was actually a euphemism. It is tempting to see the small dimensions of this room as a reflection of Mussolini's distrust or jealousy toward his former rival and *comandante* of the Fiume adventure, who in the immediate post-war years had seemed predestined to lead the motley crowd of interventionists, irredentists, Arditi, Futurists, syndicalists, nationalists and proto-fascists.

D'Annunzio's role in Italian air history – both his military actions and his contributions to the language and media profile of aviation – could only be downplayed at the EAI, certainly not left out altogether. As we saw in Chapter I.2, he was not only the "first Duce" (as Michael Leeden put it) but above all the original *primo pilota*, Italy's first and leading pilot, who together with the Futurists played a considerable part in the Fascist coding of the airplane and the flier, and in the nexus of aviation and palingenetic myth.[270] The exhibition could not avoid presenting him as a kind of "John the Baptist" of the rise of Fascist Italy[271] – although ironically the "D'Annunzio Hall" was a dead end leading nowhere.

The visitor then returned through the "sanctum," crossed the "passage at arms" (10), and entered the "Hall of Aviation in the Great War." This had been designed by Mario Sironi, and it alone contained eight airplanes, including a huge Caproni bomber and the sad remnants of the fighter flown by Francesco Baracca, Italy's leading ace. If one then went through the room devoted to airships, one would reach the "Aviation and Fascism Hall" (13), which covered the years from 1919 to 1922, the Fiume air wing and "the achievements of Fascism in aviation."[272]

This part of the exhibition further clarified the temporal order to which the Fascists aspired. The climax of the First World War was followed by a yawning gap; the heroism and self-sacrifice of Italian fliers had led to nothing, because the lack of fighting spirit on the part of the impotent Liberal government had broken the wing. There had been no fliers among the Liberals, and only the Fascists had kept the spirit of aviation alive. According to the EAI catalogue:

Fascism spreads and gains ground! It expresses the rebellion against the weaknesses and cowardice of those prepared for any kind of agreement. [...] It is the expression of a faith that becomes deeper and tougher, the greater the danger and the harder the struggle; Fascism permeates Italian life and assembles the energies of all rebels and builders. Mussolini, the creator of this new faith, the soul, the condottiere, the head who harmonizes and disciplines this enthusiasm and fervour, understands the desperation of aviation fans and the sorrowful state of the Italian air fleet. Being an aviation fanatic himself, he encourages

---

[270] See Leeden, *The First Duce*.

[271] On the British journalist Sisley Huddleston and his characterization of D'Annunzio as the "John the Baptist" of Fascism, see Woodhouse, *Gabriele D'Annunzio*, p. 3.

[272] CEAI, p. 127.

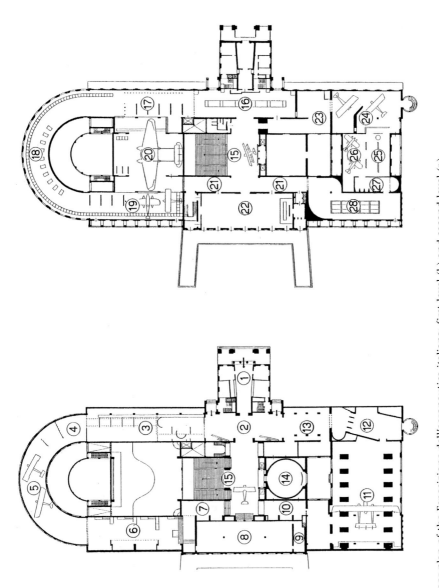

Figure 1.4 Floor plan of the Esposizione dell'aeronautica italiana: first level (l.) and second level (r.).
Note: CEAI, pp. 8–9.

340 Fascism, Aviation and Mythical Modernity

and emboldens, spurs others on, metes out punishment strictly and justly; he is not alone, but from then on he becomes the complete expression of the fliers' revolt and the determination to rebuild. He flies and flies and flies, bearing witness to a faith and electrifying a growing phalanx of supporters.[273]

Mussolini had many roles: he was variously presented as statesman, revolutionary, warrior, Roman, farmer, sportsman and racer, but also as flier and *primo pilota*, the first pilot guiding the state. (Figure 1.5)[274] In his book *Mussolini aviatore*, Guido Mattioli explained what it meant to describe Mussolini as a flier, and why this was emphasized in the EAI exhibition:

> Anyone who dares to take up flying knows that he faces a new path in the training of his mind. This is why Benito Mussolini began to fly so early. [...] The great games in history are won with moral and physical courage. Those who are afraid do not take action, create no facts, and achieve nothing. They do not fly, cannot fly.[275]

Then follows the passage from which we already quoted in the Introduction:

> No machine requires as much concentration of the human mind, as much human will power, as the flying machine does. The pilot really knows what it means to govern. Hence there appears to be a necessary, inner spiritual affinity between aviation and fascism. Every aviator is a born fascist. When he took the controls for the first time, Mussolini must have physically experienced the will to govern.[276]

The exhibition tried to make visitors grasp this link between aviation and fascism, between flying a plane and governing a state. Mattioli devoted more than 100 pages of his book to this "inner spiritual affinity," and the many accounts of the

---

[273] Ibid., p. 130.
[274] On the iconography of the "Duce," see Simonetta Falasca-Zamponi, "Mussolini's Self-Staging," in Deutsches Historisches Museum (ed.), *Kunst und Propaganda im Streit der Nationen 1930–1945*, pp. 88–95, and *The Fascist Spectacle*, pp. 42–88; Jost Philipp Klenner, "Mussolini und der Löwe. Aby Warburg und die Anfänge der politischen Ikonographie," *Zeitschrift für Ideengeschichte* 1/2007, pp. 83–98; and Clemens Zimmerman, "Das Bild Mussolinis. Dokumentarische Formungen und die Brechungen medialer Aufmerksamkeit," in Gerhard Paul (ed.), *Visual History. Ein Studienbuch*, Göttingen 2006, pp. 225–42. On Mussolini as *primo pilota*, see Gerard Silk, "Il primo pilota. Mussolini, Fascist Aeronautical Symbolism and Imperial Rome," in Claudia Lazzaro/Roger J. Crum (eds.), *Donatello among the Blackshirts. History and Modernity in the Visual Culture of Fascist Italy*, Ithaca, NY 2005, pp. 67–81.
[275] Mattioli, *Mussolini aviatore*, pp. 15–16.
[276] Ibid., p. 2.

Figure 1.5 Thayaht's *Il grande nocchiere*, and the Duce as pilot in Guido Mattioli's *Mussolini aviatore*
Note: Thayaht, *Il grande nocchiere*. Reproduced with the kind permission of the Wolfsonian Regional Foundation for Culture and Spectacle, Genoa. The portrait of Mussolini decorated the title page of Mattioli's *Mussolini aviatore*.

Duce as aviator, as well as the 1939 painting by the Futurist Ernesto Michaelles (Thayaht), give expression to it. (Figure 1.5)

In his 1939 aeropainting *Il grande nocchiere* [The Great Helmsman], Thayaht presents an allegorical vision of the Duce. He has broken the European chains, and three Savoia-Marchetti hydroplanes in the center of the picture are flying upward, perhaps to rise over the trenches suggested by lines of barbed wire in the top right. Mussolini, the bold, steely aviator and captain, is leading Italy to new shores. The Fascists did not only try to profit from the popularity of aviation and to adorn themselves with its "modernity"; they drew a more extensive analogy between Fascism and aviation, as the display at the end of the first level underlined.

The first level of the exhibition culminated in the "Icarus Hall" (14), a two-story cylindrical space, at the far end of which a statue of Icarus attached to the wall was looking and striving heavenward. In the center of the room, a blue spiral wound its way up to the ceiling. According to the catalog, this was the only room with a symbolic character. The spatial *Gesamtkunstwerk* made an unambiguous statement, and this was confirmed by a quotation from D'Annunzio beneath

Figure 1.6 The cylindrical "Icarus Hall"

the statue: "A limit to strength? There is no limit to strength. A limit to courage? There is no limit to courage. A limit to suffering? There is no limit to suffering. I tell you, *nec plus ultra* [no farther than here] is the most abominable curse against God and man."[277] Strongman swanking was as typical of D'Annunzio as the quotation of ancient wisdom: the words *nec plus ultra* supposedly adorned the Pillars of Hercules (today's Straits of Gibraltar), marking the limits of the known world and warning seafarers not to venture beyond them (Figure 1.6).[278]

---

[277] CEAI, p. 135. Cf. the memoir dedicated to Eleonora Duse covering the years from 1896 to 1907: Gabriele D'Annunzio, "Le Faville del maglio. Il venturiero senza ventura e altri studii del vivere inimitabile," in *Prose di ricerca*, vol. 1 [orig. 1924], pp. 1069–656; here p. 1231.

[278] In fact, no ancient source refers to such an inscription at the Pillars of Hercules. Pindar mentions the Pillars and presents them as a "metaphor for restraint and prudence": see Earl Rosenthal, "Plus Ultra, Non plus Ultra, and the Columnar Device of Emperor Charles V," *Journal of the Warburg and Courtauld Institutes* 34/1971, pp. 204–28; here p. 211.

D'Annunzio was probably referring to the 26th Canto of Dante's *Inferno*, where Odysseus dares to disobey the *nec plus ultra* and sail beyond the known world.

Io e' compagni eravam vecchi e tardi
quando venimmo a quella foce stretta
dov' Ercule segnò li suoi riguardi
acciò che l'uom più oltre non si metta. [...]
Le miei compagni fec' io sì aguti, con questa orazion picciola, al cammino,
Che a pena poscia li avrei ritenuti;
E volta nostra poppa nel matino,
De' remi facemmo ali al folle volo,
Sempre acquistando dal lato mancino.[279]

The relevance of D'Annunzio's dictum at the EAI was clear enough. Not even the heavens were a barrier for Fascism, which had reawakened the Icarean spirit of the Italians. The new Fascist man ventured beyond known limits and aimed ever higher; his motto was *plus ultra*.

At the exhibition, D'Annunzio's vision of the Icarean Superman was coupled with Futurist enthusiasm for technology and the cult of speed and war to produce a tangible *Gesamtkunstwerk*. But considerable reinterpretation and rewriting of the Icarus myth lay behind this total artwork. In the Fascist version the fall of Daedalus' son was not a result of hubris or arrogance, but rather the sacrifice necessary for the new man to be born and the New Italy to come into being.

The very layout of the EAI legitimized suffering, sacrifice and death, by associating them with renewal of the sacralized nation. The First World War thereby acquired a meaning, and the fallen – the burnt offerings – were integrated into the history of the rise of Fascism as one of its preconditions. Anyone who aimed higher, beyond the known limits of the world, anyone who wished for national resurrection, had to accept sacrifices. To sacrifice oneself for the nation had to be the basis for action. As we saw in Chapter II.2, D'Annunzio's aviator was a hybrid of Icarus and Christ. A Christian coding of the Icarus figure turned him into a crucified savior, and the airplane (its shadow resembling that of the "wood of sacrifice and redemption") into the cross of salvation.[280]

Having completed the circuit on the first level, the visitor passed the airplane with which D'Annunzio made his wartime flight over Vienna – a relic capable of

---

[279] "And I and my companions were already old and slow, / when we approached the narrows / where Hercules set up his boundary stones / that men might heed and never reach beyond. / [...] I spurred my comrades with this brief address / to meet the journey with such eagerness / that I could hardly, then, have held them back; / And having turned our stern toward morning, / we made wings out of our oars in a wild flight / and always gained upon our left-hand side." Dante, *Inferno*, a verse translation by Allen Mandelbaum, Berkeley, CA 1980, pp. 243–44.
[280] D'Annunzio, "La fede nell'aviazione italiana."

"producing religious effects." He then climbed the staircase, copying the upward movement of the airplane and the whole of Italy,[281] until he came to the plane in which Mussolini crashed at Arcore in early March 1921, when he was practicing for his pilot's license (15/2). The wall behind this object displayed a quotation from the Duce: "I am proud to be a flier. I boosted this pride by flying at a time when very few were flying; I crashed because I set out to become a pilot at 37, and of course I went on flying after I fell."[282] In doing this, Mussolini not only substantiated himself as aviator but affirmed his youth and future promise, as well as his spirit of self-sacrifice. Italy's rebirth began with Mussolini's airplane, with Mussolini as *primo pilota*. What followed on the second level of the Palazzo d'Arte was a story of superlative deeds and amazing scientific, technological and sporting achievements – the fulfillment through Fascism of a glorious national destiny that the liberal order had blocked off and socialism had threatened to undermine.

This success story of Fascist modernity encompassed new airports (16), research in aerodynamics (17) and meteorology (21), the development of civilian air travel and airmail (28), and records set by Italians in long-distance flight and air speed (19 and 26). An important role was also assigned to collective flights (20), especially the transatlantic *Crociera del Decennale* (22) undertaken by Italo Balbo and his "winged century," on the tenth anniversary of the founding of the Regia Aeronautica.

Balbo's "mass flight" of 1933, heading for the Century of Progress International Exposition in Chicago, was a huge propaganda victory for the Fascist regime that the EAI organizers exploited to the hilt at the end of the visit. It marked the provisional highpoint of Italy's conquest of the skies:

> Great exhibitions are not born by chance. They are inevitably sent away at the end of a cycle. In every field, they owe their beginning to a unique historical event. The Milan exhibition in general pays tribute to Italian aviation, a splendid organism that combines the genius, heroism, and will of a race of high achievers. But the exhibition could take shape only when the world, thanks to an incomparable act of heroism, had to recognize the clear superiority of the winged tricolor synthesizing the virtues of Mussolini's Italy. The *Crociera del Decennale* is that act of heroism. That is why the exhibition began with Leonardo and closes with Balbo, "the man of the Atlantic." Aviation does not end here, but a circuit is completed.[283]

Balbo shaped this magnificent air force on the Duce's instructions, above all by toughening "the spirit without which men are automata and machines are bits of metal." The *Crociera del Decennale*, however, was not only a glorious exploit in aviation:

> No, it is not possible to separate the feat of aeronautic heroism from its political echo or from the powerful impact that the legendary formation had upon

---

[281] Korff, "Vom Verlangen, Bedeutungen zu sehen," p. 95.

[282] CEAI, p. 137.

[283] Ibid., p. 177.

Italian emigrés. [...] So, the *Crociera del Decennale* does not only end an epoch of modern aviation. It is also one of Italy's most illustrious exploits – an Italy that has again become a nation of heroes and scientists, workers and artists, men of thought and men of action.[284]

Balbo's second transatlantic "mass flight" was indeed a huge success. When he and his men landed on Lake Michigan, they were acclaimed by a hundred thousand people and fêted for the next four days. The City of Chicago even named a street after Balbo – it is still there – and President Roosevelt received him in the White House. On July 21, 1933, the City of New York put on one of its tickertape parades for the Fascist heroes, and in the evening Balbo conveyed the greetings of Mussolini's Italy to 60,000 (200,000 in his own estimate) Italian Americans in Madison Square Garden:[285]

Be proud that you are Italian, [...] because you are the love and pride of the Duce – you who are faithful and fruitful, you who have the spirit and patience of the founders of Rome! Mussolini has put an end to the age of humiliations: to be Italian is an honorable title. Fifty million Italians will continue to journey across land, seas, and skies, but under the sign of Rome and on the orders of a great leader: Italy is no longer plebeian but is the army of civilization en route in the world.[286]

Balbo and his "winged century" ended the mutual acclaim with *Giovinezza* (Youth), the Fascist anthem that the Arditi had sung in the First World War and during the Fiume expedition:

Giovinezza, Giovinezza,
Primavera di bellezza
Per la vita, nell'asprezza
Il tuo canto squilla e va!
E per Benito Mussolini,
Eia eia alalà
E per la nostra Patria bella,
Eia eia alalà.
[Youth, Youth,
Spring of beauty,
In the hardship of life
Your song rings and goes!
And for Benito Mussolini,
Eia eia alalà
And for our beautiful Fatherland,
Eia eia alalà.][287]

---

[284] Ibid.
[285] See Balbo, *La Centuria alata*, p. 300; and Segrè, *Italo Balbo*, p. 247.
[286] Balbo, *La Centuria alata*, p. 298.
[287] For the full text of *Giovinezza*, in which D'Annunzio's war cry "eia eia alalà" functioned as a refrain, see Stanislao Pugliese (ed.), *Italian Fascism and Antifascism. A Critical Anthology*,

According to Balbo, the crowd inside and outside was of one voice, and its song was carried on "invisible radio waves to the fair shores of Italy."[288] In song everyone partook in the myth, in "the spring of beauty," and thus in the renewed national community.

Balbo's mass flights were the most striking and popular symbol of the new, heroic Fascist community that included visitors to the exhibition. This community, gathered "under the sign of Rome and on the orders of a great leader," pointed to the future and to a realm of eternity that also encompassed past generations of the dead. Balbo had already emphasized this on March 28, 1933, the tenth anniversary of the founding of the Italian air force, when 4,000 "blue centurions" paraded through the streets of Rome and past the Duce in the ruins of Diocletian's stadium. There Balbo addressed Mussolini as follows:

Duce! [...] six and a half years ago you gave me the command [of the air force], and today, after hard work under your indications and direct orders, I can proudly declare that the young air force of yesteryear has overcome its imperfection and defects. The hyperindividualist spirit that diminished the military character of the force has been torn up by the roots. [...] We have had hundreds of dead in ten years of tough daily work, but now the dead have left their dark graves for the radiant skies above the imperial Palatine. They want to appear before you, together with the living, as a recompense for the duty they have fulfilled. And I present them to you. They have no face and no body, but they fit among the ranks of comrades who have remained alive to fight the good fight. They are our trust in the destiny of Italy's wing.[289]

The Fascist regime attempted to root out from the whole society that "hyper individualist spirit" that went hand in hand with the liberal order and modernity.[290]

By choosing Balbo and his "winged century" as their heroes and holding their virtues up as examples to be followed, the Italian Fascists too sought to shape a "nation of fliers." Balbo's German counterpart, Hermann Göring, spelled out what this meant in his speech of June 24, 1934: "Today in National Socialist Germany," it was important that

no one runs around as a lone wolf. In the end, the people combines in a unity that is the only source of strength for future achievements. In these times, it goes without saying in German aviation too that the common goal, the common achievement, the communal will should prevail over the individual.[291]

Manchester 2001, pp. 54ff. And on its different versions and their varying success, see Bosworth, *Mussolini's Italy*, pp. 197ff.

[288] Balbo, *La Centuria alata*, p. 299.

[289] Italo Balbo's speech of March 28, 1933 on the tenth anniversary of the founding of the Regia Aeronautica, printed in *L'Ala d'Italia*, April 1933, pp. 8–15; here pp. 11–13.

[290] See Chapter II.3 above.

[291] Göring, "Der alte Fliegergeist lebt," p. 121.

Nor did Göring's analogy between flying and fascist community stop there:

> This year's *Deutschlandflug* [annual flying competition] showed what a wonderful success it has been to give German aviation this new, more solid form of association; discipline and the will to leadership are alive once more. This *Deutschlandflug* should be proof that the individual [...] works in this discipline in the spirit of a joint task. It should prove to the German *Volk* that German aviation has attained its old level, that it is imbued with the old spirit, even if it works on different soil. [...] Young Germany should be brought up in the same passion for fliers, so that the German nation becomes a nation of fliers.[292]

The EAI sought to transmit this communal spirit of "discipline" and "will to leadership" to society as a whole, to turn "transcendentally homeless" individualists into a "nation of fliers" with a common goal, will and faith. The individual was nothing, the *Volk* or nation everything. Giovanni Giurati, the PNF secretary and founder of the Gioventù fascista (the Fascist youth movement), issued the watchword for the individual: *credere, obbedire, combattere* [believe, obey, fight].[293] Individuals were expected to fuse with the nation and to sacrifice themselves for it: society should change into community, and the "flying swords" of old into flying swarms.

Balbo and his "swarms" of transatlantic pilots symbolized this transformation, and the EAI gave expression to it within society. Furthermore, the exhibition helped to inscribe the new temporal order into the new Fascist community; it allowed ordinary people to walk through the rooms and experience the new ordering of time, the rebirth of a nation after its humiliating defeat induced by the bourgeois-liberal order. Through the aesthetic and "religious" experience of the EAI and its tangible representation of myth, the visitor stepped out of profane time into sacred time. If he internalized the *plus ultra*, making it his own motto, he too could become one of the longed-for flying supermen and contribute to the new ascent of the nation. By integrating into his personal narrative the Fascist interpretations and tropes, the death-defying, sacrificial virtues of the sanctified hero, he merged with the nation and became the bearer of the myth of renewal and rebirth – the myth that was supposed to forge a meaningful, orderly whole out of the disorienting, contingent and chaotic world.

---

[292] Ibid., p. 122.
[293] See Tracy H. Koon, *Believe, Obey, Fight. Political Socialization of Youth in Fascist Italy, 1922–1943*, Chapel Hill, NC 1985.

2

Fascism and Mythical Modernity

Der Adler of April 18, 1939, an issue produced with the collaboration of the Reich air ministry, took up *inter alia* the approaching 50th birthday of the "Führer." The prolific writer Peter Supf contributed an article entitled "Waffe seiner Idee" [Force of His Idea]. The "Führer of the German Nation" had a "deep connection with aviation and aviation with him."[294] Already during the Kapp Putsch in March 1920, Hitler – who had never flown before – had the idea of chartering a plane to take him to Berlin as quickly as possible.

> With acute perception, Adolf Hitler recognized the airplane as a hugely dynamic instrument that enabled him to wage the political struggle with unprecedented speed and hitherto unsuspected intensity. – These innumerable political flights, as well as the later ones of the "flying chancellor," helped to spread aviation in Germany and strengthened confidence in its safety.

"With equal clarity," Supf continued, Adolf Hitler saw "the significance of the airplane as a weapon and as a foreign policy instrument." It was thanks to the "Führer" that Hermann Göring was placed at the head of German aviation – a man "whose extraordinary energy was alone capable, in a perilous situation, of organizing the creation of the new Luftwaffe in the shortest possible time."

After Hitler had proclaimed the "freedom of the German people to defend itself" on March 1, 1935, the new Luftwaffe took shape "with a speed that astonished the world" and spread "its protective wings over Germany." In foreign policy, it scored major successes in Hitler's hands during the "historic events of the past few months": the *Anschluss* of Austria, the annexation of the Sudetenland and later Bohemia and Moravia, and the swallowing up of Lithuania's Klaipeda Region (known in German as Memelland). "He" even provided for an offspring of the Luftwaffe: the National Socialist Flying Corps (NSFK). "Without the Führer," so ran Supf's central argument, "there would be no German air force; without the

---

[294] Peter Supf, "Waffe seiner Idee," *Der Adler*, No. 5, April 18,1939, unpaginated. This is the source also of the following quotations.

Führer German aviation would never have undergone this limitless growth that has been astounding the world with its successes."

"The prodigious change in Germany's destiny" had taken place "on a smaller scale but with the same dynamism" in the field of aviation. The revolution initiated by National Socialism had gripped all members of society, beginning with the "Führer"; the same spirit that had inspired the change in the big picture also ruled at lower levels. Supf also pointed to a link between the "Führer" and aviation that has already been discussed *mutatis mutandis* in regard to Mussolini and Fascism. It is worth quoting him at some length:

> But also at a deeper level, aviation has a natural affinity with the Führer. It is the fanatical belief out of which it was born, the heroic spirit that got it under way, the broad vision that is peculiar to it. It is the way in which it stands above things. It is its remoteness from everything small and petty. It is its thinking in different frameworks and with different measures. It is a different tempo, which creates a new sense of time. It is the greater responsibility resulting from its special kind of fervor. It is the voice of eternity clearly audible in the profound solitude of elevation. – The spirit of flying requires a different way of thinking and feeling. From that results a different way of judging and acting. The genius of the German *Volk* has always pressed beyond the confines of unnatural, untimely boundaries and obligations; it has always been most potent where it has relied on itself and followed its inner inspiration, always been strongest where it was closest to nature; and it has always drawn its deepest wisdom from the eternal. The flying spirit, however, has become the genius of our *Volk*; it is the spirit of flying time, of our own time. It is the spirit that operates in Adolf Hitler and his mighty work.[295]

All these *Führer* qualities mentioned by Supf – the association with "the eternal," with the "German genius," its "standing above things," the "new sense of time" – appear also in Leni Riefenstahl's major propaganda film of 1935: *Triumph des Willens* [Triumph of the Will].

A title at the beginning of the film reads: "On September 5, 1934, / 20 years after the outbreak of the world war, / 16 years after the onset of Germany's suffering, / 19 months after the beginning of German rebirth, / Adolf Hitler flew again to Nuremberg to review the army of his loyal followers."[296] We then see, through the front window of a Junkers 52, an ocean of clouds opening up behind the nose cylinder. To the instrumental strains of "Wach auf, es nahet gen den Tag" [Wake up, the day is nigh], from Wagner's *Mastersingers of Nuremberg*, the images suggest that the "Führer" is approaching Nuremberg as an emissary from heaven. The new dawn has come, and Hitler is appearing from the air like a

---

[295] Ibid.

[296] *Triumph des Willens*, 1934–35. Production: Reichsparteitagsfilm der L. R. Studio-Film, Berlin. Scenario and direction: Leni Riefenstahl. Music: Herbert Windt. Min. 2:06–2:51.

deus ex machina. He comes down among his people and saves them. He who has responsibility for "Germany's rebirth" ushers in a new time – eternity (Figure 2.1).

The First World War, the "crippled victory," and the defeat, as well as the threat and actuality of revolutions, sharpened an already major crisis of orientation and a sense of "transcendental homelessness." After the war, a widespread longing for order created the hope that salvation lay in a new order: the end of decadence and decline, the emergence of a community united around the "common good" and shared meaning, the beginning of a New Age under a new *nomos* akin to a heavenly canopy. And the "Führer" seemed to embody this hope and this new order as his plane made its descent and cast its cross-like shadow over the excited populace of Nuremberg.[297]

In his *Historical Time and Future Experience*, Peter Fritzsche noted:

> Germany's defeat in the war and the outbreak of Revolution created a deep sense of crisis which at once invalidated the political and social guidelines of the past and reconfigured the future as an open terrain for national renewal. There was no agreement on what Germany's capacities might be, but postwar intellectuals repeatedly cast themselves as intrepid explorers of new dimensions of time and space to reorient and reconceive the national subject. Born in the five years on either side of 1900, they endeavored to open up unrealized sources of politically sustainable time to carry Germans forward to a regenerate future or else they attempted to divest themselves of outmoded practices and assumptions in order to maximize their ability to repudiate the past and adapt to changing times; they excavated depth or played on the surface. Although the testimonies of observers in the 1920s are not accurate in any verifiable way, they reveal how widespread was the premise in Germany, as opposed to France and Britain, that the postwar world needed to be and indeed could be remapped.[298]

The contours of one such map have been drawn here on the basis of aviation discourse. The intended new order was originally discursive in nature; the work on language, that is, the reordering, renaming, relinking and reclassifying, was followed by the practical reordering and the gardening activity of the state, to use Zygmunt Bauman's metaphor.[299]

This book has investigated the extent to which the fascist quest for a new order issuing from the war should be understood as myth. The absolutized or sacralized nation or *Volk* at its center was the supreme value of society conceived as community, verifying its continued existence and composition, and vindicating the respective social context.[300] The narratives that featured in aviation discourse turned out to be mythical, in so far as the ultimate reference – and the ultimate

---

[297] On Hitler as savior, see Ian Kershaw, *The "Hitler Myth:": Image and Reality in the Third Reich*, New York, NY 2001. Cf. Colin Cook, "The Myth of the Aviator and the Flight to Fascism," *History Today* 53/2003, pp. 36–42.

[298] Fritzsche, "Historical Time and Future Experience in Post-war Germany," p. 141.

[299] See Zygmunt Bauman, *Modernity and Ambivalence*, Cambridge 1991, pp. 20–1 and 28–36.

[300] See Manfred Frank, *Gott im Exil*, Frankfurt/Main 1988, pp. 15ff.

Figure 2.1 Hitler, the "savior," arriving by air in Nuremberg, in Leni Riefenstahl's *Triumph of the Will*, first shown in 1935
Note: *Triumph des Willens*, min. 2:54, 4:33, 4:38, 4:42, 5:46, 5:50.

justification of the war, with its death and self-sacrifice – was this nation or *Volk*. This was the core of every context of meaning, the Archimedean point for the ordering of chaos and the repudiation of the absolutism of reality, and it was to this that the narrative continually returned. The nation or *Volk* was the yardstick for the deadly order of the "community," and the vanishing point of the politics of time or the temporal order. The fascist order has been presented as the paradigm of mythical order but what it displays paradigmatically needs to be shown as a *danger* ultimately inherent in modernity in general. For in its quest for order, modernity requires myth.

Modernity was (and is) not only defined by the awareness of order as a task. Modern consciousness is also "the suspicion or awareness of the inconclusiveness of extant order; a consciousness prompted and moved by the premonition of inadequacy, nay non-viability, of the order-designing, ambivalence-eliminating project; of the randomness of the world and contingency of identities that constitute it."[301] Modern existence is aporetic, therefore, since modern consciousness both yearns for myth and destroys it, longs for myth yet spurns it. This nexus of myth and modernity will now be clarified in relation to fascism. The concept of myth will allow us to grasp several intertwined phenomena that, though often examined in relation to fascism, have seldom been analyzed in their interconnection.

First, the concept of palingenesis, which Roger Griffin convincingly analyses as the "mythical core" of fascism and the characteristic trait of "political modernism," offers a centrally important approach to the mythical structure of modernity.[302] The case of D'Annunzio has shown that the palingenetic myth sprang from the need to give meaning to the human sacrifices of the First World War – an operation that involved transcending profane time in a sacred eternal time, or embedding sacrificial death in a "transhistorical" horizon. This transcendence and embedding were mythical in nature, and they occurred in and through myth. We have seen in Ernst Jünger and the Futurists that the fascist mythical order represented an attempt to break with the old order, to cancel the unpleasant aspects of previous history, to bring into being new values and a new man, and to inaugurate a new age symbolized by the aviator. This break with the old and establishment of the new happens in and through myth.

Second, with the help of the concept of myth, it is possible to understand the fascists' destructive and renovative efforts as a response to the crisis of reason and the resulting threat to order in general. The crisis of reason arose from the discovery of the contingency of order. Order is contingent if it is perceived as possible but not necessary. Order is contingent if it is exposed as a human design. For the fascists, nation and *Volk* appeared to be the Archimedean point of

---

[301] Zygmunt Bauman, *Modernity and Ambivalence*, Cambridge 1991, p. 9.
[302] See Roger Griffin, *The Nature of Fascism*, London 1993 [repr.] (esp. pp. 26–36), and *Modernism and Fascism. The Sense of a Beginning Under Mussolini and Hitler*, Basingstoke, UK 2007, esp. pp. 160–218.

a non-contingent order; they were "nature," and therefore partook of necessity. This conversion of history into nature happens in and through myth.[303]

Third, the concept of myth makes it possible to understand fascism as an answer to the crisis of historicism and the ensuing problem of value relativism. In and through myth, the temporary, historical and contingent are replaced with the "eternal" and "transhistorical." The mythical reference to "the sacred" converts relative historical values into timeless values.

Fourth, the concept of myth can join together two threads that have often remained separate in the interpretation of fascism.[304] It seems reasonable not only to view fascism as a political religion, and therefore a result of the sacralization of politics, but also to keep in mind the aestheticization of politics that finds expression in it.[305] Already in early Romanticism, but also in Nietzsche and Wagner, the search for myth took ancient Greek tragedy as its reference. Community-generating myth was supposed to be at once a religious *and* aesthetic work. Fascist myth satisfied the "religious" need of people at the time for meaning, value, community and transcendence. Through its objectification in public buildings, rituals, ceremonies and mass culture (e.g. the Esposizione dell'aeronautica italiana, or popular war books), it became the central element in the aesthetic-cultural generation of community. In fascism the values, norms and heroes of community were aesthetically staged and experienced as integral parts of a cult practice.[306]

Finally, the concept of myth enables us to think of fascism and modernity as something other than opposites – and indeed, to propose a way of looking at fascism in which, by virtue of its mythical structure, it is paradigmatic of the quest for order and therefore of modernity. Such a reversal threatens to become normative in its turn, condemning a once prized modernity.[307] To avert any misunderstandings that might arise, it is advisable to clarify the position taken here.

This book rests upon a distinction between "the Enlightenment project" and "the project of modernity."[308] Whereas the critical, emancipatory project of the Enlightenment aimed at "man's emergence from his self-incurred immaturity,"

---

[303] See Roland Barthes, *Mythologies*, New York 1972, p. 11.

[304] One major exception is the work of Emilio Gentile, especially his book *The Sacralization of Politics*.

[305] See Benjamin, "The Work of Art in the Age of Mechanical Reproduction."

[306] On Italy see *inter alia* Falasca-Zamponi, *Fascist Spectacle*; Fogu, *The Historic Imaginary*; Gentile, *The Sacralization of Politics*; Jeffrey T. Schnapp, *Staging Fascism. 18 BL and the Theater of Masses for Masses*, Stanford, CA 1996. And on Germany: George L. Mosse, *The Nationalization of the Masses*; and Reichel, *Der schöne Schein des Dritten Reiches*.

[307] On this danger, see Paul Nolte, "Abschied vom 19. Jahrhundert oder: Auf der Suche nach einer anderen Moderne," in Jürgen Osterhammel/Dieter Langewiesche/Paul Nolte (eds.), *Wege der Gesellschaftsgeschichte*, Göttingen 2006, pp. 103–32, esp. pp. 123ff.

[308] For Jürgen Habermas, by contrast, the two "projects" appear to converge into one. See Habermas, "Modernity – an Unfinished Project," in Maurizio Passerin d'Entrèves and Seyla Benhabib (eds.), *Habermas and the Unfinished Project of Modernity*.

and challenged people to use their own understanding for the sake of their freedom,[309] the project of modernity is understood – following Bauman – as a quest for order. The critical Enlightenment consciousness, thoroughly skeptical of any absolutism, seeks to delegitimize and historicize order; modern consciousness on the contrary seeks to absolutize contingencies and to create clarity and necessity where there are only ambiguities and possibilities. The attempts to achieve a rational world order derived from the Enlightenment project were also not exempt from this danger. When "reason" was absolutized, critical thought turned into mythical thought. The Enlightenment itself became a mythoid. Max Horkheimer and Theodor Adorno pointed out as long ago as 1947, in *Dialectic of Enlightenment*: "Myth is already enlightenment; and enlightenment reverts to myth."[310]

On the other hand, it seems impossible for man to renounce mythical thinking altogether – an impression strengthened by various contemporary forms of "religious" consciousness, from "fundamentalism" through esotericism and everyday orientation rituals to dietary shamanism. This is to assume, however, that mythical thinking should be regarded more as an anthropological constant, that a world without myth would be threatened with anomie. The idea that mythical thinking can and has been vanquished turns out to have been the legitimizing narrative of the Enlightenment. This is what Max Weber wrote in 1917: "Many old gods ascend from their graves; they are disenchanted and hence take the form of impersonal forces. They strive to gain power over our lives and again they resume their eternal struggle with one another."[311]

Why gods were required in the form of impersonal forces will be examined in the section on myth as a response to the crisis of historicism. Nietzsche asserted that, apart from critical-historical thinking, there is a need for "suprahistorical" powers. A dialectic of critical-historical and suprahistorical-mythical thought, of freedom and order, seems to characterize human existence. Our earlier discussion of Aby Warburg showed that the struggle with ever new anxiety-producing objects and events in the world is the reason for their "mythical" ordering and control. But only critical thought is capable of freeing men from that order, or – as Warburg put it – of winning Athens back from Alexandria.

The centrality of palingenetic myth

The fascist palingenetic myth is intertwined with a complex set of crises affecting liberalism and the model of progress, historicism and reason which extends to the interpretation of history and time in general. It would appear that the concept of myth can at least clear a path in this thicket.

As we saw in the last chapter, in connection with Ernst Jünger and the Italian Futurists, the fascist politics of time or new temporal order was based upon

---

[309] Immanuel Kant, "An Answer to the Question: `What Is Enlightenment?'" in Hans Reiss (ed.), *Political Writings*, 2nd edn, Cambridge 1991, p. 54.

[310] Theodor Adorno/Max Horkheimer, *Dialectic of Enlightenment* (orig. 1944), New York, 1972.

[311] Max Weber, "Science as a Vocation," in H. H. Gerth/C. Wright Mills (eds.), *From Max Weber. Essays in Sociology*, Boston, MA 1948, p. 149.

a palingenetic narrative. The mytheme of renewal, rebirth and reawakening promised an end to the profane time of decline and the beginning of a sacred, eternal time in which being had a meaning and death was overcome by the life of the nation or *Volk*. As George L. Mosse, Emilio Gentile and above all Roger Griffin have shown, fascism rested upon this vision of renewal and rebirth. And as Roger Griffin and Peter Osborne have shown, the vision of renewal is a founding trope of modernist chronopolitics.[312] Moreover, the palingenetic myth does not have roots only in Christian modes of thought; myths of renewal are "the most frequent mythico-ritual scenario in the religious history of humanity."[313]

Mircea Eliade stressed man's longing to overcome the meaninglessness of profane time and to "return home" in a renewed sacred time. "Archaic societies" felt a "need to regenerate themselves periodically through the annulment of time," which may be described as their "antihistorical tendency."[314] "Archaic cultures" defended themselves against history

> either by periodically abolishing it through repetition of the cosmogony and a periodic regeneration of time or by giving historical events a metahistorical meaning, a meaning that was not only consoling but was above all coherent, that is, capable of being fitted into a well-consolidated system in the cosmos and man's existence had each its *raison d'être*.[315]

But can this attempt to "annihilate history" be restricted to "archaic systems?" Or is such a restriction itself an attempt to "annihilate history" by ringing in a new sacred time of reason?

For sections of Italian and German society, the outcome of the First World War and the preceding epoch of "decadence" and "degeneration" were just such a history to be annihilated through the regeneration of time. The war triggered a defense reaction that combined with tendencies of cultural pessimism and *Zivilisationskritik* already apparent before the turn of the century.[316] Particularly in societies that experienced its outcome as a defeat, the war was devoid of all meaning; the need to embed it in a "suprahistorical" horizon transcending profane events therefore became a matter of urgency. Earlier we saw how D'Annunzio gave a suprahistorical meaning to death on the battlefield and in the skies above, by reinterpreting it as the seed from which the sacred nation would be reborn. In the popular narratives of heroism, the actual wartime deaths resulting from the interplay of social, political and economic forces but also from pure chance became elements in a higher system that provided them with meaning – that is, elements in the palingenetic myth.

---

[312] See Griffin, *The Nature of Fascism*, London 2004, pp. 26–55, and *Modernism and Fascism. The Sense of a New Beginning under Mussolini and Hitler*, Basingstoke, UK 2007.

[313] Mircea Eliade, *Mephistopheles and the Androgyne. Studies in Religious Myth and Symbol*, New York 1965, p. 158; quoted from Allen, *Myth and Religion in Mircea Eliade*, p. 205.

[314] Eliade, *The Myth of the Eternal Return*, p. 85.

[315] Ibid., p. 142.

[316] See August Nitschke et al. (eds.), *Jahrhundertwende. Der Aufbruch in die Moderne 1880–1930*, vol. 1, Reinbek bei Hamburg 1990; Thomas Rohrkrämer, *Eine andere Moderne? Zivilisationskritik, Natur und Technik in Deutschland 1880–1933*, Paderborn 1999.

In France, Britain and the USA it was possible to insert the war into an emancipatory narrative with roots in the Enlightenment.[317] The war had been waged to make the world "safer for democracy," to topple the last autocracies, to make nation-building possible for oppressed peoples and to put aggressors in their place. As to the newly emerging Soviet Union, it managed to connect with the Marxist variant of the emancipatory metanarrative while at the same time establishing a palingenetic myth in which the death of the old order was followed by the rise of the age of the proletariat.[318]

In both Italy and Germany it was much more difficult for liberal or Marxist forces to insert events into these meaningful narratives which seemed to conflict with everyday experience. The fact of the war itself, the technological, rationalized butchery, the defeat and subsequent events, led to a more fundamental questioning of the emancipatory narrative of progress.[319] "Emancipation" now seemed something alien, imposed from outside, while the experience of violence and destruction, the senseless mass killing for the sake of the nation, could not be fitted into such a narrative structure of time. Indeed, the nation – which had always been seen as a goal and engine of progress – seemed to have fallen beneath its wheels in Italy and Germany. In the USA, Britain and France, it was still possible to link the nation to progress. In Germany, however, the liberal and national strands of the narrative had drifted ever farther apart in the second half of the nineteenth century, while in postwar Italy the gulf had widened because of the "lack of credibility" of the Liberal political establishment as well as distrust of the wartime Allies. In any event, the promised national "salvation" had failed to materialize, so that the purpose of the sacrifices on the altar of the fatherland had to be sought in a temporal order in which salvation was still to be achieved.

Following on from the nineteenth-century currents of cultural pessimism and civilization critique, it therefore became common to reconsider the bourgeois conception of history as "liberty unfolding." In the eyes of many, the direction now seemed to be downward, not upward. The Enlightenment and the French Revolution were no longer understood as the beginning of a better epoch, but interpreted as an uprooting estrangement, a loss of "grip on reality," an onset of decay,[320] which had led to defeat or a "crippled victory" in the war. Fascists therefore sought to "annul" that time, to silence or eliminate its bearers, and to return

---

[317] On the concept of metanarrative, see Jean-François Lyotard, *The Postmodern Condition: A Report on Knowledge*, Minneapolis, MN 1984.

[318] On the staging of the new awakening in revolutionary Russia, see *inter alia* Dietrich Beyrau, *Petrograd, 25. Oktober 1917. Die russische Revolution und der Aufstieg des Kommunismus*, Munich 2001, pp. 76ff; and Stefan Plaggenborg, *Experiment Moderne. Der sowjetische Weg*, Frankfurt/Main 2006.

[319] This occurred in the wake of the Enlightenment and the acceleration processes that began in the eighteenth century. As the narrative of progress dislodged the hitherto prevailing space of experience and horizon of expectation, it enabled events and changes to be inserted into a new interpretive context. See Reinhart Koselleck, "'Space of Experience' and 'Horizon of Expectation' – Two Historical Categories," in Koselleck (ed.), *Futures Past: On the Semantics of Historical Time*, Cambridge, MA 1985 [orig. 1979].

[320] See Griffin, *Modernism and Fascism*.

home into sacred time. Only through such purification and regeneration of the nation or *Volk* would the sacrifice acquire its real meaning.

As we have seen in the case of Ernst Jünger, further decline of the brittle old order was interpreted not as a loss but as the precondition of the new. The war had begun to sweep away the "old and rotten," the climate of "passéism," and now there needed to be a complete *tabula rasa* on which a new order free of historical ballast could be built. The attempt to annihilate history expressed itself not least in this theme, which was propagated most clearly by the Futurists. Initiated by the war, the break had to be seen through to the end. Elimination of the old order, with its "relative" values and half-truths, was to be the test bed for construction of the new and eternal. The time was ripe for renewal and for violent utopias. This renovative posture, which researchers saw for decades as reactionary because it aimed at a putative eternity, should in fact be identified with the modern order-dissolving spirit. It was an attempt to take up the ordering of time, in the face of a defeat that was experienced as chaos.

Myth as a response to the crisis of reason

D'Annunzio, the Futurists, Jünger and numerous other European revolutionaries and intellectuals, on both the left and the right, were seeking a way out from the postwar "state of emergency" – that is, in Carl Schmitt's definition, from the "suspension of the entire existing order."[321] The order was suspended because it lacked intersubjective elements beyond dispute. There was no shortage of candidates for the position of *ens realissimum*, but their very multiplicity and diversity were perceived as a symptom of the collapse of order in general. The specter of contingency was abroad, and with it the task – but also the possibility – of establishing a new order.

The voices of the few who saw a solution in "tolerance of contingency," in living with ambivalence in the "state of emergency," were drowned out by those who insisted on the need to eliminate contingency, chaos and ambiguity.[322] According to Siegfried Kracauer, the real task was "to reach the filled space of reality with an overarching transcendent meaning," "to lead a cast-out humanity back into the new-old spheres of God-filled reality," or – in Lukács' words – to overcome "transcendental homelessness."[323]

This elimination of "crisis"[324] or "state of emergency" by producing or discovering a "filled space of reality with an overarching transcendent meaning" has a historical parallel that may help to clarify the concept of myth. In the Early

---

[321] Schmitt, *Political Theology. Four Chapters on the Concept of Sovereignty*, Chicago, IL 2006, p. 12.

[322] See Makropoulos, *Haltlose Souveranität*, p. 211.

[323] Siegfried Kracauer, "Soziologie als Wissenschaft," in Siegfried Kracauer, *Schriften*, vol. 1, Frankfurt/Main 1974, p. 7, quoted from Makropoulos, *Haltlose Souveranität*, p. 201. On transcendental homelessness, see Lukács, *The Theory of the Novel*.

[324] On the concept of crisis in this context, see Michael Makropoulos, "Crisis and Contingency: Two Categories of the Discourse of Classical Modernity," *Thesis Eleven* 111/2012.

Romantics we can see forerunners who already recognized the impending danger of a "state of emergency."[325] They held that the analytic, order-dissolving spirit of the Enlightenment had led not only to the emancipation of immature citizens but, at the same time, to the destruction of social cohesion; it had not only delegitimized the order of the *Ancien Régime* but threatened the possibility of order as such. The Early Romantics wanted to overcome this danger by means of a "new mythology" rooted in the idea of freedom, which would create a "God-filled reality" at once old and new. This reference to the Early Romantics serves to clarify a certain awareness of problems and strategy for handling them that accompanied the task of order from the very beginning. It would be wrong however, to conclude from this parallel that there is a direct continuity between Romanticism and (above all German) fascism.

The idea of a new mythology first appeared in *The Earliest System-Program of German Idealism*. The author or authors of this text from 1797 – it is unclear who of Schelling, Hegel and Hölderlin actually wrote it – believed that with a new mythology it would be possible to neutralize the atomizing effects of analytic Enlightenment reason and to counterpose an organic community to the "mechanistic state" of separate individuals.[326] Going beyond justified critique of the legitimacy of feudalism and absolutism, the analytic rationality of the Enlightenment had ground down "all synthetic positivities (the certainties of faith, the divine right of kings, aristocratic privileges, even self-certainty and morality)."[327] But this raised the danger that political rule and society, indeed order as such, could no longer be legitimized. For a (stable) order rested precisely upon "synthetic formations" and required "transcendent legitimation." The new mythology was meant to

> reverse the self-abrogation of the legitimizing force of reason, by recalling that the self-destructive forces of analysis were dependent upon a founding synthesis that no longer, as before, had to be handed down in religious traditions but could be the result of solidaristic invention: the work of poets and writers. [...] Poetry (for Wagner or Nietzsche, also music) would not only pave the way for but actually be this new mythology.[328]

The "suspension of the entire existing order" (which the Early Romantics also perceived as a threat) and the quest for order had emerged as a result of the "creeping

---

[325] On Early Romanticism, see above all the work of Manfred Frank, *The Philosophical Foundations of Early German Romanticism*, Albany, NY 2004; *Der kommende Gott; Einführung in die frühromantische Ästhetik. Vorlesungen*, Frankfurt/Main 1989; *"Unendliche Annäherung." Die Anfänge der philosophischen Frühromantik*, Frankfurt/Main 1997. For an overview of Romanticism, see Rüdiger Safranski, *Romanticism: A German Affair*, Evanston, IL 2014.

[326] On the dating and authorship of this text, see Christoph Jamme/Helmut Schneider (eds.), *Mythologie der Vernunft. Hegels "ältestes Systemprogramm des deutschen Idealismus,"* Frankfurt/Main 1984.

[327] Frank, *Gott im Exil*, p. 10. On the following, see also Frank, *Der kommende Gott*, pp. 188ff.

[328] Frank, *Gott im Exil*, p. 11.

death of God" set in motion by the Enlightenment.[329] The narrative of progress or the myth of reason was able for a time to fill the semantic vacuum: that is, to take the place of "God."[330] But the Romantics already realized that reason had undermined its own foundations and was no longer capable of legitimizing itself.

Since, for the *System-Program*, "truth and goodness become sisters in the idea of beauty," the task was to reendow poetry with a "higher dignity" as the "teacher of mankind," and to bring into being a "mythology of reason."[331] We shall return in a moment to the higher dignity of poetry or aesthetics but first we must show that the Early Romantic solution to the "state of emergency" resulting from Enlightenment critique was sought in a poetic metanarrative: that is, in man's poetic-creative achievement. It was a paradoxical, if not aporetic, situation that marked the whole quest for order: the Early Romantics were looking for an absolute as the center of a noncontingent order, but that absolute could "only" be brought into being poetically. In the wake of the Enlightenment, religious and aesthetic truths were delegitimized and truths of reason were absolutized. A poetically constructed absolute was thus immediately open to the charge of contingency.[332]

The quest for order arose out of the capacity of Enlightenment thought to deconstruct and historicize myths – a capacity that by no means did away with the need for order. The only "solution" to the state of emergency therefore consisted

---

[329] See Martin Heidegger, "Nietzsche's Word 'God Is Dead'," in Martin Heidegger, *Off the Beaten Track [Holzwege]*, Cambridge 2002, pp. 157–99.

[330] See Frank, *Der kommende Gott*, p. 192: "On the one hand, science (robbed of its practical dimension) criticizes the dogmatism of traditional (e.g. religious) explanatory models and grounds its claim to universal validity upon this unflinching negativity. On the other hand, it evades critique by assuming even the function of legitimation, the actually existing asymmetrical functional relations, that it serves as a front-ranking productive force. The ideology of autonomous rationality appeared instead of the criticized myth; that is, rational procedures themselves became mythoid."

[331] "The Earliest System-Program of German Idealism," in Jon Stewart (ed.), *Miscellaneous Writings of G. W. F. Hegel*, Evanston, IL 2002, p. 111.

[332] On this paradoxical situation, see the account of Benjamin's critique of Carl Schmitt's theory of sovereignty and constructions of aesthetic totality, in Makropoulos, *Haltlose Souveränität*, p, 206, and *Modernität als ontologischer Aufnahmezustand?*, pp. 103–58. See also Frank, *Gott im Exil*, p. 23, which argues that Nietzsche's *Übermensch* differs from "previous humans" in being able to live successfully in the "state of emergency" and to overcome it as a "poetic writer of his life": "Man until now [...] has required a supra-sensuous guarantee in order to live – a supreme point of support. If he is now deprived of one, he can live only by supporting himself with values of his own creation. But self-created values differ fundamentally from religious values in that they refer back to the creator: that is, are literally relative to me, the creator; whereas the destroyed values, whose groundlessness now became apparent, seemed to exist independently of the human subject, and they were absolute ('loosened-away'), that is, valid and support-giving, only within that independence. The *Übermensch* is thus the ideal human being, who can live with this withdrawal of the absolute; he has become godlike himself, or believes in a new god that Nietzsche calls life and that the name Dionysus vouches for in his work."

in myth that was itself immediately exposed to deconstruction. This contributed to the radicalization following the First World War, since fascism could be seen as an attempt to establish a new mythology that removed the "state of emergency." Only abandonment of the quest for order, only acceptance of the impossibility of order, would have made it possible to endure the "state of emergency" and to live with contingency and ambivalence.

The efforts of Warburg and others to ward off the "absolutism of reality" with the help of science and reason were quite incapable of eliminating the "state of emergency." Indeed, they strengthened the feeling that reason offered no secure foothold. Here Max Weber's "Science as a Vocation" lecture, given in Munich on November 7, 1917, is enough to remind us that it was a major problem for intellectuals that science had been unable to legitimize itself and to provide meaning during and after the First World War. Many, especially among the young, looked for a solution – or rather, salvation – in a kind of religion: and that meant in a mythical order:

And today? Who – aside from certain big children who are indeed found in the natural sciences – still believes that the findings of astronomy, biology, physics, or chemistry could teach us anything about the *meaning* of the world? If there is any such "meaning," along what road could one come upon its tracks? If these natural sciences lead to anything in this way, they are apt to make the belief that there is such a thing as the "meaning" of the universe die out at its very roots. And finally, science as a way "to God?" Science, this specifically irreligious power? That science today is irreligious no one will doubt in his innermost being, even if he will not admit it to himself. Redemption from the rationalism and intellectualism of science is the fundamental presupposition of living in union with the divine. This, or something similar in meaning, is one of the fundamental watchwords one hears among German youth, whose feelings are attuned to religion or who crave religious experiences.[333]

And Weber went on to quote Tolstoy: "Science is meaningless because it gives no answer to our question, the only question important for us: 'What shall we do and how shall we live?' That science does not give an answer to this is indisputable."

Weber thought that science could at least help those who asked the question correctly. It is doubtful whether most of his listeners agreed with him: they and the yearly student intakes that followed them by no means dismissed science and technology *per se*, but they sought to embed them in a different "suprahistorical" context that promised to satisfy their need for meaning and their longing for a transcendental home. That context also permitted them to employ instrumental reason and "science" for the deadly misanthropic "gardening ambitions" of the Nazi state.[334] Many fled from the "disenchanted world" into the realm of myth,

---

[333] Max Weber. "Science as a Vocation," in H. H. Gerth/C. Wright Mills (eds.), *From Max Weber. Essays in Sociology*, Boston, MA 1948, pp. 141–42.

[334] See Zygmunt Bauman, *Modernity and the Holocaust*, Ithaca, NY 1989; Bauman, *Modernity and Ambivalence*, esp. pp. 26–52; and Peukert, "Die Genesis der 'Endlösung' aus dem Geist der Wissenschaft," in *Max Webers Diagnose der Moderne*, Göttingen 1989.

whose pivotal point was the nation or *Volk*, and which charged science and technology with meaning even when these were used for the killing of millions.[335]

Myth as a response to the crisis of historicism

The religious mood and hunger for spiritual experience that Weber observed among young people was merged in the fascist political "religion." The mythical order underlying this also solved a further problem of the time: "*the* or *an* achievement of myth lies in the normative sphere."[336] War, revolution, and breakup of the old order led to collapse of the values and norms of the societies in question.[337] But these values and norms had already been shaken to their foundations, even though this was not felt by the majority of people. Their validity had proven to be no more than provisional. We cannot enter here into the background to the crisis of historicism and the problem of value relativism, so we shall have to content ourselves with a brief look at the consciousness of crisis.[338]

Unless people were prepared to give up the quest for order, the only way out of the crisis of historicism was through a mythical order. Categories of thought based on a concept of historical progression had established themselves in the wake of the Enlightenment;[339] they were a product, but also a precondition, of the ongoing critique of the (rationally ungrounded) metaphysical-religious pillars of the *Ancien Régime*. It was out of the critique of those metaphysical foundations that a critique of the historical (temporary and relative) foundations of the old order had developed.

In June 1922, the theologian Ernst Troeltsch published his essay "Die Krisis des Historismus" [The Crisis of Historicism] in *Die Neue Rundschau*. Since this incisively takes up the nexus between value relativism and historicism, it is worth quoting at some length:

> The breakup of the old tablets of values became something of a catchphrase [in the course of the nineteenth century], and in the end there were no new

---

[335] See Ulrich Herbert, *Best. Biographische Studienüber Radikalismus, Weltanschauung und Vernunft, 1903–1989*, 3rd edn, Bonn 1996, pp. 51–100; and Michael Wildt, *An Uncompromising Generation: The Nazi Leadership of the Reich Security Main Office*, Madison, WI 2009.

[336] Frank, *Gott im Exil*, p. 16.

[337] Martin Geyer has analyzed this collapse in the city of Munich: see Martin H. Geyer, *Verkehrte Welt. Revolution, Inflation und Moderne, München 1914–1924*, Göttingen 1998.

[338] On the antihistoricist complex, see Doering-Manteuffel, *Mensch, Maschine, Zeit*; Anselm Doering-Manteuffel, "Suchbewegungen in der Moderne. Religion im politischen Feld der Weimarer Republik," in Friedrich Wilhelm Graf/Klaus Große Kracht (eds.), *Religion und Gesellschaft. Europa im 20. Jahrhundert*, Cologne 2007, pp. 175–202; Jaeger, *Theorietypen der Krise des Historismus*, pp. 52–70; Wolfgang Hardtwig, "Die Krise des Geschichtsbewusstseins in Kaiserreich und Weimarer Republik"; Mai, *Agrarische Transition und industrielle Krise*; Makropoulos, *Haltlose Souveränität*, and "Crisis and Contingency"; Nowak, "Die 'antihistoristische Revolution'"; Oexle, "Von Nietzsche zu Max Weber," and "Krise des Historismus – Krise der Wirklichkeit."

[339] See Reinhart Koselleck, "The Eighteenth Century as the Beginning of Modernity," in Koselleck, *The Practice of Conceptual History*, pp. 154–69; and "'Space of Experience' and 'Horizon of Expectation' – Two Historical Categories."

tablets of values. [...] The confusion came to a head in the world war, which destroyed a host of old certainties and related historical constructions, but did not establish new ones. [...] But since these are old historical values, since their development and form have been mainly brought before us by history, the crisis is at the same time a crisis of history in its innermost core.[340]

In another passage, Troeltsch further explains the nexus between value break-down and relativism and the "crisis of history in its innermost core":

> Here we see everything in the river of becoming, in endless, ever new individu-alization, in determination by the past toward an unknown future. *State, law, morality, religion, and art are dissolved in the flux of historical becoming and are eve-rywhere intelligible to us only as parts of historical developments.* [...] It [historical understanding] shatters all eternal truths, be they ecclesiastical-supernatural (and hence supremely authoritative), be they eternal truths of reason and rational constructs of state, law, society, religion, and ethics, or be they gov-ernment constraints relating to secular authority and its prevailing form. [...] Intellectual life no longer participates in fixed, immutable truths above nature and the senses; nor does it consist in elucidating the general human truths of reason or common sense against the vagaries of superstition and fantasy, or in exploring natural law and the reconstruction of state and society on its basis. Rather, it is a continuous yet ever changing current of life, in which fleeting eddies simulate an appearance of permanence and independent existence.[341]

Even reason had proven to be historically relative. It was not that universal, eternal, godlike light of truth that Enlightenment thinkers and men such as Aby Warburg had taken it to be. All is flux: that was the root of the crisis. And *a priori* the crisis could be solved only by stepping outside the current of time into a puta-tive realm of the eternal.[342]

Fascism was perceived as the solution to the problem because it seemed able to halt the flow of becoming and the dissolution of values. Fascism – one thinks here of Eliade's analyses – sought to annul history, to regenerate time and to confer on historical events a meaning above and outside history. More: it yearned for "eternal values" – eternal because they had a link with the sacred, with the eter-nal nation exempt by definition from historical becoming, or with a *Volk* equally beyond history. This stepping outside historical time and historical understanding was possible only in and through myth.

---

[340] Ernst Troeltsch, "Die Krisis des Historismus," in *Schriften zur Politik und Kulturphilosophie (1918–1923)*, vol. 15, Berlin 2002, pp. 433–55; here pp. 447–49.

[341] Ibid., pp. 437–38.

[342] See Anselm Doering-Manteuffel, "Mensch, Maschine, Zeit. Fortschrittsbewusstsein und Kulturkritik im ersten Drittel des 20. Jahrhunderts," in *Jahrbuch des Historischen Kollegs 2003*, Munich 2004; and Hardtwig, "Die Krise des Geschichtsbewusstseins in Kaiserreich und Weimarer Republik."

Again a brief look at Nietzsche's second "untimely meditation" will help to explain this nexus of myth and "permanent" values. In his primary antihistoricist text, Nietzsche argued that an excess of history led to destruction of the illusory (but existentially necessary) foundations of a community. Culture and life itself can flourish only within a mythical horizon of meaning.[343] The "capacity to feel *unhistorically*," hence in mythical thought, is "the foundation upon which alone anything sound, healthy, and great, anything truly human, can grow."

> The unhistorical is like an atmosphere within which alone life can germinate and with the destruction of which it must vanish. It is true that only by imposing limits on this unhistorical element by thinking, reflecting, comparing, distinguishing, drawing conclusions, only through the appearance within that encompassing cloud of a vivid flash of light – thus only through the power of employing the past for the purposes of life and of again introducing into history that which has been done and is gone – did man become man: but with an excess of history man again ceases to exist, and without that envelope of the unhistorical he would never have begun or dared to begin.[344]

That "atmosphere within which alone life can germinate" is history annulled by myth. Myth turns out to be existentially necessary for a mind that wishes to destroy ambivalence. Nietzsche concedes that both "the unhistorical and the historical are necessary [...] for the health of an individual, of a people, and of a culture."[345] But "the historical sense" uproots the future if it "reigns without restraint," since it leaves behind a world without illusions.[346]

Science, according to Nietzsche, "requires superintendence and supervision"; that is the only antidote to the "malady of history."[347] History must tolerate being "transformed into a work of art." For art and religion are "suprahistorical powers,"

> which lead the eye away from becoming toward that which bestows upon existence the character of the eternal and stable [...]. Science [...] sees in these two forces hostile forces: for science considers the only right and true way of regarding things, that is to say the only scientific way, as being that which sees everywhere things that have been, things historical, and nowhere things that are, things eternal.[348]

---

[343] See Hans-Georg Gadamer, "Mythos und Vernunft," in *Gesammelte Werke*, vol. 8, *Ästhetik und Poetik*, Tübingen 1993, p. 165.

[344] Nietzsche, Friedrich Nietzsche, "On the Uses and Disadvantages of History for Life (1874)," in *Untimely Meditations*, Cambridge 1983, pp. 64–5.

[345] Ibid., p. 63.

[346] Ibid., p. 95. Cf. ibid: "If a religion, for example, which is intended to be transformed into historical knowledge under the hegemony of pure historical justice, a religion which is intended to be understood through and through as an object of science and learning, will when this process is at an end also be found to have been destroyed. The reason is that historical verification always brings to light so much that is false, crude, inhuman, absurd, violent that the mood of pious illusion in which alone anything that wants to live can live necessarily crumbles away."

[347] Ibid., p. 121.

[348] Ibid., p. 120.

In common with the Early Romantics, Nietzsche pointed to the "self-destructive forces of analysis" and recalled their "dependence upon a founding synthesis."[349] Thus, only "suprahistorical" powers established in and through myth could overcome both the crisis of reason and the value relativism bound up with historicism. The basic paradox of modernity was that life required both the conscious dissolution and the conscious generation of order, powers that eliminated myths and produced freedom as well as powers that created myths and produced order.

In the aftermath of the First World War, action in accordance with time-specific, context-dependent values counted as a "state of emergency," which could be rectified only through "eternal values." Nietzsche knew that such eternal values were illusions, that everything was historically determined and subject to becoming. But for that very reason he held that man could live only within an enveloping "atmosphere," illusory though it was. Most fascists were not conscious that their mythical, "suprahistorical" order was based upon an illusion. In their case, unlike Nietzsche's, it should not be assumed that they saw the impossibility of the suprahistorical yet chose to base themselves on it, for that would not have allowed them to escape contingency. The illusory character of the eternal nation or *Volk* on which fascism rested appears – only appears, because the contents of the mind cannot be communicated historically – to have been known only to the radical intellectuals and nihilists gathered together, for example, in the Reich Security Headquarters. For a time only war and struggle counted for them as eternal principles. "When there are no longer any generally valid goals (but only interests)," wrote Ulrich Herbert regarding Werner Best's essay of 1930 *Der Krieg und das Recht*,

> "a different ethic [must] arise out of this affirmation of reality and rejection of a redemptive goal [...] than out of the teleology of other doctrines." [...] When no value exists with a claim to certainty and permanence, what remains as an ethical standard is not a content but only the form.[350]

Since all is flux, only struggle remains as an eternal principle. This rejection of absolute values and utopian ideals could not be publicly promoted; such a "heroic" idea of the contingency of human existence was reserved for the "worldview elite," whose ideology subsequently exhausted itself in fighting and killing.[351] But even that nihilist "elite" appears to have thought of the *Volk* or the

---

[349] Frank, *Gott im Exil*, p. 11.

[350] Herbert, *Best*, p. 97.

[351] See ibid., p. 100. Herbert observed that Best's essay "lacked an explicit orientation to the *Volk* as the 'highest aim in life'," but that this did not contradict his (Best's) later statements according to which he had drawn a "non-teleological and pessimistic" picture of the world "in contrast to the 'teleological and optimistic *völkisch* view of life that [he] otherwise continued to represent'." We cannot clarify here whether or not the *Volk* or its continued existence was a redemptive goal for Best and his colleagues. But it seems plausible to assume that "*Volk* preservation" or struggle for the "life" of the "community of blood and destiny" repeatedly served to justify their murderous activity and constituted a "discourse of disinhibition and self-exoneration." In this sense, even the criminals at the Reich Security Headquarters – who

"life of the community of blood and destiny" as the ultimate value – as if they too sometimes belonged among the despised "others for whom only belief in an ultimate goal makes the present struggle bearable."[352]

The "others" – who included Hitler's mystical "chief ideologue" Alfred Rosenberg – considered the nation and *Volk*, aside from any philosophical points or intellectual reflection, as quite simply the *ens realissimum*. Not only did they "believe" in *Blut* and *Volk*; they thought its truth and validity could be scientifically demonstrated.[353] They transferred the structure of myth and the essence of the transcendental to an immanent essence; a worldly reality, whether "nation," "*Volk*" or "blood," acquired the status of the supernatural, and vice versa. This transfer and merging made it possible to satisfy both the contemporary faith in science and the longing for a transcendental homeland, as well as to create an order of a mythical nature that could nevertheless incorporate the claim to "scientific truth." This dual structure mirrored the paradoxical or aporetic character of the whole of modern existence: on the one hand, a longing for order that only myth can satisfy; on the other hand, a critical attitude to myth, involving both a capacity to deconstruct it and an incapacity to merely believe in myth without simultaneously "knowing."

The reference to Nietzsche, as previously to the Early Romantics, is not meant to demonstrate a continuity with fascism, but rather to elucidate the problem that arose through the breakup of the old "divine" order. Nietzsche, like the Early Romantics, perceived this problem seismographically and, having made it the object of philosophical reflection, thought he had found the solution in myth. It would be a mistake to assume the same degree of philosophical reflection among most of his contemporaries. It is not suggested here that fascism should be understood as conscious, or even manipulative, myth-making in response to the quest for order, such as Nietzsche or the Early Romantics proposed. Rather, fascism is conceived as a response to the quest for order and an attempt to solve the shifting, ever more radical, problem that Nietzsche and the Early Romantics diagnosed for modernity. We have shown that the task of order could be handled only with a myth. For only in myth can a "suprahistorical" element be counterposed to permanent becoming, and only in myth can clarity get the better of ambivalence and contingency.

After the First World War, the longing for order was too great, chaos and the "absolutism of reality" were pressing too hard and the feasibility of order was too strong for most people to be able to live with contingency and ambivalence.[354] This "solution" presupposes insight into the misanthropic and murderous

---

thought of themselves as "coldly objective" nihilists, above all in extreme situations affecting their own and especially other people's lives – needed a higher goal, "the justification of which was no longer a matter of reflection but was given a priori and therefore cancelled the force of acquired humanitarian principles." This rather seems to suggest that fascist intellectuals too needed an absolute or sacred focus of order. See ibid., p. 528; and Helmut Lethen, *Verhaltenslehren der Kälte. Lebensversuche zwischen den Kriegen*, Frankfurt/Main 1994.

[352] Werner Best, "Der Krieg und das Recht," in Ernst Jünger (ed.), *Krieg und Krieger*, Berlin 1930, p. 152.

[353] On Alfred Rosenberg and his *Myth of the Twentieth Century*, see Ernst Piper, *Alfred Rosenberg. Hitlers Chefideologe*, Munich 2005; and Frank, *Gott im Exil*, pp. 105–30.

[354] See Bauman, *Modernity and Ambivalence*.

consequences of the quest for order, into the limits of feasibility, and into the impossibility of a closed and finished order. It is open to doubt whether the longing for order is thereby vanquished. In any event, the project of order was in some places already abandoned.

The aestheticization of politics in and through myth

Before we turn again to the complementarity of myth and modernity, let us very briefly review the aesthetic dimension of myth that has been demonstrated with reference to the works of D'Annunzio, Jünger and the Futurists, the popular accounts of aerial warfare and the Esposizione dell'aeronautica italiana. We shall see the extent to which the "suprahistorical powers" of art and religion can be understood as a unity – a political unity, moreover. Once this unity of political art and political religion in myth has been elucidated, it will be transferred to fascism and its "everyday aviation myths."

Along with the supernatural truth of religion, the truth of art had lost its power in the wake of the Enlightenment – not least in the "subjectivization of aesthetics through the Kantian critique."[355] Only that was true that could be demonstrated and scientifically proven with the means of reason; what was believed and what was found beautiful could not be grasped with the categories of truth and falsity. The truth specific to the artwork and religion is based on their distinctive mode of expression, which is mythical. The semiotic structure of myth cannot be discussed further here, since no linguistic-structural analysis has been made of the aviation narrative.[356] But we should try to clarify in passing the nexus of religion and art, and hence of sacralization and aestheticization, in myth, once again with the help of the Early Romantics and Nietzsche.[357]

The Early Romantics, as well as Nietzsche and Wagner, did not think of art and religion as either separate or private matters. Intellectuals searching for a new mythology attacked just such a "privatization," which they saw as the root cause of alienation, individualism and social fragmentation. Indeed, in their eyes, art and religion were the very opposite of the private; they ought to be a "public affair," a *res publica*. For a community could be created only through public-political art and religion. These were bearers of the *nomos* of a society: they communicated its norms and values to the community.

The Early Romantics, then, like Nietzsche and Wagner, started from the model of Greek tragic drama. Rooted in mythology, it was not the work of an individual artist, nor therefore a purely subjective and contingent product. Tragedy was "true

---

[355] Here and on "the truth of the work of art," see Hans-Georg Gadamer, *Truth and Method*, 2nd rev. edn., London 1989, pp. 42–81.

[356] See Barthes, *Mythologies*, pp. 50–73; Lévi-Strauss, "The Structural Study of Myth"; and Frank, *Der kommende Gott*, pp. 107–08.

[357] In short we may say that, for the Early Romantics too, there was not only scientific truth and its propositions regarding particular facts, but also a superior, more primal, poetic truth native to art. Only in the creative language of the poet was the world deciphered and opened up to sense experience. See Frank, *Einführung in die frühromantische Ästhetik*, pp. 16ff.

poetry that enjoyed general validity," because it sprang "from a people's spiritual homeland, from truly public life," and because it was born "out of the totality of a nation, which as such conducts itself at the same time as identity – as individual."[358] Thus, tragedy gave expression to the value beliefs of "the whole" community.

Audience participation in a tragic performance within the framework of public ritual brought about a regeneration "of the ethical totality of the people."[359] For the tragedy was where "the social synthesis of the members of the community was acted out in ritual form."[360] It was a religious and aesthetic ceremony where the community constituted itself and affirmed its values – "an enviable (lost) unity of *ritual action, artistic performance, and political representation.*"[361] Tragedy was a "representation of public life."[362] But can this perspective be transferred to fascism?

The transfer is bound to fail if it is restricted to fascist theatre and "high" culture. It is true that there are definite parallels – Nazi open-air *Thingspiel*,[363] Italian attempts to establish a "theatre of the masses for the masses"[364] – but they enjoyed little success. Yet the fascist community did have more effective religious and aesthetic ceremonies, at which it constituted itself and bolstered its norms and values.[365] The Party conferences in Nuremberg, the funerals of "heroes," the exhibition on the tenth anniversary of the Fascist revolution, or the EAI – to name but a few examples – should be seen as just such ritual aesthetic events. But in the age of the masses, the aesthetic staging and objectification of myth may also be located elsewhere.

Roland Barthes found the post-war French "myth of everyday life" in popular cultural events, newspapers and glossy magazines, movies and photographs, even automobiles. His starting point was that, in "the age of mechanical reproduction," the work of art might have lost its original ritual aura, but that mass culture was nevertheless the bearer of myth.[366] Previous chapters have shown how

---

[358] Friedrich Wilhelm Joseph Schelling, *The Philosophy of Art*; quoted [and translated] from Frank, *Der kommende Gott*, p. 200.

[359] Christoph Jamme, *"Gott an hat ein Gewand."* Grenzen und Perspektiven philosophischer *Mythos-Theorien*, Frankfurt/Main 1999, p. 274.

[360] Frank, *Gott im Exil*, p. 11.

[361] Ibid., p. 72. Emphasis in the original.

[362] Frank, *Der kommende Gott*, pp. 197ff.

[363] See Erika Fischer-Lichte, *Theatre, Sacrifice, Ritual. Exploring Forms of Political Theatre*, London 2005; Henning Eichberg, Michael Dultz/Glen W. Gadberry/Günther Rühle, *Massenspiele. NS-Thingspiel, Arbeiterweihespiel und olympisches Zeremoniell*, Stuttgart 1977; Reichel, *Der schöne Schein des Dritten Reiches*, pp. 435–47; and Rainer Stommer, *Die inszenierte Volksgemeinschaft. Die "Thing- Bewegung" im Dritten Reich*, Marburg 1983.

[364] See Günter Berghaus (ed.), *Fascism and Theater. Comparative Studies on the Aesthetics and Politics of Performance in Europe, 1925–1945*, Providence, RI 1996; and Schnapp, *Staging Fascism*.

[365] Behrenbeck, *Der Kult um die toten Helden*; Gentile, *The Sacralization of Politics*, and *The Struggle for Modernity*; Mosse, *The Nationalization of the Masses*; George L. Mosse, "Toward a General Theory of Fascism," in Mosse (ed.), *The Fascist Revolution. Toward a General Theory of Fascism*, New York 1999, pp. 1–44; Mosse, "Fascist Aesthetics and Society. Some Considerations," in ibid., pp. 45–53; Vondung, *Magie und Manipulation*.

[366] See Benjamin, "The Work of Art in the Age of Mechanical Reproduction."

cultural events may be read as a medium of socially important narratives in the normative sphere. In the mass media too, "something existing in nature or mankind was related to a sacred sphere and grounded through that connection."[367] And the popular narratives of aviator-heroes bore a strong resemblance – in their function, not their execution – to Greek tragedy or Wagnerian opera. They were bearers of myth, and the aviator-heroes conveyed certain social values and norms by imitating model characters, adopting characteristic patterns of perception and interpretation, and inscribing the narratives in their own lives.

The Early Romantics were confident that the "new mythology" had its origins in "the people." Schelling, for instance, wrote in his Würzburg lecture of 1804 on the philosophy of art: "True poetry with general validity can arise only out of the spiritual unity of a people, out of a genuinely public life – just as science and religion find their objectivity only in the spiritual and political unity of a people."[368] The mythical work for which Schelling yearned was not regarded as the poet's subjective achievement. Rather, the poet appeared as the mouthpiece of the community; he was expected to fortify it by creating, not least through folk tales and songs, the new-old myth rooted in the national language and the spirit permeating the *Volk*.

Richard Wagner, who stood in the tradition of German Romanticism as well as early French Socialism, thought he could express the people's will and build a mythical community by harking back to ancient Teutonic sagas. The poet's job, in the "art work of the future," was "consciously to take possession of the objective bond that united the people as a *de facto* community of hardship and need, in the hour of man's greatest alienation from man."[369] Wagner saw the artist as the means to bring forth the myth whose "true poet and artist," as in times of old, "was in truth the people."[370]

This study has shown that, the fascist myth was poetry that was considered "true and generally valid" and that had its roots in "the spiritual homeland of a people." To avoid misunderstandings, a word of explanation is necessary here. The popular war books made it clear how the fascist mythical order spread into society through mass media that grew as consumption articles before and during the First World War. In popular culture, as in public speeches and buildings or the elite culture of a D'Annunzio, Marinetti or Jünger, a new order was being established.

All the media we have examined – from penny novels and glossy magazines to the canvases of Futurist *aeropittura* – performed this work on myth. The creative input came not from one individual, but from a wide range extending through anonymous media workers to prominent spokesmen. The works in question constantly transmitted norms and values, social and human models. And texts became actual reality when the recipients transferred these values, norms and

---

[367] Frank, *Gott im Exil*, p. 16.

[368] Quoted [and translated] from Frank, *Der kommende Gott*, p. 200.

[369] Frank, *Gott im Exil*, p. 76.

[370] Richard Wagner, "Das Kunstwerk der Zukunft," quoted [and translated] from Frank, *Gott im Exil*.

models to their own lives and began to perceive and interpret the world accordingly. It has become clear that the figure of the aviator should be understood as an attempt to construct a new man in line with the changed conditions – a new man who would be an integral part not only of the regenerated nation but of a new order. The values already embodied in the aviator-heroes – "manliness," heroism, self-sacrifice, pioneering spirit, wish to go ever higher and further – had to be imitated if the *Volk* was to become a "nation of fliers." The hero-constructs and their post-war metamorphoses demonstrated the need for radical transformation: henceforth the individual was to be nothing, the people or nation everything. The popular media and works of high culture remolded mass society as community; the alienated cog in a machine became the fascist new man, an organic member of the collective based upon principles of exclusion. "Flying swords" became flying swarms. Seemingly chaotic disorder gave way to a hierarchical structure in which fliers were a new nobility, based on their willingness to sacrifice themselves for the community.

It would be a mistake to see nothing but manipulative propaganda in the polyphonous "work on myth," as Benjamin (in his artwork essay) and Horkheimer and Adorno (in their writings on the culture industry) all did.[371] Fascist myth did not simply spread outward from a centrally controlled machinery. Its production and dissemination were more variegated than the Marxist thesis would suggest, with its monolithic, manipulative (late capitalist) apparatus of fascist propaganda generating a false appearance to dupe the proletariat. Our study of aviation discourse has shown that central tropes and speech rules of the fascist myth had already been established during the First World War, in relative autonomy from state-directed opinion formation.

Amateur writers and dilettantes, ever more fired up with grassroots nationalism, produced a mass of gung-ho pro-war "literature" in which killing others and laying down one's life for the nation acquired a semblance of meaning.[372] This "literary mobilization" was spontaneous. It was not initiated from above but grew out of "a people's spiritual homeland," out of discourses and order in the making. The speakers in question endeavored to structure the reality of the First World War, and in doing this they reverted to social categories of interpretation that they updated and changed. They were certainly influenced by propaganda, but the key to what they said was its function in coming to terms with the frightening reality of a war that seemed devoid of meaning. The appropriation of underlying discursive structures to interpret this new and threatening reality led to a change of those structures, to a change of myth and order. Semantic networks and rules of evidence were gradually modified. Existing associations – between sacrifice and rebirth, for instance – were enlisted to interpret the reality of war and thereby transposed to another discursive context. If, as a result, reality was

---

[371] Benjamin, "The Work of Art in the Age of Mechanical Reproduction," pp. 241–42; Adorno/Horkheimer, *Dialectic of Enlightenment.*

[372] See *inter alia* Fries, "Deutsche Schriftsteller im Ersten Weltkrieg"; and Schneider, "Zur deutschen Kriegsliteratur im Ersten Weltkrieg."

made meaningful, this led to a multiplication of similar statements – hence to their inscription into the discourse. State censorship then sanctioned, marginalized or promoted the statements available for interpretation of the reality of war. This led to the disappearance of contrary discourses and to the inscription of further elements into the discourse, so that little by little a new mythical order was constructed. For a statement to count as true and meaningful, it more and more often had to be associated with the nation or *Volk*, and only with the nation or *Volk*.

Within this perspective, the mythification process turns out to involve the marginalization of certain interpretive categories and the establishment or potentiation of others. Myth may also be seen as the establishment of a discursive hegemony: that is, it may be regarded as the outcome of a reduction of apparently meaningful statements, which can all be traced to an Absolute that tolerates none other beside it. That which is "intersubjectively beyond dispute" – that is, the sacred core of the myth with which the statements are associated – should be understood as the product of the interpretive practice. To be beyond dispute is to count as the guarantee of meaning: that is, to be capable of making the world appear ordered.

We should mention an essential aspect of this discursive practice that recalls the link with aestheticization: that is, the possibility of participation. Of course, this does not refer to political participation in the sense of democratic popular sovereignty, but rather to participation in "meaning" – which is perfectly consistent with Benjamin's observation that "fascism sees its salvation in giving the masses not their right, but instead a chance to express themselves."[373] Mass participation was limited to the semantic community, whether through "uplifting" reading of a hero's biography, identification with the protagonists of a book, wonder at a Futurist painting, or attendance at a mass event such as the EAI.[374] In establishing such possibilities, fascists seem to have satisfied the need for meaning of a disoriented society. Participation in mythical meaning, elimination of "transcendental homelessness," evidently weighed more heavily in the balance than any prospect of participation in a democratic political process.

It would certainly be wrong to suppose that all sectors and strata of society actively produced mythical meaning. The speakers in the discourse came mainly from upper and middle strata; they were articulate and capable of making their voice heard because they occupied suitable speaker positions. The fascist power apparatus excluded speakers who represented a deviant opinion, terrorizing them and sidelining any statements they managed to make. But the "passive" reproduction and updating of mythical meaning was a phenomenon that stretched right across society and rested upon a need felt in every layer. In this respect, the idea that the fascist elite did not actually believe in the myth but had a purely

---

[373] Benjamin, "The Work of Art in the Age of Mechanical Reproduction," p. 241.
[374] George L. Mosse, "Fascism and the French Revolution," in Mosse, *The Fascist Revolution. Toward a General Theory of Fascism*, New York 1999, pp. 69–93; here p. 73.

instrumental relationship to it is mistaken.[375] The fascists and future rulers were equally affected by the longing for order and a transcendental home. Drawing upon a number of existing traditions, they formulated a response to that longing that had an integrative as well as an exclusive effect. What George L. Mosse calls their "cultural revolution" rested upon a popular consensus.[376]

Fascist myth grew out of the space of experience and horizon of expectation of the national collective.[377] Myth both established and expressed the "sacred": it was thus at once originator and bearer of the sacred Realissimum. Sacralization and aestheticization of politics were united in myth, since it not only constituted "the sacred" but made it perceptible to the senses. The unity by means of which the fascists sought to produce community was lost through the analytic separation of sacralization and aestheticization. The unity of the aesthetic and sacral aspects of myth was as clear in the Futurists and the EAI as it was in popular war books or in D'Annunzio.

D'Annunzio was not only priest and prophet of the nation and the war; he was always an artist as well. He brought a certain character type to the political stage of the twentieth century, remaining a reference point for Mussolini and Hitler by virtue of his "charismatic leader" qualities and his skillful use of *mise en scène*.[378] Another aspect seems decisive, however: the fusion of art and reality. Reality or history appeared malleable material to these artist-politicians, and so too did man. In their eyes it was possible to mold a new man and to reshape society into a "nation of fliers." That they had no doubts on this score emerges from a speech given by Mussolini in June 1940. At a time when the military weakness and inadequate preparation and weaponry of the Italian armed forces were becoming apparent, he directed his rage against the "material" at his disposal. But he did not doubt his skill and know-how or what they were capable of achieving: "It is the right material that I lack. Even Michelangelo needed marble to make his statues. If he had only had clay, he would have been no more than a potter."[379] Modernity as a longing for order and a task of order finds one of its clearest expressions in this view of reality.

The complementarity of myth and modernity

On January 4, 1889, the day he broke down in Turin, Nietzsche wrote to his friend and "discoverer" Georg Brandes: "After you had discovered me, it was no trick to find me; the difficulty now is to lose me…"[380] The underlying dilemma of modernity, its

---

[375] See Gentile, *The Struggle for Modernity*, pp. 77ff.; and Saul Friedländer, *Reflections of Nazism. An Essay on Kitsch and Death*, Bloomington, IN 1993.

[376] See Mosse, "Toward a General Theory of Fascism," p. 3; and Modris Eksteins, *Rites of Spring: The Great War and the Birth of the Modern Age*, New York, 2000 [repr.], p. 315.

[377] Reinhart Koselleck, "'Space of Experience' and 'Horizon of Expectation' – Two Historical Categories," in Koselleck (ed.), *Futures Past: On the Semantics of Historical Time*, Cambridge, MA 1985 [orig. 1979].

[378] See Mosse, "The Poet and the Exercise of Political Power," pp. 87–103; and Vogel-Walter, *D'Annunzio*.

[379] Benito Mussolini, speech of June 21, 1940, quoted [and translated] from Emilio Gentile, "L' 'uomo nuovo' del fascismo," p. 260.

[380] Friedrich Nietzsche to Georg Brandes, January 4, 1889, in Friedrich Nietzsche, *Sämtliche Briefe*, p. 573. See also Martin Heidegger, *What Is Called Thinking?*, New York 1968, p. 53.

paradoxical starting position, is best expressed in these few words. Zygmunt Bauman, in his *Modernity and Ambivalence*, relates the quotation to the discovery of contingency. Once man had discovered the randomness of order, its lack of necessity, the idea could no longer be eliminated. All order now had the defect that it could be possible but did not have to be necessary. In spite or because of this, order became the essential feature of modernity, its obsession. We cannot here go into the nexus between the discovery of contingency and the "death of God," between perception of the world as a chaos to be put in order and the emergence of order as a task. But since the idea is central for an understanding of the complementarity of myth and modernity, a few words of explanation are required.[381]

It was Nietzsche who most clearly grasped and described the "state of emergency" that fascists sought to remedy. He also strove to remedy the "state of emergency" by means of a mythical order. But in Nietzsche there is also the germ of the "postmodern" perspective in which modernity has been viewed here[382] – a perspective that may be summed up as involving insight into the impossibility of order. This dichotomy of Nietzsche's offers an initial explanation of the inner turmoil characteristic of modernity as a whole.

In Heidegger's work on Nietzsche published in 1961, which is based on earlier studies in the 1930s and 1940s, a section entitled "European Nihilism" highlights a problem underlying the longing for order.[383] Nietzsche's saying "God is dead" does not refer only to the Christian God.[384] Rather, the Christian God

> stands for the "transcendent" [*das Übersinnliche*] in general in its various meanings – for "ideals" and "norms," "principles" and "rules," "ends" and "values,"

---

[381] The concept of complementarity appears most appropriate for the relationship between myth and modernity, and at the same time for the observation/perception of that relationship, in which the two terms have the appearance of being mutually exclusive. Complementarity also refers to the principle introduced by Niels Bohr in 1927 into quantum physics, which is often explained with the example of wave and particle dualism – that is, with the discovery that atomic particles display two contradictory properties. Since they appear as both particles and waves, different measurement procedures are necessary for their observation. Carl Friedrich von Weizsäcker explains as follows the complementarity of the measurement procedures or research approaches: "The complementarity consists in the fact that they cannot be used simultaneously, yet both have to be used." See "Komplementarität und Logik," in Carl Friedrich von Weizsäcker, *Zum Weltbild der Physik*, 7th edn, Stuttgart 1958, pp. 281–331; here p. 284.

[382] On Nietzsche and the onset of modernity, see Jürgen Habermas, *The Philosophical Discourse of Modernity. Twelve Lectures*, Cambridge, MA 1987; Gianni Vattimo, *The End of Modernity. Nihilism and Hermeneutics in Postmodern Culture*, Baltimore, MD 1988; and Wolfgang Welsch, *Unsere postmoderne Moderne*, 3rd edn, Weinheim 1991.

[383] See Martin Heidegger, *Nietzsche*, vol. IV, San Francisco 1991, pp. 1–196. On the relationship between Heidegger and Nietzsche, see *inter alia* Hans-Helmuth Gander (ed.), *"Verwechselt mich vor allem nicht!" Heidegger und Nietzsche*, Frankfurt/Main 1994. On Heidegger's own enthusiasm for National Socialism and his search for a "another beginning," see Philippe Lacoue-Labarthe, *Heidegger, Art, and Politics. The Fiction of the Political*, Cambridge, MA 1990; and Morat, *Von der Tat zur Gelassenheit*, pp. 160–76.

[384] See also Heidegger, "Nietzsche's Word 'God Is Dead'."

which are set "*above*" the being, in order to give being as a whole a purpose, an order, and – as it is succinctly expressed – "meaning." Nihilism is that historical process whereby the dominance of the "transcendent" becomes null and void, so that all being loses its worth and meaning.[385]

Nietzsche recognized that, with the "death of God," not only must the old order collapse, but any order rooted in the transcendent or absolute had become impossible. God's "death" meant liquidation of the transcendent and of metaphysics as such – hence liquidation too of the previous "meaning" and "purpose" that man had given to the world. The world lost its purpose and its value.[386] We cannot discuss here the causes and consequences of this process. It is a phenomenon of the *longue durée*, at the provisional "end" of which came the shaking of the foundations of Western thought and the beginning of the attempt to "rewrite" modernity.[387]

In any event, modernity *qua* consciousness of order as a task has proved to be an interlude. It appears wedged between a "premodern" and a "postmodern" consciousness, and shifts between "critical," order-dissolving thought and "mythical," order-generating thought.[388] And although neither mythical nor critical thought can be fixed to a particular epoch – each found context-dependent forms of expression that answered to diverging needs[389] – the interplay between the two appears to be characteristic particularly of modernity. A major reason for this is that the interplay was (and is) always an object of reflection; after all, it was one of the legitimizing narratives of the Enlightenment to have overcome mythical thinking.

It remains to be said that in addition to describing the longing for order and its satisfaction in myth, Nietzsche grasped and proclaimed the impossibility of myth in the modern world.[390] He further radicalized the Enlightenment by practicing the critique of critique, so that a veritable abyss opened up before him. The fundamental crisis of Western rationality that revealed itself to Nietzsche may here be perspectively "foreshortened," in such a way that myth represents at once a necessity and an impossibility. Contingency had been discovered; the trick now was to lose it.

---

[385] Heidegger, *Nietzsche*, vol. IV, p. 4.

[386] Ibid., pp. 30–5.

[387] For two diverging views of this process, see Habermas, *The Philosophical Discourse of Modernity*, and Jean-François Lyotard, "Rewriting Modernity," in Lyotard, *The Inhuman. Reflections on Time*, Stanford, CA 1991, pp. 24–35.

[388] On critical thought, see Michel Foucault, "What is Critique?," in Foucault, *The Politics of Truth*, Los Angeles, CA 1997, pp. 41–82.

[389] In this respect, the dialectic of freedom and order may be described as an anthropological constant.

[390] See Wolfgang Lange, "Tod ist bei den Göttern immer nur ein Vorurteil. Zum Komplex des Mythos bei Nietzsche," in Karl Heinz Bohrer (ed.), *Mythos und Moderne. Begriff und Bilder einer Rekonstruktion*, Frankfurt/Main 1983, pp. 111–37; and Peter Pütz, "Der Mythos bei Nietzsche," in Helmut Koopmann (ed.), *Mythos und Mythologie in der Literatur des 19. Jahrhunderts*, Frankfurt/Main 1979, pp. 251–62.

The perils resulting from the contingency of order ended in an aporia: modernity needs the order of myth, but is no longer capable of myth-making; it craves and is susceptible to myths, but is at the same time critical of them. Modernity is modernity, with its longing for order, by virtue of its capacity to question, historicize and invalidate myths. At the same time its quest for order makes it turn to myth, since only myth has the power to create order and to satisfy the longing for it. This is where the complementarity of myth and modernity proves to lie. Modernity excludes the validity of myth yet continues to strive after it.

The Enlightenment led to radical questioning of the religious-metaphysical order, which became increasingly vulnerable as a result. Many restricted its proper sphere to private life. The "meaning" bound up with it became fragile and open to challenge, or even died out altogether. But this did not do away with the longing for an order of equivalent force. Indeed, the longing for order grew out of the breakdown of order: it was a result of the discovery of its contingency. Bauman puts it like this:

> The discovery that order was *not natural* was discovery of *order as such*. The *concept* of order appeared in consciousness only simultaneously with the *problem* of order, of order as a matter of design and action, order as an obsession. [...] Order as a problem emerged in the wake of the ordering flurry, as a reflection on ordering practices. Declaration of the "non-naturalness of order" stood for an order already coming out of hiding.[391]

The task of designing a new order thus resulted from the success of critical thought in dissolving the old mythical order based on "God." Critical, historical thought was the midwife but also the gravedigger of order, for it did not spare any vision. Only much later did it seem to become thinkable to dispense with any all-encompassing order based on an absolute and to try living with ambivalence and contingency.[392] But even that attitude was capable of becoming mythoid and legitimation-oriented.

With the discovery of the "non-naturalness of order" went the discovery of its impossibility. Nevertheless, or perhaps for that very reason, order became an obsession. For although order was confronted with contingency from the start, it was impossible to relinquish the idea of order (if indeed that is ever a possibility).[393] It was, precisely, "no trick" to find contingency, but an impos-

---

[391] Bauman, *Modernity and Ambivalence*, p. 6.

[392] According to Heidegger, even Nietzsche's thought was metaphysical – although it was precisely he who proclaimed the impossibility of metaphysics after the "death of God." See Heidegger, *Nietzsche*, vol. IV, pp. 185–250.

[393] It seems useful to ascribe *acceptance* of the impossibility of order to "postmodern" consciousness, and *discovery* of its impossibility to modern consciousness (since it accompanied the discovery of the non-naturalness of order). This is possible if the distinction between "modern" and "postmodern" is thought of in terms of complementarity rather than a temporal sequence, so that even if modern and postmodern consciousness appear mutually exclusive they are both possible in one and the same epoch and one and the same person.

sibility to lose it. The transfer of properties from a transcendent to an immanent sphere only intensified the problem. The absolute status of the super-sensuous was carried over to a worldly entity, whether that was reason, the nation, the race or the proletariat. What took the place of the old "God," then, was simply another, supposedly more perfect, god or "idol," whose divine status remained hidden behind a secular façade. The old myth was supplanted or complemented with new myths, in most cases concealed beneath a scientific mantle. The old order thus functioned as both model and counter-model for the order that was to be established. The impossibility of order was either not perceived at all or suppressed by the "scientific verifiability" of the sphere of immanence.

So long as the breakup was seen as "suspension of the entire existing order," the problem could be solved only with a myth grounded on "the sacred." For only myth with a suprahistorical foundation could eliminate contingency and ambivalence and produce meaning. To renounce myth would have meant giving up order as a task. But that requires not only an understanding of the impossibility of order, but also a refusal to accept the inhuman "side effects" of order – and a willingness to live with ambivalence. Only then is it possible to forgo a "grand narrative."[394] The renunciation of myth presupposes that one acquiesces in, or positively affirms, the "state of emergency," and that one tolerates living with ambivalence.

From this point of view, modernity changes into a deadly interlude. It does have at its disposal the instrument that could enable it to destroy order and to produce "freedom." But it has not developed the skill and know-how required to handle the instrument. The modern mind strives to destroy myths, but is at the same time susceptible to them, since it conceives of freedom as absence of order, as chaos and anomie. The modern mind longs for a necessary, pregiven "natural" order and at the same time goes into raptures over the discovery of contingency. It prides itself on the "Copernican revolution," and the feasibility and design of the world permit it to engage in fantasies of omnipotence. At the same time, the modern mind recoils from the idea that the world is nothing more than a human design, for that is not capable of giving the stability it longs for.

The radicalization inherent in modernity derives not least from this paradoxical starting point, which finds its *pharmakon* only in myth. Myth was the antidote and the toxin of modernity.[395] In this too lies the complementarity of myth and modernity.

> We tell ourselves what that world was not, what it did not contain, what it did not know, what it was unaware of. That world would hardly have recognized itself in our descriptions. It would not understand what we are talking about. It would not have survived such understanding. The moment of understanding would have been the sign of its approaching death. And it was. Historically,

---

[394] See Jean-François Lyotard, *The Postmodern Condition: A Report on Knowledge*, Minneapolis, MN 1984.
[395] On the concept of *pharmakon*, see Jacques Derrida, "Plato's Pharmacy," in Derrida, *Dissemination*, Chicago, IL 1983, pp. 61–172.

this understanding was the last sigh of the passing world, and the first sound of new-born modernity.[396]

For Zygmunt Bauman, the discovery of contingency was that "last sigh of the passing world." But the discovery of the impossibility of order is also the sign of an end. The "trick" will be not to lose the discovery.

---

[396] Bauman, *Modernity and Ambivalence*, p. 5.

Bibliography

Primary sources

Published materials

Anonymous: *Aby M. Warburg zum Gedächtnis*, Darmstadt 1929.

Anonymous: "Aeropitturae aviazione," in: *L'Ala d'Italia*, November 1937, pp. 44–5.

Anonymous: "Andere Berichte," in: *Schwäbische Kronik*, August 6, 1908, No. 362.

Anonymous: *Der Luftkrieg 1914–1915. Unter Verwendung von Feldpostbriefen und Berichten von Augenzeugen dargestellt von einem Flugtechniker*, Leipzig 1915.

Anonymous: "I bimbi, le ali e il domani," in: *L'Ala d'Italia*, July 1931, pp. 519–20.[1]

Anonymous: "Über unsere Luftflotte," in: *Die Luftflotte* VII/1915, No. 2, pp. 10–11.

Anonymous: "Unsere Feldgrauen beim Lesen. Erfahrungen und Erlebnisse eines Batallions-Bücherwarts," in: *Börsenblatt für den deutschen Buchhandel*, Yr. 83, No. 190, August 17, 1916, pp. 1086–87.

Anonymous: "Ventennale," in: *L'Ala d'Italia*, March 15–31, 1939, p. 3.

Anonymous: "Vereinsprogramm des Deutschen Luftflottenvereins," in: *Illustrierte Aeronautische Mitteilungen* 12/1908, No. 19, pp. 579–81.

Anonymous: "Was verlangen die Feldgrauen zum Lesen?," in: *Börsenblatt für den deutschen Buchhandel*, Yr. 84, No. 144, June 23, 1917.

Aviaticus: *Boelcke. Der Held der Lüfte. Ein deutsches Heldenleben*, Chemnitz n.d.

Balbo, Italo: *Da Roma a Odessa. Sui cieli dell'Egeo e del Mar Nero*, Milan 1929.

Balbo, Italo: *La centuria alata*, Milan 1934.

Balbo, Italo: *L'aeronautica italiana. Realizzazioni e proposti. Discorso pronunciato alla Camera dei deputati sul "Bilancio dell'aeronautica" il 23 Marzo 1928*, Rome 1928.

Balbo, Italo: *Stormi d'Italia sul mondo*, Milan 1934.

Balbo, Italo: *Stormi in volo sull'oceano*, Milan 1931.

Balla, Giacomo et al.: "Manifesto of Aeropainting" [orig. 1932], in: Ball et al. (eds.), *Futurism. An Anthology*, New Haven, CT 2009, pp. 283–86.

Barbusse, Henri: *Under Fire* [orig.1916], trans. by Robin Buss, New York, NY 2003.

Benjamin, Walter: *Illuminations*, rev. edn, New York, NY 1969.

Best, Werner: "Der Krieg und das Recht," in: Ernst Jünger (ed.), *Krieg und Krieger*, Berlin 1930, pp. 135–61.

Binswanger, Otto: *Die Pathologie und Therapie der Neurasthenie*, Jena 1896.

Bley, Wulf: "Vorwort," in: Wulf Bley (ed.), *Deutschland zur Luft*, Stuttgart 1936.

Bloch, Ernst: *Erbschaft dieser Zeit*, Zurich 1935.

Boccioni, Umberto: "Appunti per un diario," in: *Altri inediti e apparati critici*, ed. by Zeno Birolli, Milan 1972, pp. 60–3.

Boelcke, Oswald: *Hauptmann Bölckes Feldberichte. Mit einer Einleitung von der Hand des Vaters*, Gotha 1916.

Bragaglia, Anton Giulio: "Aeroteatro," in: *L'Ala d'Italia*, February 1935, pp. 65–71.

Buddecke, Hans Joachim: *El Schahin (Der Jagdfalke). Aus meinem Fliegerleben*, Berlin 1918.

Bülow, Hilmer Freiherr von: "Vorwort," in: Giulio Douhet, *Luftherrschaft*, Berlin 1935, pp. 5–9.

Carlyle, Thomas: *On Heroes, Hero-Worship, and the Heroic in History*. Notes and Introduction by Michael K. Goldberg [orig.1841], Berkeley, CA 1993.

---

[1] The uneven reference style for *L'Ala d'Italia* is due to its varying frequency and numbering.

Cassirer, Ernst: "Eidos und Eidolon. Das Problem des Schönen und der Kunst in Platons Dialogen," in: *Vorträge der Bibliothek Warburg*, vol. 2, Leipzig 1924, pp. 1–27.

Cassirer, Ernst: *The Myth of the State* [orig.1946], New Haven, CT 1963.

Cassirer, Ernst: "Worte zur Beisetzung von Prof. Dr. Aby M. Warburg," in: Anon. (ed.), *Aby M. Warburg zum Gedächtnis*, Darmstadt 1929, unpaginated.

Ciano, Galeazzo: *Diario 1937–1943*, ed. by Renzo De Felice, Milan 1980.

Constantini, Vincenzo: "Aviazione e arte," in: *L'Ala d'Italia*, June–July 1934, pp. 61–5.

Contini, Luigi: *La signoria degli stormi*, Milan 1933.

Cutry, Francesco: "Bilancio morale dell'esposizione azzura," in: *L'Ala d'Italia*, January 1935, pp. 14–16.

D'Annunzio, Gabriele: *Das Feuer*, Munich 1900.

D'Annunzio, Gabriele: *Forse che si forse che no*, Milan 1910.

D'Annunzio, Gabriele: "Il fuoco" [orig.1900], in: D'Annunzio, *Prose di romanzi*, vol. 2, ed. by Niva Lorenzini, Milan 1989, pp. 195–518.

D'Annunzio, Gabriele: "I mistici della guerra. Morte di frate Ginepro" [orig.1918], in: D'Annunzio, *Scritti giornalistici 1889–1938*, vol. 2, Milan 2003, pp. 769–79.

D'Annunzio, Gabriele: "Il primo olocausto" [orig.1919], in: D'Annunzio, *L'Urna inesausta*, in: D'Annunzio, *Prose di ricerca*, vol. 1, Milan 2005, pp. 935–1068, pp. 984–86.

D'Annunzio, Gabriele: "Il saluto di d'Annunzio agli aviatori prima della battaglia" [orig.1917], in: D'Annunzio, *Scritti giornalistici 1889–1938*, vol. 2, Milan 2003, pp. 693–94.

D'Annunzio, Gabriele: "Il volo di D'Annunzio su Trento" [orig. 1915], in: D'Annunzio, *Scritti giornalistici 1889–1938*, vol. 2, Milan 2003, pp. 686–89.

D'Annunzio, Gabriele: "La fede nell'aviazione italiana" [orig. 1918], in: D'Annunzio, *Scritti giornalistici 1889–1938*, vol. 2, Milan 2003, pp. 736–38.

D'Annunzio, Gabriele: "L'Ala d'Italia è liberata" [orig. 1919], in: D'Annunzio, *Il sudore di sangue*, in: D'Annunzio, *Prose di ricerca*, vol. 1, Milan 2005, pp. 741–932, pp. 879–94.

D'Annunzio, Gabriele: "L'ala sul mare" [orig. 1903], in: D'Annunzio, *Laudi del cielo, del mare, della terra e degli eroi, Libro Terzo Alcyone*, ed. by Egidio Bianchetti, Milan 1956.

D'Annunzio, Gabriele: "La preghiera di Aquileia" [orig. 1918], in: D'Annunzio, *Il libro ascetico della giovane Italia*, in: D'Annunzio, *Prose di ricerca*, vol. 1, Milan 2005, pp. 411–737, pp. 600–03.

D'Annunzio, Gabriele: "La preghiera di Sernaglia" [orig. 1918], in: D'Annunzio, *Il libro ascetico della giovane Italia*, in: *Prose di ricerca*, vol. 1, Milan 2005, pp. 411–737, pp. 593–99.

D'Annunzio, Gabriele: "Le Faville del maglio. Il venturiero senza ventura e altri studii del vivere inimitabile" [orig. 1924], in: D'Annunzio, *Prose di ricerca*, vol. 1, Milan 2005, pp. 1069–656.

D'Annunzio, Gabriele: "Le solenni esequie del tenente Miraglia. Il discorso di Gabriele D'Annunzio" [orig. 1915], in: D'Annunzio, *Scritti giornalistici 1889–1938*, vol. 2, Milan 2003, pp. 869–70.

D'Annunzio, Gabriele: "Notturno" [orig. 1921], in: D'Annunzio, *Prose di ricerca*, vol. 1, Milan 2005, pp. 161–410.

D'Annunzio, Gabriele: "Orazione per la sagra dei Mille" [orig. 1915], in: D'Annunzio, *Scritti Giornalistici 1889–1938*, vol. 2, Milan 2003, pp. 675–85.

D'Annunzio, Gabriele: "Parole dette nell'Ateneo genovese il VII di Maggio, ricevendo in dono dagli studenti una targa d'oro" [orig. 1915], in: D'Annunzio, *Per la più grande Italia*, in: D'Annunzio, *Prose di ricerca*, vol. 1, Milan 2005, pp. 7–157, pp. 27–30.

D'Annunzio, Gabriele: "Parole di G. D'Annunzio dopo il volo su Vienna" [orig. 1918], in: D'Annunzio, *Scritti giornalistici 1889–1938*, vol. 2, Milan 2003, pp. 757–62.

D'Annunzio, Gabriele: *Per la più grande Italia*, Milan 1915.

D'Annunzio, Gabriele: *Prose di ricerca*, vol. 1, ed. by Annamaria Andreoli and Giorgio Zanetti, Milan 2005.

D'Annunzio, Gabriele: *Scritti giornalistici 1889–1938*, vol. 2, ed. and intro. by Annamaria Andreoli, Milan 2003.

D'Annunzio, Gabriele: *Taccuini*, ed. by Enrica Bianchetti and Roberto Forcella, Milan 1965.

D'Annunzio, Gabriele: "Tacitum Robur. Parole dette in una cena di compagni, all'alba del XXV maggio MCMXV", in: D'Annunzio, *Per la più grande Italia* [orig. 1915], in: D'Annunzio, *Prose di ricerca*, vol. 1, Milan 2005 [orig. 1915], pp. 7–157, pp. 61–5.

D'Annunzio, Gabriele: "Voci della riscossa. Alle reclute del '99" [orig. 1917], in: D'Annunzio, *Scritti giornalistici 1889–1938*, vol. 2, Milan 2003, pp. 709–17.

De Felice, Renzo/*Mariano*, Emilio (eds.): *Carteggio D'Annunzio – Mussolini (1918–1938)*, Milan 1971.

Deutscher Luftflottenverein (ed.): *Das fliegende Schwert. Wesen, Bedeutung und Taten der deutschen Luftflotte in Wort und Bild*, Oldenburg i. Gr. 1917.

Douhet, Giulio: *Il dominio dell'aria. Saggio sull'arte della guerra aerea, con un'appendice contenente nozioni elementari di aeronautica*, Rome 1921.

Eddelbüttel, F. W.: *Artillerieflieger*, Dresden 1918.

Eichler, Friedrich Ernst (ed.): *Kreuz wider Kokarde, Jagdflüge des Leutnants Ernst Udet*, Berlin 1918.

Euringer, Richard: *Fliegerschule 4. Buch der Mannschaft*, Hamburg 1929.

Faaß, (no firstname): "Lesestoff fürs Feld," in: *Börsenblatt für den deutschen Buchhandel*, Yr. 84, No. 189, August 15, 1917, pp. 969–70.

Folgore, Luciano: "Volare," in: *L'Ala d'Italia* 20/1939, April 1–15, 1939, p. 63.

Fondazione Bernocchi/Palazzo dell'arte Milano (eds.): *Esposizione dell'Aeronautica Italiana. Giugno – Ottobre 1934 – XII*, official catalog, Milan 1934.

Frank, Hans: *Im Angesicht des Galgens. Deutung Hitlers und seiner Zeit auf Grund eigener Erlebnisse und Erkenntnisse. Geschrieben im Nürnberger Justizgefängnis*, Neuhaus bei Schliersee 1955[2].

Freud, Sigmund: *Reflections on War and Death* [orig. 1915], New York, NY 1918.

Funk, (no firstname): *Unsere Luftwaffe. Jahrbuch des Luftfahrerdank e.V. 1917*, Leipzig n.d.

Göring, Hermann: "Der alte Fliegergeist lebt. Rede zum Abschluss des Deutschlandfluges am 24. Juni 1934," in: Erich Gritzbach (ed.), *Hermann Göring. Reden und Aufsätze*, Munich 1938, pp. 121–24.

Göring, Hermann: "Kameradschaft, Pflichterfüllung und Opferbereitschaft. Ansprache an 1000 Fliegerleutnants am Tage ihrer Vereidigung in Berlin am 20. Mai 1936," in: Erich Gritzbach (ed.), *Hermann Göring. Reden und Aufsätze*, Munich 1938, pp. 226–44.

Gottschalk, Rudolf Oskar: *Boelcke. Deutschlands Fliegerheld. Schilderung seines Lebensweges und seiner Heldentaten im Luftkampf*, Leipzig n.d. (circa. 1916/17).

Grey, Edward: *Twenty-Five Years, 1892–1916*, vol. 2, London 1925.

Hackenberger, Willi: *Deutschlands Eroberung der Luft. Die Entwicklung deutschen Flugwesens anhand von 315 Wirklichkeitsaufnahmen*. With a foreword by Hellmuth Hirth, Siegen 1915.

Haller, Hanns: *Der Flieger von Rottenburg*, Bayreuth 1939.

Hein, Alfred: "Zum Gedächtnis des Rittmeisters Freiherr von Richthofen," in: Jean Paul, *Rittmeister Manfred Freiherr von Richthofen. Ein Lebensbild nach Zeitungsberichten*, Diesdorf bei Gäbersdorf 1918, pp. 5–6.

Heine, Heinrich: "The Baths of Lucca", in: Heinrich Heine, *Travel Pictures*, New York, NY 2008.

Heydemarck, Georg: *Doppeldecker "C 666." Als Flieger im Westen*, Berlin 1916.

Hitler, Adolf: *Mein Kampf*, vol. 1, *Eine Abrechnung*, Munich 1933[19].

Hitler, Adolf: *Mein Kampf*, vol. 2, *Die nationalsozialistische Bewegung*, Munich 1933[19].

Hoeppner, Ernst von: *Deutschlands Krieg in der Luft. Ein Rückblick auf die Entwicklung und die Leistungen unserer Heeres-Luftstreitkräfte im Weltkriege*, Leipzig 1921.

Immelmann, Max: *Meine Kampfflüge. Selbsterlebt und selbsterzählt*, Berlin 1916.

Imwolde, K.: "Was liest der Frontsoldat?," in: *Börsenblatt für den Deutschen Buchhandel*, Yr. 84, No. 39, February 16, 1917, pp. 149–50.

Jean, Paul: "Selina oder über die Unsterblichkeit der Seele" [orig.1827] in: Jean Poul, *Sämtliche Werke*, vol. 6, ed. by Norbert Miller, Munich 1996, pp. 1105–236.

Jung, Edgar Julius: *Die Herrschaft der Minderwertigen. Ihr Zerfall und ihre Ablösung durch ein neues Reich*, Berlin 1930[2].

Jünger, Ernst: *Das Abenteuerliche Herz. Aufzeichnungen bei Tag und Nacht*, Berlin 1929; *The Adventurous Heart. Figures and Cappricios*, 2nd edn, Candor, NY 2012.

Jünger, Ernst: *Das Wäldchen 125. Eine Chronik aus den Grabenkämpfen 1918*, Berlin 1925; *Copse 125: A Chronicle from the Trench Warfare of 1918*, New York, NY 1988.

Jünger, Ernst: *Der Arbeiter. Herrschaft und Gestalt*, 3rd edn, Hamburg 1932 [orig. 1932].

Jünger, Ernst: *Der Kampf als inneres Erlebnis*, Berlin 1922.

Jünger, Ernst: "Der Kampf um das Reich, Vorwort" [orig. 1929], in: Jünger, *Politische Publizistik. 1919–1933*, ed. by Sven Olaf Berggötz, Stuttgart 2001, pp. 527–36.

Jünger, Ernst: "Der neue Nationalismus" [orig. 1927], in: Jünger, *Politische Publizistik. 1919–1933*, ed. by Sven Olaf Berggötz, Stuttgart 2001, pp. 285–91.

Jünger, Ernst: "Der neue Typ des deutschen Menschen" [orig. 1926], in: Jünger, *Politische Publizistik. 1919–1933*, ed. by Sven Olaf Berggötz, Stuttgart 2001, pp. 167–72.

Jünger, Ernst: "Die Reaktion" [orig. 1925], in: Jünger, *Politische Publizistik. 1919–1933*, ed. by Sven Olaf Berggötz, Stuttgart 2001, pp. 119–25.

Jünger, Ernst: "Die totale Mobilmachung" [orig. 1930], in: Jünger, *Politische Publizistik. 1919–1933*, ed. by Sven Olaf Berggötz, Stuttgart 2001, pp. 558–82; "Total Mobilization," in: Richard Wolin (ed.), *The Heidegger Controversy*, Cambridge, MA 1993.

Jünger, Ernst: "Die Tradition" [orig. 1925], in: Jünger, *Politische Publizistik. 1919–1933*, ed. by Sven Olaf Berggötz, Stuttgart 2001, pp. 125–31.

Jünger, Ernst: *Feuer und Blut. Ein kleiner Ausschnitt aus einer großen Schlacht* [orig. 1925], Berlin 1929.

Jünger, Ernst: *In Stahlgewittern. Aus dem Tagebuch eines Stoßtruppführers* [orig. 1920], Berlin 1930[12].

Jünger, Ernst: "Kriegsausbruch 1914," in: Ernst Jünger, *Sämtliche Werke. Erste Abteilung, vol. 1, Tagebücher 1, Der Erste Weltkrieg*, ed. by Norbert Miller, Stuttgart 1978, pp. 539–45.

Jünger, Ernst: *Storm of Steel*, trans. by Michael Hofmann, New York, NY 2003.

Jünger, Ernst: "Vorwort," in: Ernst Jünger (ed.), *Luftfahrt ist not!*, Leipzig 1928, pp. 9–13.

Jünger, Friedrich Georg: "Krieg und Krieger," in: Ernst Jünger (ed.), *Krieg und Krieger*, Berlin 1930, pp. 51–67.

Jünger, Friedrich Georg: "Manfred von Richthofen," in: Ernst Jünger (ed.), *Die Unvergessenen*, Berlin 1928, pp. 279–86.

Kafka, Franz: Kafka, "The Aeroplanes at Brescia," in: Kafka, *The Transformation and Other Stories*, trans. by Malcolm Pasley, New York, NY 1992, pp. 1–10.

Killinger, Erich: *Die Abenteuer des Ostseefliegers*, Berlin 1917.

Killinger, Erich: "Die Abenteuer des Ostseefliegers," in: *Berliner Illustrirte Zeitung*, Yr. 26, No. 11, March 18, 1917, pp. 149–51.

Koerber, Adolf Victor von: *Deutsche Heldenflieger*, Bielefeld n.d. (circa 1916/1917).

Koerber, Adolf Victor von: *Luftkreuzer im Kampf*, Leipzig 1916.

König, Karl: *Fritz der Flieger. Ein neues Kriegslesebuch für die Unter- und Mittelstufe*, Strasbourg 1917.

Kracauer, Siegfried: "Soziologie als Wissenschaft," in: Siegfried Kracauer, *Schriften*, vol. 1, ed. by Inka Mülder-Bach Frankfurt/Main 1974.

Kracauer, Siegfried: "The Little Shopgirls Go To the Movies" [orig. 1927], in: Siegfried Kracauer, *The Mass Ornament: Weimar Essays*, Cambridge, MA 1975, pp. 291–306.

Kracauer, Siegfried: *The Salaried Masses: Duty and Distraction in Weimar Germany* [orig. 1929], New York, NY 1998.

Krimer: *Aviatori*, Florence 1935.

Lamszus, Wilhelm: *Das Menschenschlachthaus. Bilder vom kommenden Krieg*, Hamburg 1912.

Laredo de Mendoza, Saverio: *Gabriele D'Annunzio. Aviatore di guerra*, Milan 1930.

Le Bon, Gustave: *Psychology of Crowds* [French orig. 1895], Southampton 2012.

Lessing, Theodor: *Geschichte als Sinngebungd es Sinnlosen oder die Geburt der Geschichte aus dem Mythos*, Munich 1927[4].

Lessing, Theodor: *Geschichte als Sinngebung des Sinnlosen oder die Geburt der Geschichte aus dem Mythos* [orig. 1919], Munich 1983.

Levi, Carlo: *Christ Stopped at Eboli* [orig. 1945], New York, NY 2006.

Loele, Kurt: Rezension zu Süersen, Elisabeth: *Die Stellung der Militär- und Zivilbehörden zur Schundliteratur*, Berlin 1916 and *Tessendorf*, Wilhelm: *Die Kriegsschundliteratur und ihre Bekämpfung. Mit einem Verzeichnis empfehlenswerter Schriften*, Halle 1916; in: *Börsenblatt für den Deutschen Buchhandel*, Yr. 84, No. 1, January 2, 1917, pp. 3–4.

Longoni, Attilio: *Fascismo ed aviazione. Gli aviatori nella rivoluzione fascista*, Milan 1931.
Luebke, Anton: *Hauptmann Boelcke. Ein Gedenkblatt für den ruhmbedeckten Heldenflieger*, Leipzig n.d. (circa 1916).
Lukács, Georg: *The Theory of the Novel. A Historico-Philosophical Essay on the Forms of Great Epic Literature* [orig. 1920], London 1971.
Marinetti, Filippo T.: "Guerra sola igiene del mondo" [orig.1915], in: Marinetti, *Teoria e invenzione futurista*, ed. by Luciano De Maria, Milan 1983, pp. 233–341.
Marinetti, Filippo T.: "I diritti artistici propugnati dai futuristi italiani. Manifesto al Governo fascista" [orig.1924], in: *Futurismo e fascismo*, in: Marinetti, *Teoria e invenzione futurista*, ed. by Luciano De Maria, Milan 1983, pp. 489–572, pp. 562–69.
Marinetti, Filippo T.: *La battaglia di Tripoli*, Milan 1912.
Marinetti, Filippo T.: "Let's Murder the Moonshine" [orig. 1909], in: Marinetti, *Selected Writings*, ed. by R. W. Flint, New York, NY 1972, pp. 45–54.
Marinetti, Filippo T.: "Lettera aperta al futurista MacDelmarle" [orig.1913], in: Marinetti, *Teoria e invenzione futurista*, ed. by Luciano De Maria, Milan 1983, pp. 91–4.
Marinetti, Filippo T.: "Prefazione futurista a 'Revolverate' di Gian Pietro Lucini," [orig. 1909] in: Marinetti, *Teoria e invenzione futurista*, ed. by Luciano De Maria, Milan 1983, pp. 27–33.
Marinetti: Filippo T.: *Selected Writings*, ed. by R. W. Flint, New York, NY 1972.
Marinetti, Filippo T.: *Taccuini 1915–1921*, ed. by Alberto Bertoni, Bologna 1987.
Marinetti, Filippo T.: "Technical Manifesto of Futurist Literature" [orig. 1912], in: Lawrence Rainey/Christine Poggi/Laura Wittman (eds.), *Futurism. An Anthology*, New Haven, CT 2009, pp. 119–25.
Marinetti, Filippo T.: "The Founding and Manifesto of Futurism" [orig. 1909], in: Marinetti, *Selected Writings*, ed. by R. W. Flint, New York, NY 1972, pp. 39–44.
Marinetti, Filippo T.: "The New Religion-Morality of Speed" [orig. 1916], in: Lawrence Rainey/Christine Poggi/Laura Wittman (eds.), *Futurism. An Anthology*, New Haven, CT 2009, pp. 224–28.
Marinetti, Filippo T./*Azari*, Fedele: *Primo dizionario aereo italiano*, Milan 1929.
Marinetti, Filippo T./*Fillia*: "Manifesto of Futurist Sacred Art" [orig. 1931], in: Lawrence Rainey/Christine Poggi/Laura Wittman (eds.), *Futurism. An Anthology*, New Haven, CT 2009, pp. 286–88.
Marinetti, Filippo T./*Fillia*: *The Futurist Cookbook* [orig. 1932], San Francisco, CA 1989.
Marquard, Alfred: "Wesen und Bedeutung der Luftwaffen," in: Deutscher Luftflottenverein (ed.), *Das fliegende Schwert. Wesen, Bedeutung und Taten der deutschen Luftflotte in Wort und Bild*, Oldenburg i. Gr. 1917, p. 15.
Marx, Karl/*Engels*, Friedrich: "Manifesto of the Communist Party," in: Marx, *Political Writings*, vol. 1, *The Revolutions of 1848*, ed. by David Fernbach, Harmondsworth 1973.
Mattioli, Guido: *Mussolini aviatore e la sua opera per l'aviazione*, Rome 1935/36.
Meinecke, Friedrich: *Cosmopolitanism and the National State* [orig. 1908], Princeton, NJ 1970.
Meyer, Friedrich Albert: *Immelmann und Boelcke. Deutsche Helden der Luft*, Warendorf i. Westfalen n.d.
Ministero dell'aeronautica (ed.): *L'Aviazione negli scritti e nella parola del Duce*, Rome 1937.
Moeller van den Bruck, Arthur: *Das Dritte Reich*, Berlin 1926.
Moeller van den Bruck, Arthur: *Das ewige Reich*, vol. 1, *Die politischen Kräfte* [orig.1923], ed. by Hans Schwarz, Breslau 1933.
Müller, Leonhard (ed.): *Fliegerleutnant Heinrich Gontermann*, Barmen n.d. [circa 1918/1919].
Müller, Wulf Dieter: *Ernst Jünger. Ein Leben im Umbruch der Zeit*, Berlin 1934.
Mussolini, Benito: *The Doctrine of Fascism* [orig. 1932], New York, NY 2006.
Mussolini, Benito: "The Political and Social Doctrine of Fascism," in: Mussolini, *My Autobiography*, Mineola, NY 2006, pp. 227–40.
Mussolini, Benito: "Volare!," in: *Il Popolo d'Italia*, August 20, 1919, quoted from Ministero dell'Aeronautica (ed.): *L'aviazione negli scritti e nella parola del Duce*, Rome 1937, p. 31.
Neumann, Franz: *Behemoth: The Structure and Practice of National Socialism*, New York, NY 1942.

Neumann, Georg Paul: *Die deutschen Luftstreitkräfte im Weltkriege. Unter Mitwirkung von 20 Offizieren und Beamten der Heeres- und Marine-Luftfahrt*, Berlin 1920.

Nieden, Johann: *Heldentum im Weltkriege 1914. Berichte von Heldentaten*, Strasbourg 1914.

Nietzsche, Friedrich: "On the Uses and Disadvantages of History for Life (1874)," in: Nietzsche, *Untimely Meditations*, Cambridge 1983, pp. 57–123.

Nietzsche, Friedrich: *Sämtliche Briefe. Kritische Studienausgabe*, vol 8, January 1887–January 1889, Munich 2003 [repr.].

Nietzsche, Friedrich: *The Gay Science*, trans. by Walter Kaufmann, New York, NY 1974.

Nietzsche, Friedrich: *Thus Spake Zarathustra*, trans. by Adrian del Caro, New York, NY 2006.

Orano, Paolo: *Balbo*, Rome 1940.

Ortega y Gasset, José: *The Revolt of the Masses* [orig. 1930], New York, NY 1957.

Panofksy, Erwin: "A.Warburg," in: *Repertorium für Kunstwissenschaft* LI/1930, pp. 1–4.

Panofksy, Erwin: "Idea." *Ein Beitrag zur Begriffsgeschichte der älteren Kunsttheorie*, Leipzig 1924.

Pavese, Cesare: *The Political Prisoner* [orig.1949], London 1955.

Petersen, Julius: *Die Sehnsucht nach dem Dritten Reich in deutscher Sage und Dichtung*, Stuttgart 1934.

Pfeifer, Karl: "1914–1933," in: *Völkischer Beobachter*, Yr. 46, No. 214, August 2, 1933.

Pinder, Wilhelm: *Das Problem der Generationen in der Kunstgeschichte Europas*, Berlin 1926.

Plüschow, Gunther: *Die Abenteuer des Fliegers von Tsingtau*, Berlin 1916.

Plüschow, Gunther: "Die Abenteuer des Fliegers von Tsingtau. Meine Erlebnisse in drei Erdteilen," in: *Berliner Illustrirte Zeitung*, Yr. 25, No. 50, December 10, 1916, pp. 757–60.

Remarque, Erich Maria: *Im Westen nichts Neues*, Cologne 2004 [orig.1928]; *All Quiet on the Western Front*, New York, NY 1987.

Remarque, Erich Maria: "Rezension," in: *Sport im Bild*, No. 12, June 2, 1928.

Richthofen, Manfred von: *Der rote Kampfflieger*, Berlin 1917; *The Red Fighter Pilot. The Autobiography of the Red Baron*, St. Petersburg, FL 1997.

Richthofen, Manfred von: *Der rote Kampfflieger. Eingeleitet und ergänzt v. Bolko von Richthofen. Mit einem Vorwort von Reichsminister Hermann Göring*, Berlin 1933.

Richthofen, Manfred von: *Sein militärisches Vermächtnis*, ed. by Kriegswissenschaftliche Abteilung der Luftwaffe, Berlin 1938.

Samuleit, Paul: "Kriegsschundliteratur. Vortrag, gehalten in der öffentlichen Versammlung der Zentralstelle zur Bekämpfung der Schundliteratur zu Berlin am 25. März 1916, Berlin 1916," in: *Börsenblatt für den Deutschen Buchhandel*, Yr. 83, No. 156, July 8, 1916.

Schäfer, Emil: *Vom Jäger zum Flieger. Tagebuchblätter und Briefe*, Berlin 1918.

Schmitt, Carl: *Political Theology. Four Chapters on the Concept of Sovereignty*, Chicago, IL 2006.

Schramm, Wilhelm von: "Schöpferische Kritik des Krieges. Ein Versuch," in: Ernst Jünger (ed.), *Krieg und Krieger*, Berlin 1930, pp. 31–49.

Schwarte, Max: *Der Krieg der Zukunft*, Leipzig 1931.

Schwarz, Hans: *Die Wiedergeburt des heroischen Menschen. Eine Langemarck-Rede vor der Greifswalder Studentenschaft am 11. November 1928*, Berlin 1930.

Seehofer, Herbert: "Das erwachte Berlin marschiert," in: *Völkischer Beobachter*, Yr. 46, No. 31, January 31, 1933.

Seehofer, Herbert: "Das Gesetz, nach dem sie angetreten," in: *Völkischer Beobachter*, Yr. 46, No. 214, August 2 1933.

Silone, Ignazio: *The Abruzzo Trilogy* [orig.1930–1940], South Royalton, VT 2000.

Soldan, George: *Der Mensch und die Schlacht der Zukunft*, Oldenburg 1925.

Sombart, Werner: *Händler und Helden. Patriotische Besinnungen*, Munich 1915.

Sorel, Georges: *Reflections on Violence* [orig. 1906], New York, NY 1975.

Spengler, Oswald: *The Decline of the West*, vol. 1 [orig. 1926], New York, NY 2000.

Supf, Peter: "Waffe seiner Idee," in: *Der Adler*, No. 5, April 18, 1939, unpaginated.

Tessendorf, Wilhelm: "Die Kriegsschundliteratur und ihre Bekämpfung. Mit einem Verzeichnis empfehlenswerter Schriften," Halle 1916, in: *Börsenblatt für den Deutschen* Buchhandel, Yr. 84, No. 1, January 2, 1917, pp. 3–4.

Tucholsky, Kurt: *Gesamtausgabe*, vol. 10, *Texte 1928*, ed. by Ute Maack, Reinbek bei Hamburg 2001.

Valentiner, Max: *300000 Tonnen versenkt! Meine U-Boots-Fahrten*, Berlin 1917.

Warburg, Aby: "Airship and Submarine in the Medieval Imagination" (1913), in: Warburg, *The Renewal of Pagan Antiquity. Contributions to the cultural history of the European Renaissance*, Los Angeles, CA 1999, pp. 333–42.

Warburg, Aby: "A Lecture on Serpent Ritual," in: *Journal of the Warburg Institute*, 2/April 1939, No. 4, pp. 277–92.

Warburg, Aby: "Italian Art and International Astrology in the Palazzo Schifanoia, Ferrara" (1912), in: Warburg, *The Renewal of Pagan Antiquity. Contributions to the Cultural History of the European Renaissance*, Los Angeles, CA 1999, pp. 563–91.

Warburg, Aby: "Pagan and Antique Prophecy in Words and Images in the Age of Luther," in: Warburg, *The Renewal of Pagan Antiquity. Contributions to the cultural history of the European Renaissance*, Los Angeles, CA 1999, pp. 597–667.

Warburg, Aby: *Schlangenritual. Ein Reisebericht*, Berlin 1988 [orig.1923].

Warburg, Aby: Tagebuch der Kulturwissenschaftlichen Bibliothek Warburg mit Einträgen von Gertrud Bing und Fritz Saxl, in: Warburg, *Gesammelte Schriften. Studienausgabe, Siebte Abteilung*, vol. 7, ed. by Karen Michels and Charlotte Schoell Glass, Berlin 2001.

Weber, Max: "Science as a Vocation," in: H. H. Gerth/C. Wright Mills (eds.) *From Max Weber. Essays in Sociology*, Boston, MA 1948, pp. 129–56.

Weddigen, Otto: *Deutschlands Luftkrieg und Helden-Flieger, 1914–1917*, Regensburg 1918.

Wegener, Georg: *Der Wall von Eisen und Feuer*, vol. 2, *Champagne – Verdun – Somme*, Leipzig 1918.

Wegener, Georg: *Der Wall von Eisen und Feuer*, vol. 3, *Die beiden letzten Jahre*, Leipzig 1920.

Wegener, Georg: *Der Wall von Eisen und Feuer. Ein Jahr an der Westfront*, Leipzig 1915.

Weizsäcker, Carl Friedrich von: "Komplementarität und Logik," in: Carl Friedrich von Weizsäcker, *Zum Weltbild der Physik*, Stuttgart 1958, pp. 281–331.

Werner, Johannes (ed.): *Briefe eines deutschen Kampffliegers an ein junges Mädchen*, Leipzig 1930.

Wildenbruch, Ernst von: "Deutschland und die Welt," in: Adolf Wasner (ed.), *Rittmeister Manfred Freiherr von Richthofen. Ein Lebensbild nach Zeitungsberichten*, Diesdorf bei Gäbersdorf 1918, pp. 87–8.

Wlotzka, Alfred: "Ikaros – Richthofen," in: Adolf Wasner (ed.), *Rittmeister Manfred Freiherr von Richthofen. Ein Lebensbild nach Zeitungsberichten*, Diesdorf bei Gäbersdorf 1918, S. 45.

Zweig, Arnold: *Education before Verdun* [German orig. 1935], New York, NY 1963.

Zweig, Stefan: *The World of Yesterday: An Autobiography*, New York, NY 1943.

Archive materials

Archive of the Deutsches Museum, Munich (ADM)
LR 00086
LR 00087
LR 00092
LR 000728
LR 000728-01
LR 000728-02
LR 00802
LR 05209
LR 05211
LR 05245
LR 05349
LR 05350
LR 05351
LR 05355

LR 05381
LR 05395
LR 05416
LR 05434
LR 05437
LR 05438
LR 05468
LR 05469
LR 05488–05492
NL 063 / 013

Archive of the Deutsches Technik Museum, Berlin (DTM)
NL 045
NL 151

Main Archive of the Bavarian State, Dept. IV War Archives (Bayerisches Hauptstaatsarchiv,
  Abt. IV Kriegsarchiv), Munich (BayHStA-KA)
DV
Iluft 34 u. 38
Iluft 56 u. 57
Iluft 86
Iluft 211
Iluft 212
Preuß. Geh. XV 29
R 496

Federal Archive, Film Archives (Bundesarchiv-Filmarchiv), Berlin (BA-FA)
Triumph des Willens, 1934/35, Produktion: Reichsparteitagfilm der L.R. Studio-Film, Berlin;
  Drehbuch und Regie: Leni Riefenstahl; Musik: Herbert Windt.

Federal Archive, Military Archive (Bundesarchiv-Militärarchiv), Freiburg (BA-MA)
MSg 1 / 74
MSg1 / 788
MSg 1 / 1907
MSg 170
RL 4 / 313

Archive of the Ullstein Publisher (Ullstein Verlagsarchiv), Berlin (UVA)
Akte Killinger
Akte Plüschow
Akte von Richthofen
Akte Valentiner

University Archive Tübingen (Universitätsarchiv Tübingen) (UAT)
117/291a
167/172
443/31

Warburg Institute Archive, London (WIA)
III.86
III. 99
FC
GC
KBW Akquisitionsbuch

Periodicals

Berliner Illustrirte Zeitung
Börsenblatt für den deutschen Buchhandel
Corriere della Sera
Der Adler
Der deutsche Sportflieger
Der Flieger
Deutsche Flugillustrierte
Deutsche Luftfahrerzeitschrift
Deutsche Luftwacht
Die Luftflotte
Ikarus
L'Ala d'Italia
L'Aquilone
L'Illustrazione Italiana
Luftwaffe
Schwäbische Kronik
Völkischer Beobachter

Secondary sources

Adamson, Walter L.: *Avant-Garde Florence. From Modernism to Fascism*, Cambridge, MA 1993.
Adamson, Walter L.: "The Culture of Italian Fascism and the Fascist Crisis of Modernity. The Case of il Selvaggio," in: JCH 30/1995, pp. 555–75.
Adorno, Theodor W./*Horkheimer*, Max: *Dialectic of Enlightenment* (orig. 1944), New York, NY 1972.
Afflerbach, Holger: "Vom Bündnispartner zum Kriegsgegner. Ursachen und Folgen des italienischen Kriegseintritts im Mai 1915," in: Johannes Hürter/Gian Enrico Rusconi (eds.), *Der Kriegseintritt Italiens im Mai 1915*, Munich 2007, pp. 53–69.
Agnese, Gino: *Marinetti. Una vita esplosiva*, Milan 1990.
Alegi, Gregory: "'L'arma fascistissima:' il falso mito dell'Aeronautica come preferita del regime," in: Massimo Ferrari (ed.), *Le ali del ventennio. L'aviazione italiana dal 1923 al 1945. Bilanci storiografici e prospettive di giudizio*, Milan 2005, pp. 111–54.
Allen, Douglas: *Myth and Religion in Mircea Eliade*, New York, NY 2002.
Aly, Götz: *Hitler's Beneficiaries. Plunder, Racial War, and the Nazi Welfare State*, New York, NY 2006.
Aly, Götz/*Heim*, Susanne: *Architects of Annihilation. Auschwitz and the Logic of Destruction*, Princeton, NJ 2003.
Anderson, Benedict: *Imagined Communities. Reflections on the Origin and Spread of Nationalism*, London 2006 [repr.].
Angehrn, Emil: *Die Überwindung des Chaos. Zur Philosophie des Mythos*, Frankfurt/Main 1996.
Angster, Julia: *Erdbeeren und Piraten. Die Royal Navy und die Ordnung der Welt 1770–1860*, Göttingen 2012.
Antliff, Mark: *Avant-Garde Fascism. The Mobilization of Myth, Art, and Culture in France, 1909–1939*, Durham, NC 2007.
Arendt, Hannah: *The Origins of Totalitarianism*, New York, NY 1951.
Arfè, Gaetano: *Storia del socialismo italiano 1892–1926*, Turin 1992.
Asendorf, Christoph: *Super Constellation. Flugzeug und Raumrevolution. Die Wirkung der Luftfahrt auf Kunst und Kultur der Moderne*, Vienna 1997.
Asholt, Wolfgang/*Fähnders*, Walter (eds.): *Der Blick vom Wolkenkratzer. Avantgarde – Avantgardekritik – Avantgardeforschung*, Amsterdam 2000.
Assmann, Jan: *Cultural Memory and Early Civilization. Writing, Remembrance, and Political Imagination*, New York, NY 2011.

Baberowski, Jörg: "Über die schöne Schwierigkeit, Geschichte zu schreiben," in: *Frankfurter Allgemeine Zeitung*, July 29, 2009, p. N3.

Baberowski, Jörg/Doering-Manteuffel, Anselm: *Ordnung durch Terror. Gewaltexzesse und Vernichtung im nationalsozialistischen und im stalinistischen Imperium*, Bonn 2006.

Bach, Martin: *Luftfahrtindustrie im Ersten Weltkrieg. Mobilisierung und Demobilisierung der britischen und deutschen Luftfahrtindustrie im Ersten Weltkrieg*, Allershausen 2003.

Ballerini, Luci: "Introduzione," in: Marinetti, FilippoT., *Mafarka il futurista*, Milan 2003, pp. VII–XLVIII.

Barner, Wilfried/*Detken*, Anke/*Wesche*, Jörg (eds.), *Texte zur modernen Mythentheorie*, Stuttgart 2003.

Bärsch, Claus-Ekkehard: *Die politische Religion des Nationalsozialismus. Die religiöse Dimension der NS-Ideologieinden Schriften von Dietrich Eckart, Joseph Goebbels, Alfred Rosenberg und Adolf Hitler*, Munich 1998.

Barth, Boris: *Dolchstoßlegenden und politische Desintegration. Das Trauma der deutschen Niederlage im Ersten Weltkrieg 1914–1933*, Düsseldorf 2003.

Barth, Boris: "Weder Bürgertum noch Adel – Zwischen Nationalstaat und kosmopolitischem Geschäft. Zur Gesellschaftsgeschichte der deutsch-jüdischen Hochfinanz vor dem Ersten Weltkrieg," in: *GG* 25/1999, pp. 94–122.

Barthes, Roland: "The Death of the Author," in: Roland Barthes, *Image – Music – Text*, New York, NY 1977, pp. 142–48.

Barthes, Roland: *Mythologies*, New York, NY 1972.

Bartov, Omer: "Man and the Mass. Reality and the Heroic Image in War," in: *History and Memory* 2/1989, pp. 99–122.

Bauerkämper, Arnd: *Der Faschismus in Europa 1918–1945*, Stuttgart 2006.

Bauman, Zygmunt: *Community. Seeking Safety in an Insecure World*, Cambridge 2001.

Bauman, Zygmunt: *Liquid Modernity*, Cambridge 2000.

Bauman, Zygmunt: *Modernity and Ambivalence*, Cambridge 1991.

Bauman, Zygmunt: *Modernity and the Holocaust*, Ithaca, NY 1989.

Baumgarth, Christa: *Geschichte des Futurismus*, Reinbek bei Hamburg 1966.

Bavaj, Riccardo: *Die Ambivalenz der Moderne im Nationalsozialismus. Eine Bilanz der Forschung*, Munich 2003.

Becker, Jörg: "Bibliographie zum Thema 'Krieg und Medien,'" in: Ulrich Albrecht/Jörg Becker (eds.), *Medien zwischen Krieg und Frieden*, Baden-Baden 2002, pp. 267–79.

Behrenbeck, Sabine: *Der Kult um die toten Helden. Nationalsozialistische Mythen, Riten und Symbole*, Vierow bei Greifswald 1996.

Behrenbeck, Sabine: "Heil," in: Étienne François/Hagen Schulze (eds.), *Deutsche Erinnerungsorte*, vol. 3, Munich 2001, pp. 310–27.

Behrens, Mathias: "'Politische Religion' – eine Religion? Bemerkungen zum Religionsbegriff," in: Hans Maier/Michael Schäfer (eds.), *Totalitarismus und Politische Religion*, vol. 2, Paderborn 1997, pp. 249–69.

Behringer, Wolfgang/*Ott-Koptschalijski*, Constance: *Der Traum vom Fliegen. Zwischen Mythos und Technik*, Frankfurt/Main 1991.

Beil, Christine: *Der ausgestellte Krieg. Präsentationen des Ersten Weltkriegs 1914–1939*, Tübingen 2004.

Belting, Hans: *Bild und Kult. Eine Geschichte des Bildes vor dem Zeitalter der Kunst*, Munich 1990.

Ben-Ghiat, Ruth: *Fascist Modernities. Italy, 1922–1945*, Berkeley, CA 2004 [repr.].

Benjamin, Walter: "Theses on the Philosophy of History," in: Walter Benjamin, *Illuminations*, New York, NY 1969[2], pp. 253–64.

Benjamin, Walter: "The Work of Art in the Age of Mechanical Reproduction," in: Walter Benjamin, *Illuminations*, New York, NY 1968, pp. 217–51.

Berger, Peter L.: *The Sacred Canopy. Elements of a Sociological Theory of Religion*, Garden City, NY 1967.

Berger, Peter L./*Luckmann*, Thomas: *The Social Construction of Reality*, Baltimore, MD 1966.

Berggötz, Sven Olaf: "Nachwort," in: Ernst Jünger, *Politische Publizistik 1919 bis 1933*, ed. by Sven Olaf Berggötz, Stuttgart 2001, pp. 834–69.

Berghaus, Günter: "Bibliography. A Futurist Reference Shelf," in: Günter Berghaus (ed.), *International Futurismin Arts and Literature*, pp. 487–596.

Berghaus, Günter (ed.): *Fascism and Theatre*. *Comparative Studies on the Aesthetics and Politics of Performance in Europe, 1925–1945*, Providence, RI 1996.

Berghaus, Günter: *Futurism and Politics*. *Between Anarchist Rebellion and Fascist Reaction, 1909–1944*, Providence, RI 1996.

Berg, Nicolas: *Luftmenschen*. *Zur Geschichte einer Metapher*, Göttingen 2008.

Berliner Geschichtswerkstatt (ed.): *Alltagskultur, Subjektivität und Geschichte*. *Zur Theorie von Alltagsgeschichte*, Münster 1994.

Bernhard, Thomas: *My Prizes*, New York, NY 2010.

Bessel, Richard (ed.): *Fascist Italy and Nazi Germany*. *Comparisons and Contrasts*, Cambridge 2000.

Betts, Paul: "The New Fascination with Fascism. The Case of Nazi Modernism," in: *JCH* 37/2002, pp. 541–58.

Bevc, Tobias: *Kulturgeneseals Dialektikvon Mythosund Vernunft*. *Ernst Cassirer und die Kritische Theorie*, Würzburg 2005.

Beyrau, Dietrich: *Petrograd, 25. Oktober 1917*. *Die russische Revolution und der Aufstieg des Kommunismus*, Munich 2001.

Binder, Hartmut: *Mit Kafka in den Süden*. *Eine historische Bilderreise in die Schweiz und zu den oberitalienischen Seen*, Prague 2007.

Bing, Gertrud: "Aby M. Warburg. Vortrag," in: Dieter Wuttke (ed.), *Aby M. Warburg*. *Ausgewählte Schriften und Würdigungen*, Baden-Baden 1980, pp. 455–64.

Blake, William: *The Marriage of Heaven and Hell*. With an Introduction and Commentary by Geoffrey Keynes, London 1975.

Blank, Ralf: "Strategischer Luftkrieg gegen Deutschland 1914–1918,": http: //www. ersterweltkrieg. clio-online. de/_Rainbow/documents/einzelne/Luftkrieg14_181.pdf, accessed on November 8, 2009.

Blessing, Ralph: *Der mögliche Frieden*. *Die Modernisierung der Außenpolitik und die deutsch-französischen Beziehungen 1923–1929*, Munich 2008.

Blumenberg, Hans: *Höhlenausgänge*, Frankfurt/Main 1989.

Blumenberg, Hans: *Paradigms for a Metaphorology*, Ithaca, NY 2010.

Blumenberg, Hans: *Work on Myth*, Cambridge, MA 1985.

Bödeker, Hans-Erich (ed.): *Begriffsgeschichte, Diskursgeschichte, Metapherngeschichte*, Göttingen 2002.

Böhme, Hartmut: *Fetischismus und Kultur*. *Eine andere Theorie der Moderne*, Reinbek bei Hamburg 2006.

Bohrer, Karl Heinz (ed.): *Mythos und Moderne*. *Begriff und Bild einer Rekonstruktion*, Frankfurt/Main 1983.

Bollenbeck, Georg: *Bildung und Kultur*. *Glanz und Elend eines deutschen Deutungsmusters*, Frankfurt/Main 1994.

Bonacker, Thorsten/*Reckwitz* Andreas (eds.): *Kulturen der Moderne*. *Soziologische Perspektiven der Gegenwart*, Frankfurt/Main 2007.

Borscheid, Peter: *Das Tempo-Virus*. *Eine Kulturgeschichte der Beschleunigung*, Frankfurt/Main 2004.

Bosworth, Richard J. B.: *Mussolini's Italy*. *Life Under the Dictatorship 1915–1945*, London 2006 [repr.].

Braun, Emily: *Mario Sironi and Italian Modernism*. *Art and Politics Under Fascism*, Cambridge 2000.

Braun, Emily: "Sironi in Context," in: Roger Griffin (ed.), *Fascism. Critical Concepts in Political Science*, vol. 3, *Fascism and Culture*, London 2004, pp. 225–48.

Breuer, Stefan: *Anatomie der konservativen Revolution*, Darmstadt 1993.

Breuer, Stefan: *Bürokratie und Charisma*. *Zur politischen Soziologie Max Webers*, Darmstadt 1994

Breuer, Stefan: "Das Charisma der Nation," in: Stefan Breuer (ed.), *Bürokratie und Charisma*. *Zur politischen Soziologie Max Webers*, Darmstadt 1994, pp. 110–43.

Breuer, Stefan: "Das Charisma des Führers," in: Stefan Breuer (ed.), *Bürokratie und Charisma. Zur politischen Soziologie Max Webers*, Darmstadt 1994, pp. 144–75.

Breuer, Stefan: *Die Völkischen in Deutschland. Kaiserreich und Weimarer Republik*, Darmstadt 2008.

Breuer, Stefan: *Grundpositionen der deutschen Rechten 1871–1945*, Tübingen 1999.

Breuer, Stefan: *Nationalismus und Faschismus. Frankreich, Italien und Deutschland im Vergleich*, Darmstadt 2005.

Breuer, Stefan: *Ordnungen der Ungleichheit – die deutsche Rechte im Widerstreit ihrer Ideen 1871–1945*, Darmstadt 2001.

Brogini-Künzi, Giulia: *Italien und der Abessinienkrieg 1935/36. Kolonialkrieg oder Totaler Krieg*, Paderborn 2006.

Bruch, Rüdiger vom: "'Der Zug der Millionen.' Massengesellschaft im Aufbruch," in: August Nitschke/Gerhard A. Ritter/Detlev J. K. Peukert/Rüdiger vom Bruch (eds.), *Jahrhundertwende. Der Aufbruch in die Moderne 1880–1930*, vol. 1, Reinbek bei Hamburg 1990, pp. 92–120.

Bruendel, Stefan: *Volksgemeinschaft oder Volksstaat. Die "Ideen von 1914" und die Neuordnung Deutschlands im Ersten Weltkrieg*, Berlin 2003.

Brunner, Otto/Conze, Werner/Koselleck, Reinhart (eds.): *Geschichtliche Grundbegriffe. Historisches Lexikon zur politisch-sozialen Sprache in Deutschland*, 8 vols., Stuttgart 1972ff.

Buchner, Bernd: *Um nationale und republikanische Identität. Die deutsche Sozialdemokratie und der Kampf um die politischen Symbole in der Weimarer Republik*, Bonn 2001.

Budisavljević, Bojan: "D'Annunzios Torpedowesen. Instrument der Vorsehung und Geschicke des Meeres im Seekrieg um Fiume herum," in: Hans Ulrich Gumbrecht/Friedrich Kittler/ Bernhard Siegert (eds.), *Der Dichter als Kommandant. D'Annunzio erobert Fiume*, Munich 1996, pp. 227–59.

Budraß, Lutz: *Flugzeugindustrie und Luftrüstung in Deutschland 1918–1945*, Düsseldorf 1998.

Bürger, Peter: *Theory of the Avant-Garde*, Minneapolis, MN 1984.

Burkert, Walter: *Homo Necans. The Anthropology of Ancient Greek Sacrificial Ritual and Myth*, Berkeley, CA 1986.

Burleigh, Michael: *Earthly Powers. Religion and Politics in Europe from the Enlightenment to the Great War*, London 2005.

Burleigh, Michael: "National Socialism as a Political Religion," in: *Totalitarian Movements and Political Religions* 1/2000, pp. 1–26.

Burleigh, Michael: *The Third Reich. A New History*, London 2000.

Buschmann, Nikolaus/Carl, Horst: "Zugänge zur Erfahrungsgeschichte des Krieges. Forschung, Theorie, Fragestellung," in: Nikolaus Buschmann/Horst Carl (eds.), *Die Erfahrung des Krieges. Erfahrungsgeschichtliche Perspektiven von der Französischen Revolution bis zum Zweiten Weltkrieg*, Paderborn 2001, pp. 11–26.

Castan, Joachim: *Der Rote Baron. Die ganze Geschichte des Manfred von Richthofen*, Stuttgart 2007.

Cervantes Saavedra, Miguel de: *The Adventures of Don Quixote* [Penguin edn], trans. by J. M. Cohen, Harmondsworth 1950.

Chakrabarty, Dipesh: *Provincializing Europe. Postcolonial Thought and Historical Difference*, Princeton, NJ 2000.

Chartier, Roger: *Forms and Meanings. Texts, Performances, and Audiences from Codex to Computers*, Philadelphia, PA 1995.

Chartier, Roger: "Laborers and Voyagers. From the Text to the Reader," in: *Diacritics* 22/1992, pp. 49–61.

Chartier, Roger: "Texts, Prints, Readings," in: Lynn Hunt (ed.), *The New Cultural History*, Berkeley, CA 1989, pp. 154–75.

Chernow, Ron: *The Warburgs. The Twentieth-Century Odyssey of a Remarkable Jewish Family*, New York, NY 1994.

Chiantera-Stutte, Patricia: *Von der Avantgarde zum Traditionalismus. Die radikalen Futuristen im italienischen Faschismus von 1919 bis 1931*, Frankfurt/Main, NY 2002.

Chickering, Roger/Förster, Stig (eds.): *Great War, Total War. Combat and Mobilization on the Western Front, 1914–1918*, New York, NY 2000.

Christadler, Marieluise: *Kriegserziehung im Jugendbuch. Literarische Mobilmachung in Deutschland und Frankreich vor 1914*, Frankfurt/Main 1978.

Ciuffoletti, Zeffiro: *Storia del PSI. Vol. I. Le origini e l'età giolittiana*, Rome 1992.

Conze, Eckart: "Die Idee eines Neuadels in den Gesellschaftsvorstellungen der SS," in: Eckart Conze/Monika Wienfort (eds.), *Adel und Moderne. Deutschland im europäischen Vergleich im 19. und 20. Jahrhundert*, Cologne 2004, pp. 151–76.

Cook, Colin: "The Myth of the Aviator and the Flight to Fascism," in: *History Today* 53/2003, pp. 36–42.

Cornelißen, Christoph: "Ein ständiges Ärgernis? Die Moderne in der (west-)deutschen Geschichtsschreibung," in: Ute Schneider/Lutz Raphael (eds.), *Dimensionen der Moderne. Festschrift für Christof Dipper*, Frankfurt/Main 2008, pp. 235–48.

Curami, Andrea: "I primi passi dell'industria aeronautica italiana," in: Paolo Ferrari (ed.), *La grande guerra aerea 1915–1918. Battaglie – industrie – bombardamenti – assi – aeroporti*, Valdagno 1994, pp. 97–139.

Curami, Andrea: "L'Aeronautica italiana dalle origini ai giorni nostri," in: Piero Del Negro (ed.), *Guida alla storia militare italiana*, Naples 1997, pp. 191–209.

Dahrendorf, Ralf: *Society and Democracy in Germany*, New York, NY 1979.

Daniel, Ute (ed.): *Augenzeugen. Kriegsberichterstattung vom 18. zum 21. Jahrhundert*, Göttingen 2006.

Dann, Otto (ed.): *Vereinswesen und bürgerliche Gesellschaft in Deutschland*, Munich 1984.

De Berti, Raffaele: "Lo sguardo dall'alto. Percorsi incrociati tra cinema e aeropittura," in: Annamaria Andreoli/Giovanni Caprara/Elena Fontanella (eds.), *Volare! Futurismo, aviomania, tecnica e cultura italiana del volo 1903–1940*, Rome 2003, pp. 175–80.

De Felice, Renzo: *D'Annunzio politico*, Rome 1978.

De Felice, Renzo: *Le interpretazioni del fascismo*, Rome 2001 [repr.].

De Grand, Alexander J.: *The Italian Nationalist Association and the Rise of Fascism in Italy*, Lincoln, NE 1978.

Demetz, Peter: *Die Flugschau von Brescia. Kafka, D'Annunzio und die Männer, die vom Himmel fielen*, Vienna 2002.

Demetz, Peter: *Worte in Freiheit. Der italienische Futurismus und die deutsche literarische Avantgarde 1912–1934*, Munich 1990.

Derrida, Jacques: "Plato's Pharmacy," in: Derrida, *Dissemination*, Chicago, IL 1983, pp. 61–172.

Desai, Ashok Valji: *Real Wages in Germany 1871–1913*, Oxford 1968.

Diaz-Bone, Rainer: "Zur Methodologisierung der Foucaultschen Diskursanalyse," in: *Historical Social Research* 31/2006, pp. 243–74.

Dickie, John: *Delizia! The Epic History of the Italians and their Food*, London 2008 [repr.].

Didi-Huberman, Georges: *L'image survivante. Histoire de l'art et temps des fantômes selon Aby Warburg*, Paris 2002.

Diner, Dan: "Geschichte der Juden. Paradigma einer europäischen Geschichtsschreibung," in: Diner, *Gedächtniszeiten. Über jüdische und andere Geschichten*, Munich 2003, pp. 246–62.

Dipper, Christof: "Die 'Geschichtlichen Grundbegriffe.' Von der Begriffsgeschichte zur Theorie der historischen Zeiten," in: *HZ* 270/2000, pp. 281–308.

Dipper, Christof: "Zwischen 'Historikerstreit' und der Debatte über 'Nationalsozialismus und die Moderne'," in: Gertraud Diendorfer/Gerhard Jagschitz/Oliver Rathkolb (eds.), *Zeitgeschichte im Wandel*, Innsbruck 1998, pp. 110–21.

Dipper, Christof/*Hudemann*, Rainer/*Petersen*, Jens (eds.): *Faschismus und Faschismen im Vergleich. Wolfgang Schieder zum 60. Geburtstag*, Cologne 1998.

Dipper, Christof/Schieder, Wolfgang (eds.): *Faschismus und Gesllschaft in Italien. Staat – Wirtschaft – Kultur*, Cologne 1998.

Doering-Manteuffel, Anselm: *Die antihistoristische Revolution im ersten Drittel des 20. Jahrhunderts. Eine Fallstudie zu den Spielräumen und Erkenntnismöglichkeiten von "Ideengeschichte,"* unpublished typescript.

Doering-Manteuffel, Anselm: "Konturen von Ordnung in den Zeitschichten des 20. Jahrhunderts," in: Thomas Etzemüller (ed.), *Die Ordnung der Moderne. Social Engineering im 20. Jahrhundert*, Bielefeld 2009, pp. 41–64.

Doering-Manteuffel, Anselm: "Mensch, Maschine, Zeit. Fortschrittsbewusstsein und Kulturkritik im ersten Drittel des 20. Jahrhunderts," in: *Jahrbuch des Historischen Kollegs 2003*, Munich 2004, pp. 91–119.

Doering-Manteuffel, Anselm: "Suchbewegungen in der Moderne. Religion im politischen Feld der Weimarer Republik," in: Friedrich Wilhelm Graf/Klaus Große Kracht (eds.), *Religion und Gesellschaft. Europa im 20. Jahrhundert*, Cologne 2007, pp. 175–202.

Doering-Manteuffel, Anselm/*Raphael*, Lutz: *Nach dem Boom. Perspektiven auf die Zeitgeschichte seit 1970*, Göttingen 2008.

Durkheim, Emile: *The Elementary Forms of the Religious Life*, New York, NY 1965 [orig. 1915].

Eatwell, Roger: "On Defining the 'Fascist Minimum'. The Centrality of Ideology," in: *Journal of Political Ideologies* 1/1996, pp. 303–19.

Echternkamp, Jörg/*Müller*, Sven Oliver (eds.): *Die Politik der Nation. Deutscher Nationalismus in Krieg und Krisen 1760–1960*, Munich 2002.

Ehrlicher, Hanno: *Die Kunst der Zerstörung. Gewaltphantasien und Manifestationspraktiken europä is cher Avantgarden*, Berlin 2001.

Eichberg, Henning/Michael Dultz/Glen W. Gadberry/Günther Rühle: *Massenspiele. NS-Thingspiel, Arbeiterweihespiel und olympisches Zeremoniell*, Stuttgart 1977.

Eisenstadt, Shmuel N.: *Die Vielfalt der Moderne*, Weilerswist 2000.

Eisenstadt, Shmuel N.: "Multiple Modernities," in: *Daedalus* 129/2000, pp. 1–29.

Eksteins, Modris: "*All Quiet on the Western Front* and the Fate of a War," in: *JCH* 15/1980, pp. 345–66.

Eksteins, Modris: "The Fate of the Film *All Quiet on the Western Front*," in: *Central European History* 13/1980, pp. 345–66.

Eksteins, Modris: *Rites of Spring. The Great War and the Birth of the Modern Age*, New York, NY 2000 [repr.].

Eley, Geoff: "Liberalism, Europe and the Bourgeoisie 1860–1914," in: David Blackbourn/Richard J. Evans (eds.), *The German Bourgeoisie. Essays on the Social History of the German Middle Class from the Late Eighteenth to the Early Twentieth Century*, London 1991, pp. 293–317.

Eley, Geoff: "The British Model and the German Road. Rethinking the Course of German History Before 1914," in: David Blackbourn/Geoff Eley (eds.), *The Peculiarities of German History. Bourgeois Society and Politics in Nineteenth-Century Germany*, Oxford 1984, pp. 37–155.

Eliade, Mircea: *The Myth of the Eternal Return. Cosmos and History*, Princeton, NJ 2005.

Eliade, Mircea: *Patterns in Comparative Religion*, Lincoln, NE 1996.

Eliade, Mircea: *The Quest. History and Meaning in Religion*, Chicago, IL 1969.

Eliade, Mircea: *The Sacred and The Profane. The Nature of Religion*, Orlando, FL 1959.

El-Tayeb, Fatima: *Schwarze Deutsche. Der Diskurs um "Rasse" und nationale Identität 1890–1933*, Frankfurt/Main 2001.

Engelsing, Rolf: "Die Perioden der Lesergeschichte in der Neuzeit," in: *Archiv für Geschichte des Buchwesens* 10/1969, pp. 945–1002.

Erdmann, Eva/*Forst*, Rainer/*Honneth*, Axel (eds.): *Ethos der Moderne. Foucaults Kritik der Aufklärung*, Frankfurt/Main 1990.

Ernst, Wolfgang: "Museale Kristallisation: *Il Vittoriale degli Italiani*," in: Hans Ulrich Gumbrecht/Friedrich Kittler/Bernhard Siegert (eds.), *Der Dichter als Kommandant. D'Annunzio erobert Fiume*, Munich 1996, pp. 309–20.

Esposito, Fernando: "Warburg und D'Annunzio Antipoden? Oder warum Athen immer wieder aus Alexandrien zurückerobert sein will," in: Gottfried Korff (ed.), *Kasten 117. Aby Warburg und der Aberglaube im Ersten Weltkrieg*, Tübingen 2007, pp. 301–23.

Falasca-Zamponi, Simonetta: *The Fascist Spectacle. The Aesthetics of Power in Mussolini's Italy*, Berkeley, CA 2000 [repr.].

Falasca-Zamponi, Simonetta: "Mussolini's Self-Staging," in: Deutsches Historisches Museum (ed.), *Kunst und Propaganda im Streit der Nationen 1930–1945*, Berlin 2007, pp. 88–95.

Falkenhausen, Susanne von: *Der Zweite Futurismus und die Kunstpolitik des Faschismus in Italien von 1922–1943*, Frankfurt/Main 1979.

Faulstich, Werner (ed.), *Image, Imageanalyse, Imagegestaltung. 2. Lüneburger Kolloquium zur Medienwissenschaft*, Bardowick 1992.

Ferrari, Massimo (ed.): *Le ali del ventennio. L'aviazione italiana dal 1923 al 1945. Bilanci storiografici e prospettive di giudizio*, Milan 2005.

Ferrari, Massimo: "La stampa aeronautica italiana in epoca fascista," in: Ferrari (ed.), *Le ali del Ventennio, L'aviazione italiana dal 1923 al 1945. Bilanci storiografici e prospettive di giudizio*, Milan 2005, pp. 31–110.

Ferrari, Paolo (ed.): *La grande guerra aerea 1915–1918. Battaglie – industrie – bombardamenti – assi – aeroporti*, Valdagno 1994.

Ferretti, Silvia: *Cassirer, Panofsky, and Warburg. Symbol, Art, and History*, New Haven, CT 1989.

Filk, Christian: *Im Bann der Live-Bilder. Krisenkommunikation, Kriegsberichterstattung und Mediensprache im Informationszeitalter. Studien nach dem Ende der Ost/West-Konfrontation*, Siegen 2006.

Fischer, Fritz: *From Kaiserreich to the Third Reich. Elements of Continuity in German History, 1871–1945*, Boston, MA 1986.

Fischer-Lichte, Erika: *Theatre, Sacrifice, Ritual. Exploring Forms of Political Theatre*, London 2005.

Fogu, Claudio: "Fascism and Historic Representation. The Garibaldian Celebrations," in: *JCH* 31/1996, pp. 317–45.

Fogu, Claudio: *The Historic Imaginary. Politics of History in Fascist Italy*, Toronto 2003.

Formigoni, Guido: *L'Italia dei cattolici. Fede e nazione dal Risorgimento alla Repubblica*, Bologna 1998.

Förster, Stig: "Das Zeitalter des totalen Krieges, 1861–1945. Konzeptionelle Überlegungen für einen historischen Strukturvergleich," in: *Mittelweg 36* 8/1999, pp. 12–29.

Förster, Stig: "Einleitung," in: Stig Förster (ed.), *An der Schwelle zum Totalen Krieg. Die militärische Debatte über den Krieg der Zukunft 1919–1939*, Paderborn 2002, pp. 15–36.

Förster, Stig/*Nagler*, Jörg (eds.): *On the Road to Total War. The American Civil War and the German Wars of Unification, 1861–1871*, New York, NY 1997.

Forsyth, Douglas: *The Crisis of Liberal Italy. Monetary and Financial Policy 1914–1922*, New York, NY 1993.

Foucault, Michel: *The Archaeology of Knowledge*, London 1972.

Foucault, Michel: "The Confession of the Flesh" (1977), interview, in: Foucault, *Power/Knowledge. Selected Interviews and Other Writings*, ed. by Colin Gordon), New York, NY 1980, pp. 194–228.

Foucault, Michel: *Discipline and Punish* [1975], New York, NY 1995.

Foucault, Michel: *The Order of Things. An Archeology of the Human Sciences*, New York, NY 1994.

Foucault, Michel: "What is Critique?" in: Foucault, *The Politics of Truth*, Los Angeles, CA 1997, pp. 41–82.

Frank, Manfred: "Das Motiv des 'kalten Herzens' in der Romantisch-symbolischen Dichtung," in: Frank, *Kaltes Herz, unendliche Fahrt, neue Mythologie. Motiv-Untersuchungen zur Pathogenese der Moderne*, Frankfurt/Main 1989, pp. 11–49.

Frank, Manfred: *Der kommende Gott. Vorlesungen über die Neue Mythologie*, Frankfurt/Main 1982.

Frank, Manfred: *Einführung in die frühromantische Ästhetik. Vorlesungen*, Frankfurt/Main 1989.

Frank, Manfred: *Gott im Exil. Vorlesungen über die Neue Mythologie*, Frankfurt/Main 1988.

Frank, Manfred: *The Philosophical Foundations of Early German Romanticism*, Albany, NY 2004.

Frank, Manfred: *"Unendliche Annäherung. "* Die Anfänge der philosophischen Frühromantik, Frankfurt/Main 1997.

Frank, Manfred: *What Is Neostructuralism?* Minneapolis, MN 1989.

Fraschetti, Alessandro: *La prima organizzazione dell'aeronautica militare in Italia dal 1884 al 1925*, Rome 1986.

Frei, Norbert: *1968. Jugendrevolte und globaler Protest*, Bonn 2008.

Frei, Norbert: "'Volksgemeinschaft.' Erfahrungsgeschichte und Lebenswirklichkeit der Hitler-Zeit," in: Frei, *1945 und wir. Das Dritte Reich im Bewusstsein der Deutschen*, Munich 2005, pp. 107–28.

Frei, Norbert: "Wie modern war der Nationalsozialismus?," in: *GG* 19/1993, pp. 367–87.

Frevert, Ute: "Das Militär als 'Schule' der Männlichkeit. Erwartungen, Angebote, Erfahrungen im 19. Jahrhundert," in: Ute Frevert (ed.), *Militär und Gesellschaft im 19. und 20. Jahrhundert*, Stuttgart 1997, pp. 145–73.

Frevert, Ute: *Ehrenmänner. Das Duell in der bürgerlichen Gesellschaft*, Munich 1991.

Frevert, Ute: "Herren und Helden. Vom Aufstieg und Niedergang des Heroismus im 19. und 20. Jahrhundert," in: Richard van Dülmen (ed.), *Erfindung des Menschen. Schöpfungsträume und Körperbilder 1500–2000*, Vienna 1998, pp. 323–44.

Friedländer, Saul: *Reflections of Nazism. An Essay on Kitsch and Death*, Bloomington, IN 1993.

Fries, Helmut: "Deutsche Schriftsteller im Ersten Weltkrieg," in: Wolfgang Michalka (ed.), *Der Erste Weltkrieg. Wirkung, Wahrnehmung, Analyse*, Munich 1994, pp. 825–48.

Fries, Helmut: *Die große Katharsis. Der Erste Weltkrieg in der Sicht deutscher Dichter und Gelehrter*, vol. 1, *Die Kriegsbegeisterung von 1914: Ursprünge – Denkweisen – Auflösung*, Konstanz 1994.

Fries, Helmut: *Die große Katharsis. Der Erste Weltkrieg in der Sicht deutscher Dichter und Gelehrter*, vol. 2, *Euphorie – Entsetzen – Widerspruch: die Schriftsteller 1914–1918*, Konstanz 1995.

Fritzsche, Peter: *A Nation of Fliers. German Aviation and the Popular Imagination*, Cambridge, MA 1992.

Fritzsche, Peter: *Germans into Nazis*, Cambridge, MA 1998.

Fritzsche, Peter: "Historical Time and Future Experience in Postwar Germany," in: Wolfgang Hardtwig (ed.), *Ordnungen in der Krise. Zur politischen Kulturgeschichte Deutschlands 1900–1933*, Munich 2007, pp. 141–64.

Fritzsche, Peter: "Nazi Modern," in: *Modernism/Modernity* 3/1996, pp. 1–21.

Fritzsche, Peter/*Hellbeck*, Jochen: "The new man in Stalinist Russia and Nazi Germany," in: Michael Geyer/Sheila Fitzpatrick (eds.), *Beyond Totalitarianism. Stalinism and Nazism Compared*, New York, NY 2009, pp. 302–41.

Fröschle, Ulrich/*Mottel*, Helmut: "Medientheoretische und mentalitätsgeschichtliche Probleme filmhistorischer Untersuchungen. Fallbeispiel *Apocalypse Now*," in: Bernhard Chiari/Matthias Rogg/Wolfgang Schmidt (eds.), *Krieg und Militär im Film des 20. Jahrhunderts*, Munich 2003, pp. 107–40.

Fuhrmann, Manfred (ed.): *Terror und Spiel. Probleme der Mythenrezeption*, Munich 1971.

Funck, Marcus: "Bereit zum Krieg? Entwurf und Praxis militärischer Männlichkeit im preußisch-deutschen Offizierskorps vor dem Ersten Weltkrieg," in: Karen Hagemann/Stefanie Schüler-Springorum (eds.), *Heimat-Front. Militär und Geschlechterverhältnisse im Zeitalter der Weltkriege*, Frankfurt/Main 2002, pp. 69–90.

Funck, Marcus: *Feudales Kriegertum und militärische Professionalität. Adel und Bürgertum in den preußisch-deutschen Offizierskorps 1860–1935*, Berlin 2005.

Funck, Marcus: "Schock und Chance. Der preußische Militäradel in der Weimarer Republik zwischen Stand und Profession," in: Heinz Reif (ed.), *Adel und Bürgertum in Deutschland*, vol. 2, *Entwicklungslinien und Wendepunkte im 20. Jahrhundert*, Berlin 2001, pp. 127–72.

Funck, Marcus: "Vom Höfling zum soldatischen Mann. Varianten und Umwandlungen adeliger Männlichkeit zwischen Kaiserreich und Nationalsozialismus," in: Eckart Conze/Monika Wienfort (eds.), *Adel und Moderne. Deutschland im europäischen Vergleich im 19. und 20. Jahrhundert*, Cologne 2004, pp. 205–35.

Fussell, Paul: *The Great War and Modern Memory*, New York, NY 2000 [repr.].

Gadamer, Hans-Georg: "Mythos und Vernunft," in: Gadamer, *Gesammelte Werke*, vol. 8, Ästhetik und Poetik, Tübingen 1993, pp. 163–69.

Gadamer, Hans-Georg: *Truth and Method*, 2nd rev. edn, London 1989.

Gander, Hans-Helmuth (ed.): *"Verwechselt mich vor allem nicht!" Heidegger und Nietzsche*, Frankfurt/Main 1994.

Gay, Peter: *Freud, Jews, and Other Germans: Masters and Victims in Modernist Culture*, New York, NY 1978.

Gay, Peter: *Weimar Culture: The Outsider as Insider*, Harmondsworth 1974.

Geinitz, Christian: "The First Air War against Noncombatants. Strategic Bombing of German Cities in World War I," in: Roger Chickering/Stig Förster (eds.), *Great War, Total War. Combat and Mobilization on the Western Front, 1914–1918*, New York, NY 2000, pp. 207–25.

Gellately, Robert: *Lenin, Stalin, and Hitler. The Age of Social Catastrophe*, London 2007.

Gellner, Ernest: *Nations and Nationalism*, Oxford 1990 [repr.].

Genette, Gérard: *Palimpsests: Literature in the Second Degree*, Lincoln, NE 1997.

Genette, Gérard: *Paratexts: Thresholds of Interpretation*, New York, NY 1997.

Gennep, Arnold van: *Rites of Passage* [orig. 1909], Chicago, IL 1961.

Gentile, Emilio: "The Conquest of Modernity: From Modernist Nationalism to Fascism," in: *Modernism/Modernity* 1/1994, pp. 55–87.

Gentile, Emilio: "Der Faschismus. Eine Definition zur Orientierung," in: *Mittelweg 36* 16/2007, pp. 81–99.

Gentile, Emilio: "Der Liktorenkult," in: Christof Dipper/Rainer Hudemann/Jens Petersen (eds.), *Faschismus und Faschismen im Vergleich. Wolfgang Schieder zum 60. Geburtstag*, Cologne 1998, pp. 247–61.

Gentile, Emilio: "Die Sakralisierung der Politik," in: Hans Maier (ed.), *Wege in die Gewalt. Die modernen politischen Religionen*, Frankfurt/Main 2000, pp. 166–82.

Gentile, Emilio: *Fascismo di pietra*, Rome 2007.

Gentile, Emilio: "Fascism, Totalitarianism and Political Religion. Definitions and Critical Reflections on Criticism of an Interpretation," in: *Totalitarian Movements and Political Religions*, 5/2004, pp. 326–75.

Gentile, Emilio: "Il futurismo e la politica. Dal nazionalismo modernista al fascismo (1909–1920)," in: Renzo *Felice* (ed.), *Futurismo, cultura e politica*, Turin 1988, pp. 105–59.

Gentile, Emilio: *La Grande Italia. Il mito della nazione nel XX secolo*, Rome 2006.

Gentile, Emilio: *"La nostra sfida alle stelle." Futuristi in politica*, Rome 2009.

Gentile, Emilio: *Le religioni della politica. Fra democrazie e totalitarismi*, Rome 2001.

Gentile, Emilio: "'L'uomo nuovo' del fascismo. Riflessioni su un esperimento totalitario di rivoluzione antropologica," in: Gentile, *Fascismo. Storia e interpretazione*, Rome 2005, pp. 235–64.

Gentile, Emilio: "The Myth of National Regeneration in Italy. From Modernist Avant-Garde to Fascism," in: Matthew Affron/Mark Antliff (eds.), *Fascist Visions. Art and Ideology in France and Italy*, Princeton, NJ 1997, pp. 25–45.

Gentile, Emilio: *The Origins of Fascist Ideology 1918–1925*, New York, NY 2005 [repr].

Gentile, Emilio: "Political Futurism and the Myth of the Italian Revolution," in: Günter Berghaus (ed.), *International Futurism in Arts and Literature*, Berlin 2000, pp. 1–14.

Gentile, Emilio: "Political Religion. A Concept and its Critics – A Critical Survey," in: *Totalitarian Movements and Political Religion* 6/2005, pp. 19–32.

Gentile, Emilio: *The Sacralization of Politics in Fascist Italy*, Cambridge, MA 1996.

Gentile, Emilio: *Storia del partito fascista, 1919–1922. Movimento e milizia*, Rome 1989.

Gentile, Emilio: *The Struggle for Modernity. Nationalism, Futurism and Fascism*, Westport, CT 2003.

Gerstner, Alexandra (ed.): *Der Neue Mensch. Utopien, Leitbilder und Reformkonzepte zwischen den Weltkriegen*, Frankfurt/Main 2006.

Gerstner, Alexandra: *Neuer Adel. Aristokratische Elitekonzeptionen zwischen Jahrhundertwende und Nationalsozialismus*, Darmstadt 2008.

Geuss, Raymond: "Die wirkliche und eine andere Moderne. Ordnungsstiftende Phantasie im *Don Quijote*," in: *Mittelweg 36* 14/2005, pp. 49–67.

Geyer, Martin H.: *Verkehrte Welt, Revolution, Inflation und Moderne, Munich 1914–1924*, Göttingen 1998.

Geyer, Michael: *Deutsche Rüstungspolitik 1860–1980*, Frankfurt/Main 1984.

Geyer, Michael/*Fitzpatrick*, Sheila (eds.): *Beyond Totalitarianism. Stalinism and Nazism Compared*, Cambridge 2009.

Gibelli, Antonio: *La grande guerra degli Italiani, 1915–1918*, Milan 2001².

Gibelli, Antonio: "'Letteratura di illetterati' nella Grande Guerra. Lineamenti di un percorso storiografico," in: Massimo Bacigalupo/Roberto De Pol (eds.), *Grande Guerra e letteratura*, Genoa 1997, pp. 37–50.

Gibelli, Antonio: *L'officina della guerra. La grande guerra e le trasformazioni del mondo mentale*, Turin 1991.

Gibellini, Pietro: "Il volo di Icaro," in: Gibellini (ed.), *Logos e Mythos. Studi su Gabriele D'Annunzio*, Florence 1985, pp. 119–32.

Giesen, Bernhard: *Triumph and Trauma*, Boulder, CO 2004.

Giovannini, Elio: *L'Italia massimalista. Socialismo e lotta sociale e politica nel primo dopoguerra italiano*, Rome 2001.

Girard, René: *Violence and the Sacred*, Baltimore, MD 1979.

Gleason, Abbott: *Totalitarianism. The Inner History of the Cold War*, New York, NY 1995.

Goebel, Stefan: *The Great War and Medieval Memory. War, Remembrance and Medievalism in Britain and Germany, 1914–1940*, Cambridge 2007.

Goldstein, Laurence: *The Flying Machine and Modern Literature*, Bloomington, IN 1986.

Gollbach, Michael: *Die Wiederkehr des Weltkrieges in der Literatur. Zu den Frontromanen der späten Zwanziger Jahre*, Kronberg Ts. 1978.

Gombrich, Ernst H.: *Aby Warburg. An Intellectual Biography*, Oxford 1986.

Gooch, John: *Mussolini and his Generals. The Armed Forces and Fascist Foreign Policy 1922–1940*, Cambridge 2007.

Graf, Friedrich Wilhelm: "Alter Geist und neuer Mensch. Religiöse Zukunftserwartungen um 1900," in: Ute Frevert (ed.), *Das Neue Jahrhundert. Europäische Zeitdiagnosen und Zukunftsentwürfe um 1900*, Göttingen 2000, pp. 185–228.

Graf, Rüdiger: *Die Zukunft der Weimarer Republik. Krisen und Zukunftsaneignungen in Deutschland 1918–1933*, Munich 2008.

Griffin, Roger (ed.): *Fascism. Critical Concepts in Political Science*, vol. 1., *The Nature of Fascism*, London 2004.

Griffin, Roger: "'I am no longer human. I am a Titan. A God!' The Fascist Quest to Regenerate Time," in: Roger Griffin (ed.), *A Fascist Century. Essays*, New York, NY 2008, pp. 3–23.

Griffin, Roger: "Introduction," in: Roger Griffin (ed.), *International Fascism. Theories, Causes and the New Consensus*, London 1998, pp. 1–20.

Griffin, Roger: *Modernism and Fascism. The Sense of a New Beginning under Mussolini and Hitler*, Basingstoke 2007.

Griffin, Roger: "Modernity Under the New Order: The Fascist Project for Managing the Future," in: Griffin, *A Fascist Century. Essays*, New York, NY 2008, pp. 24–45.

Griffin, Roger: *The Nature of Fascism*, London 1993 [repr.].

Griffin, Roger: "The Primacy of Culture. The Current Growth (or Manufacture) of Consensus within Fascist Studies," in: *JCH* 37/2002, pp. 21–43.

Groehler, Olaf: *Geschichte des Luftkrieges 1910 bis 1970*, Berlin (East) 1970.

Grosse, Jürgen: *Kritik der Geschichte. Probleme und Formen seit 1800*, Tübingen 2006.

Guerrini, Irene/*Pluviano*, Marco: *Francesco Baracca una vita al volo. Guerra e privato di un mito dell'aviazione*, Udine 2000.

Gumbrecht, Hans Ulrich: *Dimensionen und Grenzen der Begriffsgeschichte*, Munich 2006.

Gumbrecht, Hans Ulrich: "I redentori della vittoria. Über Fiumes Ort in der Genealogie des Faschismus," in: Hans Ulrich Gumbrecht/Friedrich Kittler/Bernhard Siegert (eds.), *Der Dichter als Kommandant. D'Annunzio erobert Fiume*, Munich 1996, pp. 83–15.

Gumbrecht, Hans Ulrich: "Modern, Modernität, Moderne," in: Otto Brunner/Werner Conze/Reinhart Koselleck (eds.), *Geschichtliche Grundbegriffe. Historisches Lexikon zur politisch-sozialen Sprache in Deutschland*, vol. 4, Stuttgart 1978, pp. 93–131.

Gumbrecht, Hans Ulrich/*Kittler* Friedrich/*Siegert*, Bernhard (eds.): *Der Dichter als Kommandant. D'Annunzio erobert Fiume*, Munich 1996.

Günther, Hans: *Der sozialistische Übermensch. M. Gorkij und der sowjetische Heldenmythos*, Stuttgart 1993.

Habeck, Mary R.: "Technology in the First World War. The View from Below," in: Jay Winter/Geoffrey Parker/Mary R. Habeck (eds.), *The Great War and the Twentieth Century*, New Haven, CT 2000, pp. 99–131.

Habermas, Jürgen: "Modernity – an Unfinished Project," in: Maurizio Passerin d'Entrèves/Seyla Benhabib (eds.), *Habermas and the Unfinished Project of Modernity. Critical Essays on The Philosophical Discourse of Modernity*, Cambridge 1996, pp. 38–58.

Habermas, Jürgen: *The Philosophical Discourse of Modernity. Twelve Lectures*, Cambridge, MA 1987.

Habermas, Jürgen: *Theory of Communicative Action*, vol. 1, *Reason and the Rationalization of Society*, Frankfurt/Main 1981, pp. 225–366.

Haffner, Sebastian: *Defying Hitler. A Memoir*, New York, NY 2003.

Härmänmaa, Marja: *Un patriota che sfidò la decadenza. F. T. Marinetti e l'idea dell'uomo nuovo fascista 1929–1944*, Helsinki 2000.

Harmening, Dieter: *Wörterbuch des Aberglaubens*, Stuttgart 2005.

Hebekus, Uwe: Ästhetische Ermächtigung. *Zum politischen Ort der Literatur im Zeitraum der Klassischen Moderne*, Munich 2009.

Heffen, Annegret: *Der Reichskunstwart. Kunstpolitik in den Jahren 1920–1930. Zu den Bemühungen um eine offizielle Reichskunstpolitik in der Weimarer Republik*, Essen 1986.

Heidegger, Martin: "The Age of the World Picture" (1938), in: *Off the Beaten Track [Holzwege]*, Cambridge 2002, pp. 57–85.

Heidegger, Martin: *Being and Time*, New York, NY 1962.

Heidegger, Martin: "Einblick in das, was ist," in: Heidegger, *Bremer und Freiburger Vorträge. Gesamtaugabe*, vol. 79, ed. by Petra Jaeger, Frankfurt/Main 1994, pp. 3–77.

Heidegger, Martin: *Einleitung in die Philosophie. Freiburger Vorlesung Wintersemester 1928/29*, Heidegger, *Gesamtausgabe*, vol. 27, ed. by Otto Saame and Ina Saame, Frankfurt/Main 1996.

Heidegger, Martin: "Letter on Humanism," in: Heidegger, *Basic Writings*, rev. edn, New York, NY 1993, pp. 213–66.

Heidegger, Martin: *Nietzsche*, vols. I and II, paperback edn, San Francisco, CA 1991.

Heidegger, Martin: *Nietzsche*, vols. III and IV, paperback edn, San Francisco, CA 1991.

Heidegger, Martin: "Nietzsche's Word 'God Is Dead,'" in: Heidegger, *Off the Beaten Track [Holzwege]*, Cambridge 2002, pp. 157–99.

Heidegger, Martin: "The Question Concerning Technology," in: Heidegger, *Basic Writings*, rev. edn, New York, NY 1993, pp. 213–66.

Heidegger, Martin: *What is Called Thinking?*, New York, NY 1968.

Heidegger, Martin: *Zu Ernst Jünger, Gesamtausgabe*, vol. 90, ed. by Peter Trawny, Frankfurt/Main 2004.

Heise, Joachim S.: "Sprache im Dienste des Völkerringens. Linguistische Perspektiven zum Ersten Weltkrieg," in: *Krieg und Literatur –War and Literature*, vol. V/1999, pp. 37–54.

Hellbeck, Jochen: *Revolution on My Mind. Writing a Diary under Stalin*, Cambridge, MA 2006.

Herbert, Ulrich: "Arbeiterschaft im 'Dritten Reich.' Zwischenbilanz und offene Fragen," in: *GG* 15/1989, pp. 320–60.

Herbert, Ulrich: *Best. Biographische Studien über Radikalismus, Weltanschauung und Vernunft, 1903–1989*, Bonn 1996.

Herbert, Ulrich: "Europe in High Modernity. Reflections on a Theory of the 20th Century," in: *JMEH* 5/2007, pp. 5–21.

Herf, Jeffrey: "Der nationalsozialistische Technikdiskurs. Die deutschen Eigenheiten des reaktionären Modernismus," in: Wolfgang Emmerich/Carl Wege (eds.), *Der Technikdiskurs in der Hitler-Stalin-Ära*, Stuttgart 1995, pp. 72–93.

Herf, Jeffrey: "The Engineer as Ideologue. Reactionary Modernists in Weimar and NaziGermany," in: *JCH* 19/1984, S. 631–48.

Herf, Jeffrey: *Reactionary Modernism. Technology, Culture, and Politics in Weimar and the Third Reich*, Cambridge, MA 1984.

Hermand, Jost: *Der alte Traum vom neuen Reich. Völkische Utopien und Nationalsozialismus*, Weinheim 1955.

Hettling, Manfred: "Eine anstrengende Affäre. Die Sozialgeschichte des Bürgertums," in: Sven Oliver Müller/Cornelius Torp (eds.), *Das Deutsche Kaiserreich in der Kontroverse*, Göttingen 2009, pp. 219–32.

Hewitt, Andrew: *Fascist Modernism. Aesthetics, Politics, and the Avant-Garde*, Stanford, CA 1993.

Hildermeier, Manfred: "Revolution und Kultur. Der 'neue Mensch' in der frühen Sowjetunion," in: *Jahrbuch des Historischen Kollegs* 1996, Munich 1997, pp. 51–67.

Hinz, Manfred: *Die Zukunft der Katastrophe. Mythische und rationalistische Geschichtstheorie im italienischen Futurismus*, Berlin 1985.

Hirschfeld, Gerhard: "The Battle of the Somme 1916," in: Gerhard Hirschfeld/Gerd Krumeich/Irina Renz (eds.), *Scorched Earth. The Germans on the Somme 1914–1918*, Barnsley 2009, pp. 63–81.

Hobsbawm, Eric J.: *The Age of Extremes: The Short Twentieth Century, 1914–1991*, London 1994.

Hobsbawm, Eric J.: *The Age of Revolution. Europe 1789–1848*, London 1988 [repr.].

Hobsbawm, Eric J.: *Nations and Nationalism since 1780. Programme, Myth, Reality*, Cambridge 1990.

Hobsbawm, Eric J./*Ranger*, Terence (eds.): *The Invention of Tradition*, Cambridge 1983.

Hobson, Rolf: *Imperialism at Sea. Naval Strategic Thought, the Ideology of Sea Power, and the Tirpitz Plan, 1875–1914*, Boston, MA 2002.

Hochgeschwender, Michael: *Amerikanische Religion. Evangelikalismus, Pfingstlertum und Fundamentalismus*, Frankfurt/Main 2007.

Hoeres, Peter: *Der Krieg der Philosophen. Die deutsche und die britische Philosophie im Ersten Weltkrieg*, Paderborn 2004.

Höhler, Sabine: *Luftfahrtforschung und Luftfahrtmythos. Wissenschaftliche Ballonfahrt in Deutschland 1880–1910*, Frankfurt/Main 2001.

Homepage of the Bundeswehr Jagdgeschwader 71 "Richthofen," accessed on August 18, 2010: http: //www. luftwaffe. de-/portal/a/luftwaffe/org/luftm/jg71.

Homepage of D'Annunzio's Vittoriale degli italiani, accessed on August 14, 2009: http: // www. vittoriale.it.

Horkheimer, Max: "The Jews in Europe," in: Stephen Eric Bonner/Douglas Mackay Kellner (eds.), *Critical Theory and Society. A Reader*, New York, NY 1989, pp. 77–94.

Hunt, Lynn: *Measuring Time, Making History*, Budapest 2008.

Hüppauf, Bernd (ed.): *Ansichten vom Krieg. Vergleichende Studien zum Ersten Weltkrieg in Literatur und Gesellschaft*, Königstein Ts. 1984.

Hüppauf, Bernd: "Das Schlachtfeld als Raum im Kopf," in: Steffen Martus/Martina Münkler/ Werner Röcke (eds.), *Schlachtfelder. Codierung von Gewalt im medialen Wandel*, Berlin 2003, pp. 207–34.

Hüppauf, Bernd: "Das Unzeitgemäße der Avantgarden. Die Zeit, Avantgarden und die Gegenwart," in: Wolfgang Asholt/Walter Fähnders (eds.), *Der Blick vom Wolkenkratzer. Avantgarde – Avantgardekritik – Avantgardeforschung*, Amsterdam 2000, pp. 547–82.

Hüppauf, Bernd: "Räume der Destruktion und Konstruktion von Raum. Landschaft, Sehen, Raum und der Erste Weltkrieg," in: *Krieg und Literatur* III/1991, pp. 105–23.

Hüppauf, Bernd: "Schlachtenmythen und die Konstruktion des 'Neuen Menschen,'" in: Gerhard Hirschfeld/Gerd Krumeich (eds.), *Keiner fühlt sich hier mehr als Mensch ... Erlebnis und Wirkung des Ersten Weltkrieges*, Essen 1993, pp. 43–84.

Huxley, Aldous: *The Doors of Perception*, London 1954.

Ingold, Felix Philipp: *Literatur und Aviatik. Europäische Flugdichtung 1909–1927*, Frankfurt/ Main 1980.

Isenghi, Mario: *Giornali di trincea, 1915–1918*, Turin 1977.

Isenghi, Mario: *Il mito della grande guerra*, Bologna 2002.

Isenghi, Mario: *Le guerre degli italiani. Parole, immagini, ricordi 1848–1945*, Bologna 2005.

Isenghi, Mario: *L'Italia del fascio*, Florence 1996.

Isenghi, Mario/*Rochat*, Giorgio: *La grande guerra*, Milan 2004 [repr.].

Jaeger, Friedrich: "Theorietypen der Krise des Historismus," in: Wolfgang Bialas/Gérard Raulet (eds.), *Die Historismusdebatte in der Weimarer Republik*, Frankfurt/Main 1996, pp. 52–70.

Jamme, Christoph: *Einführung in die Philosophie des Mythos. Neuzeit und Gegenwart*, Darmstadt 1991.

Jamme, Christoph: *"Gott an hat ein Gewand." Grenzen und Perspektiven philosophischer Mythos-Theorien der Gegenwart*, Frankfurt/Main 1999.

Jamme, Christoph/*Schneider*, Helmut (eds.): *Mythologie der Vernunft. Hegels "ältestes Systemprogramm des deutschen Idealismus,"* Frankfurt/Main 1984.

Janz, Oliver: "Monumenti di carta. Le pubblicazioni in memoria dei caduti della Prima Guerra Mondiale," in: Oliver Janz/Fabrizio Dolci (eds.), *Non Omnis Moriar. Gli opuscoli di necrologi o per i caduti italiani nella Grande Guerra*, Rome 2003, pp. 11–44.

Janz, Oliver: "Zwischen privater Trauer und öffentlichem Gedenken. Der burgerliche Gefallenenkult in Italien während des Ersten Weltkrieges," in: *GG* 28/2002, pp. 554–73.

Jean, Yaron: "'Mental Aviation' – Conquering the Skies in the Weimar Republic," in: *Tel Aviver Jahrbuchfür deutsche Geschichte* 28/1999, pp. 429–58.

Jesse, Eckhard (ed.): *Totalitarismusim 20. Jahrhundert. Eine Bilanz der internationalen Forschung*, Baden-Baden 19992.

Josting, Petra: *Der Jugendschrifttums-Kampf des Nationalsozialistischen Lehrerbundes*, Hildesheim 1995.

Kaelble, Hartmut: "Wie feudal waren die deutschen Unternehmer im Kaiserreich? Ein Zwischenbericht," in: Richard Tilly (ed.), *Beiträge zur quantitativen vergleichenden Unternehmensgeschichte*, Stuttgart 1985, pp. 148–71.

Kant, Immanuel: "An Answer to the Question: 'What Is Enlightenment?'," in: Hans Reiss (ed.), *Political Writings*, 2nd edn, Cambridge 1991, pp. 54–60.

Kater, Michael H.: "Generationskonflikt als Entwicklungsfaktor in der NS-Bewegung vor 1933," in: *GG* 11/1985, pp. 217–43.

Keegan, John: *The First World War*, New York, NY 1999.

Kehrt, Christian: *Moderne Krieger. Die Technikerfahrungen deutscher Militärpiloten 1910–1945*, Paderborn 2010.

Kehrt, Christian: "'Schneid, Takt und gute Nerven.' Der Habitus deutscher Militärpiloten und Beobachter im Kontext technisch strukturierter Handlungszusammenhänge, 1914–1918," in: *Technikgeschichte* 72/2005, pp. 177–201.

Kennan, George F.: *The Decline of Bismarck's European Order. Franco-Russian Relations, 1875–1890*, Princeton, NJ 1981.

Kennedy, Paul: *The Rise and Fall of British Naval Mastery*, London 2004³.

Kennett, Lee: *The First Air War. 1914–1918*, New York, NY 1991.

Kern, Stephen: *The Culture of Time and Space, 1880–1918*, London 1983.

Kershaw, Ian: *The "Hitler Myth": Image and Reality in the Third Reich*, Oxford 2001.

Kershaw, Ian: *Hitler 1936–1945: Nemesis*, New York, NY 2000.

Kester, Bernadette: *Film Front Weimar. Representations of the First World War in German Films of the Weimar Period (1919–1933)*, Amsterdam 2003.

Ketelsen, Uwe K.: "Ernst Jüngers *Der Arbeiter*. Ein faschistisches Modernitätskonzept," in: Ketelsen (ed.), *Literatur und Drittes Reich*, Schernfeld 1992, pp. 258–85.

Kiesel, Helmuth: *Ernst Jünger. Die Biographie*, Munich 2007.

Kittler, Friedrich: "Il fiore delle truppe scelte," in: Hans Ulrich Gumbrecht/Friedrich Kittler/Bernhard Siegert (eds.), *Der Dichter als Kommandant. D'Annunzio erobert Fiume*, Munich 1996, pp. 205–25.

Kittstein, Ulrich: *"Mit Geschichte will man etwas." Historisches Erzählen in der Weimarer Republik und im Exil (1918–1945)*, Würzburg 2006.

Klee, Ernst: *Das Kulturlexikon zum Dritten Reich. Wer war was vor und nach 1945*, Frankfurt/Main 2007.

Klenner, Jost Philipp: "Mussolini und der Löwe. Aby Warburg und die Anfänge der politischen Ikonographie," in: *Zeitschrift für Ideengeschichte* 1/2007, pp. 83–98.

Klinger, Cornelia: "Die Utopie der Versöhnung von Kunst und Leben. Die Transformation einer Idee im 20. Jahrhundert. Vom Staat als Kunstwerk zum life-style des Individuums," in: Klinger/Wolfgang (eds.), *Das Jahrhundert der Avantgarden*, Munich 2004, pp. 211–45.

Klinkhammer, Lutz: "Der Partisanenkrieg der Wehrmacht 1941–1944," in: Rolf-Dieter Müller/Hans-Erich Volkmann (eds.), *Die Wehrmacht. Mythos und Realität*, Munich 1999, pp. 815–36.

Koch, Lars: *Der Erste Weltkrieg als Medium der Gegenmoderne. Zu den Werken von Walter Flex und Ernst Jünger*, Würzburg 2006, pp. 287–330.

Köhler, Karl: "Organisationsgeschichte der Luftwaffe von den Anfängen bis 1918," in: Militärgeschichtliches Forschungsamt (ed.), *Handbuch zur deutschen Militärgeschichte 1648–1939*, vol. 3, Part V, Munich 1979, pp. 283–311.

König, Wolfgang: *Wilhelm II. und die Moderne. Der Kaiser und die technisch-industrielle Welt*, Paderborn 2007.

Koon, Tracy H.: *Believe, Obey, Fight. Political Socialization of Youth in Fascist Italy, 1922–1943*, Chapel Hill, NC 1985.

Köppen, Manuel: *Das Entsetzen des Beobachters. Krieg und Medien im 19. und 20. Jahrhundert*, Heidelberg 2005.

Korff, Gottfried: "Vom Verlangen, Bedeutungen zu sehen," in: Ulrich Borsdorf/Heinrich Theodor Grütter/Jörn Rüsen (eds.), *Die Aneignung der Vergangenheit. Musealisierung und Geschichte*, Bielefeld 2004, pp. 81–103.

Koselleck, Reinhart: "The Eighteenth Century as the Beginning of Modernity," in: Koselleck, *The Practice of Conceptual History. Timing History, Spacing Concepts*, Stanford, CA 2002, pp. 154–69.

Koselleck, Reinhart: "Einleitung," in: Otto Brunner/Werner Conze/Reinhart Koselleck (eds.), *Geschichtliche Grundbegriffe. Historisches Lexikon zur politisch-sozialen Sprache in Deutschland*, vol. 1, Stuttgart 1972, pp. XIII–XXVII.

Koselleck, Reinhart: "Fortschritt," in: Otto Brunner/Werner Conze/Reinhart Koselleck (eds.), *Geschichtliche Grundbegriffe. Historisches Lexikon zur politisch-sozialen Sprache in Deutschland*, vol. 2, Stuttgart 1975, pp. 351–423.

Koselleck, Reinhart: *Futures Past. On the Semantics of Historical Time*, Cambridge, MA 1985.

Koselleck, Reinhart: "Neuzeit – Remarks on the Semantics of Modern Concepts of Movement," in: Koselleck, *Futures Past. On the Semantics of Historical Time*, Cambridge, MA 1985, pp. 222–54.

Koselleck, Reinhart: "On the Anthropological and Semantic Structure of Bildung," in: Koselleck, *The Practice of Conceptual History. Timing History, Spacing Concepts*, Stanford, CA 2002, pp. 170–207.

Koselleck, Reinhart: *The Practice of Conceptual History. Timing History, Spacing Concepts*, Stanford, CA 2002.

Koselleck, Reinhart: "'Space of Experience' and 'Horizon of Expectation' – Two Historical Categories," in: Koselleck, *Futures Past. On the Semantics of Historical Time*, Cambridge, MA 1985, pp. 267–88.

Koselleck, Reinhart: "Transformations of Experience and Methodological Change," in: Koselleck. *The Practice of Conceptual History. Timing History, Spacing Concepts*, Stanford, CA 2002, pp. 45–83.

Koslowski, Peter: *Der Mythos der Moderne. Die dichterische Philosophie Ernst Jüngers*, Munich 1991.

Kroll, Frank Lothar: *Utopie als Ideologie. Geschichtsdenken und politisches Handeln im Dritten Reich*, Paderborn 1998.

Krumeich, Gerd (ed.): *Versailles 1919. Ziele, Wirkung, Wahrnehmung*, Essen 2001.

Kühne, Thomas: *Kameradschaft. Die Soldaten des nationalsozialistischen Krieges und das 20. Jahrhundert*, Göttingen 2006.

Kühne, Thomas (ed.): *Männergeschichte–Geschlechtergeschichte. Männlichkeit im Wandel der Moderne*, Frankfurt/Main 1996.

Kunze, Rolf-Ulrich: *Nation und Nationalismus*, Darmstadt 2005.

Laak, Dirk van: *Gespräche in der Sicherheit des Schweigens. Carl Schmitt in der politischen Geistesgeschichte der frühen Bundesrepublik*, Berlin 1993.

Laak, Dirk van: "Planung. Geschichte und Gegenwart des Vorgriffs auf die Zukunft," in: *GG* 34/2008, pp. 305–26.

Lacoue-Labarthe, Philippe: *Heidegger, Art, and Politics. The Fiction of the Political*, Cambridge, MA 1990.

Lakoff, George/Johnson, Mark: *Metaphors We Live By*, Chicago, IL 1980.

Landwehr, Achim: *Geschichte des Sagbaren. Einführung in die historische Diskursanalyse*, Tübingen 2001.

Langewiesche, Dieter: "Bildungsbürgertum und Liberalismus im 19. Jahrhundert," in: Jürgen Kocka (ed.), *Bildungsbürgertum im 19. Jahrhundert. vol. IV. Politischer Einfluß und gesellschaftliche Formation*, Stuttgart 1989, pp. 95–121.

Langewiesche, Dieter: *Liberalismus in Deutschland*, Frankfurt/Main 1988.

Langewiesche, Dieter: *Nation, Nationalismus und Nationalstaat in Deutschland und Europa*, Munich 2000.

Langewiesche, Dieter: "Nation, Nationalismus, Nationalstaat. Forschungsstand und Forschungsperspektiven," in: *NPL* 40/1995, pp. 190–236.

Lange, Wolfgang: "Tod ist bei den Göttern immer nur ein Vorurteil. Zum Komplex des Mythos bei Nietzsche," in: Karl Heinz Bohrer (ed.), *Mythos und Moderne*, Frankfurt/Main 1983, pp. 111–37.

Latham, Michael: *Modernization as Ideology. American Social Science and "Nation Building" in the Kennedy Era*, Chapel Hill, NC 2000.

Latour, Bruno: *We Have Never Been Modern*, Cambridge, MA 1993.

Latzel, Klaus: *Vom Sterben im Krieg. Wandlungen in der Einstellung zum Soldatentod vom Siebenjährigen Krieg bis zum II. Weltkrieg*, Warendorf 1988.

Lazzaro, Claudia/Crum, Roger J. (eds.): *Donatello among the Blackshirts. History and Modernity in the Visual Culture of Fascist Italy*, Ithaca, NY 2005.

Leeden, Micheal A.: *The First Duce*, Baltimore, MD 1977.

Leed, Eric J.: *No Man's Land, Combat and Identity in World War I*, Cambridge 1979.

Leendertz, Ariane: *Ordnung schaffen. Deutsche Raumplanung im 20. Jahrhundert*, Göttingen 2008.

Leonhard, Jörn: *Liberalismus. Zur historischen Semantik eines europäischen Deutungsmusters*, Munich 2001.

Lepsius, Mario Rainer: "Nation und Nationalismus in Deutschland," in: Lepsius (ed.), *Interessen, Ideen und Institutionen*, Opladen 1990, pp. 232–46.

Lethen, Helmut: *Cool Conduct. The Culture of Distance in Weimar Germany*, Berkeley, CA 2002.

Lévi-Strauss, Claude: *Rasse und Geschichte*, Frankfurt/Main 1972.

Lévi-Strauss, Claude: "The Structural Study of Myth," in: Lévi-Strauss, *Structural Anthropology*, New York, NY 1963, pp. 206–31.

Levsen, Sonja: *Elite, Männlichkeit und Krieg. Tübinger und Cambridger Studenten 1900–1929*, Göttingen 2006.

Linder, Ann. P.: *Princes of the Trenches. Narrating the German Experience of the First World War*, Columbia, SC 1996.

Linz, Juan J.: "Der religiöse Gebrauch der Politik und/oder der politische Gebrauch der Religion. Ersatz-Ideologie gegen Ersatz-Religion," in: Hans Maier (ed.), *Totalitarismus und Politische Religion, Vol. 1*, Paderborn 1996, pp. 129–54.

Lipp, Anne: *Meinungslenkung im Krieg. Kriegserfahrungen deutscher Soldaten und ihre Deutung 1914–1918*, Göttingen 2003.

Loewenstein, Bedrich: *Der Fortschrittsglaube. Geschichte einer europäischen Idee*, Göttingen 2009.

Longerich, Peter: *Heinrich Himmler*, Oxford/New York 2012.

Longoni, Maurizio: "Gli 'assi' sul fronte italiano," in: Paolo Ferrari (ed.), *La grande guerra aerea 1915–1918. Battaglie – industrie – bombardamenti – assi – aeroporti*, Valdagno 1994, pp. 291–322.

Löwith, Karl: *My Life in Germany Before and After 1933 – A Report*, Chicago, IL 1994.

Lowry, Stephen: "Star," in: Hans-Otto Hügel (ed.), *Handbuch Populäre Kultur. Begriffe, Theorien und Diskussionen*, Stuttgart 2003, pp. 441–45.

Lyotard, Jean-François: *The Postmodern Condition: A Report on Knowledge*, Minneapolis, MN 1984.

Lyotard, Jean-François: "Rewriting Modernity," in: Lyotard, *The Inhuman: Reflections on Time*, Stanford, CA 1991, pp. 24–35.

Maase, Kaspar: *Grenzenloses Vergnügen. Der Aufstieg der Massenkultur 1850–1970*, Frankfurt/Main 1997.

Maase, Kasper: "'Schundliteratur' und Jugendschutz im Ersten Weltkrieg – Eine Fallstudie zur Kommunikationskontrolle in Deutschland," in: kommunikation@gesellschaft 3/2002, accessed on August 7, 2008: http: //www. uni- frankfurt. de/fb03/K. G/B3_2002_Maase. pdf.

Maase, Kaspar/*Kaschuba*, Wolfgang (ed.): *Schund und Schönheit. Populäre Kultur um 1900*, Cologne 2001.

Maier, Charles S.: *Recasting Bourgeois Europe. Stabilization in France, Germany and Italy in the Decade After World War I*, Princeton, NJ 1975.

Maier, Hans: *Politische Religionen. Die totalitären Regime und das Christentum*, Freiburg 1995.

Maier, Hans (ed.): *Totalitarianism and Political Religions. Concepts for the Comparison of Dictatorships: Theory and History of Interpretation*, 3 vols., New York, NY 2004–2007.

Maier, Hans (ed.): *Wege in die Gewalt. Die modernen politischen Religionen*, Frankfurt/Main 2000.

Mai, Gunther: "Agrarische Transition und industrielle Krise. Anti-Modernismus in Europa in der ersten Hälfte des 20. Jahrhunderts," in: *JMEH* 4/2006, pp. 5–37.

Makropoulos, Michael: "Crisis and Contingency: Two Categories of the Discourse of Classical Modernity," *Thesis Eleven* 111/2012, pp. 9–18.

Makropoulos, Michael: "Haltlose Souveränität. Benjamin, Schmitt und die Klassische Moderne in Deutschland," in: Manfred Gangl/Gérard Raulet (ed.), *Intellektuellendiskurse in der Weimarer Republik*, Darmstadt 1994, pp. 197–211.

Makropoulos, Michael: *Modernität als ontologischer Ausnahmezustand? Walter Benjamins Theorie der Moderne*, Munich 1989.

Malinowski, Stephan: "'Führertum' und 'Neuer Adel.' Die Deutsche Adelgenossenschaft und der Deutsche Herrenklub in der Weimarer Republik," in: Heinz Reif (ed.), *Adel und Bürgertum in Deutschland II. Entwicklungslinien und Wendepunkte im 20. Jahrhundert*, Berlin 2001, pp. 173–211.

Malinowski, Stephan: "Ihr liebster Feind. Die deutsche Sozialgeschichte und der preußische Adel," in: Sven OliverMüller/Cornelius Torp (eds.), *Das Deutsche Kaiserreich in der Kontroverse*, Göttingen 2009, pp. 203–18.

Malinowski, Stephan: *Vom König zum Führer. Sozialer Niedergang und politische Radikalisierung im deutschen Adel zwischen Kaiserreich und NS-Staat*, Berlin 2003².

Mann, Michael: *Fascists*, Cambridge 2004.

Martschukat, Jürgen: "Geschichte schreiben mit Foucault – eine Einleitung," in: *Martschukat* (ed.), *Geschichte schreiben mit Foucault*, Frankfurt/Main 2002, pp. 7–26.

Marwedel, Rainer: *Theodor Lessing 1872–1933. Eine Biographie*, Darmstadt 1987.

Marx, Karl: "The Eighteenth Brumaire of Louis Bonaparte," in: *Political Writings*, vol. 2, *Surveys from Exile*, ed. by David Fernbach, Harmondsworth 1973, pp. 143–249.

Mason, Timothy W.: "Italy and Modernization: A Montage," in: *History Workshop Journal* 25/1998, pp. 127–47.

Mason, Timothy W.: *Nazism, Fascism and the Working Class*, Cambridge 1995.

Mason, Timothy W.: "The Origins of the Law on the Organization of National Labour of 20 January 1934. An Investigation into the Relationship between 'Archaic' and 'Modern' Elements in Recent German History," in: Mason (ed.), *Nazism, Fascism and the Working Class*, Cambridge 1995, pp. 77–103.

Massignani, Alessandro: "La grande guerra: un bilancio complessivo," in: Paolo Ferrari (ed.), *L'Aeronautica italiana. Una storia del Novecento*, Milan 2004, pp. 267–99.

Massignani, Alessandro: "La guerra aerea sul fronte italiano," in: Paolo Ferrari (ed.), *La grande guerra aerea 1915–1918. Battaglie – industrie – bombardamenti – assi – aeroporti*, Valdagno 1994, pp. 17–55.

Mattioli, Aram: *Experimentierfeld der Gewalt. Der Abessinienkrieg und seine internationale Bedeutung 1935–1941*, Zurich 2005.

McEwan, Dorothea: "Ein Kampf gegen Windmühlen. Warburgs pro-italienische publizistische Initiative," in: Gottfried Korff (ed.), *Kasten 117. Aby Warburg und der Aberglaube im Ersten Weltkrieg*, Tübingen 2007, S. 135–63.

McEwan, Dorothea: "IDEA VINCIT – 'Die siegende, fliegende Idea.' Ein künstlerischer Auftrag von Aby Warburg," in: Sabine Flach/Inge Münz-Koenen/Marianne Streisand (eds.), *Der Bilderatlas im Wechsel der Künste und Medien*, Munich 2005, pp. 121–51.

Mergel, Thomas: "Die Modernisierungstheorie auf dem Weg zu einer Theorie der Moderne," in: Thomas Mergel/Thomas Welskopp (eds.), *Geschichte zwischen Kultur und Gesellschaft. Beiträge zur Theoriedebatte*, Munich 1997, pp. 203–232.

Meriggi, Marco: "Soziale Klassen, Institutionen und Nationalisierung im liberalen Italien," in: *GG* 26/2000, pp. 201–18.

Merton, Robert K.: "The Ethos of Science," in: Piotr Sztompka (ed.), *On Social Structure and Science*, Chicago, IL 1996, pp. 267–76.

Meschnig, Alexander: *Der Wille zur Bewegung. Militärischer Traum und totalitäres Programm. Eine Mentalitätsgeschichte vom Ersten Weltkrieg zum Nationalsozialismus*, Bielefeld 2008.

Metzler, Gabriele: "'Geborgenheit im gesicherten Fortschritt.' Das Jahrzehnt von Planbarkeit und Machbarkeit," in: Matthias Frese/Julia Paulus/Karl Teppe (eds.), *Demokratisierung und gesellschaftlicher Aufbruch. Die sechziger Jahre als Wendezeit der Bundesrepublik*, Paderborn 2003, pp. 777–97.

Michaud, Eric: *The Cult of Art in Nazi Germany*, Stanford, CA 2004.

Michels, Karen: *Aby Warburg. Im Bannkreis der Ideen*, Munich 2007.

Michels, Karen/*Schoell-Glass*, Charlotte: "Aby Warburg et les timbres en tant que document culturel," in: *Protée* 20/2002, pp. 85–94.

Mohler, Armin: *Die konservative Revolution in Deutschland 1918–1932. Ein Handbuch*, Stuttgart 1999.

Mommsen, Hans: "Die Auflösung des Bürgertums seit dem späten 19. Jahrhundert," in: Jürgen Kocka (ed.), *Bürger und Bürgerlichkeit im 19. Jahrhundert*, Göttingen 1987, pp. 288–315.

Mommsen, Hans: "Nationalsozialismus und Modernisierung," in: *GG* 21/1995, pp. 391–402.

Morat, Daniel: *Von der Tat zur Gelassenheit. Konservatives Denken bei Martin Heidegger, Ernst Jünger und Friedrich Georg Jünger 1920–1960*, Göttingen 2007.

Morrow, John H. Jr.: *German Air Power in World War I*, Lincoln, NE 1982.

Morrow, John H. Jr.: *The Great War in the Air. Military Aviation from 1909 to 1921*, Washington, DC 1993.

Morrow, John H. Jr.: "Knights of the Sky. The Rise of Military Aviation," in: Frans Coetzee/Marylin Shevin-Coetzee (eds.), *Authority, Identity and the Social History of the Great War*, Providence, RI 1995, pp. 305–24.

Mosse, George L.: *The Crisis of German Ideology. Intellectual Origins of the Third Reich*, London 1966 [repr.].

Mosse, George L.: "Das deutsch-jüdische Bildungsbürgertum," in: Koselleck (ed.), *Bildungsbürgertum im 19. Jahrhundert. vol. II. Bildungsgüter und Bildungswissen*, Stuttgart 1990, pp. 168–80.

Mosse, George L.: *Fallen Soldiers. Reshaping the Memory of the World Wars*, New York, NY 1990.

Mosse, George L.: "Fascism and the Avant Garde," in: Mosse, *The Fascist Revolution. Toward a General Theory of Fascism*, New York, NY 1999, pp. 137–55.

Mosse, George L.: "Fascist Aesthetics and Society: Some Considerations," in: Mosse, *The Fascist Revolution. Toward a General Theory of Fascism*, New York, NY 1999, pp. 45–53.

Mosse, George L.: *The Fascist Revolution. Towards a General Theory of Fascism*, New York, NY 1999.

Mosse, George L.: "Futurismo e culture politiche in Europa: una prospettiva globale," in: Renzo De Felice (ed.), *Futurismo, cultura e politica*, Turin 1988, pp. 13–31.

Mosse, George L.: *The Image of Man. The Creation of Modern Masculinity*, Oxford 1998 [repr.].

Mosse, George L.: *Jüdische Intellektuelle in Deutschland. Zwischen Religion und Nationalismus*, Frankfurt/Main 1992.

Mosse, George L.: "The Knights of the Sky and the Myth of the War Experience," in: Robert A. Hinde/Helen E. Watson (eds.), *War: A Cruel Necessity? The Bases of Institutionalized Violence*, London 1995, pp. 132–42.

Mosse, George L.: *The Nationalization of the Masses. Political Symbolism and Mass Movements in Germany from the Napoleonic Wars through the Third Reich*, New York, NY 1975.

Mosse, George L.: "The Poet and the Exercise of Political Power: Gabriele D'Annunzio," in: Mosse, *Masses and Man. Nationalist and Fascist Perceptions of Reality*, New York, NY 1980, pp. 87–103.

Mosse, George L.: "The Political Culture of Italian Futurism: A General Perspective," in: *Journal of Contemporary History*, CH XXV, April–July 2–3, 1990, pp. 253–68.

Mosse, George L.: "Toward a General Theory of Fascism," in: Mosse, *The Fascist Revolution. Toward a General Theory of Fascism*, New York, NY 1999, pp. 1–44.

Müller, Hans-Harald: *Der Krieg und die Schriftsteller. Der Kriegsroman der Weimarer Republik*, Stuttgart 1986.

Münkler, Herfried: *Die Deutschen und ihre Mythen*, Berlin 2009.

Müller, Sven Oliver: *Die Nation als Waffe und Vorstellung, Nationalismus in Deutschland und Großbritannien im Ersten Weltkrieg*, Göttingen 2002.

Müller, Sven Oliver: "Die umkämpfte Nation. Legitimationsprobleme im kriegführenden Kaiserreich," in: Jörg Echternkamp/Sven Oliver Müller (eds.), *Die Politik der Nation. Deutscher Nationalismus in Krieg und Krisen*, Munich 2002, pp. 149–72.

Münz, Christoph: "'Wohin die Sprache nicht reicht…' Sprache und Sprachbilder zwischen Bilderverbot und Schweigebot," in: Bettina Bannasch/Almuth Hammer (eds.), *Verbot der Bilder – Gebot der Erinnerung. Mediale Repräsentationen der Schoah*, Frankfurt/Main 2004, pp. 147–66.

Naumann, Michael: *Strukturwandel des Heroismus. Vom sakralen zum revolutionären Heldentum*, Königstein Ts. 1984.

Neitzel, Sönke: "Zum strategischen Mißerfolg verdammt? Die deutschen Luftstreitkräfte in beiden Weltkriegen," in: Bruno Thoß/Hans-Erich Volkmann (eds.), *Erster Weltkrieg – Zweiter Weltkrieg. Ein Vergleich. Krieg, Kriegserlebnis, Kriegserfahrung in Deutschland*, Paderborn 2002, pp. 167–92.

Niethammer, Lutz (ed.): *"Die Jahre weiß man nicht, wo man die heute hinsetzen soll." Faschismuserfahrungen im Ruhrgebiet*, Berlin 1983.

Nipperdey, Thomas: "Der Faschismus in seiner Epoche. Zu den Werken von Ernst Nolte zum Faschismus," in: HZ 210/1970, pp. 620–38.

Nipperdey, Thomas: *Deutsche Geschichte 1866–1918*, vol. 2, *Machtstaat vor der Demokratie*, Munich 1992.

Nitschke, August/Gerhard A. Ritter/Detlev J. K. Peukert/Rüdiger vom Bruch (eds.): *Jahrhundertwende. Aufbruch in die Moderne 1880–1930*, 2 vols., Reinbek bei Hamburg 1990.

Nolte, Ernst: *Three Faces of Fascism: Action Française, Italian Fascism, National Socialism*, New York, NY 1969.

Nolte, Ernst: *Die Krise des liberalen Systems und die faschistischen Bewegungen*, Munich 1968.

Nolte, Paul: "Abschied vom 19. Jahrhundert oder Auf der Suche nach einer anderen Moderne," in: Jürgen Osterhammel/Dieter Langewiesche/Paul Nolte (eds.), *Wege der Gesellschaftsgeschichte*, Göttingen 2006, pp. 103–32.

Nolte, Paul: *Die Ordnung der deutschen Gesellschaft. Selbstentwurf und Selbstbeschreibung im 20. Jahrhundert*, Munich 2000.

Nolzen, Armin/Reichardt, Sven (eds.): *Faschismus in Italien und Deutschland. Studien zu Transfer und Vergleich*, Göttingen 2005.

Nowak, Kurt: "Die 'antihistoristische Revolution.' Symptome und Folgen der Krise historischer Weltorientierung nach dem Ersten Weltkrieg in Deutschland," in: Horst Renz/Friedrich Wilhelm Graf (eds.), *Umstrittene Moderne. Die Zukunft der Neuzeit im Urteil der Epoche Ernst Troeltschs*, Gütersloh 1987, pp. 133–71.

Nusser, Peter: *Trivialliteratur*, Stuttgart 1991.

Oexle, Otto Gerhard: "Krise des Historismus – Krise der Wirklichkeit. Eine Problemgeschichte der Moderne," in: Oexle (ed.), *Krise des Historismus – Krise der Wirklichkeit. Wissenschaft, Kunst und Literatur*, Göttingen 2007, pp. 11–116.

Oexle, Otto Gerhard: "Von Nietzsche zu Max Weber: Wertproblem und Objektivitätsforderung der Wissenschaft im Zeichen des Historismus," in: Oexle (ed.), *Geschichtswissenschaft im Zeichen des Historismus. Studien zu Problemgeschichten der Moderne*, Göttingen 1996, pp. 73–94.

Oppelt, Ulrike: *Film und Propaganda im Ersten Weltkrieg. Propaganda als Medienrealität im Aktualitäten- und Dokumentarfilm*, Stuttgart 2002.

Osborne, Peter: *The Politics of Time. Modernity and the Avantgarde*, London 1995.

Overmans, Rüdiger: "Kriegsverluste," in: Gerhard Hirschfeld/Gerd Krumeich/Irina Renz (eds.), *Enzyklopädie Erster Weltkrieg*, Paderborn 2003, pp. 663–6.

Overy, Richard J.: *The Dictators. Hitler's Germany and Stalin's Russia*, London 2004.

Paetzold, Heinz: *Ernst Cassirer. Von Marburg nach New York. Eine philosophische Biographie*, Darmstadt 1995.

Palmer, Scott W.: *Dictatorship of the Air. Aviation Culture and the Fate of Modern Russia*, Cambridge 2006.

Patel, Kiran Klaus: *"Soldaten der Arbeit." Arbeitsdienste in Deutschland und den USA 1933–1945*, Göttingen 2003.

Paxton, Robert O.: *The Anatomy of Fascism*, London 2005 [repr.].

Paxton, Robert O.: "The Five Stages of Fascism," in: *Journal of Contemporary History* CH 70/1998, pp. 1–23.

Payne, Stanley G.: *A History of Fascism 1914–1945*, London 1997².

Payne, Stanley G.: *Fascism. Comparison and Definition*, Madison, WI 1980.

Peukert, Detlev J. K.: "Die Genesis der 'Endlösung' aus dem Geist der Wissenschaft," in: Peukert (ed.), *Max Webers Diagnose der Moderne*, Göttingen 1989, pp. 102–21.

Peukert, Detlev J. K.: "Die Unordnung der Dinge. Michel Foucault und die deutsche Geschichtswissenschaft," in: François Ewald/Bernhard Waldenfels (eds.), *Spiele der Wahrheit. Michel Foucaults Denken*, Frankfurt/Main 1991, pp. 320–39.

Peukert, Detlev J. K.: *Max Webers Diagnose der Moderne*, Göttingen 1989.

Peukert, Detlev J. K.: "'Mit uns zieht die neue Zeit...' Jugend zwischen Disziplinierung und Revolte," in: August Nitschke/Gerhard A. Ritter/Detlev J. K. Peukert/Rüdiger vom Bruch (eds.), *Jahrhundertwende. Der Aufbruch in die Moderne 1880–1930*, vol. 1, Reinbek bei Hamburg 1990, pp. 176–202.

Peukert, Detlev J. K.: *Inside Nazi Germany: Conformity, Opposition, and Racism in Everyday Life*, New Haven, CT 1987.

Peukert, Detlev J. K.: *The Weimar Republic. The Crisis of Classical Modernity*, London 1991.

Piper, Ernst: *Alfred Rosenberg. Hitlers Chefideologe*, Munich 2005.

Plaggenborg, Stefan: *Experiment Moderne. Der sowjetische Weg*, Frankfurt/Main 2006.

Planert, Ute: "Kulturkritik und Geschlechterverhältnis. Zur Krise der Geschlechterordnung zwischen Jahrhundertwende und 'Drittem Reich'," in: Wolfgang Hardtwig (ed.), *Ordnungen in der Krise. Zur politischen Kulturgeschichte Deutschlands 1900–1933*, Munich 2007, pp. 191–214.

Plato: "Politeia," in: Plato, *Werke in acht Bänden*, translated into German by Friedrich Schleirmacher, vol. 4, ed. by Gunther Eigler, Darmstadt 1990².

Poggi, Christine: *Inventing Futurism. The Art and Politics of Artificial Optimism*, Princeton, NJ 2009.

Poser, Hans: "Zeit und Ewigkeit. Zeitkonzepte als Orientierungswissen," in: Hans Michael Baumgartner (ed.), *Das Rätsel der Zeit. Philosophische Analysen*, Freiburg 1996², pp. 17–50.

Potempa, Harald: *Die Königlich-Bayerische Fliegertruppe 1914–1918*, Frankfurt/Main 1997.

Prinz, Michael/*Zitelmann*, Rainer: *Nationalsozialismus und Modernisierung*, Darmstadt 1991.

Procacci, Giovanna: *Dalla rassegnazione alla rivolta. Mentalità e comportamenti popolari nella Grande Guerra*, Rome 1999.

Prümm, Karl: *Die Literatur des Soldatischen Nationalismus der 20er Jahre. Gruppenideologie und Epochenproblematik*, 2 vols., Kronberg Ts. 1974.

Pugliese, Stanislao (ed.): *Italian Fascism and Antifascism. A Critical Anthology*, Manchester 2001.

Puschner, Uwe: *Die völkische Bewegung im wilhelminischenmKaiserreich. Sprache – Rasse – Religion*, Darmstadt 2001.

Puschner, Uwe/*Schmitz*, Walter/*Ulbricht*, Justus H. (eds.): *Handbuch zur "VölkischenBewegung" 1871–1918*, Munich 1996.

Pütz, Peter: "Der Mythos bei Nietzsche," in: Helmut Koopmann (ed.), *Mythos und Mythologie in der Literatur des 19. Jahrhunderts*, Frankfurt/Main 1979, pp. 251–62.

Rabinbach, Anson: "Nationalsozialismus und Moderne. Zur Technik-Interpretation im Dritten Reich," in: Wolfgang Emmerich/Carl Wege (eds.), *Der Technikdiskurs in der Hitler-Stalin-Ära*, Stuttgart 1995, pp. 94–113.

Radkau, Joachim: *Das Zeitalter der Nervosität. Deutschland zwischen Bismarck und Hitler*, Munich 1998.

Radkau, Joachim: *Max Weber: A Biography*, Cambridge 2009; *Max Weber. Die Leidenschaft des Denkens*, Munich 2005.

Radkau, Joachim: *Technik in Deutschland. Vom 18. Jahrhundert bis zur Gegenwart*, Frankfurt/ Main 1989.

Raphael, Lutz: "Die Verwissenschaftlichung des Sozialen als methodische und konzeptionelle Herausforderung für eine Sozialgeschichte des 20. Jahrhunderts," in: *GG* 22/1996, pp. 165–93.

Raphael, Lutz: "Ideen als gesellschaftliche Gestaltungskraft im Europa der Neuzeit: Bemerkungen zur Bilanz eines DFG-Schwerpunktprogramms," in: Lutz Raphael/Heinz-Elmar Tenorth (eds.), *Ideen als gesellschaftliche Gestaltungskraft im Europa der Neuzeit. Beiträge für eine erneuerte Geistesgeschichte*, Munich 2006, pp. 11–27.

Raphael, Lutz: "Ordnungsmuster der 'Hochmoderne?' Die Theorie der Moderne und die Geschichte der europäischen Gesellschaften im 20. Jahrhundert," in: Lutz Raphael/Ute Schneider (eds.), *Dimensionen der Moderne. Festschrift für Christof Dipper*, Frankfurt/Main 2008, pp. 73–91.

Rastelli, Achille: "I bombardamenti sulle città," in: Paolo Ferrari (ed.), *La grande guerra aerea 1915–1918. Battaglie – industrie – bombardamenti – assi – aeroporti*, Valdagno 1994, pp. 183–249.

Raulff, Ulrich: "Der aufhaltsame Aufstieg einer Idee. 'Idea vincit:' Warburg, Stresemann und die Briefmarke," in: *Vorträge aus dem Warburg-Haus*, vol. 6, Berlin 2002, pp. 125–62.

Raulff, Ulrich: "Nachwort," in: Aby Warburg, *Schlangenritual. Ein Reisebericht*, Berlin 1988, pp. 63–94.

Raulff, Ulrich: "Von der Privatbibliothek des Gelehrten zum Forschungsinstitut. Aby Warburg, Ernst Cassirer und die neue Kulturwissenschaft," in: *GG* 23/1997, pp. 28–43.

Raulff, Ulrich: *Wilde Energien. Vier Versuche zu Aby Warburg*, Göttingen 2003.

Reichardt, Rolf: "Einleitung," in: Rolf Reichardt/Eberhard Schmitt (eds.), *Handbuch politisch-sozialer Grundbegriffe in Frankreich 1680–1820*, vols. 1/2, Munich 1985, pp. 39–148.

Reichardt, Sven: "Bourdieu für Historiker? Ein kultursoziologisches Angebot an die Sozialgeschichte," in: Thomas Mergel/Thomas Welskopp (eds.), *Geschichte zwischen Kultur und Gesellschaft. Beiträge zur Theoriedebatte*, Munich 1997, pp. 71–93.

Reichardt, Sven: *Faschistische Kampfbünde. Gewalt und Gemeinschaft im italienischen Squadrismus und in der deutschen SA*, Cologne 2002.

Reichardt, Sven: "Neue Wege der vergleichenden Faschismusforschung," in: *Mittelweg 36* 16/2007, pp. 9–25.

Reichardt, Sven: "Praxeologie und Faschismus. Gewalt und Gemeinschaft als Elemente eines praxeologischen Faschismusbegriffs," in: Karl H. Hörning/Julia Reuter (eds.), *Doing Culture. Neue Positionen zum Verhältnis von Kultur und sozialer Praxis*, Bielefeld 2004, pp. 129–53.

Reichardt, Sven: "Was mit dem Faschismus passiert ist. Ein Literaturbericht zur internationalen Faschismusforschung seit 1990," Part 1, in: *NPL* XLIX/2004, pp. 385–406.

Reichel, Peter: *Der schöne Schein des Dritten Reiches. Gewalt und Faszination des deutschen Faschismus*, Hamburg 2006 [repr.].

Reichherzer, Frank: "'Alles ist Front!' Wehrwissenschaften in Deutschland und die Bellifizierung der Gesellschaft vom Ersten Weltkrieg bis in den Kalten Krieg," Paderborn 2012.

Reichherzer, Frank: "'Das Wehr-Denken ist deutsch, nationalsozialistisch.' Zum Verhältnis von wehrwissenschaftlichem Denken und nationalsozialistischer Ideologie in der Zwischenkriegszeit," in: Käte Meyer-Drawe/Kristin Platt (eds.), Wissenschaft im Einsatz, Munich 2007, pp. 243–67.

Reif, Heinz: "Einleitung," in: Reif (ed.), Adel und Bürgertum in Deutschland, vol. 1, Entwicklungslinien und Wendepunkte im 19. Jahrhundert, Berlin 2000, pp. 7–27.

Reimann, Aribert: Der große Krieg der Sprachen. Untersuchungen zur historischen Semantik in Deutschland und England zur Zeit des Ersten Weltkrieges, Essen 2000.

Reimann, Bruno W.: "...die Feder durch das Schwert ersetzen..." Ernst Jüngers politische Publizistik 1923–1933, Marburg 2001.

Reinicke, Helmut: "Zeppelin, Karl May und die deutschen Auffahrten nach Dschinnistan," in: Dieter R. Bauer/Wolfgang Behringer (eds.), Fliegen und Schweben. Annäherung an eine menschliche Sensation, Munich 1997, pp. 317–43.

Reulecke, Jürgen: "Neuer Mensch und neue Männlichkeit. Die 'junge Generation' im ersten Drittel des 20. Jahrhunderts," in: Jahrbuch des Historischen Kollegs 2001, Munich 2002, pp. 109–38.

Riall, Lucy: Garibaldi. Invention of a Hero, New Haven, CT 2007.

Rieger, Bernhard: Technology and the Culture of Modernity in Britain and Germany, 1890–1945, Cambridge 2005.

Robert, David D.: "How Not To Think about Fascism and Ideology, Intellectual Antecedents and Historical Meaning," in: JCH 35/2000, pp. 185–211.

Rochat, Giorgio: Gli arditi della grande guerra. Origini, battaglie e miti, Milan 1981.

Rochat, Giorgio: Italo Balbo. Aviatore e ministro dell'aeronautica 1926–1933, Ferrara 1979.

Rochat, Giorgio: Italo Balbo. Lo squadrista, l'aviatore, il gerarca, Turin 2003.

Roeck, Bernd: Der junge Aby Warburg, Munich 1997.

Roeck, Bernd: "Psychohistorie im Zeichen Saturns. Aby Warburgs Denksystem und die moderne Kulturgeschichte," in: Wolfgang Hardtwig/Hans-Ulrich Wehler (eds.), Kulturgeschichte Heute, Göttingen 1996, pp. 231–54.

Rohkrämer, Thomas: "Antimodernism, Reactionary Modernism and National Socialism. Technocratic Tendencies in Germany, 1890–1945," in: Contemporary European History 8/1999, pp. 29–50.

Rohkrämer, Thomas: Der Militarismus der "kleinen Leute." Die Kriegervereine im Deutschen Kaiserreich 1871–1914, Munich 1990.

Rohkrämer, Thomas: "Die Verzauberung der Schlange. Krieg, Technik und Zivilisationskritik beim frühen Ernst Jünger," in: Wolfgang Michalka (ed.), Der Erste Weltkrieg. Wirkung, Wahrnehmung, Analyse, Munich 1994, pp. 848–74.

Rohkrämer, Thomas: Eine andere Moderne? Zivilisationskritik, Natur und Technik in Deutschland 1880–1930, Paderborn 1999.

Rohkrämer, Thomas: "Kult der Gewalt und Sehnsucht nach Ordnung – Ernst Jünger und der soldatische Nationalismus in der Weimarer Republik," in: Sociologus 51/2001, pp. 28–48.

Rosa, Hartmut: Social Acceleration. A New Theory of Modernity, New York, NY 2013.

Rosenbaum, Eduard/Sherman, Ari J.: Das Bankhaus M. M. Warburg & Co., Hamburg 1978².

Rosenthal, Bernice Glatzer: New Myth, New World. From Nietzsche to Stalinism, University Park, PA 2002.

Rosenthal, Earl: "Plus Ultra, Non plus Ultra, and the Columnar Device of Emperor Charles V," in: Journal of the Warburg and Courtauld Institutes 34/1971, pp. 204–28.

Rother, Rainer (commissioned by the Deutsches Historisches Museum): Der Weltkrieg 1914–1918. Ereignis und Erinnerung, Berlin 2004.

Rusconi, Gian Enrico: "Das Hasardspiel des Jahres 1915. Warum sich Italien für den Eintritt in den Ersten Weltkrieg entschied," in: Hürter, Johannes/Gian Enrico Rusconi (eds.), Der Kriegseintritt Italiens im Mai 1915, Munich 2007, pp. 13–52.

Russo, Antonella: *Il fascismo in mostra. Storia fotografica della società Italiana*, Rome 1999.

Safranski, Rüdiger: *Romanticism: A German Affair*, Evanston, IL 2014.

Salaris, Claudia: *aero… futurismo e mito del volo*, Rome 1985.

Salaris, Claudia: *Alla festa della rivoluzione. Artisti e libertari con D'Annuzio a Fiume*, Bologna 2002.

Salaris, Claudia: *Artecrazia. L'avanguardia futurista negli anni del fascismo*, Florence 1992.

Salaris, Claudia: *Storia del futurismo. Libri, giornali, manifesti*, Rome 1985.

Salewski, Michael (ed.): *Die Deutschen und die See. Studien zur deutschen Marinegeschichte des 19. und 20. Jahrhunderts*, Stuttgart 2002.

Salierno, Vito: *D'Annunzio e i suoi editori*, Milan 1987.

Sarasin, Philipp: "Geschichtswissenschaft und Diskursanalyse," in: Sarasin, *Geschichtswissenschaft und Diskursanalyse*, Frankfurt/Main 2003, pp. 10–60.

Sarasin, Philipp: "Subjekte, Diskurse, Körper. Überlegungen zu einer diskursanalytischen Kulturgeschichte," in: Wolfgang Hardtwig/Hans-Ulrich Wehler (eds.), *Kulturgeschichte Heute*, Göttingen 1996, pp. 131–64.

Schapiro, Jacob Salwyn: "Thomas Carlyle, Prophet of Fascism," in: *JMH* XVII/1945, pp. 97–115.

Schenda, Rudolf: *Die Lesestoffe der kleinen Leute. Studien zur populären Literatur im 19. und 20. Jahrhundert*, Munich 1976.

Schieder, Wolfgang: "Die Geburt des Faschismus aus der Krise der Moderne," in: Christof Dipper (ed.), *Deutschland und Italien 1860–1960*, Munich 2005, pp. 159–79.

Schieder, Wolfgang: "Die NSDAP vor 1933. Profil einer faschistischen Partei," in: *GG* 19/1993, pp. 141–54.

Schieder, Wolfgang (ed.): *Faschismus als soziale Bewegung. Deutschland und Italien im Vergleich*, Hamburg 1976.

Schieder, Wolfgang: *Faschistische Diktaturen. Studien zu Italien und Deutschland*, Göttingen 2008.

Schildt, Axel: "NS-Regime, Modernisierung und Moderne. Anmerkungen zur Hochkonjunktur einerandauernden Diskussion," in: Dan Diner/Frank Stern (eds.), *Nationalsozialismus aus heutiger Perspektive*, Göttingen 1994, pp. 3–22.

Schilling, René: *"Kriegshelden." Deutungsmuster heroischer Männlichkeit in Deutschland 1813–1945*, Paderborn 2002.

Schivelbusch, Wolfgang: *Three New Deals. Reflections on Roosevelt's America, Mussolini's Italy, and Hitler's Germany, 1933–1939*, New York, NY 2006.

Schivelbusch, Wolfgang: *The Culture of Defeat. On National Trauma, Mourning, and Recovery*, New York, NY 2003.

Schivelbusch, Wolfgang: *The Railway Journey. The Industrialization of Time and Space in the 19th Century*, Berkeley, CA 1986.

Schlager, Claudia: "Zwischen Feindesliebe und Erbfeindschaft. Deutsche und französische Katholiken im Ersten Weltkrieg," in: Reinhard Johler/Freddy Raphaël/Claudia Schlager/Patrick Schmoll (eds.), *Zwischen Krieg und Frieden. Die Konstruktion des Feindes*, Tübingen 2009.

Schluchter, Wolfgang: *The Rise of Western Rationalism. Max Weber's Developmental Theory*, Berkeley, CA 1981.

Schmidt, Anne: *Belehrung – Propaganda – Vertrauensarbeit. Zum Wandel amtlicher Kommunikationspolitik in Deutschland 1914–1918*, Essen 2006.

Schmidt-Bergmann, Hansgeorg: "*Mafarka le Futuriste* – F. T. Marinettis literarische Konstruktion des futuristischen Heroismus," in: Filippo T. Marinetti, *Mafarka der Futurist. Afrikanischer Roman*, Munich 2004, pp. 261–84.

Schmidt-Bergmann, Hansgeorg: *Futurismus. Geschichte, Ästhetik, Dokumente*, Reinbek bei Hamburg 1993.

Schnapp, Jeffrey T.: *Anno X. La Mostra della Rivoluzione fascista del 1932*, Pisa 2003.

Schnapp, Jeffrey T.: "Epic Demonstrations. Fascist Modernity and the 1932 Exhibition of the Fascist Revolution," in: Richard Joseph Golsan (ed.), *Fascism, Aesthetics, and Culture*, Hanover, NH 1992, pp. 1–37.

Schnapp, Jeffrey T.: "Mostre," in: Deutsches Historisches Museum (ed.), *Kunst und Propaganda im Streit der Nationen 1930–1945*, Berlin 2007, pp. 78–87.

Schnapp, Jeffrey T.: "Propeller Talk," in: *Modernism/Modernity* 1/1994, pp. 153–78.

Schnapp, Jeffrey T.: *Staging Fascism. 18BL and the Theater of Masses for Masses*, Stanford, CA 1996.

Schneider, Jost: *Sozialgeschichte des Lesens. Zur historischen Entwicklung und sozialen Differenzierung der literarischen Kommunikation in Deutschland*, Berlin 2004.

Schneider, Thomas F.: "'Die Meute hinter Remarque.' Zur Rezeption von *Im Westen nichts Neues* 1928–1930," in: *Jahrbuch zur Literatur der Weimarer Republik* 1/1995, pp. 143–70.

Schneider, Thomas F.: "'Krieg ist Krieg schließlich.' Erich Maria Remarque: *Im Westen nichts Neues* (1928)," in: Thomas F. Schneider/Hans Wagener (eds.), *Von Richthofen bis Remarque: Deutschsprachige Prosa zum 1. Weltkrieg*, Amsterdam 2003, pp. 217–32.

Schneider, Thomas F.: "Endlich die 'Wahrheit' über den Krieg. Zu deutscher Kriegsliteratur," *Text und Kritik* 124/1994, pp. 38–51.

Schneider, Thomas F.: "Zur deutschen Kriegsliteratur im Ersten Weltkrieg," in: Thomas F. Schneider (ed.), *Kriegserlebnis und Legendenbildung. Das Bild des "modernen" Krieges in Literatur, Theater, Photographie und Film*, 3 vols., Osnabrück 1999, pp. 101–14.

Schneider, Thomas F.: "Zwischen Wahrheitsanspruch und Fiktion. Zur deutschen Kriegsliteratur im Ersten Weltkrieg," in: Rolf Spilker/Bernd Ulrich (eds.), *Der Tod als Maschinist. Der industrialisierte Krieg 1914–1918. Eine Ausstellung des Museums für Industriekultur Osnabrück im Rahmen des Jubiläums "350 Jahre Westfälischer Friede" 17. 5.–23. 8. 1998*, Bramsche 1998, pp. 142–53.

Schneider, Thomas F./*Heinemann*, Julia/*Hischer*, Frank/*Kuhlmann*, Johanna/*Puls*, Peter: *Die Autoren und Bücher der deutschsprachigen Literatur zum 1. Weltkrieg 1914–1939. Ein bio-bibliographisches Handbuch*, Göttingen 2008, pp. 7–14.

Schneider, Thomas F./*Heinemann*, Julia/*Hischer*, Frank/*Kuhlmann*, Johanna/*Puls*, Peter: "Einleitung," in: Thomas F. Schneider/Julia Heinemann/Frank Hischer/Johanna Kuhlmann/Peter Puls (eds.) *Die Autoren und Bücher der deutschsprachigen Literatur zum 1. Weltkrieg 1914–1939. Ein bio-bibliographisches Handbuch*, Göttingen 2008, pp. 7–14.

Schneider, Thomas F./*Wagner*, Hans: "Einleitung," in: Thomas F. Schneider/Hans Wagner (eds.), *Von Richthofen bis Remarque. Deutschsprachige Prosa zum I. Weltkrieg*, Amsterdam 2003, pp. 11–16.

Schneider, Ute: "Spurensuche: Reinhart Koselleck und die 'Moderne'," in: Ute Schneider/Lutz Raphael (eds.), *Dimensionen der Moderne. Festschrift für Christof Dipper*, Frankfurt/Main 2008, pp. 61–71.

Schneider, Uwe (ed.): *Krieg der Geister. Erster Weltkrieg und literarische Moderne*, Würzburg 2000.

Schoell-Glass, Charlotte: *Aby Warburg und der Antisemitismus. Kulturwissenschaft als Geistespolitik*, Frankfurt/Main 1998.

Schoenbaum, David: *Hitler's Social Revolution. Class and Status in Nazi Germany*, New York, NY 1966.

Schoeps, H. Julius: "Der ungeliebte Außenseiter. Zum Leben und Werk des Philosophen und Schriftstellers Theodor Lessing," in: Walter Grab/H. Julius Schoeps (eds.), *Juden in der Weimarer Republik*, Stuttgart 1986, pp. 200–17.

Schöttler, Peter: "Sozialgeschichtliches Paradigma und historische Diskursanalyse," in: Jürgen Fohrmann/Harro Müller (eds.), *Diskurstheorien und Literaturwissenschaft*, Frankfurt/Main 1988, pp. 159–99.

Schüler-Springorum, Stefanie: *Krieg und Fliegen. Die Legion Condor im Spanischen Bürgerkrieg*, Paderborn 2010.

Schüler-Springorum, Stefanie: "Vom Fliegen und Töten. Militärische Männlichkeit in der deutschen Fliegerliteratur, 1914–1939," in: Karen Hagemann (ed.), *Heimat-Front*, Frankfurt/Main 2001, pp. 208–33.

Schwartz, Peter J.: "Aby Warburgs Kriegskartothek. Vorbericht einer Rekonstruktion," in: Gottfried Korff (ed.), *Kasten 117. Aby Warburg und der Aberglaube im Ersten Weltkrieg*, Tübingen 2007, pp. 39–69.

Scirocco, Alfonso: *Garibaldi. Battaglie, amori, ideali di un cittadino del mondo*, Rome 2001.

Scott, James C.: *Seeing Like a State. How Certain Schemes to Improve the Human Condition Have Failed*, New Haven, NJ 1998.

Scuccimara, Luca: "Era fascista," in: Victoria De Grazia/Sergio Luzzatto (eds.), *Dizionario del fascismo*, vol. 1, Turin 2002, pp. 480–81.

Scudiero, Maurizio: "Die Metamorphosen des Futurismus, von der futuristischen Rekonstruktion des Universums zur mechanischen Kunst. Die Kunst tritt ins Leben ein," in: Ingo Bartsch/Maurizio Scudiero (eds.), *...auch wir Maschinen, auch wir mechanisiert!... Die zweite Phase des italienischen Futurismus 1915–1945*, Bielefeld 2002, pp. 15–29.

Segeberg, Harro: *Literatur im Medienzeitalter. Literatur, Technik und Medien seit 1914*, Darmstadt 2003.

Segeberg, Harro: *Literatur im technischen Zeitalter. Von der Frühzeit der deutschen Aufklärung bis zum Beginn des Ersten Weltkrieges*, Darmstadt 1997.

Segeberg, Harro: "Regressive Modernisierung. Kriegserlebnis und Moderne-Kritik in Ernst Jüngers Frühwerk," in: Segeberg (ed.), *Vom Wert der Arbeit. Zur literarischen Konstitution des Wertkomplexes "Arbeit" in der deutschen Literatur (1770–1930)*, Tübingen 1991, pp. 338–78.

Segrè, Claudio G.: *Italo Balbo. A Fascist Life*, Berkeley, CA 1987.

Seubold, Günter: "Martin Heideggers Stellungnahme zu Jüngers 'Arbeiter' im Spiegel seiner Technikkritik," in: Friedrich Strack (ed.), *Titan Technik, Ernst und Friedrich Georg Jünger über das technische Zeitalter*, Würzburg 2000, pp. 119–32.

Sheehan, James J.: *German Liberalism in the Nineteenth Century*, Chicago, IL 1978.

Showalter, Dennis E.: "Mass Warfare and the Impact of Technology," in: Roger Chickering/Stig Förster (eds.), *Great War, Total War. Combat and Mobilization on the Western Front, 1914–1918*, New York, NY 2000, pp. 73–93.

Sieg, Ulrich: *Jüdische Intellektuelle im Ersten Weltkrieg. Kriegserfahrungen, weltanschauliche Debatten und kulturelle Neuentwürfe*, Berlin 2001.

Sieg, Ulrich: *Jüdische Intellektuelle und die Krise der bürgerlichen Welt im Ersten Weltkrieg*, Stuttgart 2000.

Siegert, Bernhard: "L'Ombra della macchina alata. Gabriele D'Annunzios *renovatio imperii* im Licht der Luftkriegsgeschichte 1909–1940," in: Hans Ulrich Gumbrecht/Friedrich Kittler (eds.), *Der Dichter als Kommandant. D'Annunzio erobert Fiume*, Munich 1996, pp. 261–306.

Siegfried, Detlef: *Der Fliegerblick. Intellektuelle, Radikalismus und Flugzeugproduktion bei Junkers 1914–1934*, Bonn 2001.

Silk, Gerard: "Il primo pilota. Mussolini, Fascist Aeronautical Symbolism and Imperial Rome," in: Claudia Lazzaro/Roger J. Crum (eds.), *Donatello among the Blackshirts. History and Modernity in the Visual Culture of Fascist Italy*, Ithaca, NY 2005, pp. 67–81.

Soeffner, Hans-Georg: "Flying Moles (Pigeon-Breeding Miners in the Ruhr District). The Totemistic Enchantment of Reality and the Technological Disenchantment of Longing," in: Soeffner (ed.), *The Order of Rituals. The Interpretation of Everyday Life*, New Brunswick, NJ 1997, pp. 95–116.

Soldani, Simonetta/Turi, Gabriele: "Introduzione," in: Soldani/Turi (eds.), *Fare gli italiani. Scuola e cultura nell'Italia contemporanea*, vol. 1, *La nascista dello Stato nazionale*, Bologna 1993, pp. 9–34.

Söllner, Alfons/Walkenhaus, Ralf/Wieland, Karin (eds.): *Totalitarismus. Eine Ideengeschichte des 20. Jahrhunderts*, Berlin 1997.

Spackman, Barbara: *Fascist Virilities. Rhetoric, Ideology, and Social Fantasy in Italy*, Minneapolis, MN 1996.

Spackman, Barbara: "Mafarka and Son: Marinetti's Homophobic Economics," in: *Modernism/Modernity* 1/1994, pp. 89–107.

Spilker, Rolf/Ulrich, Bernd (eds.): *Der Tod als Maschinist. Der industrialisierte Krieg 1914–1918. Eine Ausstellung des Museums für Industriekultur Osnabrück im Rahmen des Jubiläums "350 Jahre Westfälischer Friede" 17. 5.–23. 8. 1998*, catalog, Bramsche 1998.

Stagl, Justin: Stagl, "Immanenz und Transzendenz – ethnologisch," in: Jan Assmann/ Rolf Trauzettel (eds.), *Tod, Jenseits und Identität. Perspektiven einer kulturwissenschaftlichen Thanatologie*, Freiburg 2002, pp. 562–74.

Stambolis, Barbara: *Mythos Jugend. Leitbild und Krisensymptom. Ein Aspekt der politischen Kultur im 20. Jahrhundert*, Schwalbach a. Ts. 2003.

Stephenson, Jill: "Inclusion. Building the National Community in Propaganda and Practice," in: Jill Caplan (ed.), *Nazi Germany*, Oxford 2008, pp. 99–121.

Sternhell, Zeev/*Sznajder*, Mario/*Asheri*, Maia: *Die Entstehung der faschistischen Ideologie. Von Sorel zu Mussolini*, Hamburg 1999.

Stevenson, David: *1914–1918. The History of the First World War*, London 2004.

Stimili, Davide (ed.): *Ludwig Binswanger. Aby Warburg. La guarigione infinita. Storia clinica di Aby Warburg*, Vicenza 2005.

Stockhausen, Tilmann von: *Die Kulturwissenschaftliche Bibliothek Warburg. Architektur, Einrichtung und Organisation*, Hamburg 1992.

Stolleis, Michael: *Geschichte des öffentlichen Rechts in Deutschland*, vol. 3, *Staats- und Verwaltungsrechtswissenschaft in Republik und Diktatur, 1914–1945*, Munich 1999.

Stommer, Rainer: *Die inszenierte Volksgemeinschaft. Die "Thing-Bewegung" im Dritten Reich*, Marburg 1983.

Stone, Marla Susan: "Staging Fascism. The Exhibition of the Fascist Revolution," in: JCH 28/1993, pp 215–43.

Stone, Marla Susan: *The Patron State. Culture and Politics in Fascist Italy*, Princeton, NJ 1998.

Stoneman, Mark R.: "Bürgerliche und adlige Krieger. Zum Verhältnis zwischen sozialer Herkunft und Berufskultur im wilhelminischen Armee-Offizierskorps," in: Heinz Reif (ed.), *Adel und Bürgertum in Deutschland*, vol. 2, *Entwicklungslinien und Wendepunkte im 20. Jahrhundert*, Berlin 2001, S. 25–63.

Strack, Friedrich: *Titan Technik. Ernst und Friedrich Georg Jünger über das technische Zeitalter*, Würzburg 2000.

Süß, Dietmar/*Süß*, Winfried: "'Volksgemeinschaft' und Vernichtungskrieg. Gesellschaft im nationalsozialistischen Deutschland," in: Dietmar Süß/Winfried Süß (eds.), *Das "Dritte Reich." Eine Einführung*, Munich 2008, pp. 79–100.

Syon, Guillaume de: *Zeppelin! Germany and the Airship, 1900–1939*, Baltimore, MD 2002.

Tager, Michael: "Myth and Politics in the Works of Sorel and Barthes," in: *Journal of the History of Ideas* 47/1986, pp. 625–39.

Taigel, Hermann: *Louis Laiblin, Privatier. Ein schwäbischer Mäzen*, Pfullingen 2005.

Takács, Sarolta A.: "Kybele," in: Hubert Cancik/Helmuth Schneider (eds.), *Der Neue Pauly*, vol. 6, Stuttgart 1999, Sp. 950–56.

Theweleit, Klaus: *Männerphantasien*, vol. 1, *Frauen, Fluten, Körper, Geschichte*, Frankfurt/Main 1977.

Tranfaglia, Nicola: *Fascismi e modernizzazione in Europa*, Turin 2001.

Tranfaglia, Nicola: *La prima guerra mondiale e il fascismo*, Turin 1995.

Trischler, Helmuth: *Luft- und Raumfahrtforschung in Deutschland 1900–1970. Politische Geschichte einer Wissenschaft*, Frankfurt/Main 1992.

Troeltsch, Ernst: "Die Krisis des Historismus," in: Ernst Troeltsch, *Schriften zur Politik und Kulturphilosophie (1918–1923). Kritische Gesamtausgabe im Auftrag der Heidelberger Akademie der Wissenschaften*, ed. by Friedrich Wilhelm Graf/Volker Drehsen/Gangolf Hübinger/ Trutz Rendtorff, vol. 15, ed. by Gangolf Hübinger with Johannes Mikuteit, Berlin 2002, pp. 433–55.

Turchetta, Gianni: "Introduzione," in: Gabriele D'Annunzio, *Notturno*, Milan 2003[4], pp. V–XLIII.

Turner, Henry A. Jr.: "Fascism and Modernization," in: *World Politics* 24/1972, pp. 547–64.

Uka, Walter: "Idol/Ikone," in: Hans-Otto Hügel (eds.), *Handbuch populäre Kultur. Begriffe, Theorien und Diskussionen*, Stuttgart 2003, pp. 255–59.

Vattimo, Gianni: *The End of Modernity: Nihilism and Hermeneutics in Postmodern Culture*, Cambridge 1988.

Verdone, Mario/*Berghaus*, Günter: "*Vita futurista* and Early Futurist Cinema," in: Günter Berghaus (ed.), *International Futurism in Arts and Literature*, Berlin 2000, pp. 398–421.

Verhey, Jeffrey: *The Spirit of 1914. Militarism, Myth, and Mobilization in Germany*, New York, NY 2000.

Vierhaus, Rudolf: "Bildung," in: Otto Brunner/Werner Conze/Reinhart Koselleck (eds.), *Geschichtliche Grundbegriffe. Historisches Lexikon zur politisch-sozialen Sprache in Deutschland*, vol. l, Stuttgart 1972, pp. 508–51.

Vigo, Giovanni: "Gli italiani alla conquista dell'alfabeto," in: Simonetta Soldani/Gabriele Turi (eds.), *Fare gli italiani. Scuola e cultura nell'Italia contemporanea*, vol. 1, *La nascista dello Stato nazionale*, Bologna 1993, pp. 37–66.

Virilio, Paul: *Polar Inertia*, London 1999.

Virilio, Paul: *Revolutionen der Geschwindigkeit*, Berlin 1993.

Visser, Romeke: "Fascist Doctrine and the Cult of the *Romanità*," in: JCH 27/1992, pp. 5–22.

Voegelin, Eric: *The Political Religions*, in: Voegelin, *Modernity without Restraints. The Collected Works of Eric Voegelin*, vol. 5, ed. by Manfred Henningsen, Columbia, MO 2000.

Vogel, Bettina: "Guido Keller – Mystiker des Futurismus," in: Hans Ulrich Gumbrecht/Friedrich Kittler/Bernhard Siegert (eds.), *Der Dichter als Kommandant. D'Annunzio erobert Fiume*, Munich 1996, pp. 117–32.

Vogel-Walter, Bettina: *D'Annunzio – Abenteurer und charismatischer Führer*, Frankfurt/Main 2004.

Volkmann, Uwe: *Solidarität – Programm und Prinzip der Verfassung*, Tübingen 1998.

Volkov, Shulamit: "Jewish Success in Science," in: Volkov, *Germans, Jews, and Antisemites: Trials in Emancipation*, New York, NY 2006, pp. 224–47.

Vollmer, Frank: *Die politische Kultur des Faschismus. Stätten totalitärer Diktatur in Italien*, Cologne 2007.

Vollmer, Jörg Friedrich: *Imaginäre Schlachtfelder. Kriegsliteratur in der Weimarer Republik. Eine literatursoziologische Untersuchung*, http: //www. diss. fu-berlin. de/diss/receive/FUDISS_thesis_000000001060, accessed on August 6, 2008.

Vondung, Klaus: *Magie und Manipulation. Ideologischer Kult und Politische Religion des Nationalsozialismus*, Göttingen 1971.

Vondung, Klaus: "Probleme einer Sozialgeschichte der Ideen," in: Vondung (ed.), *Das wilhelminische Bildungsbürgertum. Zur Sozialgeschichte seiner Ideen*, Göttingen 1976, pp. 5–19.

Walkenhorst, Peter: *Nation – Volk – Rasse. Radikaler Nationalismus im Deutschen Kaiserreich 1890–1914*, Göttingen 2007.

Warnke, Martin: "'Ich bin ein wissenschaftlicher Privatbankier, dessen Credit so gut ist wieder der Reichsbank.' Aby Warburg und die Warburg Bank. Vorwort," in: Karen Michels, *Aby Warburg. Im Bannkreis der Ideen*, Munich 2007, pp. 11–19.

Weber, Max: *Economy and Society*, Berkeley, CA 1968, 2 vols.

Weber, Max: *From Max Weber. Essays in Sociology*, ed. by H. H. Gerth and C. Wright Mills, Boston, MA 1948.

Weber, Max: *The Protestant Ethic and the Spirit of Capitalism* [1930], New York, NY 1992.

Weber, Max: "The 'Rationalism' of Western Civilization," in: Stephen Kalberg (ed.), *Readings and Commentaries on Modernity*, Malden, MA 2005, pp. 53–64.

Weber, Max: *The Religion of China*, rev. edn, ed. by C. Wright Mills, Boston, MA 1968; *Die Wirtschaftsethik der Weltreligionen. Konfuzianismus und Taoismus. Schriften 1915–1920. Max Weber-Gesamtausgabe, Abt. I Schriften und Reden*, vol. 19, ed. by Helwig Schmidt-Glintzer with Petra Kolonko, Tübingen 1989.

Wedemeyer-Kolwe, Bernd: "*Der neue Mensch.*" *Körperkultur im Kaiserreich und in der Weimarer Republik*, Würzburg 2004.

Wedepohl, Claudia: "'Agitationsmittel für die Bearbeitung der Ungelehrten.' Warburgs Reformationsstudien zwischen Kriegsbeobachtung, historisch-kritischer Forschung und Verfolgungswahn," in: Gottfried Korff (ed.), *Kasten 117. Aby Warburg und der Aberglaube im Ersten Weltkrieg*, Tübingen 2007, pp. 325–68.

Wehler, Hans-Ulrich: *Deutsche Gesellschaftsgeschichte*, vol. 3, *Von der "Deutschen Doppelrevolution" bis zum Beginn des Ersten Weltkrieges 1849–1914*, Munich 1995.

Wehler, Hans-Ulrich: *Deutsche Gesellschaftsgeschichte*, vol. 4, *Vom Beginn des Ersten Weltkrieges bis zur Gründung der beiden deutschen Staaten 1914–1949*, Munich 2003.

Wehler, Hans-Ulrich: *Deutsche Gesellschaftsgeschichte*, vol. 5, *Von der Gründung der beiden deutschen Staaten bis zur Vereinigung 1949–1990*, Munich 2008.

Wehler, Hans-Ulrich: *Modernisierungstheorie und Geschichte*, Göttingen 1975.

Weiler, Bernd: *Die Ordnung des Fortschritts. Zum Aufstieg und Fall der Fortschrittsidee in der "jungen" Anthropologie*, Bielefeld 2006.

Welch, David: *Germany, Propaganda and Total War, 1914–1918. The Sins of Omission*, New Brunswick, NJ 2000.

Welch, David: "Nazi Propaganda and the Volksgemeinschaft. Constructing a People's Community," in: *JCH* 39/2004, pp. 213–38.

Welsch, Wolfgang: *Unsere postmoderne Moderne*, Weinheim 1991[3].

Welskopp, Thomas: "Identität ex negativo. Der 'deutsche Sonderweg' als Metaerzählung in der bundesdeutschen Geschichtswissenschaft der siebziger und achtziger Jahre," in: Konrad H. Jarausch (ed.), *Die historische Meistererzählung. Deutungslinien der deutschen Nationalgeschichte nach 1945*, Göttingen 2002, pp. 109–39.

Wette, Wolfram (ed.): *Schule der Gewalt. Militarismus in Deutschland 1871–1945*, Berlin 2005.

White, Hayden: *Metahistory. The Historical Imagination in Nineteenth-century Europe*, Baltimore, MD 1973.

White, Hayden: "The Value of Narrativity in the Representation of Reality," in: *Critical Inquiry* 7/1980, pp. 5–27.

Wilding, Peter: "Krieg – Technik – Moderne. Die Eskalation der Gewalt im 'Ingenieur- Krieg.' Zur Technisierung des Ersten Weltkrieges," in: Petra Ernst/Sabine A. Haring/Werner Suppanz (eds.), *Aggression und Katharsis. Der Erste Weltkrieg im Diskurs der Moderne*, Vienna 2004, pp. 163–86.

Wildt, Michael: *An Uncompromising Generation. The Nazi Leadership of the Reich Security Main Office*, Madison, WI 2009.

Wildt, Michael: *Hitler's Volksgemeinschaft and the Dynamics of Racial Exclusion. Violence against Jews in Provincial Germany, 1919–1939*, New York, NY 2012.

Winkle, Ralph: *Der Dank des Vaterlandes. Eine Symbolgeschichte des Eisernen Kreuzes 1914 bis 1936*, Essen 2007.

Wippermann, Wolfgang: *Faschismustheorien. Die Entwicklung der Diskussion von den Anfängen bis heute*, Darmstadt 1997[7].

Wippermann, Wolfgang: "The Post-War German Left and Fascism," in: *JCH* 11/1976, pp. 185–219.

Wippermann, Wolfgang: *Totalitarismustheorien. Die Entwicklung der Diskussion von den Anfängen bis heute*, Darmstadt 1998.

Wittgenstein, Ludwig: *Philosophical Investigations*, Oxford 1968.

Wittmann, Reinhard: *Geschichte des deutschen Buchhandels. Ein Überblick*, Munich 1991.

Wohl, Robert: *A Passion for Wings. Aviation and the Western Imagination 1908–1918*, New Haven, CT 1994.

Wohl, Robert: *The Generation of 1914*, Cambridge, MA 1979.

Wohl, Robert: *The Spectacle of Flight. Aviation and the Western Imagination 1920–1950*, New Haven, CT 2005.

Woller, Hans: *Geschichte Italiens im 20. Jahrhundert*, Munich 2010.

Woodhouse, John: *Gabriele D'Annunzio. Defiant Archangel*, Oxford 1998.

Wünsch, Marianne: "Ernst Jüngers *Der Arbeiter*. Grundpositionen und Probleme," in: Lutz Hagestedt (ed.), *Politik – Mythos – Kunst*, Berlin 2004, pp. 459–75.

Wuttke, Dieter: "Die Emigration der Kulturwissenschaftlichen Bibliothek Warburg und die Anfänge des Universitätsfaches Kunstgeschichte in Großbritannien," in: Horst Bredekamp/Michael Diers/Charlotte Schoell-Glass (eds.), *Aby Warburg. Akten des internationalen Symposions Hamburg 1990*, Weinheim 1991, pp. 141–63.

Zachriat, Wolf Gorch: *Die Ambivalenz des Fortschritts. Friedrich Nietzsches Kulturkritik,* Berlin 2001.

Zeidler, Manfred: *Reichswehr und Rote Armee 1920–1933. Wege und Stationen einer ungewöhnlichen Zusammenarbeit,* Munich 1994².

Ziemann, Benjamin: *War Experiences in Rural Germany, 1914–1923,* New York, NY 2007.

Zimmerman, Clemens: "Das Bild Mussolinis. Dokumentarische Formungen und die Brechungen medialer Aufmerksamkeit," in: Gerhard Paul (ed.), *Visual History. Ein Studienbuch,* Göttingen 2006, pp. 225–42.

Zimmermann, Michael: "Ausbruchshoffnung. Junge Bergleute in den Dreißiger Jahren," in: Lutz Niethammer (ed.), *"Die Jahre weiß man nicht, wo man die heute hinsetzen soll." Faschismuserfahrungen im Ruhrgebiet. Lebensgeschichte und Sozialkultur im Ruhrgebiet 1930 bis 1960,* vol. 1, Bonn 1983, pp. 97–132.

Zitelmann, Rainer: *Hitler: The Policies of Seduction,* London 1999.

Zoppi, Sergio: *Dalla rerum novarum alla democrazia cristiana di murri,* Bologna 1991.

Index

Printed and bound in the United States of America